CHEMISTRY

By the same authors
Chemistry: Calculations

By J. A. Hunt, A. Sykes and J. P. Mason
Chemistry: Experiments
Chemistry: The Teachers' and Technicians' Guide

CHEMISTRY

J. A. Hunt M.A.
Durrants School, Croxley Green

A. Sykes B.Tech., Ph.D.
Formerly of Alleyne's School, Stevenage

Longman Group UK Limited
Longman House, Burnt Mill, Harlow, Essex CM20 2JE, England
and Associated Companies throughout the World

First published 1984
Seventh impression 1987
ISBN 0 582 33108 0

Set in Monophoto Times New Roman

Produced by Longman Group (FE) Ltd
Printed in Hong Kong

Preface

This book is organized in a series of twelve themes which relate laboratory investigations both to the theory of chemistry and to the applied, social and historical aspects of the subject. The book reflects our teaching in a variety of schools and fulfils the requirements of the National Criteria for GCSE Chemistry.

This is more of a workbook than a reader. The outline experiments include sample results and suggested discussion points. We have aimed to give guidance to those who have done the experiments to help them make sense of their results and reach appropriate conclusions. We have not wanted to give too much away and spoil the spirit of investigation.

There are questions on most pages to test understanding. The questions often refer to the reference section at the end of the book. Words printed in **bold** type in the text are defined in the glossary. We have made these definitions as informal as possible, so that they aid understanding and are not just a form of words to be learned by heart.

Each chapter ends with a summary which suggests alternative ways of presenting the ideas and information in the chapter. The review questions at the end of each theme have been selected from recent public examinations, and they use ideas from the theme as a whole.

Chemistry: Experiments is a companion publication which we wrote with the help of Joy Mason of Godolphin and Latymer School. It is a file of labcards and worksheets, which can be used as the basis of a varied programme of practical work suitable for any of the GCSE syllabuses. We have found the use of labcards and worksheets very helpful when doing practical assessments. They provide opportunities for planning, manipulative skills, observation and interpretation.

We have also written *Chemistry: Calculations* which covers all the many types of numerical problems which feature in the GCSE, including atomic structure, rates of reaction, and energetics, as well as calculations based on formulae and equations. It is particularly designed for those aiming to achieve higher grades at GCSE, but it will also be a great help for those going on to A-level science courses.

Each of these publications can be used independently of the others, but taken together they provide for a complete course in chemistry for GCSE single subject and Combined Science syllabuses. Detailed guidance on the use of the publications is given in *Chemistry: The Teachers' and Technicians' Guide*. As well as giving teaching hints and technical notes, the guide shows how the ASE *Science and Technology in Society* units can be used in the context of the themes in *Chemistry*.

We are grateful to Mr L. K. Turner (Headmaster of Watford Grammar School) who encouraged us to embark on this project. Barry Nelson of North Westminster Community School read all our drafts and made many valuable suggestions for improving them. We are indebted to present and former colleagues at schools where we have taught. Finally we should like to thank those at Longman who have been involved in preparing our books for publication: Andrew Ransom, who has helped us plan the whole scheme; the designers, Randell Harris and Mick Harris; and Laurice Suess, who has had editorial responsibility for them all.

Andrew Hunt
Alan Sykes
January 1987

Contents

Preface
Safety information

Theme A Chemicals and where they come from

1 Pure chemicals 2
2 Elements and compounds 13
3 Atoms, molecules and ions 20
Review questions 29

Theme B The air

4 The composition of air 34
5 Oxygen 38
6 Burning, breathing and rusting 42
7 The air in industry 50
Review questions 56

Theme C Water

8 Water as a compound 60
9 Water as a solvent 63
10 Water in the home 66
Review questions 73

Theme D The metals

11 The physical properties and uses of metals 76
12 The chemical properties of metals 80
13 The extraction of metals from their ores 87
Review questions 93

Theme E The periodic table

14 Looking for patterns 96
15 Metals in the periodic table 99
16 Non-metals in the periodic table 103
17 Atomic structure 110
Review questions 114

Theme F Structure and bonding

18 Investigating structure 118
19 Covalent molecules and giant structures 128
20 Ions and ionic crystals 140
Review questions 144

Theme G Acids, bases and salts

21 Acids 148
22 Bases 156
23 Salts 161
24 Acid–base theories 170
Review questions 173

Theme H How much? How fast? How far?

25 How much? 176
26 How fast? 190
27 How far? 202
28 The contact process 211
Review questions 215

Theme I Chemicals from oil and gas

29 Introducing organic chemistry 218
30 Oil 223
31 Alcohols, acids and esters 229
32 Plastics 235
Review questions 241

Theme J Chemistry and food

33 Carbohydrates 244
34 Proteins 251
35 Fertilizers from ammonia 259
Review questions 267

Theme K Chemistry and energy

36 Energy and structure 270
37 Energy from fuels 277
38 Energy changes in chemical reactions 290
39 Nuclear energy 295
Review questions 307

Theme L Chemistry and electricity

40 Electrolysis 310
41 Oxidation and reduction 321
42 Electricity from chemical reactions 327
Review questions 332

Reference section

Tables of data 336
Units 344
Names of chemicals 345
Glossary 346

Index 351

Acknowledgements

We are grateful to the following Examining Bodies for permission to reproduce questions from past examination papers:

The Associated Examining Board (**AEB**); Associated Lancashire Schools Examining Board (**ALSEB**); University of Cambridge Local Examinations Syndicate (**CLES**); East Anglian Examinations Board (**EAEB**); East Midland Regional Examinations Board (**EMREB**); Joint Matriculation Board (**JMB**); University of London School Examinations Department (**L**); North West Regional Examinations Board (**NWREB**); Oxford & Cambridge Schools Examination Board (**O&C**); Oxford Delegacy of Local Examinations (**O**); The South-East Regional Examinations Board (**SEREB**); Southern Regional Examinations Board (**SREB**); Southern Universities Joint Board (**SUJB**); South Western Examinations Board (**SWEB**); Welsh Joint Education Committee (**WJEC**); The West Midlands Examinations Board (**WMEB**); Yorkshire & Humberside Regional Examinations Board (**YHREB**).

We are grateful to the following for permission to reproduce photographs and original artwork:

Airship Industries, 7.5(b); Anglian Water, 10.2; BBC, Theme A; Bowaters UK Paper Company, 16.5(c); Alan Brain, 23.10; British Aerospace, Theme D; British Alcan Aluminium, 18.16(c); British Gas, 29.5; British Leyland, 26.18; British Library, 14.1; British Oxygen, 7.2, 7.3 and 7.4; British Paper Federation, 33.10; British Petroleum, 31.9, 32.6 and 37.15; British Sugar Bureau, 1.14; British Steel, 11.1, 11.5, 13.4, 18.19(a), 18.19(c), 18.20, 18.21 and Theme K; British Tourist Authority, 18.18; Building Research Establishment, Crown Copyright, 6.12; Bureau International des Poids et Mesures, Sèvres, Theme M; Camera Press, 6.11 (Jean Regis Roustan), 7.5(a) (Norman Sklarewitz), 36.1 (Andy Kyle) and 36.9(d) (L. Smillie); W. Canning Materials, 40.11; J. Allan Cash, 6.5(a); Chloride Batteries, 42.7, 42.8 and 42.9; Chubb Fire Security, 6.5 (b and c); Crystal Structures, 19.26(b), 19.27(b) and 19.28(b); James Davis Photography, Themes B, C and H; De Beers Diamond Information Centre, 19.19 and 19.20; Documentation Française/M. Brigaud/Sodel EFL, 37.21; English Steel Forge and Engineering Corporation, 11.2(a); Ever Ready, Theme L; *Farmers Weekly*, 21.6(a) (Peter Adams), 35.10 (Keith Huggett) and 35.14; Robert Fowler, 2.1(3a); Galvanizers Association, 18.6; Geographers' A-Z Map Company, 17.23 (based on the Ordnance Survey Maps with the sanction of the Controller of Her Majesty's Stationery Office); Golden Wonder, Theme J; A. Gregory and M. Trompeteler, *Practical: Photography*, Longman, 16.6(a); Dr. Michael Hudson, Department of Chemistry, University of Reading, 19.13(a); F. A. Hughes Marine, 6.10; ICI, 16.5(b), 21.6(b), 22.9, 28.2 (Mond Division), 34.13, 35.11 and 35.12 (Agricultural Division); IMI, 40.9; Imperial War Museum, 33.11; Institute of Geological Sciences, 1.1, 1.2, Theme F, 18.1, 18.4(a), 18.11(a), 19.18, 19.26(a), 19.27(a), 19.28(a) and 23.3(b); Jaguar Cars, 26.19; JEOL UK, 3.9; JET Joint Undertaking, 37.20; Joseph Ash Galvanizing, 21.4; Keystone Press Agency, 6.13 and 6.14; Frank Lane, 8.3; London Transport, 17.2; Loughborough University of Technology, 37.7; Metal Box, 11.2(b); Meteorological Office, HMSO Crown Copyright, 22.5; Permutit Water Soft, 10.9; Professor D. C. Phillips, Department of Zoology,

University of Oxford, 18.5; Polaroid UK, 16.7(a); Press Association, 37.22; Press-Tige Pictures, 16.6(b), 18.19(b) and 21.6(c); *Revised Nuffield Chemistry Handbook for Pupils*, figure 9.30, 19.23; Rio Tinto Zinc, 13.6 and 13.7; Ann Ronan Picture Library, 4.3 and Theme E; Royal Institution, 14.2; Science Museum, 3.10; Science Photo Library, 8.4 (Martin Dohrn), 11.4 (Dr. Tony Brain), 34.9 (Dr. Jeremy Burgess) and 37.18 (NASA); Scotch Whisky Association, 1.15 and 33.12 (Anthony James); Shell UK, 2.1(1), 2.1(2), 2.1(3b), 16.5(a), 26.9, 26.18, 26.20, Theme I, 32.1 and 37.23 (Michael Sturley); Smith and Nephew, 23.3(a); Space Frontiers/NASA, 42.10; Stone Manganese Marine, 18.16(a); Thames Water, 8.1, 10.1 and 37.17; Thermit Welding GB, 12.7; United Kingdom Atomic Energy Authority, 11.3, 39.5, 39.6, 39.10, 39.14, 39.15, 39.16, 39.17, 39.18, 39.19, 39.20, 39.21 and 39.22; University of Cambridge Department of Earth Sciences, 18.2(b); Van den Berghs and Jurgens, 30.10; Vision International/Paolo Koch, 37.19; Wessex Water, 10.11; Jerry Wooldridge, Theme G. All other photographs by Longman Photographic Unit.

The photographs of experiments were taken at Alleyne's School, Stevenage. We are grateful to the following technicians for their help with the preparation: Mrs. H. A. Davies, Mrs. J. M. Urquhart and Mrs. E. A. Roberts, who built the model shown in 18.11(b). Student helpers were A. J. F. Atkinson, M. J. Lusher, M. Thorpe and G. D. H. Turner. Labels for 25.2, 41.1 and 41.2 were prepared by J. M. Innes, Head of Art. The chemicals used were lent by A. Howat of May and Baker, space-filling models by Needs Plastics Ltd and ball-and-spoke models by Griffin and George. Photographs were also taken at Crystal Structures Ltd, Bottisham, Cambridge and at Rocirc, Bishop's Stortford.

We are also grateful to the following for permission to redraw copyright material: Association for Science Education, *Chemistry and Industry*, 1975, 1.11; British Gas, 29.4; British Petroleum, 32.13 and 37.16 (*Statistical Review of World Energy*, 1981); Department of Energy, *Energy, A Key Resource*, Crown Copyright, reproduced with the permission of the Controller of Her Majesty's Stationery Office, 37.3; Halton Chemical Industry Museum, Widnes, 22.6; Heinemann Educational Books, *Science in Society Teachers' Guide*, directed by J. Lewis for the Association for Science Education, 1.10; Macmillan Publishers, *Chemistry Today*, E. S. Henderson, 1.4; Oxford University Press, 1978, *Chemistry Matters*, Richard Hart, 19.14; Shell UK, *Oil*, 30.2; Unilever Educational Publications, *Detergents*, 10.5; UK Atomic Energy Authority, 37.4.

The cover photograph was taken by Paul Brierley and shows crystals of rochelle salt (sodium potassium tartrate, $C_4H_4O_6KNa.4H_2O$) viewed in polarized light. This salt was discovered in La Rochelle, France, in the seventeenth century and used to be a constituent of health salts.

Safety information

Chemical experiments in general can be dangerous, so you should always do them carefully and under supervision. This book is not a laboratory manual – it only describes experiments in outline. To do them, you will need full instructions from your teacher, and you should follow these accurately.

Some chemicals are particularly hazardous. These have been pointed out in this book by standard hazard symbols, which are illustrated and explained below.

Explosive
These substances may explode if ignited in air or exposed to heat. A sudden shock or friction may also start an explosion.

Oxidizing
These substances may produce much heat as they react with other materials. They can create a fire risk.

Highly flammable
These are solids, liquids or gases that may easily catch fire in a laboratory under normal conditions.

Toxic
These substances are a serious risk to health, and can cause death. The chemicals may have their effects when they are swallowed or breathed in or are absorbed through the skin.

Harmful
These chemicals are less of a health risk than toxic substances, but they must still be handled with care.

Corrosive
These chemicals destroy living tissues, including eyes and skin.

Irritant
These substances are not corrosive but they can cause reddening or blistering of the skin. The effect may be immediate or it may only be observed after prolonged, or repeated, contact with the chemical.

Radioactive
The radioactive chemicals used in schools have low activity. They are normally only used by teachers, for demonstrations. They should be treated in the same way as toxic substances.

THEME A

Chemicals and where they come from

Old remedies and modern drugs are made with chemicals from rocks, plants, the sea, coal and oil

1 Pure chemicals

1.1 Chemicals from rocks

Figure 1.1 The large white areas in this lump of granite are feldspar crystals

Many important chemicals come from the ground. Sometimes the chemicals are found **pure**, but usually they are mixed with other things. Figure 1.1 shows a piece of granite. This can be seen to consist of a mixture of crystals. The feldspar, mica and quartz crystals are the minerals which make up the granite rock.

Chemists use the word 'pure' to mean a single substance not mixed with anything else. So water or oxygen gas can be pure. Air cannot be chemically pure because it is a mixture of oxygen, nitrogen and other gases.

Some rocks are made of only one mineral. An important example is limestone. It is quarried on a large scale for use in industry and agriculture (see figure 1.2). Limestone consists of calcium carbonate which is also found naturally as chalk and as marble.

Figure 1.2 A limestone quarry in South Wales

Limestone can be quarried from the surface because many hills in Britain are made of this rock. Other important rocks are found beneath the ground. In Cheshire there are large underground beds of salt. These provide the main raw material for chemical factories producing chlorine, sodium and alkalis. There is only one British mine for extracting dry rock salt, which is at Winsford. Explosives are used to break up the rock face. The lumps of rock are then crushed before being hoisted to the surface. The ground rock salt is not pure. It contains about 94 per cent salt (sodium chloride). The chief impurity is clay. The impurities do not matter because the salt is mainly used to melt ice and snow on the roads in winter. Experiment 1a shows how rock salt can be purified.

Questions

1 Which of these things are pure in the chemist's sense: natural gas, sea water, sand, wood, sugar, soil, chalk, petrol, shampoo, beer, polythene, table salt, milk, honey?

2 What does the word 'pure' mean in its everyday sense (for example, when used to describe pure air or pure honey)?

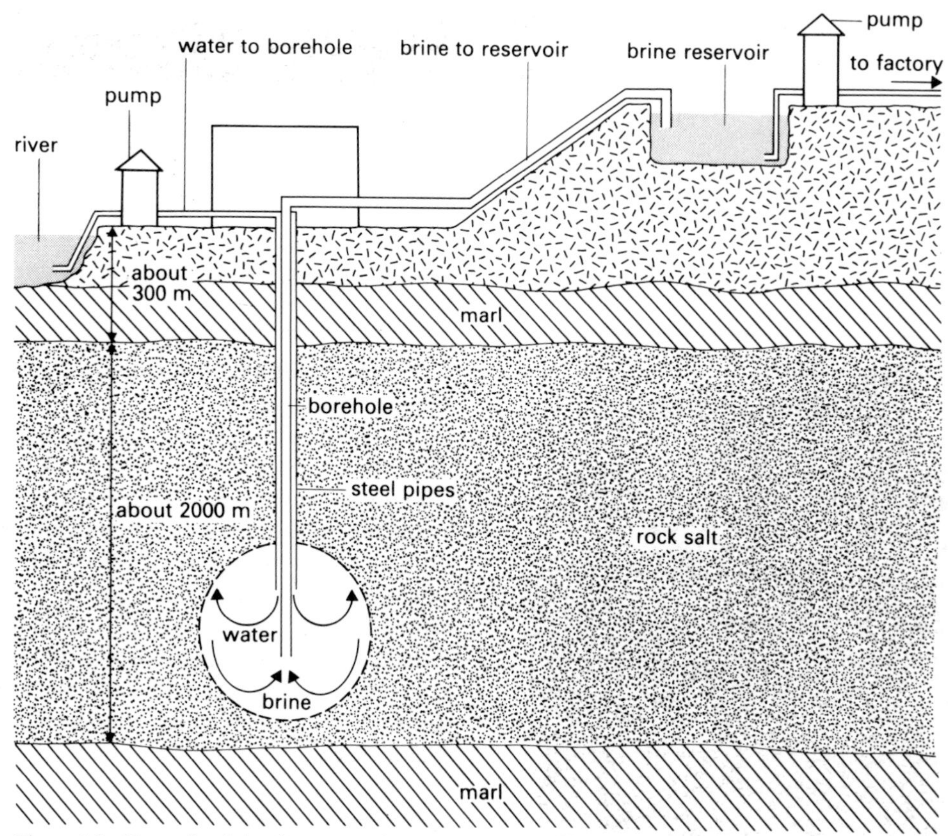

Figure 1.3 Extracting brine from an underground salt bed

The salt used for the chemical industry is not mined but is obtained by pumping water down into the rock (see figure 1.3). The salt dissolves and is carried to the surface in solution. The clay does not dissolve, and remains underground. The solution of salt in water is called brine.

A **solvent** is a liquid which can be used to dissolve things. A **solute** is a substance which dissolves in a solvent to make a solution. Brine is a solution of salt (the solute) in water (the solvent). A substance which dissolves is said to be *soluble.*

Large-scale pumping to extract salt as brine started in about 1870. Uncontrolled pumping for about sixty years resulted in widespread subsidence and flooding of land. From time to time there were disastrous collapses which destroyed whole chemical works. Nowadays pumping is planned so that the cavities in the rock are spaced out and separated by pillars of rock which prevent the surrounding land subsiding.

Salt crystals are recovered from brine by **evaporating** the water. Much energy is

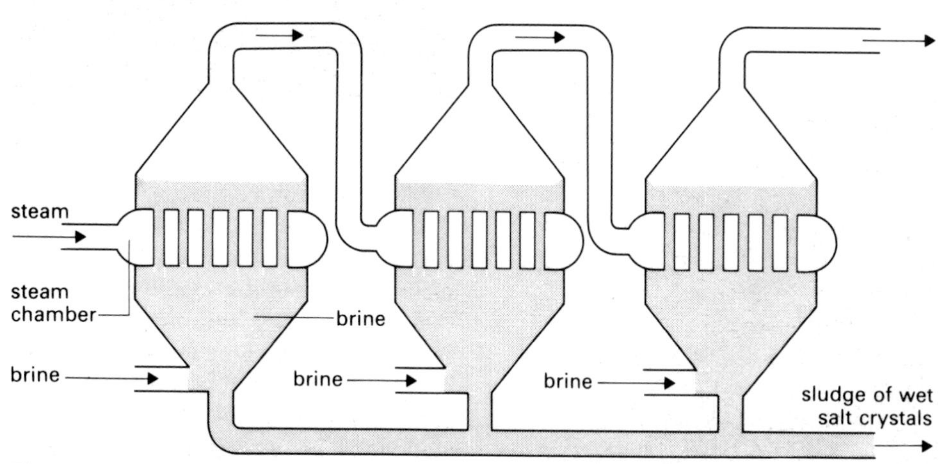

Figure 1.4 Obtaining salt crystals by steam evaporation of brine

needed to turn water into steam, and the modern industrial process is designed to be as efficient as possible in its use of energy (see figure 1.4). It uses the principle that the **boiling** point of a liquid is lowered when the pressure is lowered. Energy is saved by using the steam from the first vessel to heat the brine in the second, thus raising more steam to boil the brine in the third vessel. This is possible because a vacuum is applied to the third vessel so that the brine in it boils at a lower temperature than the brine in the second vessel, which in turn boils at a lower temperature than the brine in the first vessel. As the water evaporates the salt **crystallizes** out. The crystals are separated by **filtering** or by using centrifuges.

Experiment 1a
Pure salt from rock salt

The diagrams in figure 1.5 show how to get pure salt from rock salt in the laboratory.

Results
The insoluble impurities (clay or sand) do not dissolve in the water and are caught in the filter paper. The salt solution passes through the filter. Pure white crystals of salt appear when the water is evaporated.

Discussion
At which stages do the following processes occur: dissolving, evaporation, crystallization? Which is the solvent and which is the solute in the solution formed? Why is the evaporation completed on a steam bath? Why is this method much less efficient in its use of energy than the industrial process?

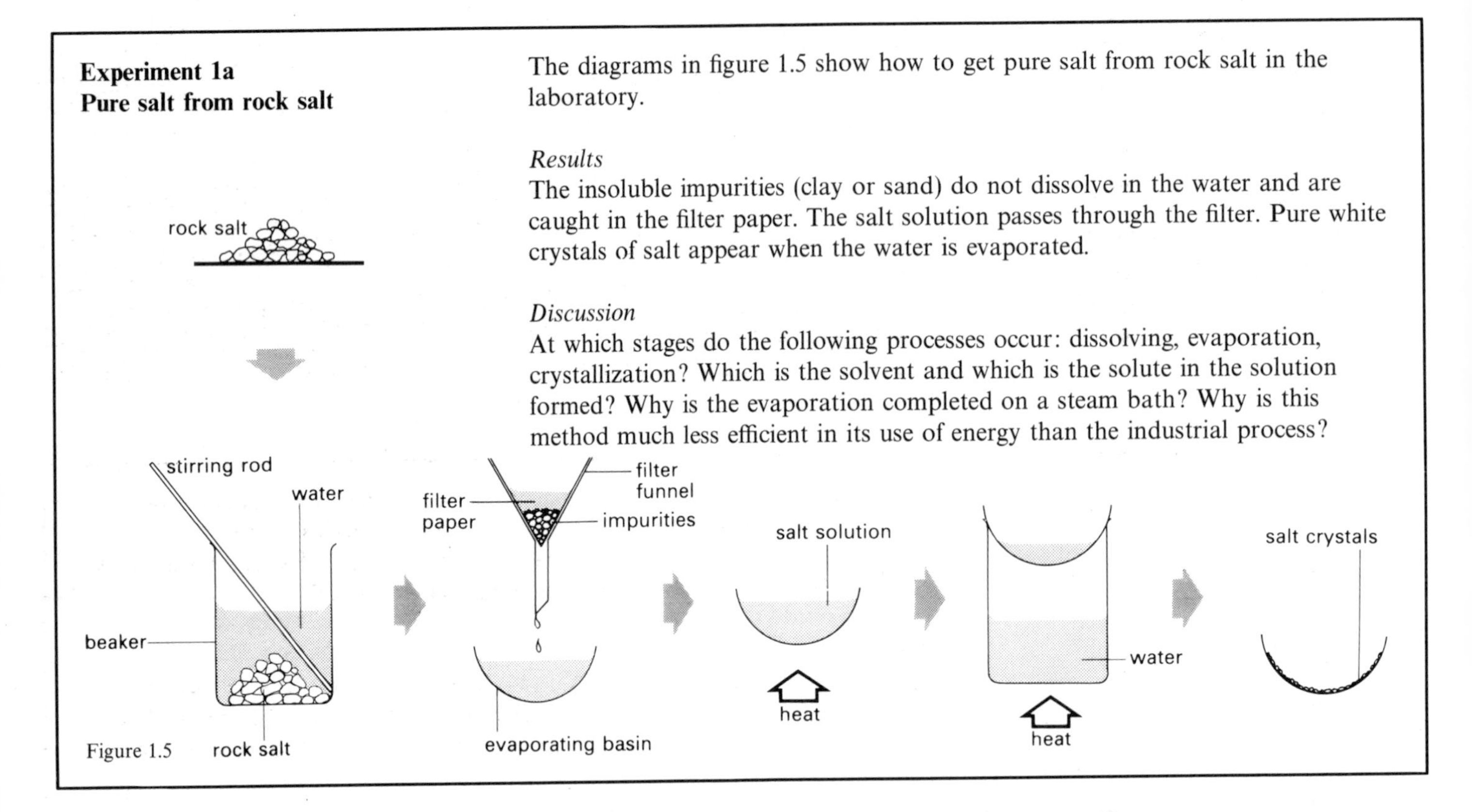

Figure 1.5

Questions

1 Which of these substances are soluble in water and which are insoluble: chalk, sugar, sand, copper, charcoal, soap, alum, copper(II) sulphate?
2 Which of these pairs of mixed solids can be separated by the method used in experiment 1a: alum and clay, copper(II) sulphate and sand, chalk and sand, salt and sugar, copper and iron, sugar and sawdust?

Sulphur is also found underground. Sulphur is an important raw material for the chemical industry because it is used to make sulphuric acid. In 1865, large underground deposits of sulphur were discovered in America under layers of quicksand. The quicksand made it impossible to mine the sulphur in the usual way. The sulphur could not be used until Herman Frasch had invented an ingenious method for **melting** the sulphur underground and then piping it to the surface while still hot.

Sulphur has a low melting point compared with other minerals. It melts at 115 °C, which is only a little above the normal boiling point of water. Under pressure, water can be heated to well above 100 °C without boiling, and thus be hot enough to melt sulphur when pumped below ground. In the Frasch process, hot compressed air is then used to force the molten sulphur to the surface.

Questions

1 In the Frasch process, why is the sulphur pumped up between the pipes carrying the hot water and the hot compressed air down to the sulphur beds?
2 Use the list of properties below to explain why the use of water in the extraction of salt is quite different from its use in the Frasch process.
Sulphur melts at 115 °C and is insoluble in water.
Salt melts at 808 °C and is soluble in water.
Sand melts at 1610 °C and is insoluble in water.

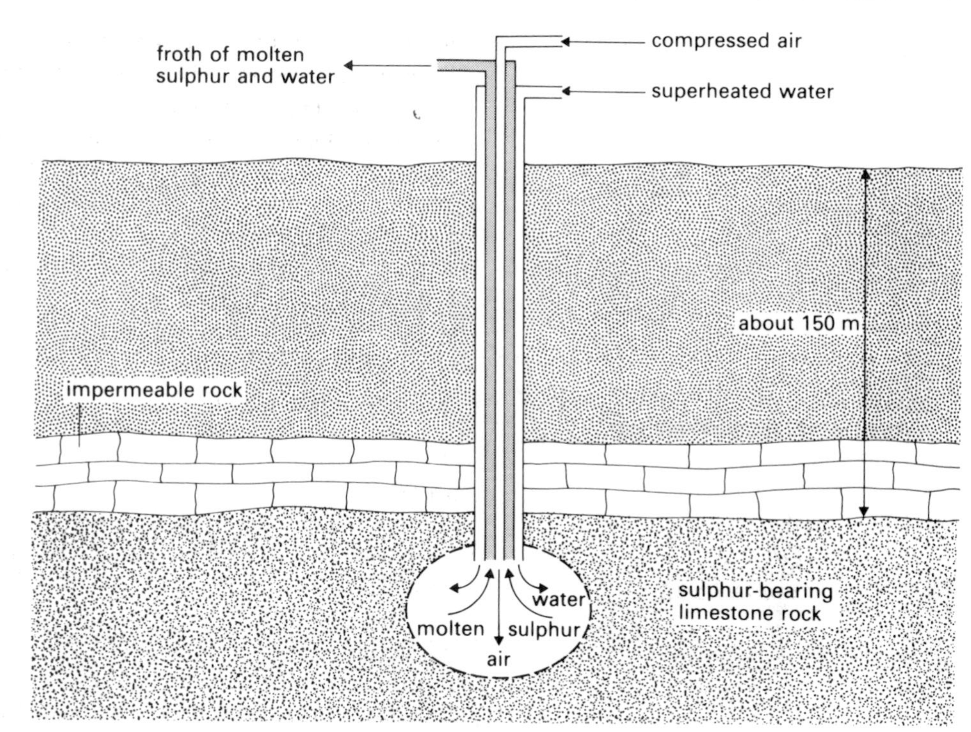

Figure 1.6 The Frasch process for obtaining sulphur

Experiment 1b
Melting stearic acid

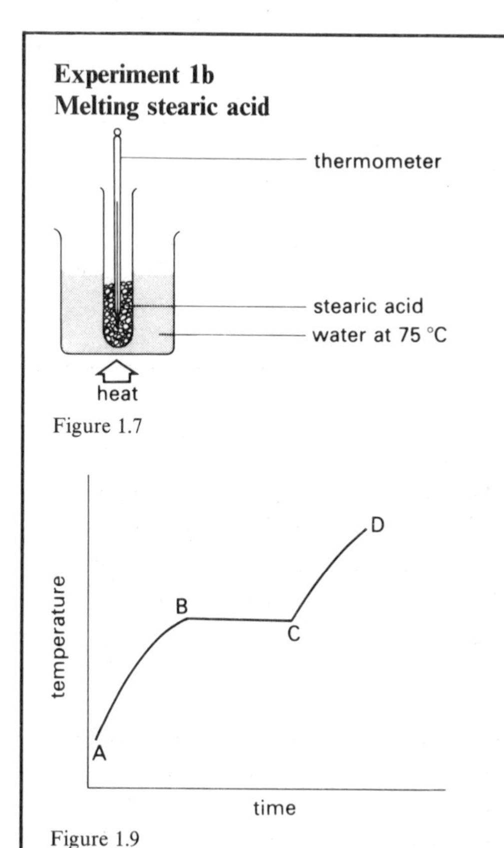

Figure 1.7

Figure 1.9

The melting of substances which melt below 100 °C can be investigated using the apparatus shown in figure 1.7. The hot water maintains a steady temperature outside the test-tube. The temperature is noted at regular intervals until some time after all the stearic acid has become liquid.

Results
Figure 1.8 gives a table of results for this experiment. Plot these results on a graph. Put the temperature scale on the vertical (y) axis and the time scale on the horizontal (x) axis. From your graph work out the melting point of stearic acid.

Time (minutes)	0	$\frac{1}{2}$	$1\frac{1}{2}$	2	$2\frac{1}{2}$	3	4	5	6	7	8	$8\frac{1}{2}$	9	$9\frac{1}{2}$	10
Temperature (°C)	19	29	40	48	53	55	55	55	55	55	55	64	70	73	74

Figure 1.8

Discussion
Figure 1.9 shows what the graph might look like in an ideal experiment. From A to B, the heat entering the tube warms up the solid. At B the solid starts to melt, and at C it is all molten. From C to D, the heat warms up the liquid. From B to C, the heat entering the tube causes no rise in temperature. The energy is used to melt the solid. Why is heat needed to melt a solid?

Chemists measure melting points for two main reasons. Each substance has its own melting point so a measured melting point can be used like a fingerprint to help to identify an unknown substance. Also, watching a substance melt can show whether or not it is pure. A pure solid melts at a definite temperature but an impure solid softens and melts over a range of temperatures, like butter on a summer day.

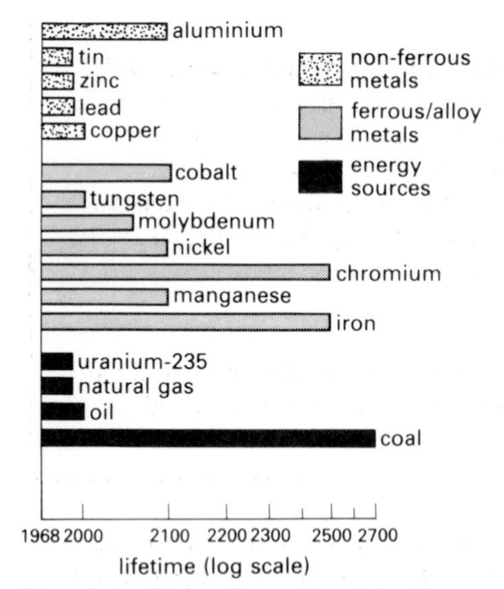

Figure 1.10 The earth's supplies of metals and other raw materials are limited and will eventually run out. This graph shows the estimated lifetimes of some resources

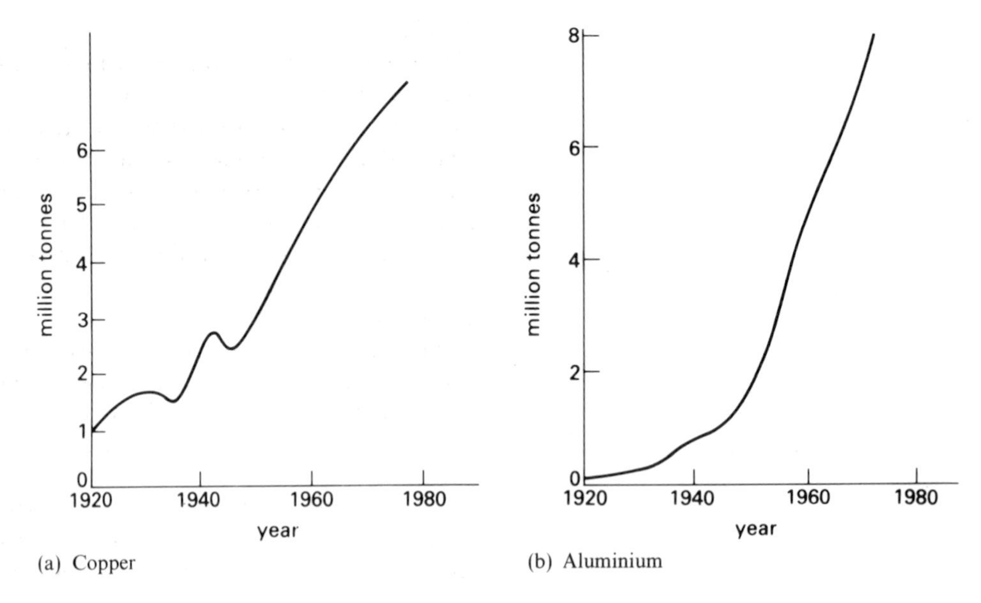

Figure 1.11 Metals are very important in modern life. These graphs show the world use of two metals in the period 1920–80

Metals come from rocks. A few metals, such as gold and copper, can be found free in nature but most metals are hidden in ores which have to be treated chemically to extract the metal. Often the amount of ore in the rock is so low that it has to be separated from the other minerals and concentrated before the metal can be extracted. The extraction of some metals is described in chapter 13.

1.2 Chemicals from the sea

Sea water is undrinkable and cannot be used for watering crops because it contains a high concentration of dissolved salts. **Freezing** is one natural process for getting pure water from sea water. When the sea freezes, only the water solidifies. The salts are left in solution. Icebergs consist of pure water in the solid state.

Distillation is the usual method for purifying water but it is expensive because of the large amount of energy needed. It is only used on a large scale for getting pure water from sea water in areas such as the Persian Gulf where there is little or no fresh water, but plenty of fuel.

Experiment 1c
Pure water from sea water

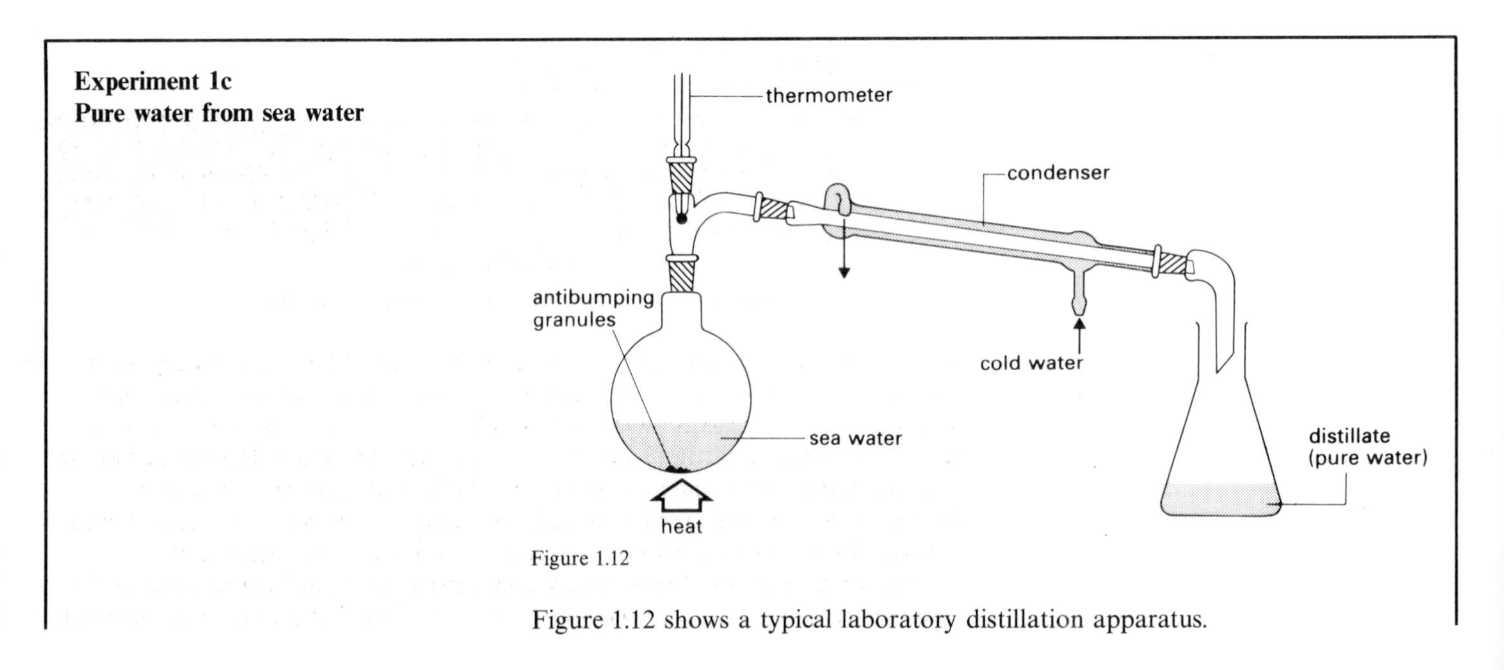

Figure 1.12

Figure 1.12 shows a typical laboratory distillation apparatus.

Results
The water boils and turns to steam in the flask. The thermometer reads 100 °C, which is the boiling point of water. In the condenser the steam is cooled, turns back into water and runs down to the collecting flask. The distillate is pure water. The 'antibumping granules' (or pieces of broken pot) in the distilling flask help to make sure that the bubbles of steam form smoothly. Without the granules the steam bubbles may form explosively, shaking the whole apparatus. This explosive formation of steam bubbles is called bumping.

Discussion
Distillation purifies the water because the dissolved salts do not evaporate, and stay in the flask. The process uses much energy because the heat supplied to evaporate the water is carried away by the cooling water in the condenser.

In some parts of the world the sea is the main source of common salt (sodium chloride). The sea is trapped in shallow pools (pans) and heat from the sun is used to evaporate the water. There is much less magnesium chloride than sodium chloride in the sea but there is enough for it to be the main source of magnesium metal. Bromine is also extracted from sea water. There is a factory for extracting bromine from sea water on the coast of Anglesey. Bromine is a dark red liquid. Substances containing bromine are used to make photographic film, drugs and additives for petrol.

1.3 Chemicals from plants

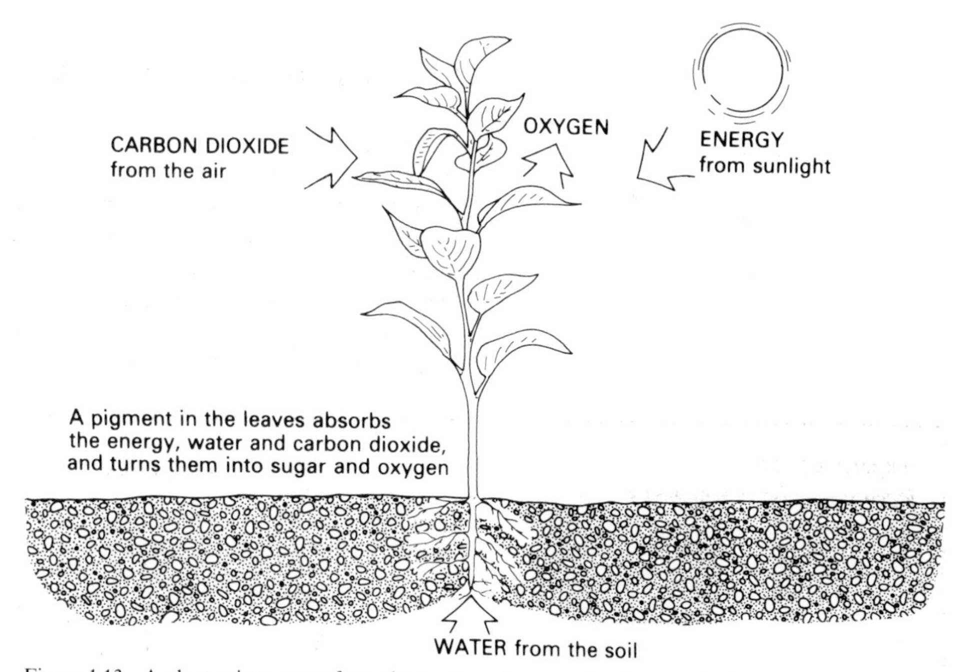

Figure 1.13 A plant using energy from the sun to make sugar by photosynthesis

Man has always obtained many chemicals from plants. Plants trap energy from the sun as they grow. This energy can be reused for growth or movement when plants are eaten as food or burnt as fuels. Drugs and dyes all came from plants before the chemical discoveries made in the last hundred years. The bark of willow trees was used to treat pain and fevers for which we would now use synthetic aspirin. Indigo has been used as a blue dye for cotton for over 5000 years. It was extracted from a plant grown as a main crop in India. More recently, the antibiotic penicillin was discovered by Alexander Fleming in 1928 as a result of a chance infection of a bacterial culture by a mould. The first plastics were made by

Questions

1 From which plant, other than beet, is sugar obtained? In which parts of the world is this plant grown?
2 Can you describe a laboratory experiment which uses seed crystals in crystallization?
3 How does a laboratory centrifuge differ from a spin drier or a centrifuge used in sugar refining?

Figure 1.14 Sugar refining: separating out the crystals using a centrifuge

making chemical changes to plant fibres. Natural rubber comes from the latex which oozes from rubber trees when the bark is cut and peeled. Oils from corn and palm fruits are used for cooking and for the manufacture of margarine and soaps.

Sugar beet is an important crop in East Anglia, Lincolnshire and the West Midlands. The roots are harvested in the autumn and sent to factories for refining. At the factory the beets are sliced and treated with hot water, which extracts the sugar as a juice. The juice is filtered to remove impurities. Next it is steam-heated to evaporate much of the water and then crystallized by further evaporation in vacuum pans. Seed crystals of icing sugar are added. These grow into bigger crystals as the water evaporates. In the end, the pan is filled with crystals and syrup. The crystals are separated from the syrup in centrifuges. Each centrifuge has a basket with holes in the side with a sieve inside. As the basket spins at high speed, the syrup is forced out through the holes while the crystals are caught in the sieve. It works on the same principle as a spin drier for clothes.

Fermentation has been used for thousands of years to convert sugar into alcohol (ethanol) and carbon dioxide. Yeast is a plant (a fungus) which can get energy from sugars as it grows during fermentation. Growth continues until all the sugar is used or until the alcohol concentration reaches about 12 per cent by mass. At this level the alcohol poisons the yeast. Nearly pure alcohol can be obtained by distilling the solution. Distillation is used to make spirits, such as whisky, rum and brandy, which contain about 35 per cent alcohol by mass.

When distillation is used to separate a mixture of liquids it is called **fractional distillation** because it splits the mixture into two or more parts (fractions). The method works because the liquids boil at different temperatures. For example, alcohol (ethanol) boils at 78 °C so it turns to a vapour more easily than water, which boils at 100 °C. Figure 1.16 shows a typical laboratory apparatus for fractional distillation. The thermometer shows the boiling point of the liquid being distilled. It reads 78 °C at the start while nearly pure alcohol distils over. At this stage any water which evaporates in the flask **condenses** in the fractionating column and drips back into the flask.

Iodine is a chemical related to bromine. It is used to make photographic film and polarizing sunglasses. Our bodies cannot grow normally unless there is a small amount of iodine in the food we eat. There is very little iodine in the sea (about 1 part in 20 million) but it is concentrated in the leaves of some seaweeds.

Figure 1.15 Copper vessels like these have been used for distilling whisky for many years

Question

Would you use simple distillation, fractional distillation or a separating funnel to separate (a) cooking oil and vinegar, (b) salt and water, (c) paraffin and petrol, (d) oil and water?

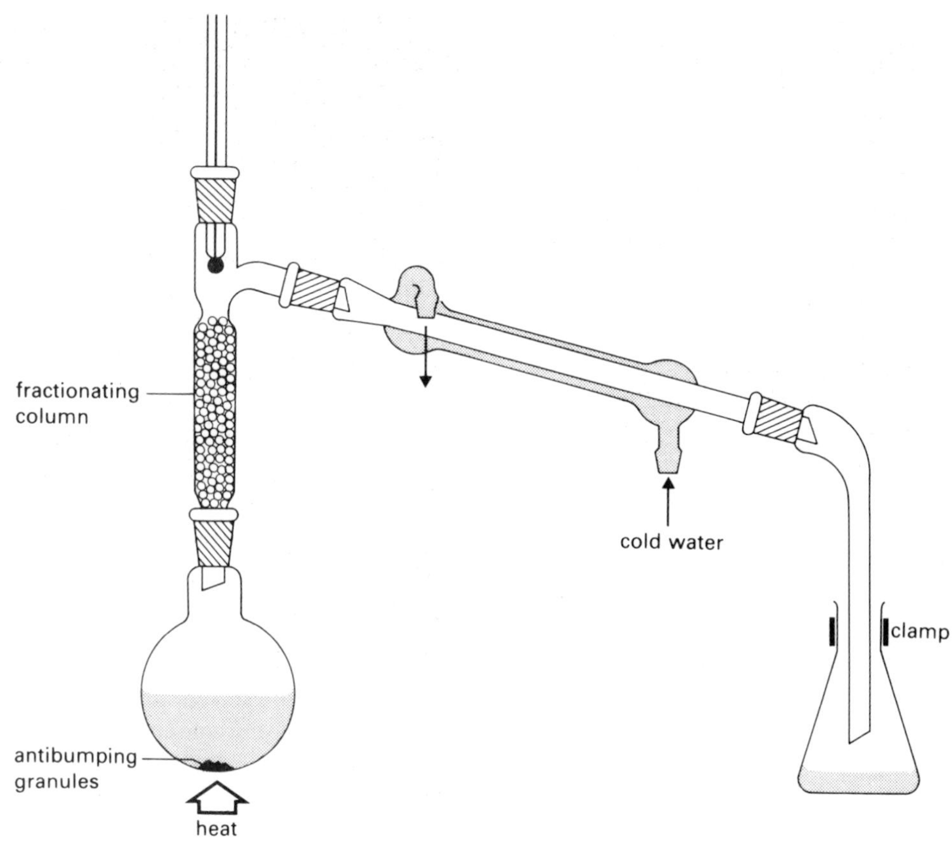

Figure 1.16 Fractional distillation

Experiment 1d
Iodine from seaweed

iodine
1,1,1-trichloroethane

hydrogen peroxide

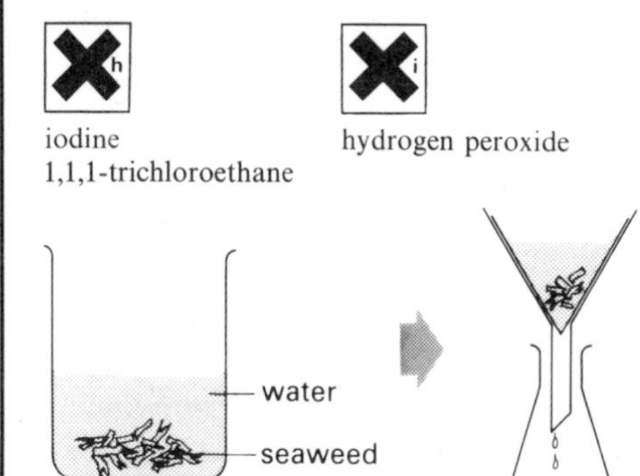

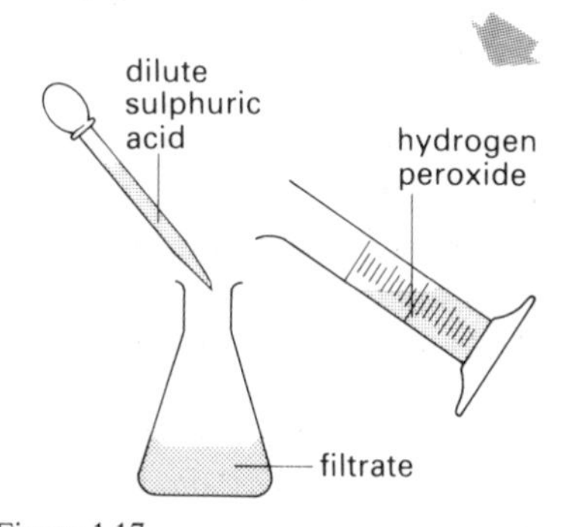

Figure 1.17

The series of diagrams in figure 1.17 illustrates the steps of the experiment.

Results

The filtrate turns yellow when the hydrogen peroxide is added. The 1,1,1-trichloroethane does not mix with water and forms a lower layer. On shaking, the water layer in the funnel becomes colourless and the 1,1,1-trichloroethane turns pale red.

Discussion

The iodine is not in its free state in the seaweed. The hydrogen peroxide releases the iodine. Iodine is more soluble in 1,1,1-trichloroethane than in water. When the separating funnel is shaken, the iodine is extracted into the solvent in which it is more soluble. Everything else stays in the water. This is an example of *solvent extraction*. The solution of iodine in 1,1,1-trichloroethane can be run off through the tap. Iodine crystals are obtained by evaporating the solvent.

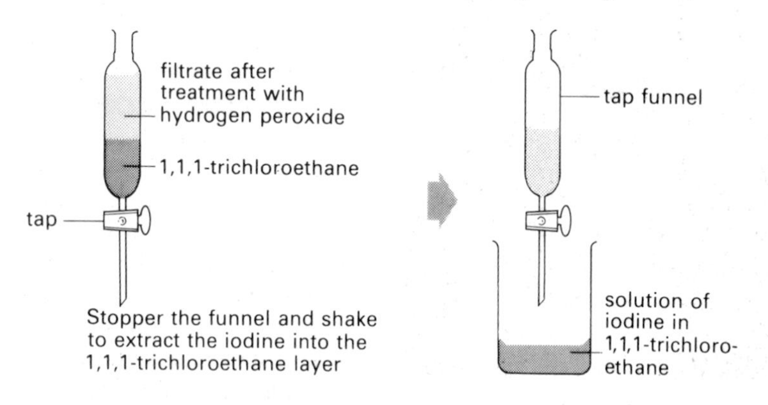

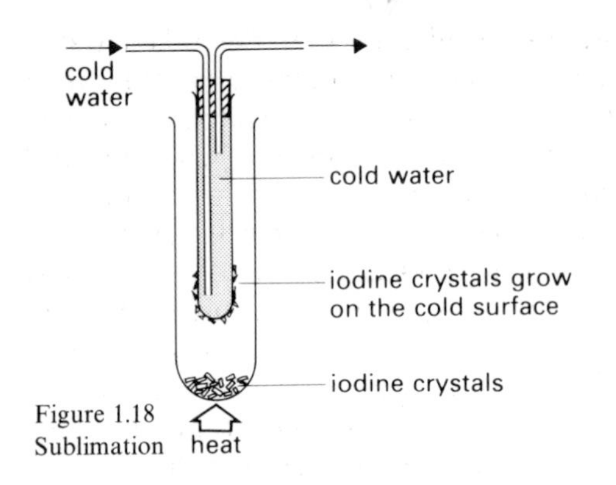

Figure 1.18
Sublimation

Iodine crystals, when warmed, turn to a beautiful violet vapour – the same colour as iodine dissolved in cyclohexane (see p. 131). The vapour condenses onto a cool surface, not as a liquid, but as a solid. The crystals which form are shiny grey and look almost metallic. This change direct from the vapour to the solid state is called **sublimation**. Sublimation can be used to purify iodine. A suitable apparatus for the purification is shown in figure 1.18.

Chlorophyll is the green pigment in the leaves of plants. It is extracted for use as a food colour. Most plant colours are easily damaged by heat, acids or alkalis and so they have to be investigated under mild conditions. **Chromatography** is used to separate plant colours.

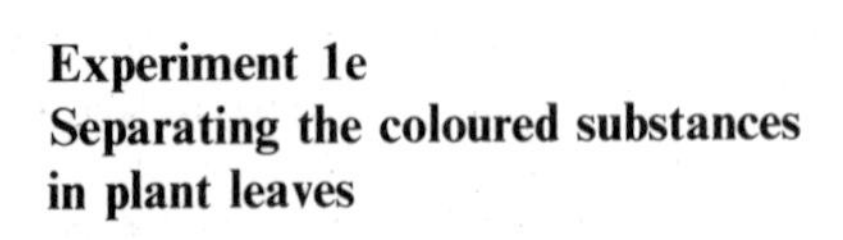

Experiment 1e
Separating the coloured substances in plant leaves

ethanol
propanone

The procedure for this experiment is described by the diagrams in figure 1.19.

Results
The solvent rises up the paper, and after a time green and yellow bands of colour are seen. The green band is chlorophyll. The yellow bands are carotenes.

Discussion
In this experiment the solvent moves while the paper is still. Colours which tend to dissolve in the solvent are carried further than colours which tend to stick to the paper. Thus the colours move at different speeds and are separated.

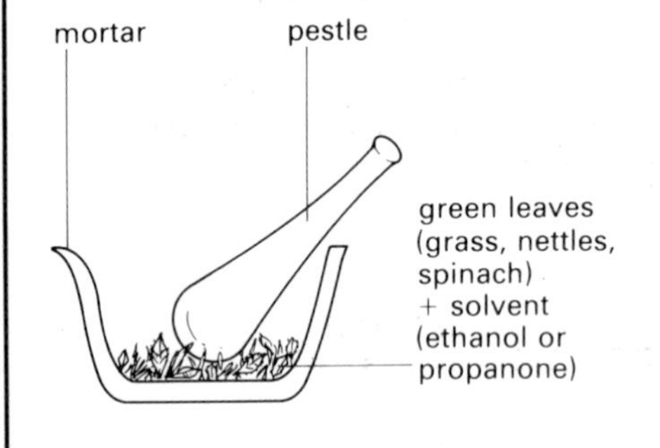

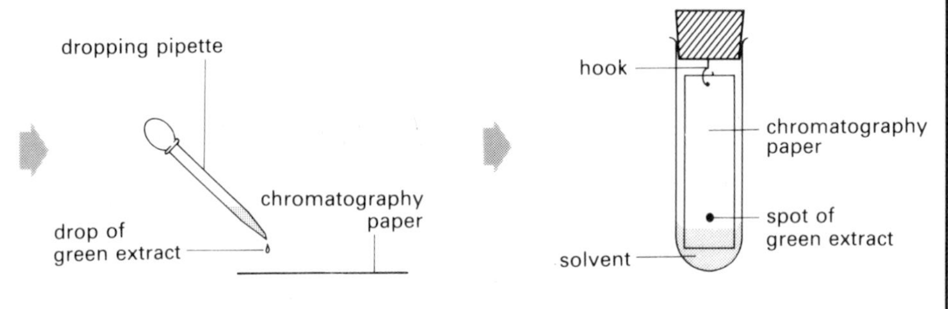

Figure 1.19

1.4 Chemicals from the air

The air is a mixture of gases. Most of the air is made of oxygen and nitrogen. Also present are carbon dioxide, the noble gases (argon, neon, helium, krypton and xenon) and varying amounts of water vapour. There are no simple methods for separating gas mixtures. The industrial separation of the gases in the air is based on the fractional distillation of liquid air. This process, and the properties and uses of the gases, are described in chapters 5 and 7.

1.5 Chemicals from coal and crude oil

Two hundred and fifty years ago, people depended almost entirely on plants, including trees, for fuels and chemicals. Plants grow slowly and need large areas of land. To meet an increasing demand for fuels, drugs, dyes, detergents and solvents, chemists have developed ways of making these things from coal and crude oil (petroleum).

These discoveries have changed the pattern of life in many parts of the world. For example, until the end of the nineteenth century, huge areas of land in Europe and Asia were used for the growth of the madder plant. The alizarin extracted from the roots of the plant was a red dye for cotton. Then chemists discovered

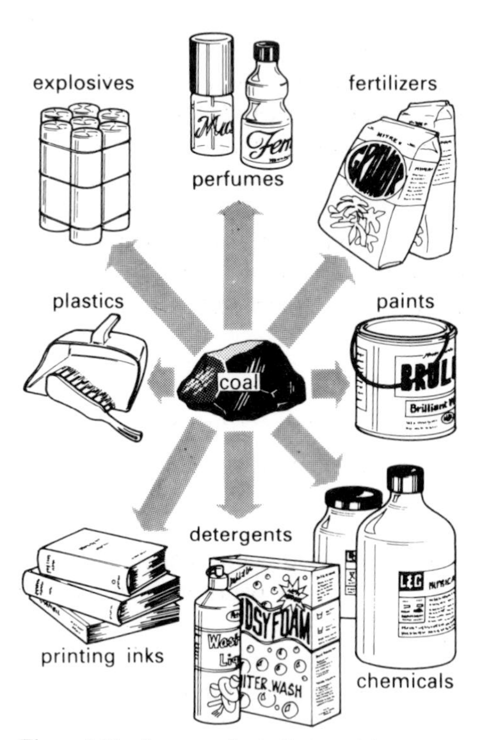

Figure 1.20 Some products that contain chemicals from coal

how to make alizarin cheaply from a chemical found in coal tar. Soon there was no market for alizarin from madder, and the farmers had to change to other crops.

Coal was formed from the remains of giant reeds, mosses and scaly trees, which grew in the swamps of the Carboniferous period about 300 million years ago. Coal can be thought of as a concentrated source of plant chemicals. Coal is used in three ways: as a fuel, as coke and as a source of chemicals. Coke is made by heating crushed coal in special ovens. About two-thirds of a tonne of coke is made from one tonne of coal. The rest is given off as gases and vapours. Useful chemicals can be condensed from the gases. Some materials made from these chemicals are shown in figure 1.20. The importance of coal is discussed further in chapter 37.

Crude oil is more convenient than coal as a source of fuels and chemicals. This is because it is a mixture of liquids, which are easier to transport, handle and process than solids. Oil is taken from wells to refineries where it is processed in three stages: separation, conversion and purification. The many liquids in oil have different boiling points, so they can be separated by fractional distillation. On an industrial scale the apparatus differs from that shown in figure 1.16. The plant is designed to run continuously and to produce a number of fractions (see figure 1.21).

Distillation takes place in a tall steel tower. The tower is hottest at the bottom and gradually cools towards the top. The crude oil is heated in a furnace and then fed in near the bottom of the column. The vapours rise up the column until they are cool enough to condense. Different substances condense at different levels and are separated. To help to separate the fractions, the column is divided by a series of trays with holes covered by bubble traps as shown in figure 1.21.

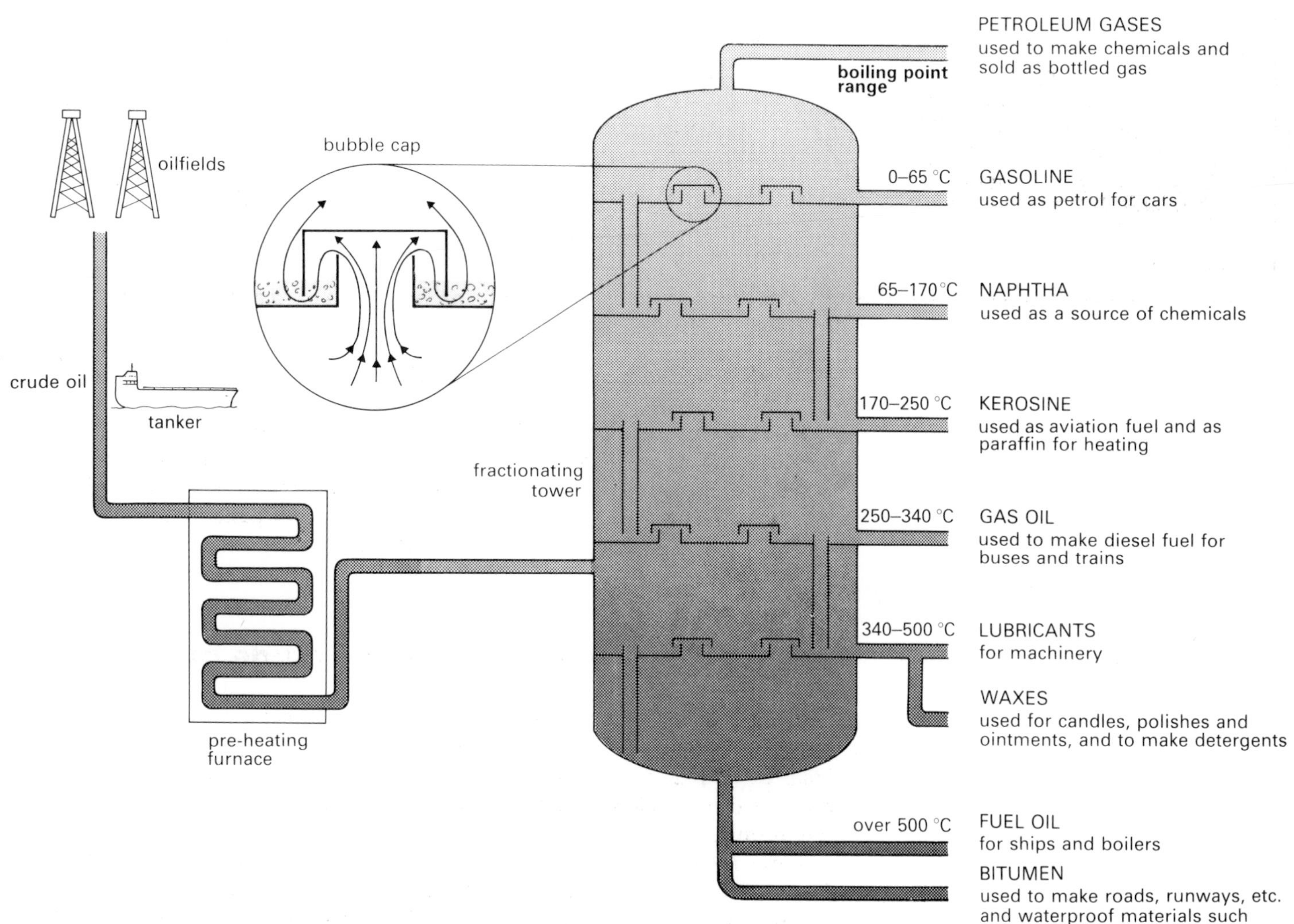

Figure 1.21 Fractional distillation of oil

Summary

1 Copy and complete the table in figure 1.22 to summarize the methods of separation mentioned in this chapter.

Method of separation	Type of mixture to be separated	Example
Filtration	An insoluble solid mixed with a liquid	Sand from a mixture of sand with salt water

Figure 1.22

The other methods listed should include distillation, fractional distillation, evaporation, sublimation, crystallization, the use of a centrifuge and the use of a separating funnel.

2 Copy and complete the diagram in figure 1.23, giving examples of the important chemicals obtained from the earth, the sea, plants, the air and from coal and oil.

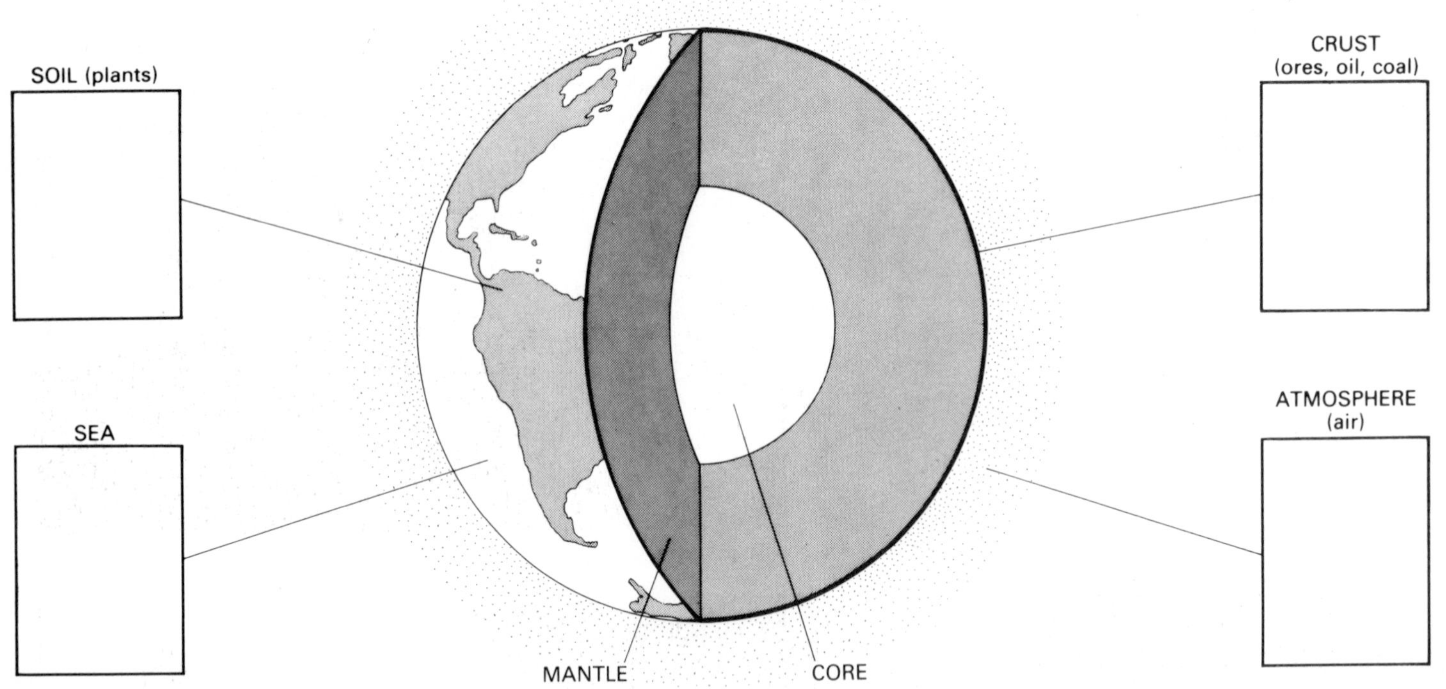

Figure 1.23

2 Elements and compounds

Figure 2.1 Chemical magic! Crude oil is refined to produce the chemicals needed to make modern materials used in such things as climbing ropes, portable boats and dry ski slopes

2.1 Chemical changes

The separations described in chapter 1 are all ways of getting pure chemicals from natural mixtures. These changes, such as melting, evaporation and dissolving, are sometimes called *temporary* (*or physical*) *changes* because they are easily reversed. No chemical magic is involved. The salt and sulphur are in the ground in an impure state. They are extracted in a pure state, but nothing new is made. Chemists can seem like magicians when they conjure new products from raw materials. Coal and oil can be made into dyes, fertilizers, medicines and plastics. Chlorine for sterilizing water comes from salt. Sulphuric acid is produced from sulphur.

Unlike magicians, chemists like to be able to understand and explain what they are doing. They have learnt that it is useful to sort substances into groups and

Question

Sort the substances listed on the right
(a) into two groups – natural and synthetic.
(b) into two groups – coloured and colourless.
(c) into three groups – solids, liquids and gases at room temperature.

patterns. A common classification is the division of substances into solids, liquids and gases at room temperature.

In the following list some of the substances are chemically simple, while others are complicated: salt, sugar, alcohol (ethanol), polythene, natural gas, starch, limestone, argon, mica, copper, gold, water, magnesium, nitrogen, bromine, oxygen, iodine, sulphur, mercury. It is impossible to tell which are which just by looking at them.

One way to find out whether a substance is simple or complicated is to try to take it apart. Chemists have to be *analysts*, trying to split things up to see what they are made of. Two forms of energy – heat and electricity – are often used in analysis. The use of heat to break things up is called **thermal decomposition**. Splitting things with electricity is called **electrolysis**.

Experiment 2a
Heating copper(II) sulphate crystals

copper(II) sulphate

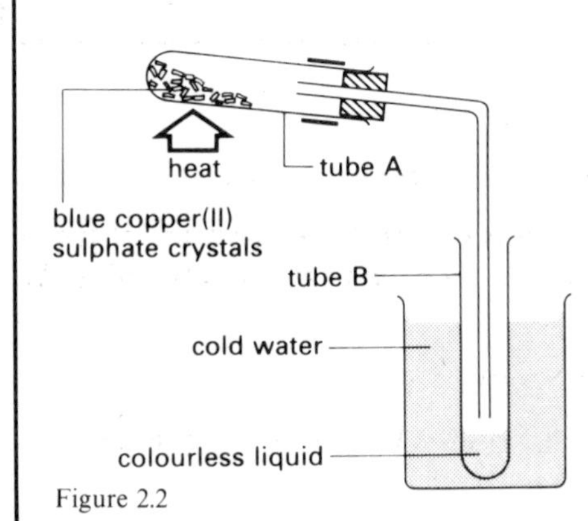

Figure 2.2

Figure 2.2 shows a simple apparatus for investigating the effect of heat on blue copper(II) sulphate crystals. Test-tube A and its contents may be weighed before and after heating.

Results
When the crystals are heated they gradually change from blue to a whitish–grey colour. At the same time a vapour is given off which condenses in tube B as a colourless liquid. The liquid collected in tube B boils at 100 °C.

Sample weighings:
Mass of tube A empty = 12.15 g
Mass of tube A + crystals before heating = 14.68 g
Mass of tube A + white solid after heating = 13.77 g

Discussion
Why is tube B surrounded by cold water? What is the liquid that condenses in tube B? Why does tube A lose mass during the experiment?

In this experiment the blue crystals are decomposed. They have been turned into a white powder and a colourless liquid. Calculate the mass of blue crystals which were heated. Also work out the mass of liquid driven off.

The results of experiment 2a can be written as a word equation.

Blue crystals(s) ⟶ White powder(s) + Colourless liquid(l)

The symbols for solid (s) and liquid (l) are examples of **state symbols** which show the physical states of the substances in the chemical change.

The colourless liquid given off when copper(II) sulphate crystals are heated is water. Before heating, the water was trapped in the crystals. This is an example of **water of crystallization**. The white powder formed by removing the water is called **anhydrous** copper sulphate. The word 'anhydrous' means 'without water'. So the word equation can be rewritten:

Blue copper(II) sulphate(s) ⟶ Anhydrous copper(II) sulphate(s) + Water(l)

(reactant: Blue copper(II) sulphate(s); products: Anhydrous copper(II) sulphate(s) + Water(l))

Question

Use the results of experiment 2a to calculate the mass of water in 100 g of the blue crystals. (This is the percentage of water of crystallization in blue copper(II) sulphate.)

The thermal decomposition of blue copper(II) sulphate is an example of a chemical reaction. In a reaction, one or more reactants changes into new products.

The decomposition of copper(II) sulphate crystals is **reversible**. When water is added to cold, anhydrous copper(II) sulphate, it turns blue again. This change can be used to test for the presence of water.

Cobalt chloride is another substance which changes colour when it loses its water of crystallization. The crystals are pink. On heating they turn blue. Cobalt chloride test papers turn blue when dried in an oven. The blue strips act as water detectors – they turn pink if moisture is present.

Experiment 2b
The electrolysis of lead(II) bromide

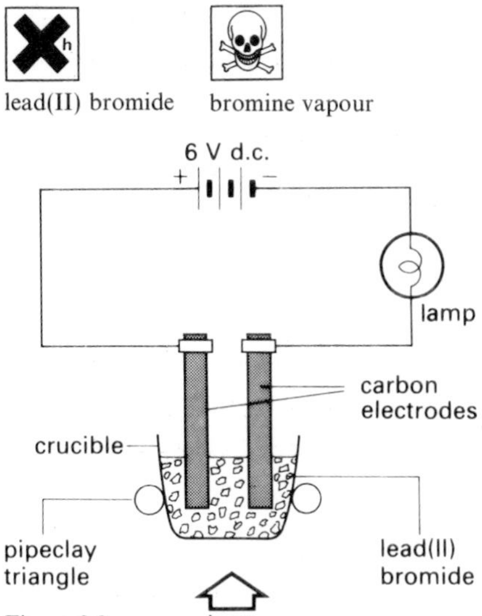

lead(II) bromide bromine vapour

6 V d.c.
lamp
carbon electrodes
crucible
pipeclay triangle
lead(II) bromide
heat

Figure 2.3

Figure 2.3 shows how the effect of electricity on lead(II) bromide can be investigated. The bulb in the circuit glows if an electric current flows. The apparatus is set up in a fume cupboard because poisonous vapours are given off during the experiment.

Results
The bulb does not light while the lead bromide is solid, because solid lead bromide does not conduct. When it is heated it melts and then the bulb lights. Bubbles of gas form round the positive rod (the electrode called the **anode**) and each bubble produces a puff of orange–brown vapour. At the end of the experiment the liquid can be poured off to show that a shiny bead of metal has collected on the bottom of the crucible.

Discussion
What could be done to show that it is the electric current and not the heat which decomposes the lead bromide? How could the apparatus be changed to show that the metal is formed at the negative rod (the electrode called the **cathode**)? The orange–brown vapour formed at the anode condenses to a red liquid at room temperature. How could the apparatus be changed to collect and condense this product?

Electrolysis splits lead bromide into two products:

Lead(II) bromide(l) ⟶ Shiny metal(l) + Orange–brown vapour(g)

The word equation shows the states of the substances at the temperature of the experiment. The shiny metal is lead and the orange–brown vapour is bromine. At room temperature the word equation is:

Lead(II) bromide(s) ⟶ Lead(s) + Bromine(l)

Thermal decomposition and electrolysis are used on a large scale in industry to

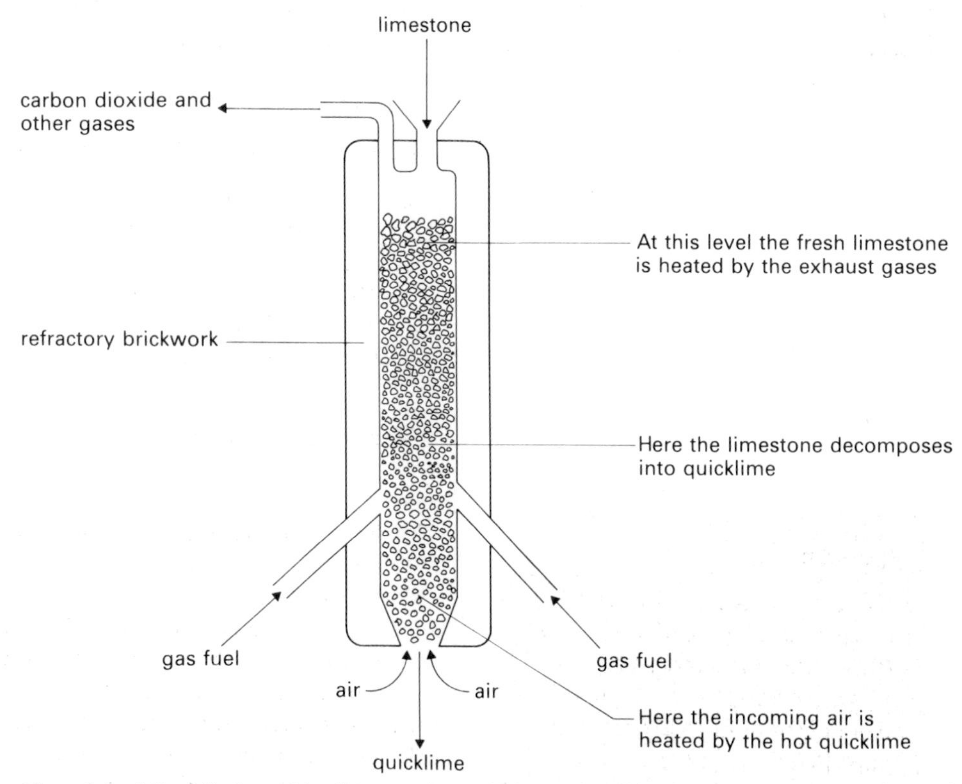

Figure 2.4 A limekiln for making limestone into quicklime

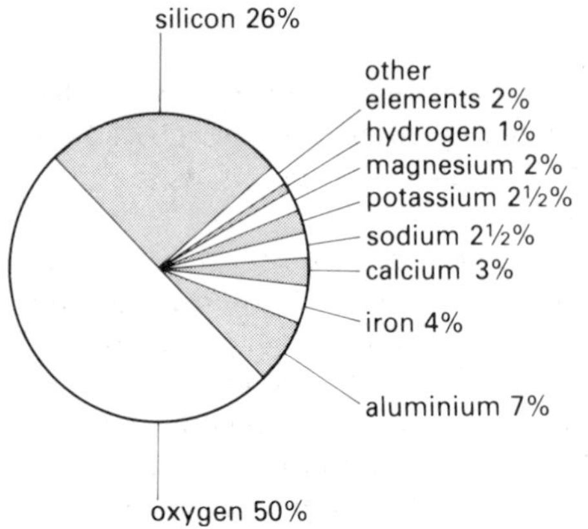

Figure 2.5 Composition of the earth's crust

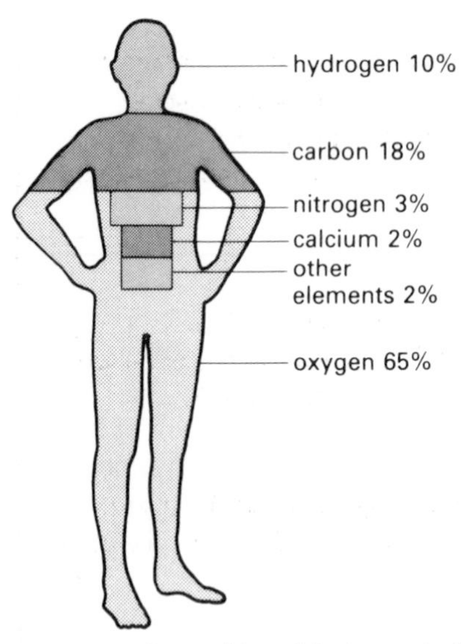

Figure 2.6 Composition of the human body

manufacture chemicals. About one-fifth of the limestone quarried is decomposed in limekilns (see figure 2.4). The word equation for the reaction in the kiln is:

Limestone(s) ⟶ Quicklime(s) + Carbon dioxide(g)

Quicklime is used in agriculture to condition the soil; it is also used to make alkalis. The uses and properties of alkalis are described in chapter 22.

Electrolysis is used to extract metals such as sodium, magnesium, calcium and aluminium from their ores. The chemistry of these processes is given in chapter 13. Other important chemicals made by electrolysis are chlorine and the alkali, sodium hydroxide (see chapter 22).

2.2 Elements

Using heat and electricity, chemists have found which substances can be decomposed and which cannot. The substances which cannot be split up in any way are called **elements**. These are the basic things from which everything is made.

The history of the discovery of the elements is an adventure story full of excitement and competition. It can be as thrilling for a scientist to make a new discovery as it is for a climber to be the first person to reach a mountain peak. In 1807, Humphry Davy investigated the electrolysis of molten potassium hydroxide. As the shiny beads of the element potassium appeared and caught fire in the air, he could not contain his joy but leapt about the room in delight. In his notes he described this as a 'capital experiment' (see figure 14.2 on p. 97).

There are over a hundred known elements, so it is surprising that astronomers think that about 90 per cent of the universe consists of the one element hydrogen. Another 9 per cent is helium, leaving only 1 per cent for all the other elements. In the night sky we see millions of stars and few planets. Stars consist mainly of hydrogen. However, the proportions of the elements in the earth, the sea and the air, and in the human body are quite different, as shown in figures 2.5 and 2.6.

Every element has been given a chemical symbol based on either its English or its Latin name. Many of the symbols are made up of two letters. The first letter is a capital and the second is small. A full list of these symbols can be found in figure 14.3 on p. 97.

Most of the elements are metals. It is usually quite easy to recognize a metal by its properties. Metals are shiny, strong, bendable and good conductors of heat and electricity. Elements which do not have these typical properties are classed as non-metals. There are only 22 non-metals. There are a few elements which are difficult to classify because they seem to have some metallic and some non-metallic properties. Examples are silicon and germanium. These elements are sometimes called metalloids. The properties of metals and non-metals are compared in figure 2.7.

Property	**Metals**	**Non-metals**
State at room temperature	All solids except mercury	Eleven gases and one liquid (bromine); the rest are solids
Melting point	Usually high	Usually low
Appearance	Shiny when polished	Dull
Effect of bending and hammering	They can be bent or hammered into shape	They snap or crumble and are brittle
Electrical conductivity	Good conductors	Poor conductors

Figure 2.7

Questions

1 What are the symbols of these elements: copper, lead, sodium, potassium, nitrogen?
2 What are the names of the elements with these symbols: Ag, Br, Cl, H, Li, N, O, P, Si, Sn, U, Xe?

2.3 Compounds

Compounds are pure substances made up of two or more elements joined together. Compounds can be decomposed by heat or electricity because they can be split up into their elements. For example, in experiment 2b, the compound lead(II) bromide splits into its elements, lead and bromine.

Joining elements to form compounds is called **synthesis**. Synthesis is a chemical reaction. The mixture of elements usually has to be heated to start the change, but once the reaction has started it will often continue on its own, giving out more heat. A chemical reaction which gives out heat is called an **exothermic** reaction.

A *mixture* has the properties of the substances in it and can be separated by the methods described in chapter 1. There is no reaction when a mixture is made.

A *compound* has new properties different from those of its elements. There is a chemical reaction when a compound forms and a chemical reaction is needed to split it up again. Heat is given out or taken in during these reactions.

Experiment 2c
Water as a product of burning

hydrogen

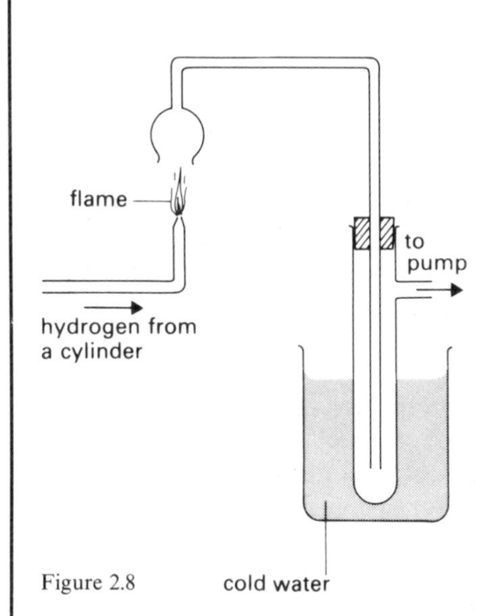

Figure 2.8

The apparatus shown in figure 2.8 is designed to show what happens when hydrogen burns in air. The hydrogen gas comes from a cylinder and reacts with oxygen in the air. The products of burning are drawn into the apparatus by a pump. The test-tube surrounded by cold water acts as a condenser.

Results
The reaction of hydrogen with oxygen gives out much energy, so there is a very hot flame. During the experiment a colourless liquid condenses in the test-tube. If the liquid in the test-tube is heated, it is found that it boils at 100 °C.

Discussion
This experiment suggests that water is a compound of hydrogen and oxygen. The two elements join in the flame and the water formed is collected in the cool test-tube. The word equation is:

Hydrogen(g) + Oxygen(g) $\longrightarrow$ Water(l)

The experiment described is not good enough, by itself, to prove that this is the correct equation. The hydrogen might have reacted with another gas in the air. The water might not be the only product. How would you redesign the apparatus to show that the hydrogen *is* reacting with oxygen? What would you do to find out whether the liquid in the cooled tube is a single substance or a mixture?

Experiment 2d
Combining carbon with oxygen

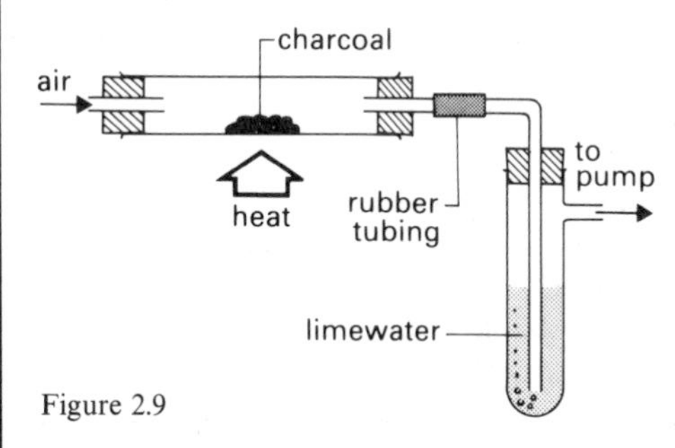

Figure 2.9

The apparatus shown in figure 2.9 can be used to investigate a simple synthesis reaction. Charcoal consists of the element carbon. It is heated in a stream of oxygen, which is also an element.

Results
Before heating, there is no reaction and the bubbles of oxygen have no effect on the limewater. On heating, the charcoal starts to glow brightly and burns away. Now the gas bubbling through the limewater makes it turn cloudy white.

Discussion
What are the signs that a chemical reaction takes place? In what ways is the gas formed different from the elements used to make it?

Questions

1 Sort the substances listed on p. 14 into two groups: elements and compounds.
2 Which elements are present in the compounds with these names: sodium chloride, potassium nitrate, zinc sulphide, magnesium bromate?

Carbon and oxygen join to form a colourless gas called carbon dioxide. Carbon dioxide is easily recognized: it is colourless and has no smell but turns limewater cloudy white.

Carbon(s) + Oxygen(g) $\longrightarrow$ Carbon dioxide(g)

The names of simple compounds end either as 'ide', 'ate' or 'ite'. The ending 'ide' shows that the compound contains only the elements mentioned in the name. Lead bromide is a compound of lead and bromine. Carbon dioxide is a compound of carbon and oxygen. The ending 'ate' or 'ite' shows that the compound contains oxygen as well as the elements mentioned in the name. The chemical name for limestone is calcium carbonate which shows that it is made from the three elements, calcium, carbon and oxygen. Copper sulphate contains copper, sulphur and oxygen.

Since the beginning of the last century, many of the new ideas in chemical theory have come from making measurements. Modern chemistry began when scientists such as Lavoisier and Dalton realized the importance of accurate weighing to discover the changes in mass which happen during chemical reactions.

Experiment 2e
Measuring the mass change when magnesium burns

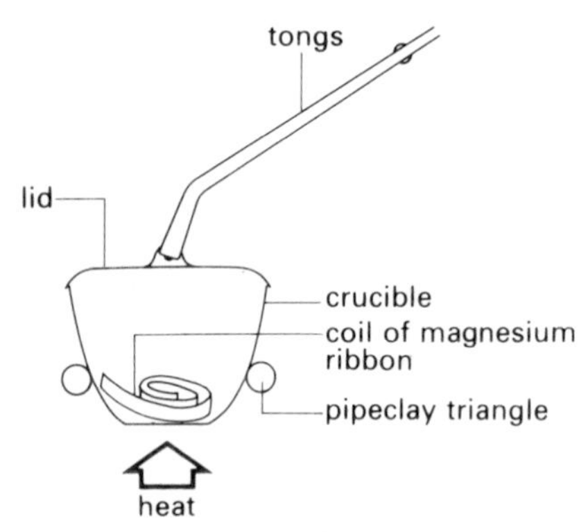

Figure 2.10

The length of magnesium ribbon is cleaned, coiled and then placed in a crucible which has been weighed. The crucible is weighed again before and after heating. The lid is needed to stop the product escaping. The lid is lifted with tongs from time to time to let in more air.

Results
The magnesium glows brightly, each time the lid is raised. It gradually changes to a greyish–white ash.

Typical results are:
Mass of crucible + lid = 23.50 g
Mass of crucible + lid + magnesium before heating = 23.68 g
Mass of crucible + lid + product after heating = 23.80 g

Discussion
How could the weighings be used to show that the reaction was complete?

The word equation for the reaction is:

Magnesium(s) + Oxygen(g) $\longrightarrow$ Magnesium oxide(s)

The table in figure 2.11 includes results from a series of experiments. Plot the results on a graph with the mass of oxygen on the vertical (y) axis and the mass of magnesium on the horizontal (x) axis. Use the measurements given above to add one more point to the graph.

Experiment number	Mass of magnesium (g)	Mass of oxygen (g)
1	0.06	0.04
2	0.15	0.10
3	0.22	0.14
4	0.30	0.20
5	0.28	0.18
6	0.10	0.06

Figure 2.11

Most simple chemical compounds have a definite composition. The straight line graph from experiment 2e shows that the mass of oxygen in the compound increases in exact proportion to the amount of magnesium. Every sample of magnesium oxide contains 60% of magnesium and 40% of oxygen by mass. This is an example of the *law of constant composition.*

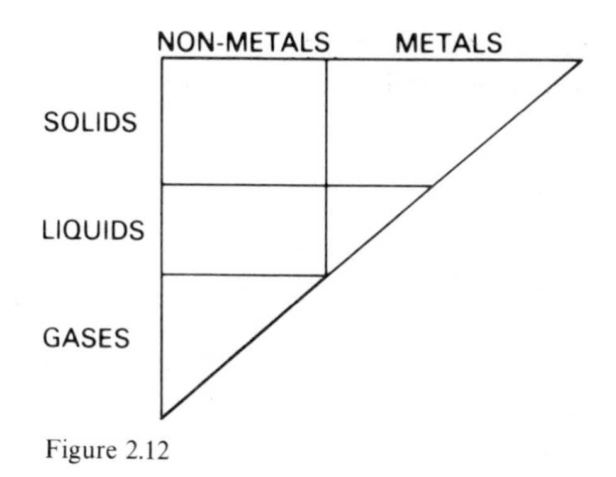

Figure 2.12

Careful measurements have shown that, if you take all the reactants and products into account, there is no change in mass during a chemical reaction. This is the basis of the term 'equation'. The total mass of the products of a reaction is equal to the total mass of the reactants. This is the *law of conservation of mass*, which states that matter is neither created nor destroyed during a chemical reaction.

Summary

1 Make a large copy of the outline table in figure 2.12. Write into the spaces in the table the symbols of elements which belong in each area. Table 1 on page 336 will help you.
2 Compounds can be made from elements by *synthesis*. Compounds can be split up by *thermal decomposition* or by *electrolysis*. Explain what the words in italics mean, using examples from this chapter. Write a word equation, with state symbols, for each chemical change you include in your summary.
3 What do chemists understand by the laws of conservation of mass and of constant composition?

3 Atoms, molecules and ions

3.1 Particles

Imagine cutting a block of iron in half, then in half again, and so on. Can you keep on cutting it in half for ever or is there a limit – a particle which cannot be broken down further?

About 2500 years ago a Greek philosopher called Democritus suggested that all matter (solids, liquids and gases) is made up of tiny, indivisible particles. We have evidence for this from simple experiments.

Bromine vapour is very heavy but it spreads throughout a gas jar of air. This is **diffusion**. Particles of bromine intermingle with particles in the air. It is difficult to explain this observation without using the idea of particles.

Questions

1 Using the idea of particles, explain why
(a) ether evaporates quickly and can be smelled all round a room.
(b) salt disappears when it is stirred into water.
(c) a purple colour spreads through water from a small crystal of potassium manganate(VII).

2 Think about the steps in making a cup of sweet tea. What do you imagine happens to the particles of water, milk, tea and sugar as the tea is made?

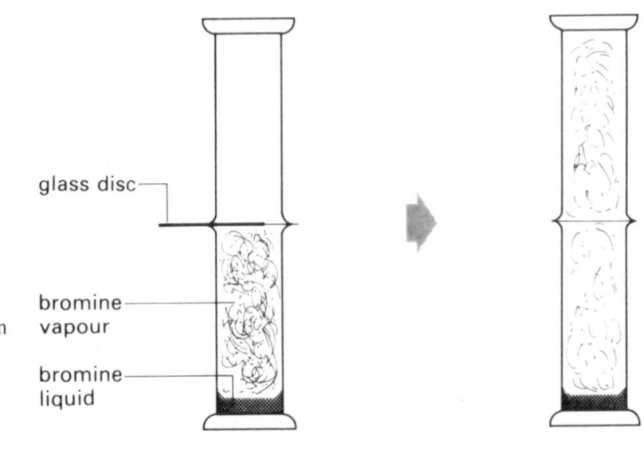

Figure 3.1 The bromine diffusion experiment: when the glass disc is carefully removed, the bromine vapour spreads out to fill both gas jars

Experiment 3a
Diffusion of ammonia and hydrogen chloride

concentrated hydrochloric acid

concentrated ammonia solution

In the tube in figure 3.2, hydrogen chloride is released from the cotton wool plug soaked in concentrated hydrochloric acid. Ammonia is released from the concentrated ammonia solution.

Results
After about ten minutes, a white ring of smoke forms about two-fifths of the way along the tube from the hydrochloric acid end.

Discussion
Ammonia and hydrogen chloride particles diffuse along the tube. When they meet, a white solid, ammonium chloride, is formed. One gas diffuses faster than the other. Which one? Why is there a difference in speed? Why does it take some time for the gases to meet? Does the air have any effect on the speed of diffusion?

cotton wool soaked with concentrated hydrochloric acid

glass tube about 1 m long

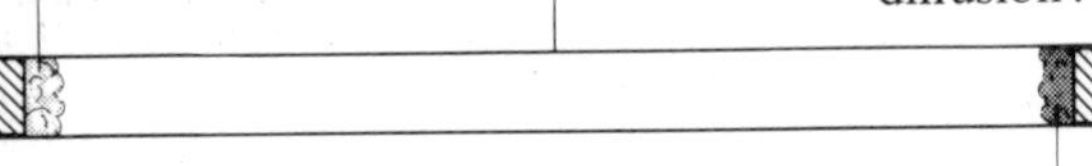

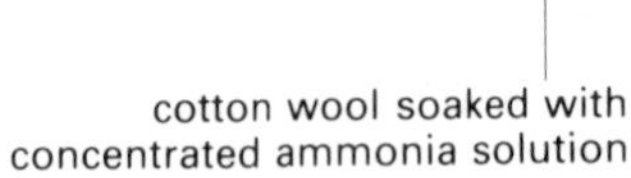

cotton wool soaked with concentrated ammonia solution

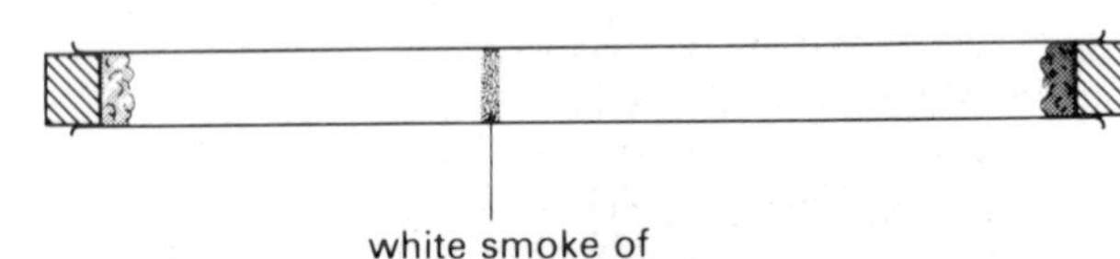

Figure 3.2

3.2 Particles in solids, liquids and gases

Substances can exist in three states – as solids, liquids or gases. From their properties we can picture how the particles behave in these states.

Solids

Solids keep their own shape, and many solids form crystals with regular shapes. Some mineral crystals are shown in figure 18.1 in chapter 18. When crystals grow the shape remains the same. Most solids are difficult to compress.

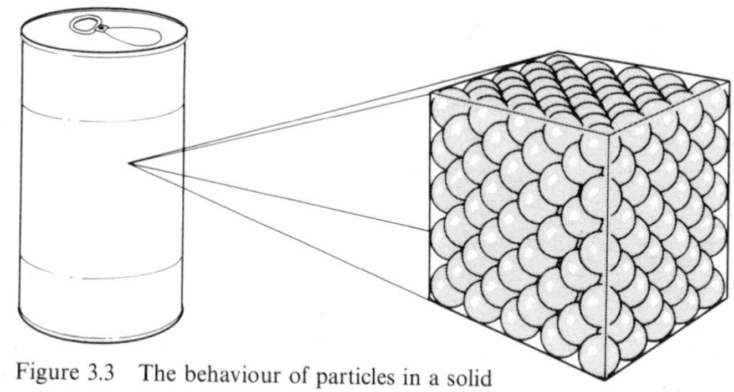

Particles in a solid are packed close together in a regular way. The particles do not move freely, but vibrate in their positions.

Figure 3.3 The behaviour of particles in a solid

Liquids

Liquids take the shape of the container they are in and are easily poured. Some liquids, like water, flow easily. Others, like oil and treacle, are thick and sticky. They are said to be **viscous**. Liquids are difficult to compress.

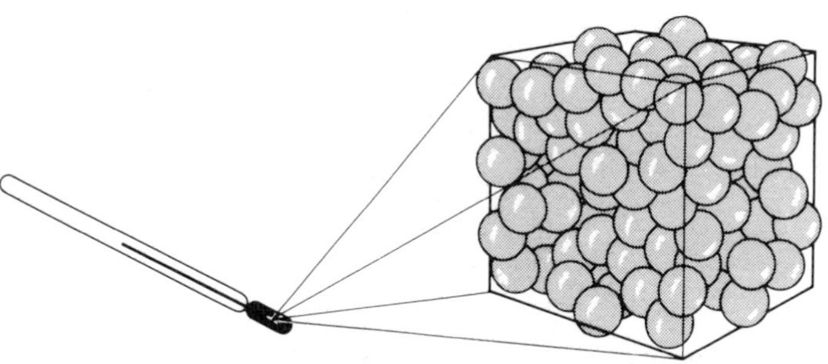

The particles in a liquid are closely packed but are free to move around, sliding past each other.

Figure 3.4 The behaviour of particles in a liquid

Gases

Gases fill the volume available and are easily compressed. They are much less dense than solids and liquids.

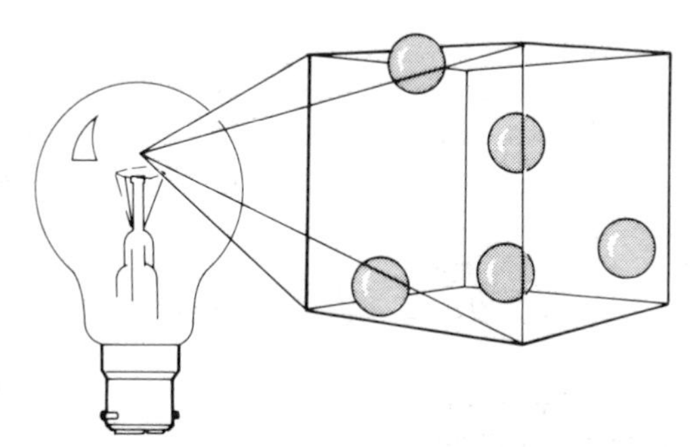

In a gas the particles are spread out, so the densities of gases are very low compared with solids and liquids. The particles move rapidly in a random manner, colliding with other particles and the vessel's walls. Pressure is caused by particles hitting the walls. Light particles move faster than heavier ones.

Figure 3.5 The behaviour of particles in a gas

Questions

1 Name some substances which you have seen as crystals. Is a sugar cube a crystal?
2 Treacle is more viscous than water. Can you suggest why?
3 Why is a liquid (the hydraulic fluid) used to operate the brakes and clutch in a car?
4 Explain why a small volume of water in a kettle can fill a kitchen with steam.
5 How does the pressure of a gas change with (a) temperature and (b) volume? Explain the changes in terms of particles.

The study of *Brownian motion* gives more evidence that particles exist and that they are in constant, random motion in liquids and gases. This was first noticed by Robert Brown, a botanist, in 1827 as he used a microscope to study pollen grains suspended in water.

Experiment 3b
Brownian motion

Brownian motion can be seen by observing smoke particles using the apparatus shown in figure 3.6.

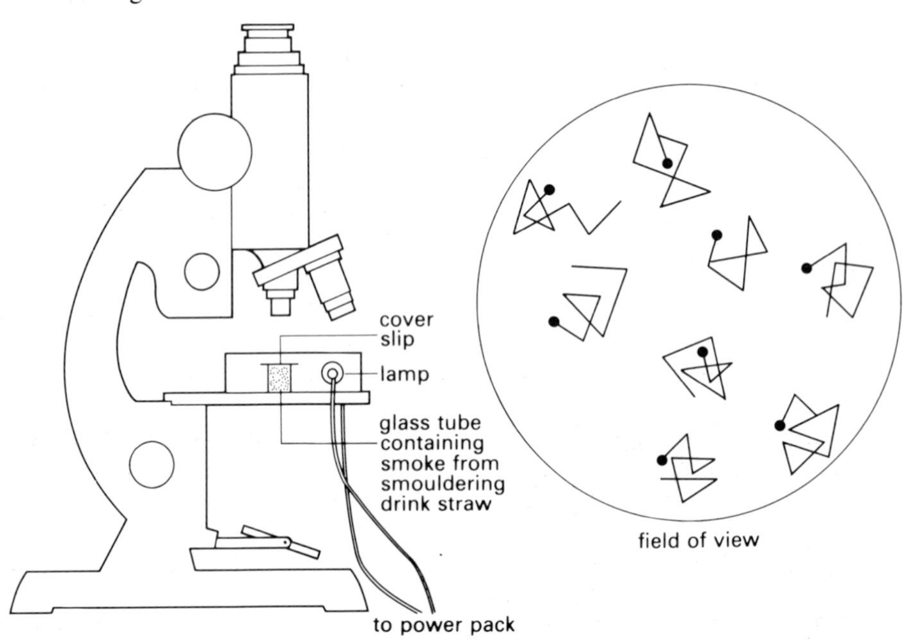

Figure 3.6

Results
Smoke particles move randomly with a jerky, haphazard movement. This is Brownian motion.

Discussion
Imagine an aerial view of a skating rink with a large, but underweight, elephant (on skates, of course) in the middle, and hundreds of small children charging around the ice in all directions. From the viewer's position the children are invisible. Inevitably, a child soon collides with the elephant, which is shunted to one side. Then a charge from the opposite side causes the elephant to change direction, and so on. How would you describe the motion of the elephant over a long period of time?

When studying smoke the visible particles correspond to the elephant, and the children to the air particles, which cannot be seen using the microscope. The movement of the visible particles suggests that invisible particles are moving about rapidly in a haphazard way and colliding with the larger particles.

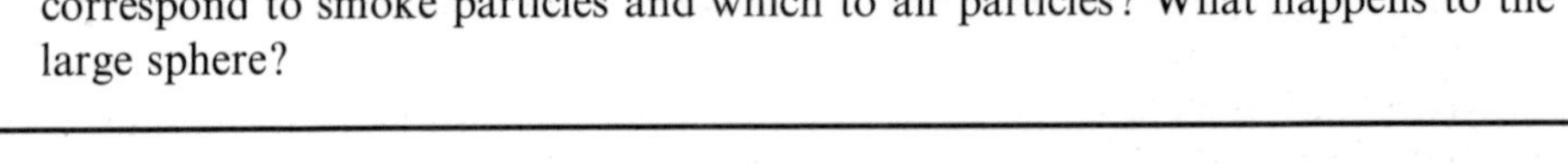

A model of Brownian motion is shown in figure 3.7. Which spheres correspond to smoke particles and which to air particles? What happens to the large sphere?

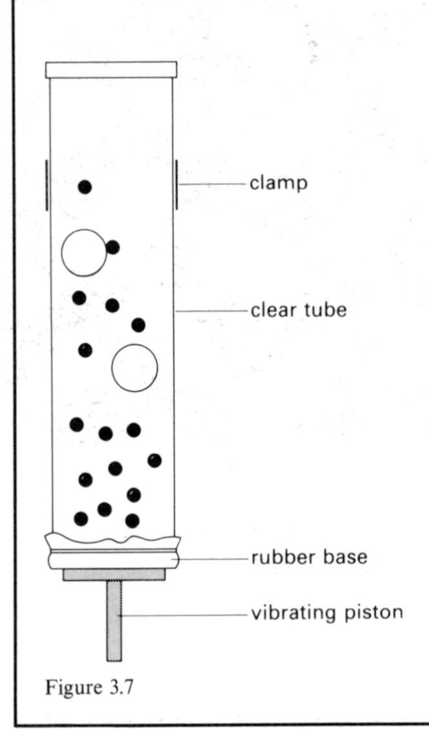

Figure 3.7

The particles which make up substances must be very small. Very small things are measured in nanometres (nm). The use of this unit is explained on p. 344.

The size of the particles in olive oil can be estimated from a simple experiment. It depends on the fact that when the oil is dropped onto water it spreads out to give a film just one particle thick.

Experiment 3c
How small are the particles in olive oil?

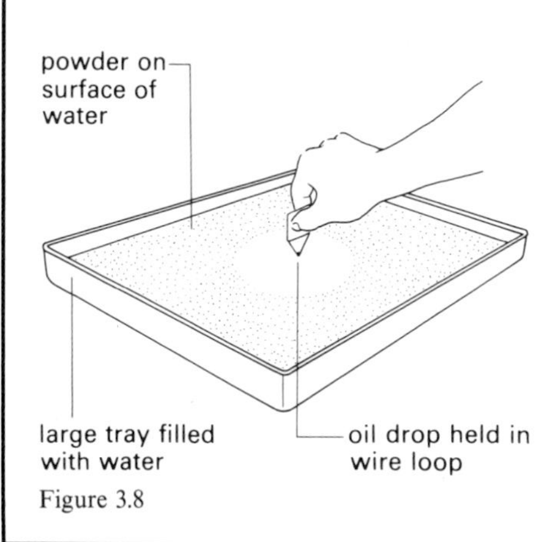

Figure 3.8

An oil drop about 0.5 mm in diameter is caught on a wire loop. An oil film is formed as it touches the powdered water surface and spreads out.

Results
Typical results are:
Diameter of oil drop = 0.5 mm
Approximate size of the oil film = 240 mm × 260 mm

Discussion
Assuming that the oil drop is a small cube with each side 0.5 mm long, what is its volume?

What is the area of the oil film in mm^2?

The volume of the film is equal to the area multiplied by the thickness. Work out the thickness of the film in mm. What is the thickness in nm?

The thickness of the film gives an idea of the length of an olive oil particle. It is about 2 nm. Each olive oil particle is, in fact, made up of several smaller particles joined together. These particles must be only a fraction of 2 nm in diameter.

Optical microscopes cannot be used to see very small particles. They can only distinguish objects about 1000 nm in diameter or larger. However, the most powerful electron microscopes developed in the last twenty years can magnify up to two million times. They have been used to observe small groups of particles (see figure 3.9). In some elements, single particles have been observed.

Figure 3.9 Very small particles can be 'seen' with electron microscopes

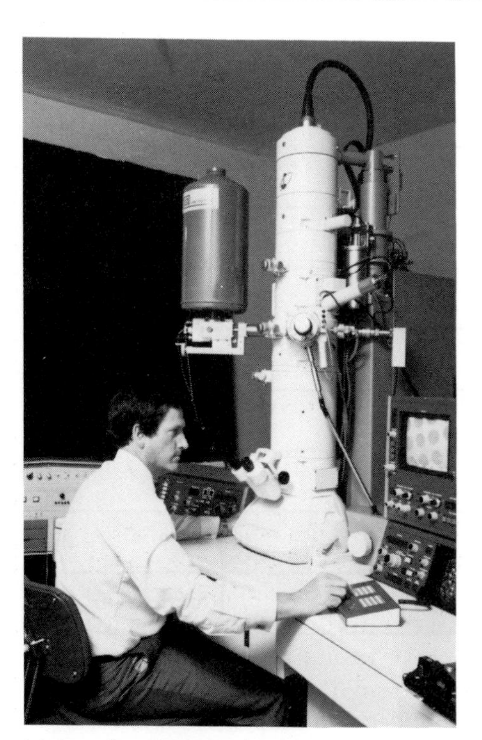

(a) A modern electron microscope

(b) An electron micrograph of a silver compound, showing individual silver particles

Questions

1 How many particles 1 nm in diameter can fit side by side inside a 1 mm division on your ruler?
2 Estimate the number of particles, each 0.1 nm in diameter, in a pin head. To make the problem easier, assume that the particles and pin head are cubes.

The particles in elements such as iron and copper are about 0.1–0.5 nm in diameter – about 100 000 times smaller than a single grain of flour!

3.3 Particles in elements

Two elements, iron and sulphur, differ in their appearance and in their chemical properties. The particles which make up the two elements are different. In 1803 John Dalton described his idea of these particles in his atomic theory of matter.

All matter, he suggested, is made up of small particles called **atoms**, which cannot be broken down. The word 'atom', or one like it – *atomos*, had been used by the Greeks 2500 years earlier. The word means 'indivisible'. Atoms cannot be created or destroyed. In Dalton's theory atoms of one element are all identical

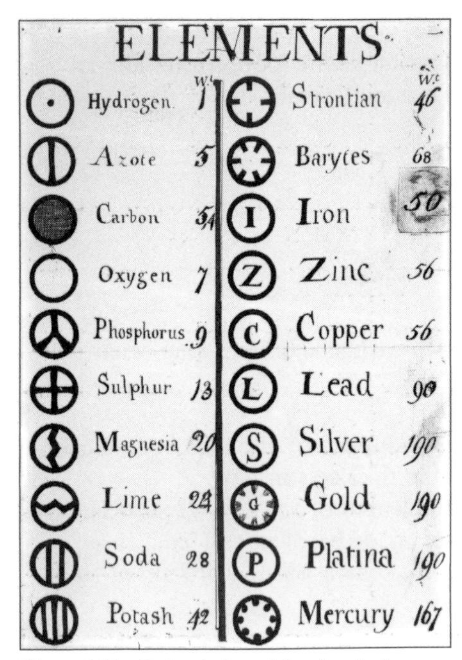

Figure 3.10 Dalton's list of the chemical elements and his symbols for them

(they have the same mass and properties), but the atoms of different elements are different. Figure 3.10 shows the symbols Dalton used for the atoms of different elements. The modern symbols are shown on p. 97.

We now know that atoms are not indivisible and that atoms of an element are not always identical (see chapter 39). But Dalton's ideas laid the foundations of modern chemistry.

All elements are made up of atoms. In some elements, particularly the metals, atoms are packed together in a regular way. Figure 18.11 in chapter 18 shows a model of a crystal of copper. When metals are melted the atoms become free to move about.

In other elements – non-metals, such as oxygen, sulphur, phosphorus, chlorine, nitrogen and bromine – small groups of atoms are joined tightly together. These groups are called **molecules**. They are the smallest particles that usually exist separately in these elements. The molecular formula is a way of describing how many atoms there are in one molecule.

An oxygen molecule consists of two oxygen atoms joined together. The molecular formula is O_2. In oxygen gas (and liquid) it is the oxygen molecules which move around freely (see figure 3.11). Some other elements which exist as molecules are shown in figure 3.12. Notice that all the common gases have diatomic molecules, i.e. they have two atoms in a molecule.

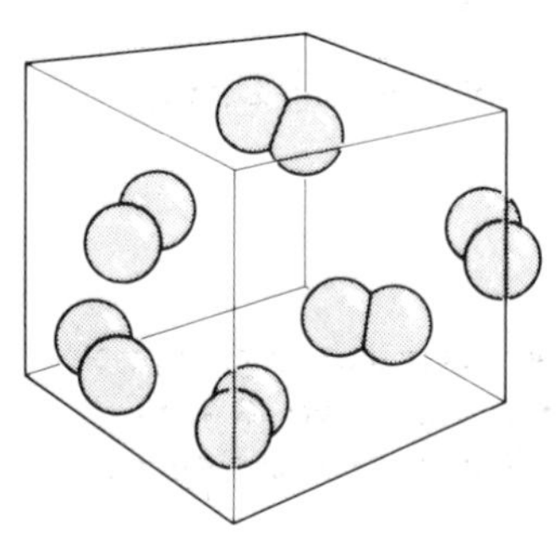

Figure 3.11 Oxygen molecules contain two atoms

Element	State at room temperature	Number of atoms in molecule	Formula
Oxygen	Gas	2	$O_2(g)$
Nitrogen	Gas	2	$N_2(g)$
Hydrogen	Gas	2	$H_2(g)$
Chlorine	Gas	2	$Cl_2(g)$
Bromine	Liquid	2	$Br_2(l)$
Iodine	Solid	2	$I_2(s)$
Phosphorus	Solid	4	$P_4(s)$
Sulphur	Solid	8	$S_8(s)$

Figure 3.12

The structure of elements is described in more detail in chapters 18 and 19.

Questions

1 Azote is listed in figure 3.10. What is the modern name for this element?
2 Which metallic element is not a solid, and so does not have a regular arrangement of atoms at room temperature?
3 Using books in the library, find out more about John Dalton.

3.4 Particles in compounds

Compounds are formed by combining two or more elements. Dalton suggested that simple numbers of atoms of each element come together. However, not all atoms combine in the same way, so we have different types of compounds.

The table in figure 3.13 shows the properties of some common compounds.

The compounds in the table can be put into two groups. The compounds in one group contain only non-metals. The compounds in the other group contain a metal combined with one, or more, non-metals.

Non-metal compounds melt and vaporize easily. Most are insoluble in water. They do not conduct electricity even if they do dissolve. They may be gases, liquids or solids at room temperature.

Compounds of metals with non-metals are solid at room temperature. They are

Question

Which of the compounds in figure 3.13 consist only of non-metals? Which consist of a metal combined with one or more non-metals? Tables 2 and 3 on pp. 337–40 may help you.

Compound	State at room temperature	Does it melt easily?	Is it soluble in water?	Does the solution conduct?
Sodium chloride	Solid	No	Yes	Yes
Calcium oxide	Solid	No	Slightly	Yes
Ethanol	Liquid	—	Yes	No
Copper sulphate	Solid	No	Yes	Yes
Methane	Gas	—	No	—
Paraffin wax	Solid	Yes	No	—
Sugar	Solid	Yes	Yes	No
Calcium sulphate	Solid	No	No	—

Figure 3.13

difficult to melt. Many are soluble in water and the solutions formed conduct electricity.

Non-metal compounds and compounds containing metals behave differently because the atoms in them combine in different ways.

Non-metal compounds

In most compounds containing only non-metals the atoms combine in groups to form molecules. This is like the non-metal elements themselves, except that in compounds different types of atoms are joined together. The molecular formula gives the number of each type of atom in the molecule.

Methane is a compound. It is the main component of natural gas. Each molecule of methane contains a carbon atom and four hydrogen atoms; the molecular formula is CH_4. Figure 3.14 shows a model of a molecule of methane. Models of other simple molecules are shown in figure 19.7 on p. 130.

Each type of atom has its own 'combining power' – the number of bonds it can form with other atoms. Some of these are given in figure 3.15. From this information the molecular formula of a compound can be predicted.

Figure 3.14 A model of a methane molecule

Element	Symbol	Combining power
Carbon	C	4
Nitrogen	N	3
Oxygen	O	2
Sulphur	S	2
Hydrogen	H	1
Chlorine	Cl	1
Bromine	Br	1

Figure 3.15

Question

For each of the following compounds, name the elements present and work out the formula: (a) water, (b) hydrogen chloride, (c) tetrachloromethane, (d) hydrogen sulphide, (e) carbon dioxide, (f) carbon disulphide. Is there a connection between the name of a compound and its formula?

Water is a compound of oxygen and hydrogen. Oxygen has a combining power of 2 and it can be represented as —O—. Hydrogen has a combining power of 1 and can be represented as H—. Two hydrogen atoms can bond to one oxygen atom:

H— —O— —H

The formula of water is H_2O.

Unfortunately, not all non-metal compound formulae can be predicted using the 'combining powers' given in figure 3.15 because some can vary. You could not predict the formula of carbon monoxide, CO, nor of sulphur dioxide, SO_2, and these have to be learnt. Fortunately, the name often gives a clue to the formula.

Metal/non-metal compounds

Compounds of a metal and one or more non-metals are **electrolytes**. They conduct electricity when molten or when dissolved in water. The electric current is carried by the movement of electrically charged particles called **ions**.

Sodium chloride consists of ions. A sodium ion is a sodium atom, but with a positive charge. It has the symbol Na^+. A chloride ion is a chlorine atom, but with a negative charge. It has the symbol Cl^-.

In a crystal of sodium chloride, sodium ions are attracted to chloride ions by electrostatic forces. Next to each sodium ion is a chloride ion, and so on. Ions pack together in a regular way, like the atoms which make up a metal structure. Figure 20.4 in chapter 20 shows a model of the structure of sodium chloride. A crystal of sodium chloride is shown on p. 117. It is cubic in shape. Does the model fit this observation?

The formula of sodium chloride is NaCl because there is one sodium ion, Na^+, for every chloride ion, Cl^-. We cannot talk about a molecule of sodium chloride.

Not all ions have single positive or negative charges like sodium and chloride ions. The formula of lead bromide is $PbBr_2$. In this compound, for every lead ion there are two bromide ions. All compounds are electrically neutral, so the charge on a lead ion must be twice that on a bromide ion. A bromide ion, like a chloride ion, has a single negative charge, Br^-, so a lead ion must have a double positive charge, Pb^{2+}.

Compound	Ions present		Formula
	Positive	Negative	
Calcium chloride A calcium ion with a 2+ charge needs two chloride ions each with a 1− charge to make an electrically neutral compound.	Ca^{2+} 2+	Cl^- Cl^- 2−	$CaCl_2$
Copper(II) sulphate The (II) indicates that the copper ion has a 2+ charge. A copper ion needs one sulphate ion which has a 2− charge.	Cu^{2+} 2+	SO_4^{2-} 2−	$CuSO_4$
Zinc nitrate In the formula the brackets around NO_3 indicate that two complete nitrate ions are needed.	Zn^{2+} 2+	NO_3^- NO_3^- 2−	$Zn(NO_3)_2$
Aluminium sulphate	Al^{3+} Al^{3+} 6+	SO_4^{2-} SO_4^{2-} SO_4^{2-} 6−	$Al_2(SO_4)_3$

Figure 3.16

Questions

1 Put the following compounds into one of two groups depending on whether they are molecular or ionic compounds: ethanol, copper oxide, lithium chloride, methane, glucose, silver bromide, naphthalene.
2 What are the formulae for the following ionic compounds: (a) potassium iodide, (b) calcium carbonate, (c) sodium sulphate, (d) calcium hydroxide, (e) aluminium chloride?

Table 8 on p. 343 gives the charges on some common ions. From this table it can be seen that

* metal ions are always positive, while non-metal ions are negative (this is true for all elements except hydrogen)
* some metals can form more than one type of ion – a Roman numeral after a name shows which ion is present
* some negative ions are compound ions containing more than one type of atom.

If the charges on ions are known, then the formulae of ionic compounds can be found because the number of positive and negative charges in a compound must be equal. The number of charges on an ion can be thought of as a 'combining power'. Figure 3.16 gives some examples.

3.5 Atoms, molecules and ions in chemical reactions

In a chemical reaction, there is no change in mass if all the reactants and products are taken into account. This, together with Dalton's suggestion that atoms cannot be created or destroyed, leads to the conclusion that in chemical reactions, atoms are simply rearranged.

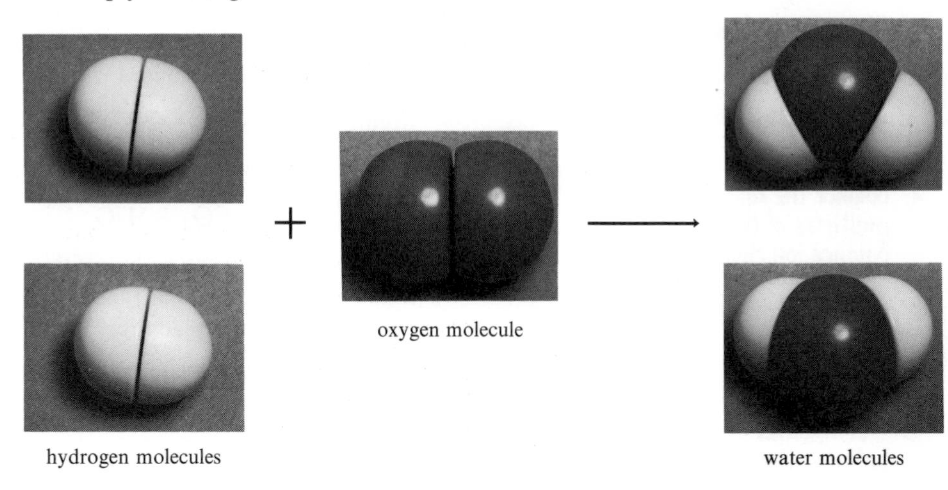

Figure 3.17 Model equation for the reaction of hydrogen with oxygen to give water

Hydrogen burns in oxygen to form molecules of water. The formula of water is H_2O. Figure 3.17 shows an 'equation' for the reaction, using models. A water molecule contains only one oxygen atom, so one oxygen molecule can give rise to two water molecules provided that there are two hydrogen molecules available to supply all the hydrogen atoms necessary. There are the same numbers of hydrogen and oxygen atoms on both sides of the equation – the atoms have simply been rearranged. The reaction can be described in shorthand using symbols and formulae. State symbols show whether the substances are gases, solids, liquids or dissolved in water.

$$2H_2(g) + O_2(g) \longrightarrow 2H_2O(g)$$

The thermal decomposition of limestone (calcium carbonate) gives quicklime (calcium oxide) and carbon dioxide (see p. 16). Calcium carbonate and calcium oxide are ionic; carbon dioxide consists of molecules. Figure 3.18 shows the model equation. The equation

$$CaCO_3(s) \longrightarrow CaO(s) + CO_2(g)$$

Questions

1 Make and draw a 'model' equation to represent the burning of methane in oxygen.
2 Write equations for the following reactions:
(a) burning magnesium in oxygen to give magnesium oxide.
(b) electrolysis of lead(II) bromide(l) to give lead and bromine.

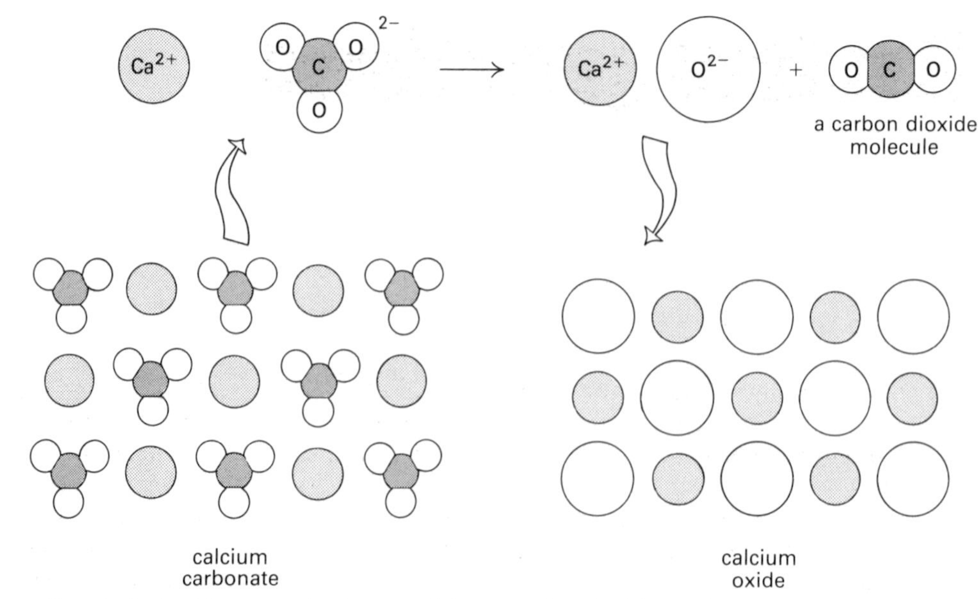

Figure 3.18 Model equation for the thermal decomposition of calcium carbonate

describes the reaction in symbols and formulae. Are there the same numbers of Ca, C and O atoms on both sides of the equation?

Writing an equation involves four steps:

* Writing a word equation for the reaction, e.g. for natural gas burning in air to give carbon dioxide and water

  Methane + Oxygen ⟶ Carbon dioxide + Water

* Representing the reactants and products using symbols and formulae

  $CH_4 + O_2 \longrightarrow CO_2 + H_2O$

* 'Balancing' the equation by writing numbers in front of the formulae so that the number of each type of atom is the same on both sides of the equation. The formulae must not be changed to balance the equation

  $CH_4 + 2O_2 \longrightarrow CO_2 + 2H_2O$

* Adding state symbols

  $CH_4(g) + 2O_2(g) \longrightarrow CO_2(g) + 2H_2O(g)$

In equations involving ions, the charges on each side of the equation must balance. These steps are summarized in figure 3.19.

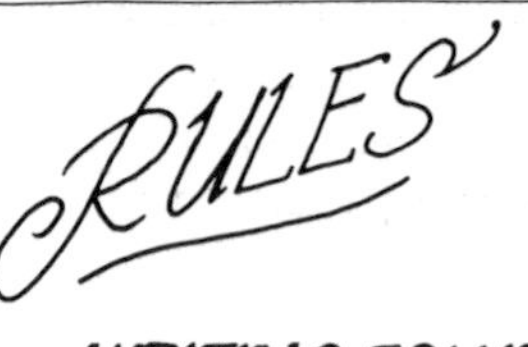

WRITING EQUATIONS

1/ Be sure the reaction takes place.
2/ Write down a 'word equation'.
3/ Write down the correct formulae for each reactant and product. (O_2, H_2, Cl_2 etc.)
4/ Balance the equation by using multiples of these formulae. Balance ion charges too!
5/ Add state symbols.

NEVER change the formula of a compound or element to balance the equation.

Figure 3.19 Writing an equation for a reaction

Summary

1 Draw diagrams to show how you picture the following processes in terms of the movement of particles: water evaporating, sugar dissolving in water, ether diffusing through air.
2 How does the idea that the particles of a gas are in rapid motion help to explain the pressure of a gas and Brownian motion?
3 What happens to atoms when they turn into molecules? Give examples of molecules of elements and molecules of compounds. Draw diagrams and give the molecular formulae.
4 What is the difference between an atom and an ion? What types of compound are ionic? Which elements form positive ions? Which elements form negative ions?

Review questions

1 Chromatography, crystallization, distillation, filtration, heating and weighing, solution, sublimation, washing. Which *one* of these processes would you use for each of the following?
(a) To obtain sand from a mixture of sand and sea water.
(b) To remove the *final traces* of sea water from the sand in part (a).
(c) To extract the red colouring from some crushed flower petals.
(d) To discover whether the red colouring from the crushed flower petals was a single colour or a mixture of colours.
(e) To obtain pure water from a salt solution.
(f) To make sure that a sample of copper(II) oxide was completely free from moisture.
(g) To purify a sample of ammonium chloride by heating.
(SREB)

2 Select the appropriate word from the following list to complete the sentences given below: condensation, decomposition, evaporation, freezing, melting, sublimation, dissolving, diffusion.
(a) A wet school yard dries in the wind because of ... .
(b) Grease spots can be removed from clothing by ... with tetrachloromethane (carbon tetrachloride).
(c) Ammonia solution spilled in a laboratory can be smelled in every corner because of ... .
(d) The process in which heated iodine turns into a gas without melting is called ... . **(WJEC)**

3 (a) For a pure substance, state what is meant by (i) melting point and (ii) boiling point. What is the effect of impurities on these *two* physical properties?
(b) (i) State what is meant by sublimation.
(ii) Give the name and formula of a substance that sublimes.
(c) Explain what is meant by 'distillation'. State briefly why the distillation of crude oil is referred to as 'fractional distillation'. **(WJEC)**

4 Propanone (acetone) may be used to extract many of the chemicals present in grass. It is flammable and has a boiling point of 56 °C (329 K). A student prepared a propanone extract of grass, but decided that it was too dilute to use.
(a) Explain how the solution could be safely concentrated.
(b) Describe the effect of putting a drop of the concentrated extract on filter paper, allowing it to dry and then spotting it at intervals with further drops of propanone.
(c) State what information this experiment gives about the nature of grass.
(d) In a similar experiment, a pupil decided to use water instead of propanone, but his attempt failed. Explain why this should be so.
(e) Name *one* other solvent apart from water and propanone and suggest a simple test or experiment to illustrate its solvent property. **(O)**

5 In a chemistry textbook published early in the last century, reference is made to *alembics* which were used 'to separate liquids of different degrees of volatility (i.e. different boiling points) and to preserve the more volatile or both of them'.

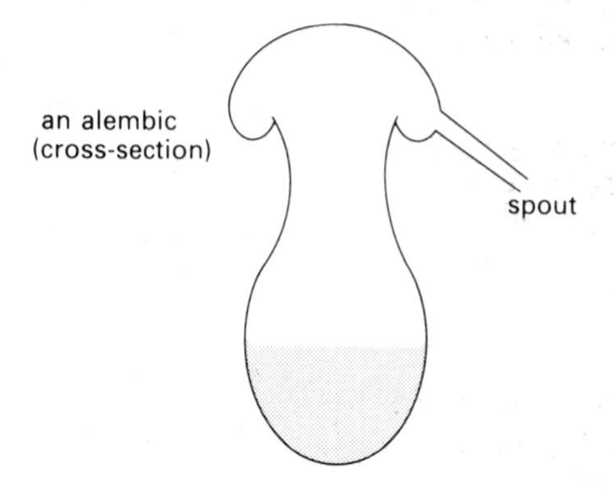

(a) Name *two* liquids which when mixed could be separated by means of an alembic.
(b) Explain clearly the principles underlying the operation of an alembic.
(c) As a piece of chemical equipment, the alembic was far from efficient. Give *two* reasons for this. **(O)**

6 (a) The following is a list of common substances: air, carbon, sodium hydroxide, chlorine, iron, sugar, sulphur, copper, sodium carbonate crystals, water, brass. From the list name
(i) *two* metallic elements,
(ii) *two* non-metallic elements,
(iii) *two* compounds,
(iv) *one* mixture.
(b) (i) Name the *four* elements contained in sodium carbonate *crystals.*
(ii) Give the formula for sodium carbonate crystals.
(iii) Give one use for sodium hydroxide other than in the laboratory. **(SREB)**

7 (a) The apparatus drawn below can be used to determine the melting point of naphthalene, which melts between 75 °C and 85 °C.

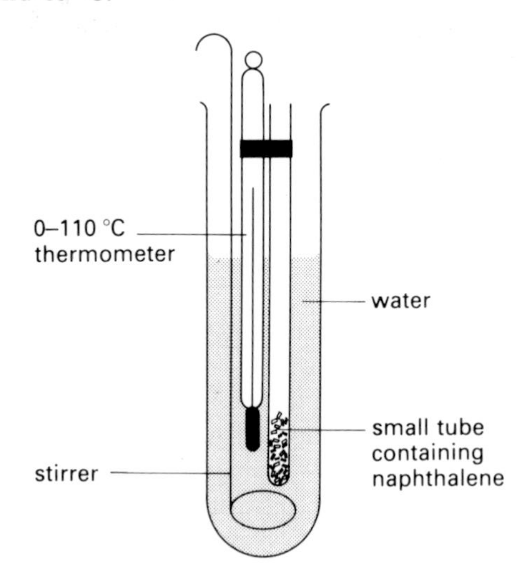

(i) Why is a stirrer necessary?
(ii) Why must the tube be heated *gently*?
(iii) How will you know when the melting point of the naphthalene has been reached?
(iv) State and explain *two* changes that would be necessary to the above apparatus in order to determine the melting point of a solid which melts between 120 °C and 125 °C.

(b) A supply of hydrogen was burned in air and the resulting gases passed through a cooled tube. A colourless liquid collected which was found to boil at 100 °C.
(i) Give the common and chemical names for the colourless liquid.
(ii) Why was it necessary to pass the resulting gases through a *cooled* tube?
(iii) How would the gases be made to pass through the cooled tube?
(iv) Draw a simple labelled diagram of the apparatus you would use to measure the boiling point of the colourless liquid.

(c) Some iron filings and some powdered sulphur were mixed together in the correct reacting proportions. The mixture was heated and after a short time the heat was removed and the mixture continued to glow. A dark grey mass was left.
(i) Why was the iron used in the form of filings and the sulphur powdered?
(ii) What was the dark grey mass which was left?
(iii) What would happen if dilute hydrochloric acid were added to the mixture of iron and sulphur? What would happen if dilute hydrochloric acid were added to the final dark grey solid?
(iv) State *one* method by which a mixture of iron and sulphur could be separated.
(v) What was the cause of the glow in the experiment described above?
(vi) What do you understand by the term 'correct reacting proportions'? **(SEREB)**

8

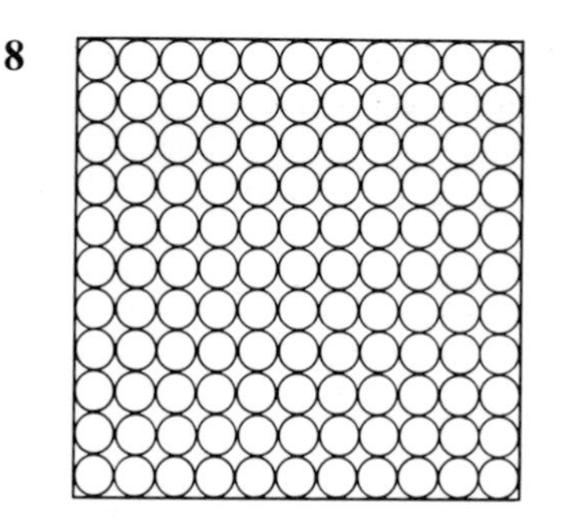

The diagram above shows the arrangement of particles in a solid element having a melting point of 40 °C and a boiling point of 180 °C. Draw diagrams to show the arrangement of the particles (a) at 62 °C and (b) at 191 °C. **(EMREB)**

9 In the boxes below, different atoms are represented by ○ and ●.

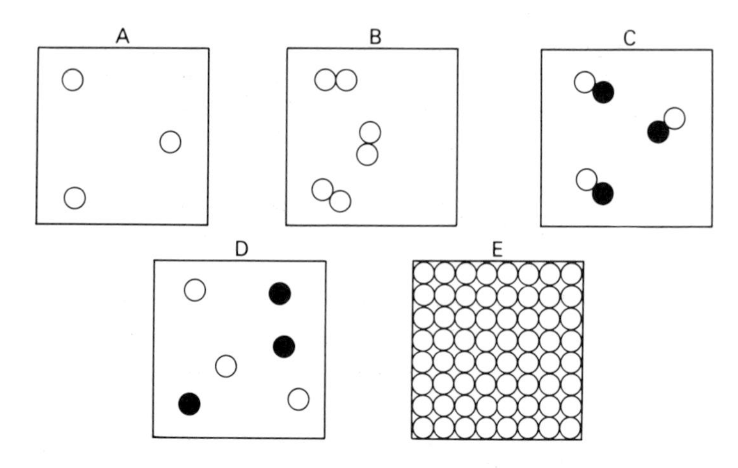

Which box represents:
(a) a mixture of gases?
(b) a gaseous compound?
(c) a solid?
(d) oxygen?
(e) a simple gas such as neon? **(EMREB)**

10 Ammonia gas reacts with hydrogen chloride gas to produce a white solid:

$$NH_3(g) + HCl(g) \longrightarrow NH_4Cl(s)$$

A long, dry glass tube was clamped horizontally. A piece of cotton wool soaked in concentrated ammonia solution and a piece of cotton wool soaked in concentrated hydrochloric acid were placed in opposite ends of the glass tube at the same time. After ten minutes a white solid ring was formed in the tube as shown below.

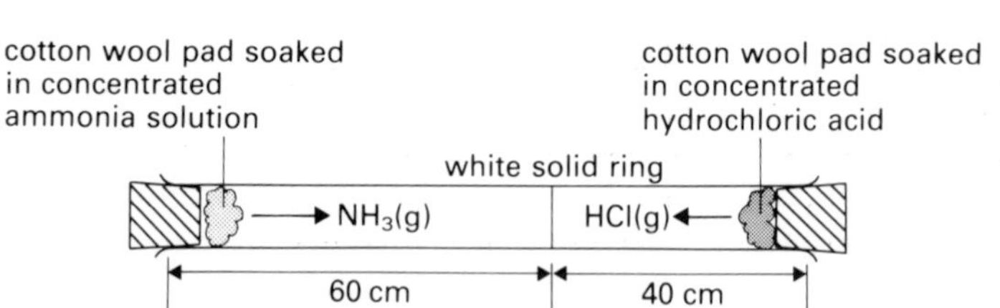

(a) Name the white solid formed inside the glass tube.

(b) Why was the white solid formed nearer to the end of the tube containing the pad soaked in concentrated hydrochloric acid?
(c) What name is given to the movement of molecules which causes the formation of this solid?
(d) Molecules of ammonia and hydrogen chloride move rapidly at room temperature. Give a reason why the formation of the white solid was so slow. **(EAEB)**

11 For producing chemicals on a large scale, cheap and abundant supplies of raw materials are required. Some examples of such raw materials are: air, fats, metal oxides, petroleum, sea water.
(a) Select *one* chemical in the production of which one or more of the above raw materials plays an essential part and describe the production of this chemical (technical details are not required). The raw material(s) must be converted into other substances.
(b) Comment briefly on the importance to society of the chemical you have selected and suggest what would be likely to happen if the raw material(s) in the list required for production of this chemical became unobtainable. **(L)**

12 Five students investigating the thermal decomposition of a compound found that a gas was evolved. They each then repeated their experiments, finding the loss in mass of the compound and measuring the volume of gas evolved at room temperature and pressure. Their results are as follows:

Student	A	B	C	D	E
Loss in mass (g)	0.060	0.032	0.107	0.083	0.090
Volume of gas (cm^3)	45	24	80	62	75

(a) Plot the results on a graph.
(b) (i) One student made an error in measuring the volume of his gas. Draw a ring around this point.
(ii) Draw a straight line through the other points.
(c) Use your graph to find the mass of $100\,cm^3$ of gas.
Below are the densities of some common gases at room temperature and pressure.

Gas	**Density (g/litre)**
Hydrogen	0.008
Nitrogen	1.16
Oxygen	1.33
Carbon dioxide	1.83

(d) State, with a reason, which gas might have been the one evolved during the experiments.
(e) Sketch an apparatus in which this experiment might have been carried out.
(f) Name *one* compound which on heating might have given off the gas you have named in (d), and write an equation for the reaction. **(JMB)**

THEME B

The air

Air for breathing and burning. Air for oxygen. Air for industry. Air for fun

4 The composition of air

Air is a very important mixture. It is all around us. There seems to be an endless supply in the atmosphere. We need it to breathe and live. We need it to burn fossil fuels to get energy.

The average person breathes in about 14 500 litres of air each day when resting. For someone doing energetic exercise, the air intake could be about 30–40 litres every minute. Only part of this air is used. The rest is breathed out unchanged. Breathed out air includes other gases formed in exchange for the part used.

About 100 000 000 litres of air is needed for the complete **combustion** of 1 tonne of coal. Again, only part of the air is used and other gases are produced in exchange. A litre of petrol needs over 10 000 litres of air to allow it to burn completely.

Many processes which occur in the air use just part of it. There seems to be an 'active' component.

Copper is a metallic element which becomes grey–black on the surface when it is heated in air. A reaction takes place between copper and part of the air.

Experiment 4a
How much air is used when copper reacts with air?

Figure 4.1 shows the apparatus used. Fresh copper wire is packed into the silica tube. One syringe contains 100 cm^3 of air. The copper is heated while the air is pushed to and fro across it.

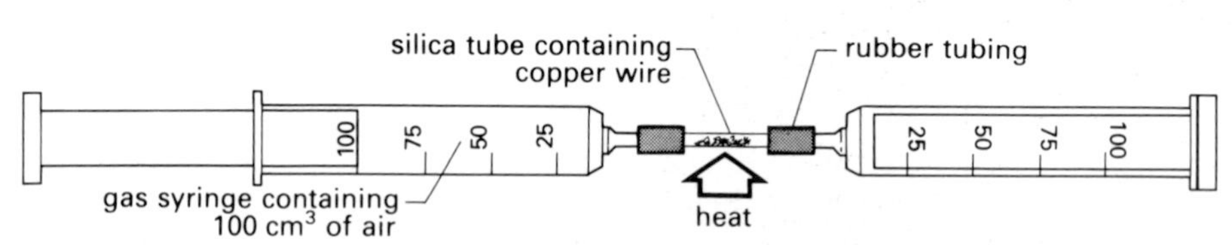

Figure 4.1

Results
As the copper becomes grey–black, the volume of air decreases. When no more change occurs, and after cooling, the volume of air left is about 79 cm^3. What volume of air is used by the copper? What percentage is this of the air available?

Discussion
Copper reacts with something in the air to give a grey–black product – a compound. An unreactive gas is left. The reaction can be written as:

$$\text{Copper} + \text{Air} \longrightarrow \text{Product} + \text{Inactive air } (79\,\%)$$

The 21 % active gas in the air is oxygen. Of the remaining 79 %, 78 % is nitrogen.

The grey–black solid is copper(II) oxide. Using the rules given in chapter 3, write a symbol equation for the reaction of copper with oxygen giving copper(II) oxide.

Would you expect any mass change in the copper during the experiment? Why?

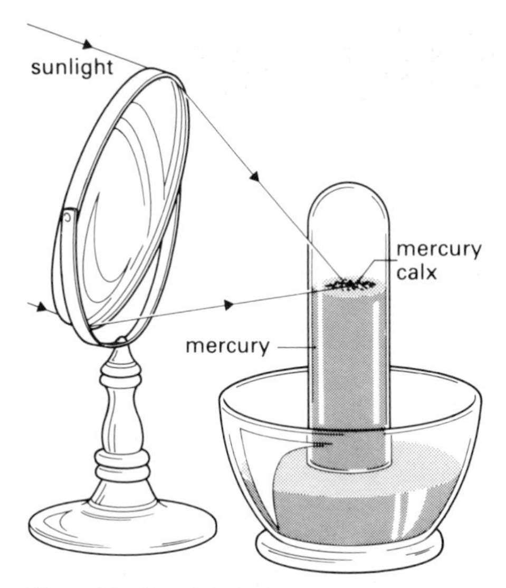

Figure 4.2 Joseph Priestley used the heat of the sun to obtain oxygen from mercury calx (mercury oxide)

Oxygen (21 %) and nitrogen (78 %) are the two main gases in the air.

Carl Scheele (1742–86), a Swedish chemist, first prepared oxygen in the laboratory in 1774 by heating potassium nitrate, but by the time he had published his findings, a British chemist, Joseph Priestley (1733–1804) had also prepared the gas by heating red mercury oxide (see figure 4.2), and had published his results. It was Antoine Lavoisier (1743–94), a great French chemist, who gave the gas its name and showed that the gas prepared by Scheele and Priestley was the same as the active gas in the air. Figure 4.3 shows Lavoisier in his laboratory. In the background is the apparatus he used for the sequence of important experiments illustrated in figure 4.4. He found that

Mercury + Air ⟶ Red solid product + Inactive air

The red product was then called mercury calx. We call it mercury oxide. The mercury had combined with oxygen from the air:

Mercury + Oxygen ⟶ Mercury oxide

In his second experiment he heated mercury oxide and obtained mercury and oxygen:

Mercury oxide ⟶ Mercury + Oxygen

Figure 4.3 Antoine Lavoisier in his laboratory

Questions

1 Draw a modern piece of apparatus which could be used to reproduce Lavoisier's experiment. Is it necessary to collect the gas over mercury?
2 Write symbol equations to describe Lavoisier's experiment.
3 Using books in the library, find out more about Scheele, Priestley and Lavoisier and their contributions to chemistry.

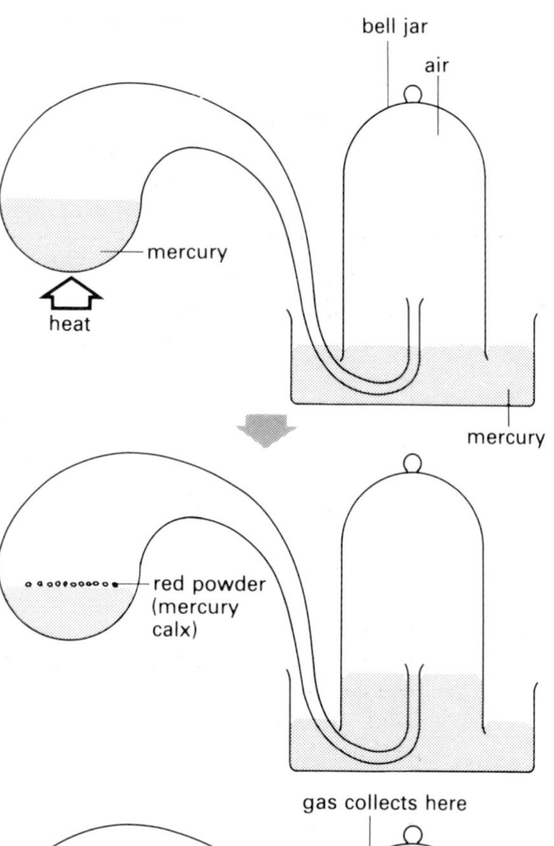

Mercury in the flask was heated over a charcoal fire. The 'delivery tube' opened into air trapped above mercury.

After heating for a long time, a red powder appeared on the mercury in the flask. In the bell jar, the mercury had risen to take up about 1/5 the volume of the bell jar and replace the active gas used up.

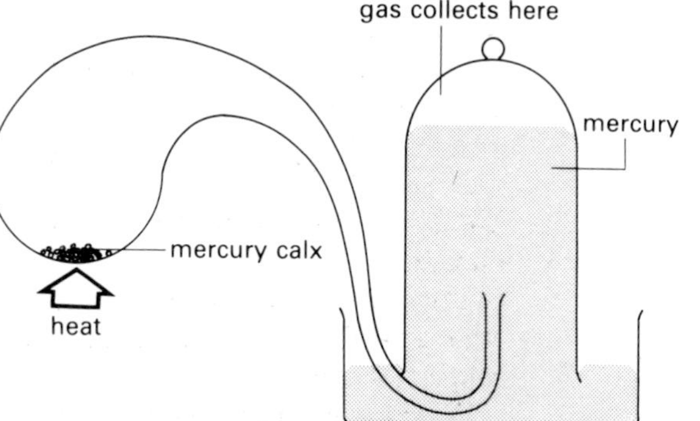

The red powder was then heated in a clean flask, with the bell jar full of mercury. The red mercury calx changed to mercury and a gas collected in the bell jar. Its volume was about 1/5 that of the bell jar. It was oxygen.

Figure 4.4 Lavoisier's experiments, which showed that the gas given off when mercury calx was heated was the same as the active gas in the air

Oxygen and nitrogen account for about 99 per cent of dry air. Experiment 4b illustrates the detection of other gases.

Experiment 4b
Detecting carbon dioxide and water vapour in the air

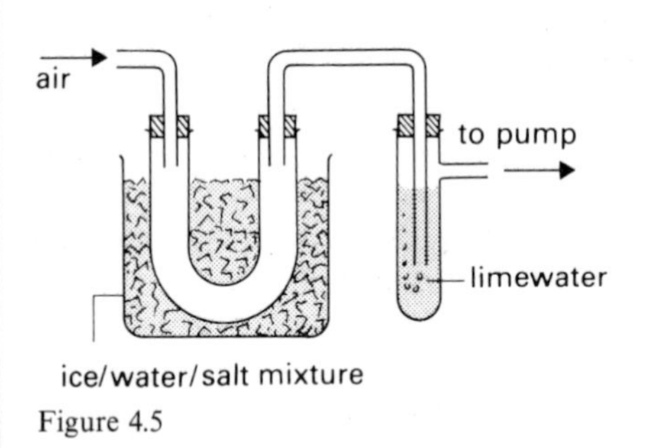

Figure 4.5

Figure 4.5 illustrates the apparatus which can be used. The U-tube is clean and dry at the start of the experiment. A gentle stream of air is drawn through the apparatus using a water pump.

Results
After about 5–10 minutes, droplets of a clear, colourless liquid condense in the U-tube, and the limewater becomes slightly cloudy.

Discussion
The positive limewater test shows that there is some carbon dioxide in the air. What tests would you carry out to show that the colourless liquid is water? Can you devise an experiment to determine the percentage of carbon dioxide and water vapour in the air?

The amount of water vapour in the air can vary from about 4% in a tropical jungle to almost 0% in desert areas. The carbon dioxide amount remains fairly constant at about 0.03%.

Often, in science, things are discovered by putting together several apparently unrelated observations or pieces of information – rather like the pieces of a jigsaw fitting together to give a complete picture. This is how another gas in the air was discovered. It was difficult to detect because it is so unreactive. There were three pieces to this jigsaw, as shown in figure 4.6. Fitting the pieces of evidence together,

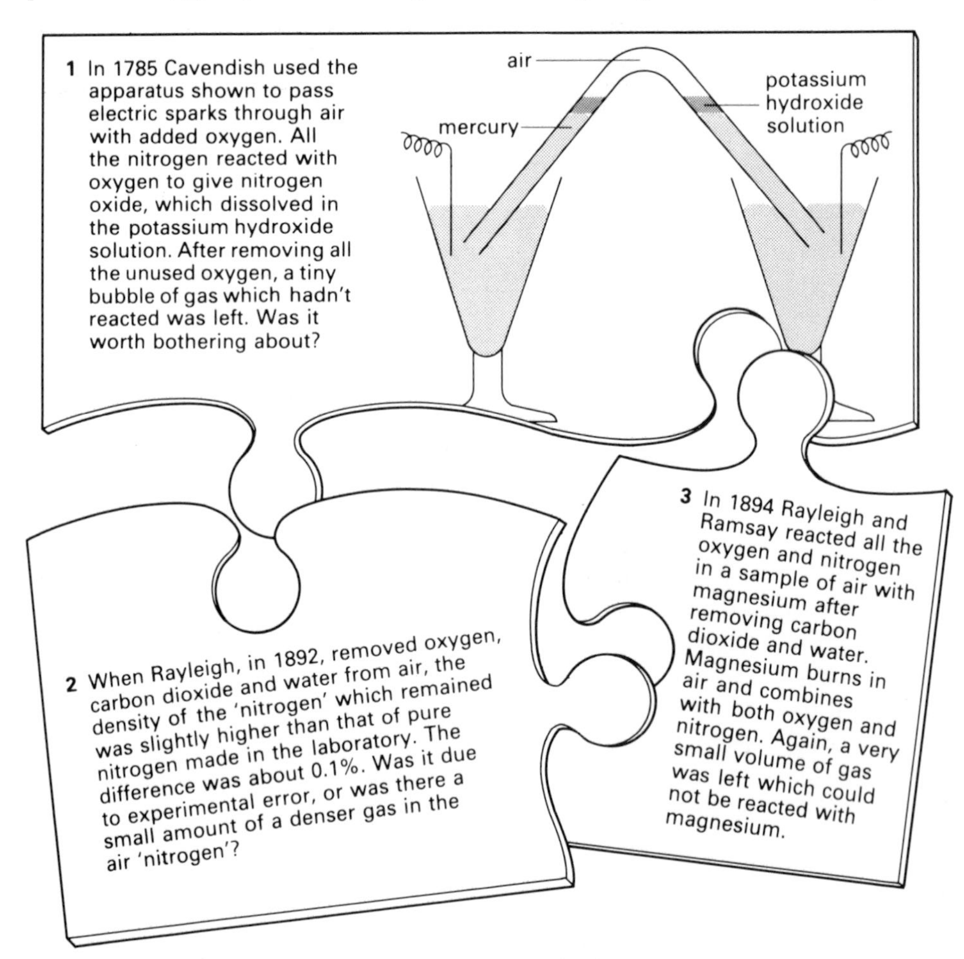

1 In 1785 Cavendish used the apparatus shown to pass electric sparks through air with added oxygen. All the nitrogen reacted with oxygen to give nitrogen oxide, which dissolved in the potassium hydroxide solution. After removing all the unused oxygen, a tiny bubble of gas which hadn't reacted was left. Was it worth bothering about?
air
mercury
potassium hydroxide solution
2 When Rayleigh, in 1892, removed oxygen, carbon dioxide and water from air, the density of the 'nitrogen' which remained was slightly higher than that of pure nitrogen made in the laboratory. The difference was about 0.1%. Was it due to experimental error, or was there a small amount of a denser gas in the air 'nitrogen'?
3 In 1894 Rayleigh and Ramsay reacted all the oxygen and nitrogen in a sample of air with magnesium after removing carbon dioxide and water. Magnesium burns in air and combines with both oxygen and nitrogen. Again, a very small volume of gas was left which could not be reacted with magnesium.

Figure 4.6 Pieces of evidence for the presence of argon in the air

Questions

1 Find out how and when the noble gases other than argon were discovered.
2 Can you suggest why the amounts of oxygen and carbon dioxide in the air might vary slightly from place to place, especially between urban and rural areas?

Rayleigh and Ramsay announced the discovery of a very unreactive gas called argon (from a Greek word meaning 'lazy'). About 0.9 per cent of the air is argon. It is more dense than nitrogen. It was only discovered because of very careful, accurate experimental work.

Argon is one of a family of colourless, unreactive gases which includes helium, neon, krypton and xenon. These were discovered a few years after argon. They have been given several names – 'the inert gases', 'the rare gases' and 'the noble gases'. It is now known that they are not all totally inert (unreactive). Compounds such as the xenon fluorides, XeF_6 and XeF_4, have been made. It is also arguable whether they are all rare! In a room 4 m × 4 m × 2.5 m, there will be about 372 litres of argon; in every deep breath we take there will be about 5 cm^3 of argon. Perhaps the title 'the noble gases' is the most appropriate.

Air is a mixture of nine gases and, while the amount of water vapour can vary from place to place, the composition of dry air remains fairly constant. A typical analysis is shown in figure 4.7.

Gas	Volume percentage in dry air
Nitrogen	78.03
Oxygen	20.99
Argon	0.93
Carbon dioxide	0.03
Neon	0.001 5
Hydrogen	0.001 0
Helium	0.000 5
Krypton	0.000 1
Xenon	0.000 008

Figure 4.7

Summary

1 The air may be thought of as being made up of nitrogen, oxygen and other gases. Draw a bar chart, or pie chart, to show the proportions of these three main parts of the air. The data you need is in figure 4.7.
2 Now consider the other gases. Using a different scale, draw a second bar chart to show the relative amounts of the gases which make up this third part of the air. Take the values from figure 4.7.

5 Oxygen

Many oxygen containing compounds can be decomposed to give oxygen. Priestley heated mercury oxide to obtain mercury and oxygen:

$$2HgO(s) \longrightarrow 2Hg(l) + O_2(g)$$

Scheele heated potassium nitrate, KNO_3. Oxygen and a solid product (potassium nitrite, KNO_2) were formed:

$$2KNO_3(s) \longrightarrow 2KNO_2(s) + O_2(g)$$

All nitrates decompose to give oxygen as one of the products when they are heated.

In the laboratory, oxygen is usually prepared by allowing a solution of hydrogen peroxide, H_2O_2, a colourless liquid, to come into contact with powdered manganese(IV) oxide (manganese dioxide), MnO_2.

Experiment 5a
Preparation and properties of oxygen

hydrogen peroxide

Figure 5.1 shows the apparatus used to obtain oxygen from hydrogen peroxide, H_2O_2. Oxygen is collected by displacing water from a gas jar or test-tube.

If dry gas is required, it can be passed through a drying agent. One suitable drying agent is silica gel, which can be used in a U-tube, as shown in figure 5.2.

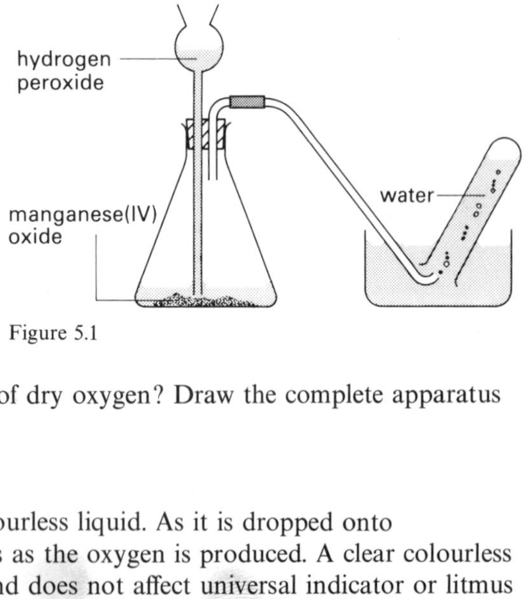

Figure 5.1

How would you collect a sample of dry oxygen? Draw the complete apparatus you would need.

Results
Hydrogen peroxide is a clear, colourless liquid. As it is dropped onto manganese(IV) oxide it effervesces as the oxygen is produced. A clear colourless gas is collected. It has no smell and does not affect universal indicator or litmus paper. When a lighted splint is lowered into the gas, the wood burns rapidly with a brighter flame. A glowing splint relights.

Discussion
Hydrogen peroxide is an unstable compound and decomposes to give water and oxygen even at room temperature:

$$2H_2O_2(aq) \longrightarrow 2H_2O(l) + O_2(g)$$

However, the reaction is slow. It is speeded up by adding manganese(IV) oxide, which acts as a **catalyst** and does not change during the reaction. Catalysts are described in more detail in chapter 26.

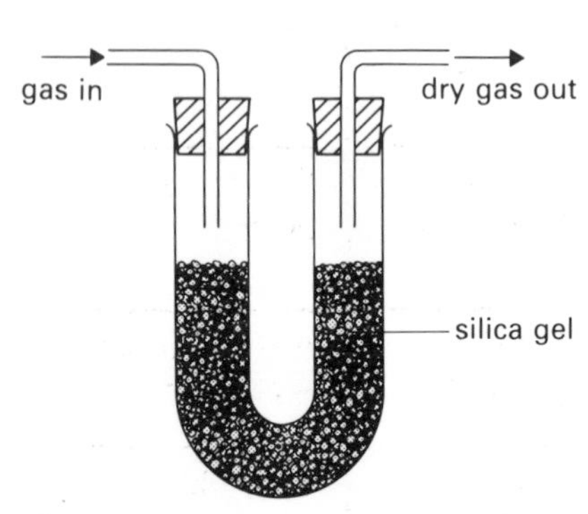

Figure 5.2

Questions

1 Potassium manganate(VII), $KMnO_4$, is a purple crystalline solid. When heated it crackles, gives off oxygen and leaves black manganese(IV) oxide.
(a) Draw the apparatus you would use to obtain oxygen from potassium manganate(VII).
(b) How would you prevent tiny crystals of potassium manganate(VII) 'spitting' out of the tube during heating?
2 How would you distinguish between three unlabelled test-tubes, one of which contains oxygen, another nitrogen, and the third, carbon dioxide?
3 Imagine a world in which the air contained 80% oxygen and 20% nitrogen. Describe some of the strange events that might occur. What effect does nitrogen have on the properties of oxygen?

The key properties of oxygen are shown in figure 5.3. Nitrogen is similar to oxygen in appearance, but chemically it is different. Its properties are also shown in figure 5.3.

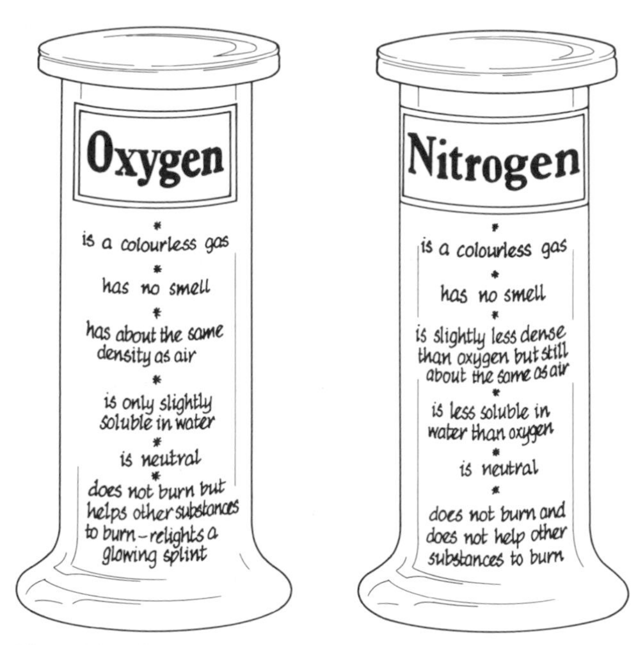

Figure 5.3 The properties of oxygen and nitrogen

Reactions that occur between elements and oxygen in the air take place more readily in pure oxygen. Some of these reactions are studied in experiment 5b.

Experiment 5b
Burning elements in oxygen

phosphorus

sodium

Using tongs or a combustion spoon, samples of elements are heated in a flame and then placed in a test-tube of oxygen (see figure 5.4). White phosphorus is *not* heated before placing in oxygen. When each reaction stops, the product is examined and water added. If it dissolves, the solution is tested with full range **indicator** paper to determine whether it is **acidic** or **alkaline**.

Results
Figure 5.5 (the table below) shows a typical set of results.

Element	Reaction with oxygen	Appearance of product	Name of product	Effect of water on product
Magnesium	Burns with bright white flame	White powder	Magnesium oxide	Dissolves slightly, pH = 9
Copper	Does not burn. Surface turns black	Black solid	Copper(II) oxide	Insoluble
Phosphorus	Burns with white flame. White smoke	White smoke	Phosphorus pentoxide	Dissolves easily, pH = 2
Carbon	Glows red hot	Colourless gas	Carbon dioxide	Probably dissolves slightly, pH $\approx$ 6
Iron	Glows red hot and sparks	Black solid	Iron(III) oxide	Insoluble

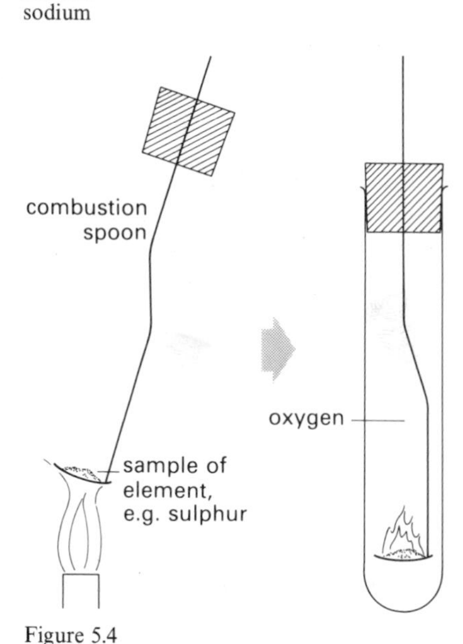

Figure 5.4

Sulphur	Burns with bright blue flame. Steamy gas with choking smell	Colourless steamy gas	Sulphur dioxide	Dissolves easily, pH = 2
Calcium	Burns with bright red flame	White powder	Calcium oxide	Mostly dissolves, pH = 10
Sodium	Burns with bright yellow–orange flame	White solid	Sodium oxide	Dissolves readily, pH = 12

Figure 5.5

Discussion

What evidence is there that chemical reactions have occurred? All the elements in the table react with oxygen. They combine with it to give compounds called oxides. Some of the elements react more readily than others.

Metal oxides behave differently from non-metal oxides. Which elements in figure 5.5 are non-metals? What do their oxides have in common? Which elements are metals? In what ways are their oxides similar?

Most elements combine with oxygen. When they do, oxides are formed. This process is called **oxidation**; we say that the element has been oxidized. The reactivity of elements with oxygen varies considerably. Some metals, like sodium and magnesium, react vigorously; others, like iron, are less reactive. Some do not combine directly with oxygen. These include gold and platinum. There is an order of reactivity among the non-metals, too.

The oxides of metals behave differently from the oxides of non-metals:

* metal oxides are always solid and if they dissolve in water, they give alkaline solutions
* non-metal oxides are usually gases and dissolve in water to give acid solutions.

Questions

1 For each of the elements in figure 5.5, give the symbol, classify it as a metal or non-metal, find the formula of the oxide from table 2 on p. 337, and write an equation for the reaction with oxygen.

2 List the metals in figure 5.5 in order of their reactivity with oxygen, with the most reactive at the top. (From the information given, some metals will 'tie' at this stage.)

3 When sodium, magnesium and calcium oxides are placed in water, they react to give hydroxides as they dissolve. Write equations to show the reactions of the oxides with water.

Question

Put the following oxides in the order you would predict for their ease of reduction: magnesium oxide, copper(II) oxide, sodium oxide, iron(III) oxide, mercury oxide.

Some of the reactive metals, such as sodium, magnesium and calcium, will also react with nitrogen. Solid compounds called nitrides are formed.

Many processes in chemistry can be reversed. The removal of oxygen is called **reduction**. Some metal oxides will give up their oxygen readily. Priestley prepared oxygen by heating mercury oxide. However, very few metal oxides behave like mercury oxide, and more drastic chemical reactions are needed to remove the oxygen. The ease of reduction depends on how readily the element combines with oxygen. It is difficult to remove the oxygen from the oxides of reactive metals like magnesium. The process of reduction is discussed fully in chapters 13 and 41.

Summary

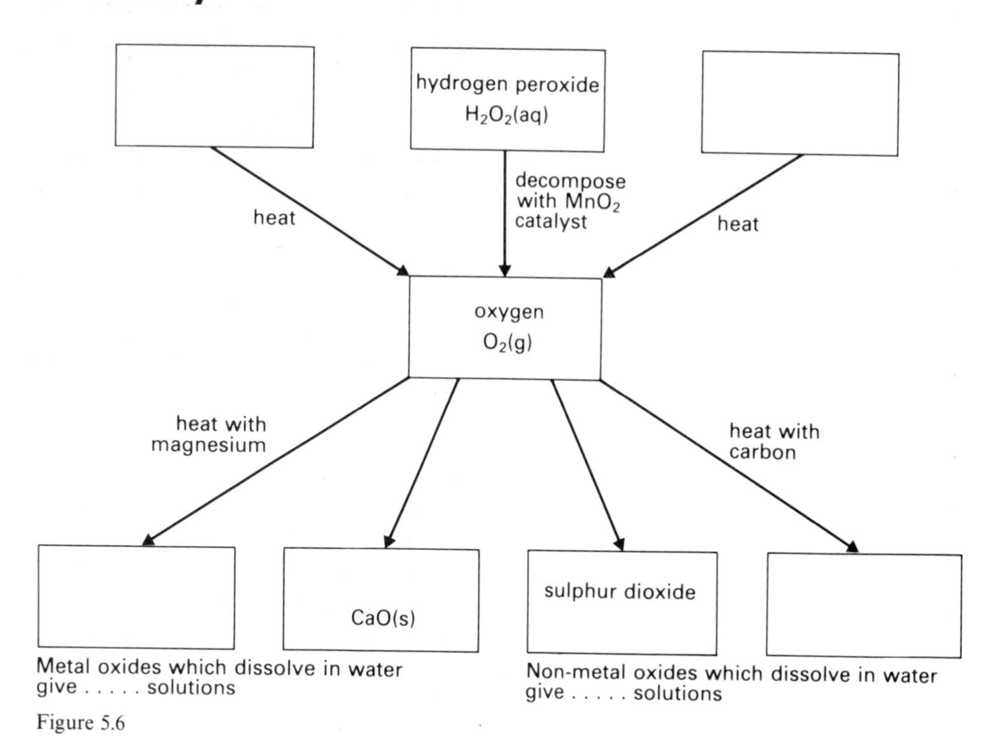

Figure 5.6

Copy and complete the diagram in figure 5.6. Write the name, formula and state symbol for each of the main products of the reactions in the boxes. Write the reaction conditions beside the arrows.

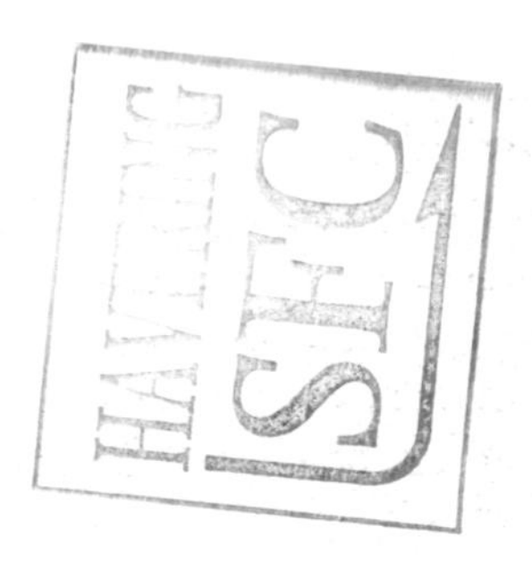

6 Burning, breathing and rusting

We burn fuels to keep warm, cook food, run cars and generate electricity. We breathe, of course, to live. Rust, we could well manage without, but is being formed all around us. Metals corrode and it seems a never-ending fight to prevent the process occurring. When iron or steel corrodes the process is called rusting. Burning, breathing and rusting are familiar processes which involve the air.

6.1 Burning

Burning (**combustion**) reactions are those in which a substance reacts rapidly, usually with oxygen. There is a flame and heat energy is released. Combustion involves oxidation. Copper and magnesium are both easily oxidized. When magnesium reacts with oxygen there is a bright white, hot flame, but with copper there is no flame. Magnesium burns but copper does not, even though the chemical reactions are very similar:

$$2Mg(s) + O_2(g) \longrightarrow 2MgO(s)$$
$$2Cu(s) + O_2(g) \longrightarrow 2CuO(s)$$

Common combustion reactions include those involving wood, paper, coal and natural gas. Other fuels such as coke, petrol and paraffin are made from the fossil fuels, coal and crude oil.

Questions

1 Why are coal and oil described as fossil fuels?
2 List other fuels in addition to those mentioned.

Experiment 6a
What are the products of burning fuels?

Figure 6.1 shows some apparatus that can be used to investigate the products formed when a candle burns. (Other fuels could also be burnt under the thistle funnel.) Air and the combustion products are drawn through the apparatus using a water pump.

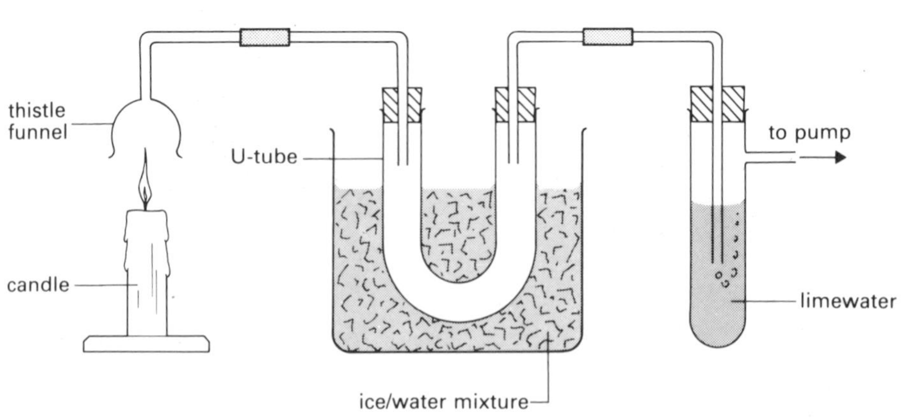

Figure 6.1

Results
In a short time the limewater becomes milky and a liquid condenses in the U-tube. The liquid turns cobalt chloride paper from blue to pink. With some fuels, such as candle wax, soot appears on the inside of the funnel.

Discussion

What are the two main products of combustion of the fuel used? When elements burn in air or oxygen, a single oxide is formed. The common fuels are compounds (or sometimes mixtures of compounds), and on burning they give mixtures of oxides. Carbon dioxide is formed if the fuel contains a carbon compound. Any hydrogen in the fuel combines with oxygen to form water. Soot (carbon) forms when the fuel is not completely burnt. Incomplete combustion occurs either when there is not enough oxygen to fully oxidize the fuel or when the flame is not hot enough.

Question

A Bunsen burner gives different flames depending on whether the air hole is open or closed.
(a) What are the differences between the two types of flame?
(b) In which type of flame is the combustion of natural gas more complete?
(c) What evidence is there for incomplete combustion in the other type of flame?

In a car engine, petrol is burnt with air in the cylinders. The carburettor allows the correct mixture of petrol and air into the cylinder. If the mixture is too 'rich', and contains too much petrol relative to air, incomplete combustion leads to carbon being formed. The inside of the exhaust pipe of a well-tuned engine is clean. A sooty exhaust indicates bad tuning. What is the effect of a 'weak' mixture?

Carbon monoxide can be formed when there is not enough oxygen for complete combustion. Like carbon dioxide, it is a colourless gas and has very little smell, but it is very poisonous. It combines very strongly with haemoglobin in the blood and prevents it from doing its normal job of carrying oxygen round the body.

Carbon monoxide burns with a flickering blue flame. It is readily oxidized to carbon dioxide:

$$2CO(g) + O_2(g) \longrightarrow 2CO_2(g)$$

Figure 6.2 shows where carbon monoxide and carbon dioxide are formed in a coke fire. Coke is almost pure carbon.

Carbon monoxide plays an important part in the blast furnace. It reduces iron oxide to iron (see chapter 13).

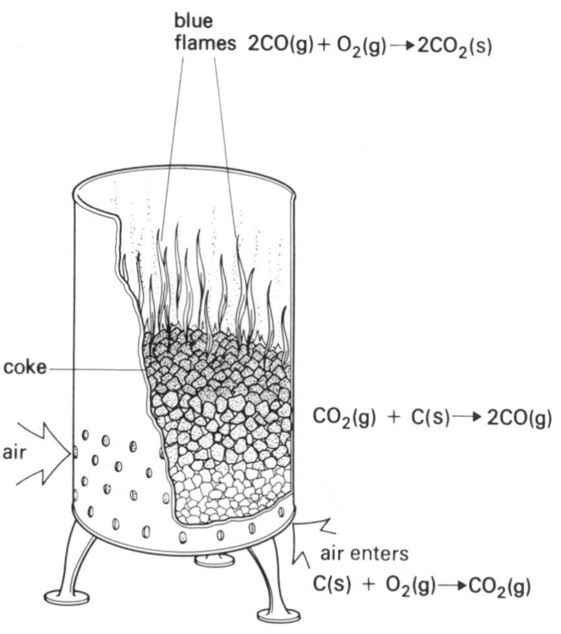

Figure 6.2 The reactions that take place in a coke fire

Carbon Dioxide
is a colourless gas
has no smell
is much denser than air
is moderately soluble in water
is slightly acidic
does not burn and does not help other substances to burn
turns limewater milky

Figure 6.3 The properties of carbon dioxide

Questions

1 Carbon monoxide can be made in the laboratory by passing carbon dioxide over hot carbon. Draw a labelled diagram of some apparatus which could be used to prepare and collect samples in test-tubes. The gas is almost insoluble in water.
2 Why is it dangerous to run a car engine in a closed garage?
3 Figure 6.3 summarizes the properties of carbon dioxide. Make a similar summary of the properties of carbon monoxide.

When a fuel burns,

- oxygen is needed
- the fuel is oxidized: fuels which contain carbon and hydrogen give carbon dioxide and water on complete combustion
- heat energy is released.

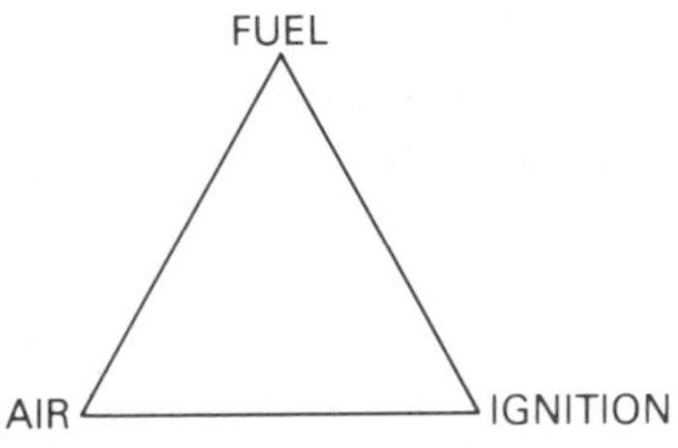

Figure 6.4 The fire triangle

Before burning can start, three components are needed. They are shown in the fire triangle, figure 6.4. If any one component is missing, a fire cannot occur.

Petrol can be stored safely in air, provided there is no naked flame or spark – a source of ignition – nearby. One spark, and a fire is created. Dry woodland can be completely destroyed by one cigarette end carelessly thrown away.

A fire can be extinguished by removing any one of the three components. Figure 6.5 shows some fire extinguishers in use. What is being removed by each type? How do they work?

(a)

(b)

(c)

Figure 6.5 Some fire extinguishers in use
(a) Using conventional hoses
(b) An airfield crash tender spraying foam
(c) Using a hand-held carbon dioxide extinguisher on a liquid fuel fire

Questions

1 Explain why
(a) bonfire fireworks should be stored in a closed metal box.
(b) car drivers should switch off their engines on a petrol station forecourt.
2 How would you deal with the following fire situations: (a) a chip pan on fire in the kitchen, (b) ethanol in a test-tube on fire in the laboratory and (c) a person whose clothes are on fire?

6.2 Breathing

Breathing involves the air. All our lives we rhythmically take in and exhale 'air' – slowly and gently while we are resting, rapidly and deeper when we are taking exercise. Any interruption to the process is a threat to life.

What do we breathe out? How does exhaled air compare with the atmospheric air we breathe in? When we breathe out onto a cold window it mists up, and we can 'see' our breath on a frosty day. These observations suggest one substance present in exhaled air.

Experiment 6b
Comparing inhaled and exhaled air

Figure 6.6 shows simple experiments which can be carried out to compare inhaled and exhaled air. To complete the comparison, the temperatures of exhaled and atmospheric air are recorded.

Results

Within 30 seconds, the limewater in (a) turns milky. What happens in (b) in the same time? Again, in a short time, a clear, colourless liquid collects in the test-tube in (c), and this can be shown to be water. What happens in (d) in the same time? The temperature of exhaled air is about 30 °C. What is normal room temperature?

Discussion

During breathing, some of the oxygen in the air is used up, and carbon dioxide, water and heat energy are produced. Typical analyses of inhaled and exhaled air are:

	Inhaled air	Exhaled air
Oxygen	21%	16%
Carbon dioxide	0.03%	4%

As you might expect, the nitrogen and other gases are unaffected.

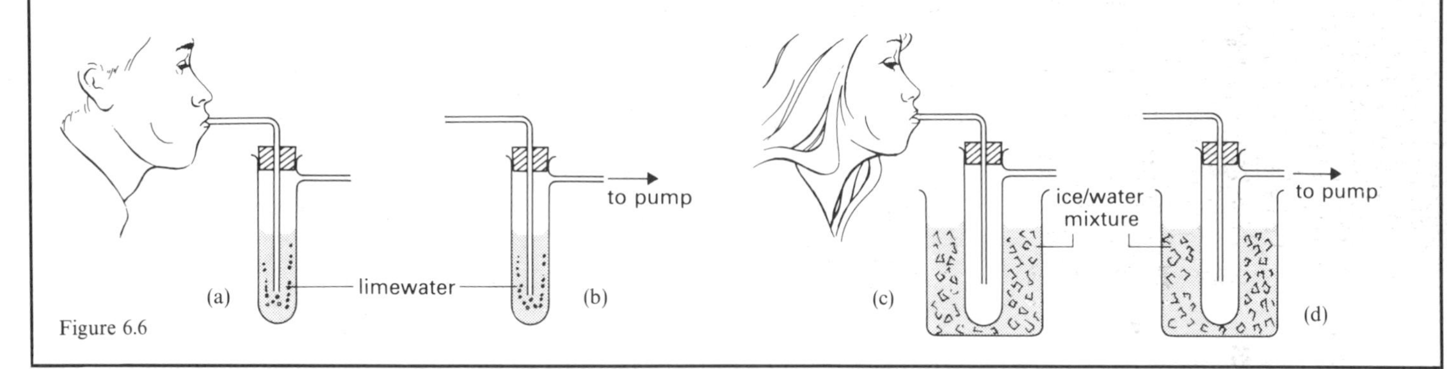

Figure 6.6

Questions

1 (a) What would you expect to happen if lit candles are lowered into gas jars of atmospheric air and exhaled air?
(b) How would you collect exhaled air?
(c) How would you extend the experiment to test air which had been breathed repeatedly?
2 Why does breathing become 'difficult', and why do you become tired, in a crowded, unventilated room?
3 Why is it not a good idea to use Bunsen burners for heating laboratories if the boiler fails in winter?
4 Why do workers such as miners, who have been in a dusty environment for many years, breathe more rapidly than people who have worked in a clean area?

During 'breathing', gas exchange occurs in the lungs in tiny sacs called *alveoli.* There are about 700 million of these. The total surface area of all the alveoli is very large – about the area of a tennis court. Inhaled air passes into the blood and, in exchange, waste gases pass from the blood into the sac to be exhaled.

Breathing is similar to burning: in both processes, oxygen is used, and carbon dioxide, water and heat energy are produced. For burning, a fuel is needed. What is the 'fuel' for breathing? The obvious answer is food.

Questions

1 Why do athletes often eat glucose sweets just before a race?
2 Why does our breathing rate increase when we do energetic exercise?

Certain types of food, particularly **carbohydrates** and fats (which we call the energy foods), are broken down in the body to provide heat energy to keep us warm and the mechanical energy to be active. Oxygen helps the breakdown of these foods. The reaction of foods with oxygen is called *respiration.* The chemical nature of the energy foods is discussed in chapter 33.

Glucose, a sugar, can be burnt in the laboratory:

$$\text{Glucose} + \text{Oxygen} \longrightarrow \text{Carbon dioxide} + \text{Water} + \text{Heat}$$

$$C_6H_{12}O_6(s) + 6O_2(g) \longrightarrow 6CO_2(g) + 6H_2O(l) + \text{Heat}$$

We call this *combustion.* In our bodies a very similar reaction occurs, but this time we call it *respiration.*

6.3 Rusting

Rust is the red–brown flaky solid which forms on iron and steel.

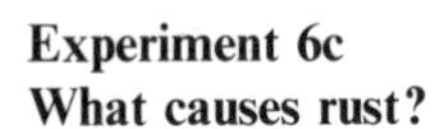

Experiment 6c
What causes rust?

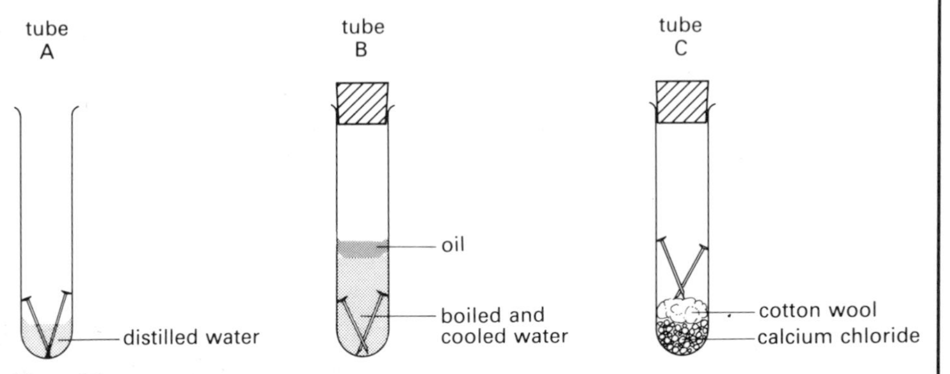

Figure 6.7

The experiment is based on the series of test-tubes shown in figure 6.7. Tube A contains nails in contact with water and air. In tube B, the nails are in distilled water which has been boiled to remove dissolved air. The layer of oil prevents air redissolving in the water. In tube C, the nails are in dry air (calcium chloride is a drying agent).

Results
Typical results after 2–3 days are shown in figure 6.8.

Tube	Nails in contact with ...	Observations
A	air and water	Rust formed. Some has flaked off and collected in the bottom of the tube
B	water only	Nails mostly shiny. A hint of rust on the tips
C	dry air	No sign of rust

Figure 6.8

Discussion
In addition to iron, what else is needed for the formation of rust? Why might there be a hint of rust in tube B?

Rusting is a process involving the air. Which of the gases in the air takes part in rusting?

Experiment 6d
How much of the air is used when iron rusts?

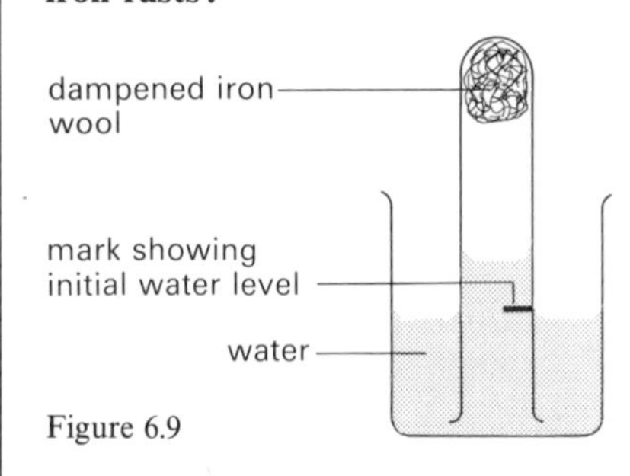

Figure 6.9

In figure 6.9 a large test-tube containing dampened iron wool stands in a beaker of water. The water level is noted at the start of the experiment and over the following 2–3 days.

Results
After two days there are signs of rust on the iron and the water level has risen. It stops rising about 26 mm up a 125 mm test-tube.

Discussion
Why has the water risen up the test-tube? What percentage of the air has been used? This proportion corresponds to the amount of oxygen in the air. So this gas, in addition to water, must also be present for rusting to occur.

Questions

1 Where in your house would an iron object rust the quickest? In which room would rusting occur least? Give reasons for your answers.
2 It has been suggested that a car which is frequently put away wet into a garage rusts more quickly than one left outside. Do you think there is any truth in this? Give reasons for your answer.
3 Design experiments to confirm that it is oxygen and not nitrogen that is needed for rusting.

Comparing rusting with burning, it seems reasonable to assume that rust is some form of iron oxide. It is hydrated iron oxide. What would you expect to happen when rust is heated?

Rusting, like burning and breathing, is an **oxidation** process. In all these processes, oxygen is used and oxides are formed.

Although iron is the only metal that rusts, others corrode or tarnish, and oxygen is usually involved in these corrosion processes, too. More details of these reactions are given in chapter 12.

Iron can be protected from rust by preventing oxygen and water getting to the surface. Painting, oiling and electroplating are all used. Electroplating is described in chapter 40.

Figure 6.10 Bars of zinc attached to a ship's hull to protect it from corrosion by sea water: you can see that the bars are badly pitted, while the hull itself is still smooth

Question

Which methods are used for protecting iron and steel used for (a) metal window frames, (b) car bumpers, (c) car bodies, (d) bicycle wheel spokes, (e) food cans, (f) bridges and (g) garden tools?

Sometimes these methods are inadequate or impractical for protecting a metal structure, e.g. an underground pipe. In these cases, blocks of zinc or magnesium are connected to the iron. The reactive metal corrodes instead of the iron. It is less expensive to replace blocks of zinc or magnesium than underground pipes or harbour piers. Figure 6.10 shows an example of this type of protection.

6.4 Air pollution

Do the windows and paintwork in your house become dirty very quickly? When it rains, do cars in your street become dirty? Do even the flowers look dusty? Are stone churches and statues in your area badly decayed? These are all symptoms of air pollution. When people first used fire for heating and cooking they began to pollute the air with soot and corrosive gases. Over a long period of time, these pollutants damage our health and property. The problems are worst in heavily industrialized areas (see figure 6.11).

Figure 6.11 Heavy industry in the Ruhr basin is polluting the air and the River Rhine

Coal and crude oil contain small amounts of sulphur compounds. When these fuels burn, the sulphur is oxidized to sulphur dioxide, which is a colourless, choking gas. Smoke containing sulphur dioxide trapped in low cloud was responsible for the very thick, deadly smogs which were common in large cities some years ago.

In areas which have been heavily polluted with sulphur dioxide, stonework is badly decayed and metals corroded. Sulphur dioxide is an acidic oxide. It dissolves in rain-water to give an acid. Its effect on stonework is shown in figure 6.12.

Pollution from sulphur dioxide is less serious now than it was in the early 1950s. Following the Clean Air Act of 1956, many areas became 'smokeless zones'. Only smokeless fuels, such as coke, could be used in these areas. In modern coal-fired power stations, high chimney stacks have reduced the concentration of sulphur dioxide at ground level, by taking polluting gases higher into the air.

Car exhaust gases pollute the air. The problems are most serious in large modern cities. Exhaust gases include nitrogen oxides, unburnt hydrocarbon fuels and lead compounds, in addition to carbon monoxide and carbon dioxide. Nitrogen oxides can have a similar effect to sulphur dioxide. Why? Perhaps the most serious pollutants are the lead compounds which may be added to petrol to make engines more efficient. When the fuel is burnt they are released into the atmosphere. If breathed in, they might accumulate in the body and cause brain damage. Around some motorways quite high concentrations of lead in the air have been found. Fortunately, in some parts of the world there are now strict laws controlling exhaust gases (see figure 6.14), and in some countries lead compounds cannot be added to petrol. Some pollutants can be removed by using a filter system and converter fitted into the exhaust pipe (see figure 26.18 on p. 198).

Many of the things we use are made of plastics. The problem of disposing of unwanted materials can be serious. For example, pvc, which is used to make such things as toys, guttering and vinyl wallpaper, releases hydrogen chloride gas when

Questions

1 In this section, four sources of air pollution are mentioned. List them and add any others you may have noticed – some may be particular local problems.
2 Several methods of reducing pollution are mentioned. List them and add more.
3 What particular pollution problems would you expect (a) near an airport, (b) near a cement works, (c) near farm land, (d) in a city and (e) near brickworks?
4 Design an experiment to compare the concentrations of solid particles in the air in different areas.
5 (a) Do you know about the types of pollutant which cigarette smoke contains?
(b) Should smoking be prohibited in all public places? Support your answer with reasons.
6 Design an experiment to show that smoke from a coal fire contains acidic gases.

Figure 6.12 Acidic rain has caused the decay of these thirteenth-century stone figures on the west front of Wells Cathedral

Figure 6.13 Car exhaust gases are a major source of pollution in cities

Figure 6.14 A Budapest policeman checking a lorry's exhaust gases for high levels of pollutants

burnt. This gas has a choking smell and is very soluble in rain-water, giving a solution of hydrochloric acid. A possible solution to the problem is the use of biodegradable plastics which rot away naturally if buried.

Summary

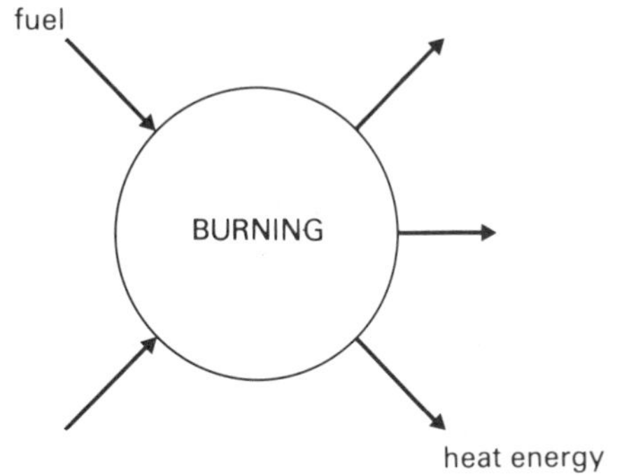

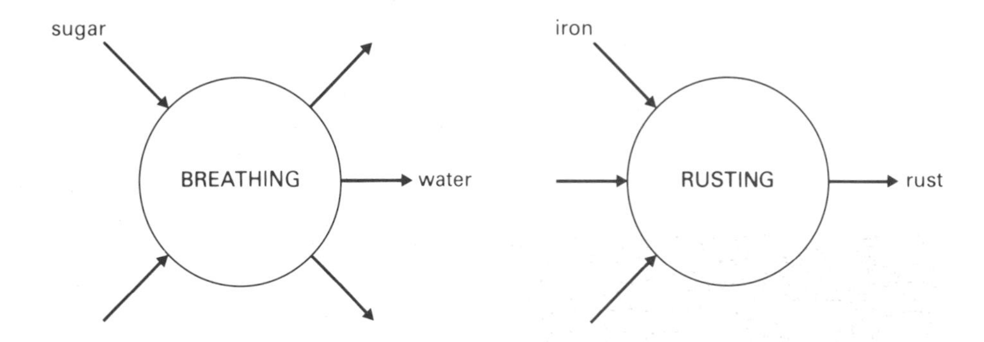

Figure 6.15

1 Copy and complete the parts of figure 6.15 to summarize what happens during burning, breathing and rusting.
2 Draw up a balance sheet to show the good things and bad things which result from burning, breathing and rusting.

7 The air in industry

Air is an important raw material in the chemicals industry. Hot air is blown into blast furnaces used for getting iron from iron ore. Air is also used in the production of sulphuric acid and nitric acid. In these examples, the air is being used as a cheap and convenient source of oxygen. These processes are described in chapters 13, 28 and 35.

For many uses, individual gases from the air are needed, and for some of these uses, the gases must be very pure. Over 100 million tonnes of pure oxygen are used throughout the world each year, and about half that amount of pure nitrogen. There is also a need for other gases, including carbon dioxide, argon, neon and other noble gases. The air is the obvious source of all these gases.

7.1 Separating the gases of the air

The gases of the air are separated by **fractional distillation** of liquid air. Fractional distillation was described in chapter 1. It depends on the different liquids in the mixture having different boiling points.

There are three stages in the industrial separation of air. A flow diagram of the process is shown in figure 7.1.

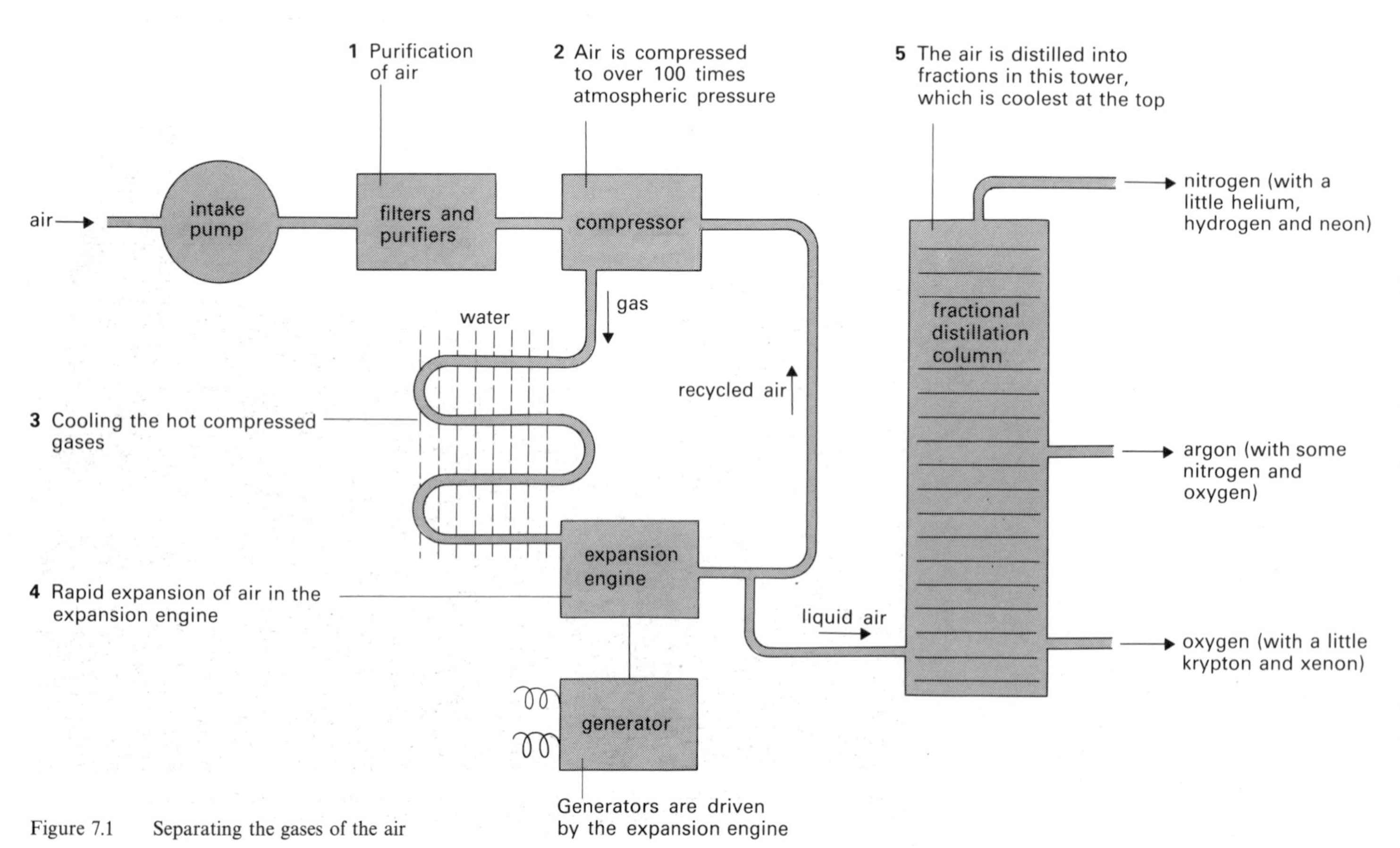

Figure 7.1 Separating the gases of the air

Questions

Look up the boiling points of the gases in the air (see figure 4.7) in tables 1 and 2 on pp. 336–9.

1 Put the gases in order of their boiling points, with the lowest one first.

2 To what temperature must air be cooled so that all the gases become liquid?

3 If liquid air is allowed to warm up, which component boils first and becomes a gas?

Purification (1)

The air is filtered to remove dust, then water and carbon dioxide are removed. (If they were not removed, they would become solid in the cooling process and block the pipes.)

Liquefaction of air (2–4)

When a gas is compressed and then allowed to expand quickly, it cools. The air is compressed to over 100 times atmospheric pressure, and as this happens it becomes hot. Still under pressure, it is cooled again by flowing water over the pipes. In the expansion engine, air expands rapidly and the temperature falls. By recycling the air, compressing it, then allowing it to expand, very low temperatures – low enough for the air to become liquid – are reached. There is an energy bonus because the expansion engine can drive generators which provide electricity for other parts of the process, such as the compressor.

Fractional distillation of liquid air (5)

Fractional distillation takes place in a rectangular tower (see figure 7.2), which is cooler inside at the top than at the bottom. Components with the lowest boiling points become gases most easily and are obtained as gases from the top of the tower. So nitrogen, with a little helium, hydrogen and neon, is obtained at the top. Oxygen and a little krypton and xenon, all with higher boiling points, are obtained from the bottom of the tower. Argon, mixed with some nitrogen and oxygen, comes out from the middle. Pure gases are not obtained from the main column. More fractional distillation processes are needed.

For many uses, the gases are compressed and stored in cylinders. Each gas cylinder has its colour code.

Oxygen and nitrogen may be liquefied and then transported and stored in large metal 'vacuum flasks'. In the steel industry, oxygen is used on such a large scale that the gas is piped direct from the air separation plant to the steel furnaces.

Question

Look at gas cylinders in your school laboratory. What are the colour codes for (a) oxygen, (b) nitrogen and (c) hydrogen? Can you find out any other colour codes? What colour are the cylinders used to supply chlorine to swimming pools?

Figure 7.2 An air separation plant. Fractionation takes place in the tall tower on the far right. You can also see the special transporters and storage tanks for liquid oxygen

7.2 Uses of the gases of the air

The uses of the gases reflect their properties.

Oxygen

Oxygen helps combustion, causes oxidation and is essential for life. Whenever a person has difficulty in breathing normally because of illness or unconsciousness (either after an accident or during and after an operation), pure, 'medical grade' oxygen, sometimes mixed with other gases, can keep life going and help the recovery of normal breathing. Divers and high altitude climbers need a supply of oxygen from cylinders.

Most steel is now made using the 'basic oxygen' steelmaking process. A blast of oxygen is passed over the surface of molten iron and speeds up the oxidation of carbon to carbon dioxide and of other impurities in the crude iron. This process is described in more detail in chapter 13.

Fuels burn better and with a hotter flame in oxygen than in air. Ethyne (which is still commonly known as acetylene) burns in oxygen to give a very hot flame. The flame is hot enough to weld and cut metals.

One of the newer uses of oxygen is in sewage treatment and for cleaning polluted rivers. Sewage is broken down naturally by micro-organisms which use the air to oxidize harmful materials to give harmless products. The process can be speeded up to treat greater amounts of sewage by pumping in oxygen. In a similar way, oxygen is used in brewing to speed up the early stages of **fermentation**.

(a)

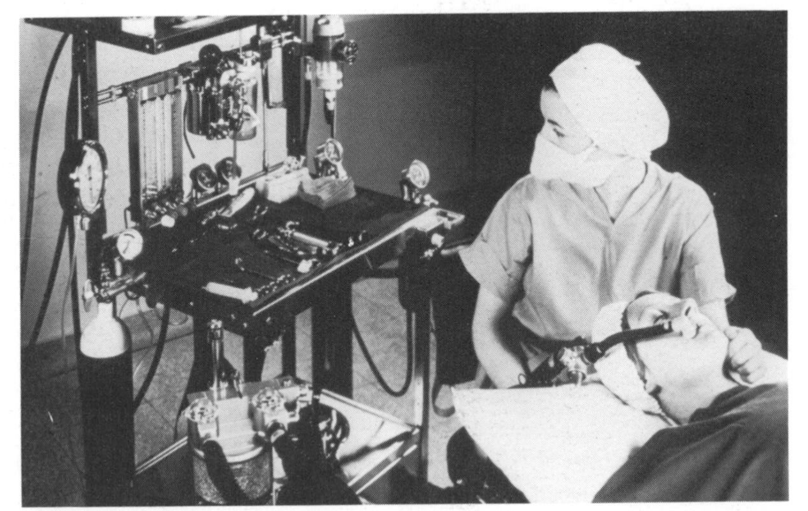

(b)

(c)

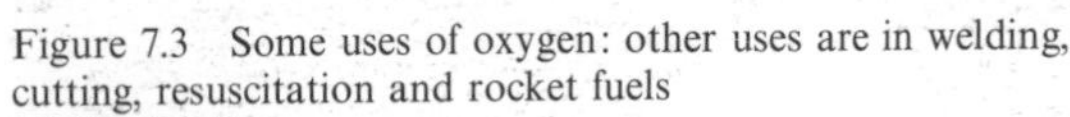

Figure 7.3 Some uses of oxygen: other uses are in welding, cutting, resuscitation and rocket fuels
(a) Steelmaking
(b) In hospitals
(c) Sewage treatment

Nitrogen

Two properties of nitrogen are important. It is very unreactive, preventing oxidation, and it has a very low boiling point, −196 °C. Nitrogen is, in a sense, the opposite of oxygen; it is used to prevent fires and to stop the filaments of electric light bulbs burning away by oxidation. Many foods such as bacon and peas are packed in nitrogen to stop them being spoiled by oxygen in the air. Coinage metals are heat treated in a nitrogen atmosphere to prevent oxidation.

At −196 °C nitrogen is a cold, unreactive liquid, ideal for quick freezing of foods and for preserving biological tissue for transplant surgery or semen for artificial insemination. It can be used for shrinking metals so that machine parts can be fitted together. Liquid nitrogen is inexpensive, and ideal for freezing the contents of pipes in which there are no valves, so that repairs can be made.

For such an unreactive gas, nitrogen has many important uses.

(a)

(b)

(c)

(d)

Figure 7.4 Some uses of liquid nitrogen: another use is for the storage of biological materials
(a) Freezing food
(b) Purging oil tankers to remove oxygen and prevent fire
(c) Shrinking metal parts so they fit together
(d) Freezing the contents of a pipeline so repairs can be made

The noble gases

Helium, neon, argon, krypton and xenon are all unreactive and they have special properties which make them particularly useful.

Figure 7.5

(a) The *Hindenburg*

(b) *Skyship 500*

Helium is very light, with a density about one-fifth that of air. The first airships were filled with hydrogen. Hydrogen is lighter than helium but has one major disadvantage in airships – it burns. Figure 7.5(a) shows the end of the *Hindenburg* in 1937. New airships like *Skyship 500* are filled with helium (see figure 7.5(b)). Hydrogen is still used for meteorological balloons.

Helium and argon are very unreactive – even less reactive than nitrogen – and can be used to protect high purity metals during heat treatment. They are much more expensive than nitrogen because there are smaller amounts in the air (most helium is, in fact, obtained from natural gas in the USA), so they are only used where very high purity is required.

Neon, krypton and xenon are used in lights of different kinds. Neon-filled advertising lights like those in Piccadilly Circus are familiar sights in large cities, while krypton and xenon are used in high intensity lamps.

Question

Which gas would you expect to be used for
(a) balloon races at the fair?
(b) photographers' flash tubes?
(c) flushing out furnaces before inflammable gases are introduced?
(d) enriching air used in the blast furnace?
(e) filling potato crisp bags?
(f) filling red indicator lamps for instrument panels?
(g) some foam fire extinguishers?
(h) doing experiments close to absolute zero (−273 °C)?
(i) mixing with natural gas to give a flame hot enough to work Pyrex glass?

Summary

Summarize the main ideas in this chapter by copying and completing figure 7.6.

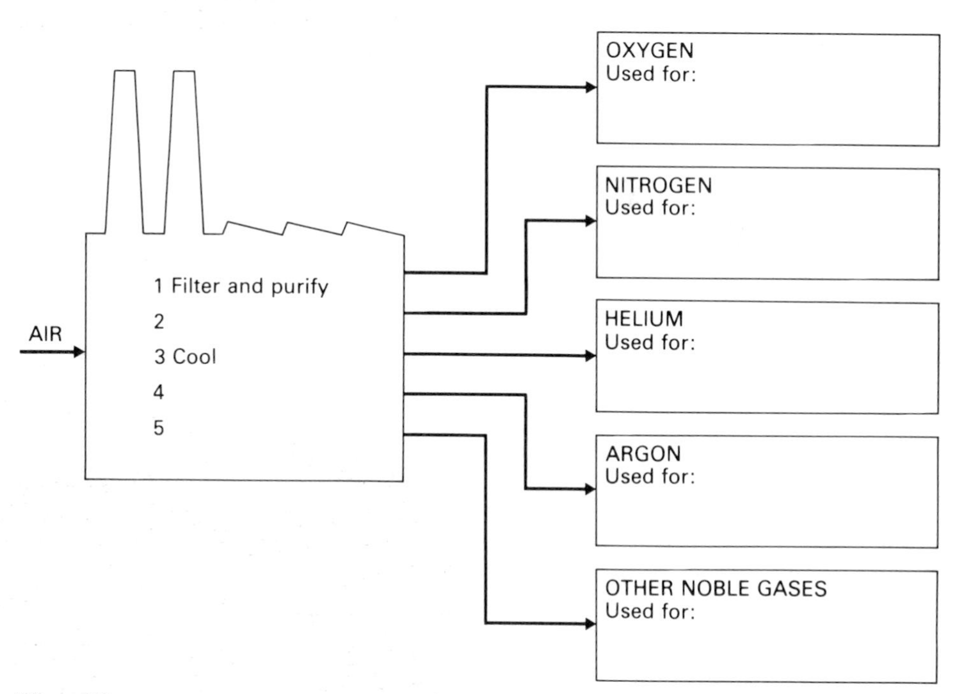

Figure 7.6

Review questions

1 (a) Name *three* gases, apart from oxygen, that are present in the atmosphere.
(b) Name *two* gases which cause pollution in the atmosphere.
(c) Give *two* important large scale uses of oxygen.
(d) White phosphorus burns if it is allowed to come into contact with the air. In the experiment below, a small piece of white phosphorus was allowed to burn slowly in 50 cm^3 of air. After some time, the phosphorus stops burning and the water is seen to have risen in the tube.

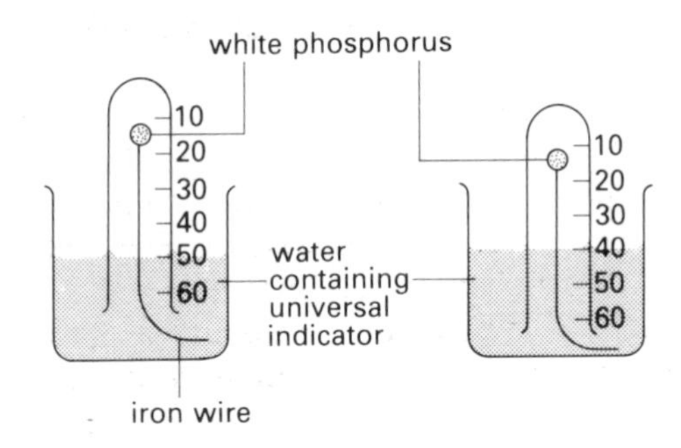

(i) Explain why the phosphorus stops burning.
(ii) Name the substance formed by some of the phosphorus during the experiment.
(iii) Use the results of the experiment shown in the diagram to calculate the percentage of oxygen present in air. Show your working.
(iv) What changes would take place in the appearance of the universal indicator during the experiment? Explain why these changes take place.
(e) Explain why the air that we breathe out contains a smaller percentage of oxygen than atmospheric air.
(f) Explain why galvanized iron rusts *more slowly* than untreated iron. **(SWEB)**

2 (a) Air is a mixture of several different gases. Name a gas in air which
(i) turns limewater milky.
(ii) supports combustion (allows things to burn in it).
(iii) is a noble gas.
(iv) condenses into a colourless liquid at room temperature.
(v) makes up approximately 78% of the air.
(b) (i) Name a gas which does not contain carbon and which seriously pollutes the air.
(ii) Name another material, other than gases, that pollutes the air. **(EMREB)**

3 (a) (i) Name *three* substances, other than oxygen and nitrogen, that are always present in the atmosphere.
(ii) Describe, with the aid of a diagram of the apparatus, an experiment by which you could demonstrate the presence in the atmosphere of *one* of these three substances.
(b) Describe, with the aid of a diagram of the apparatus, the preparation and collection of a sample of oxygen, starting with hydrogen peroxide.
(c) Outline briefly how air is liquefied. How is oxygen obtained from liquid air?
(d) State the approximate percentage of oxygen by volume in the atmosphere, and explain briefly why the figure does not vary very much. **(CLES)**

4 The figure below shows a diagram of a fire extinguisher.

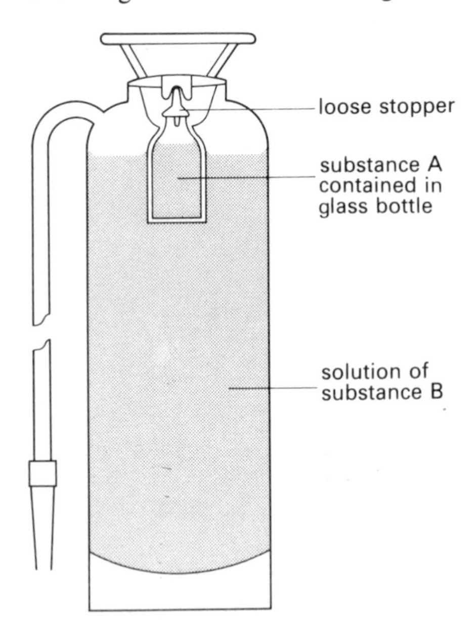

(a) Name the substances A and B shown on the diagram.
(b) Explain the working of this extinguisher.
(c) Explain the likely effect of using an extinguisher of this type on fires resulting from (i) an electrical fault, (ii) burning waste paper and (iii) spilt paraffin from an oil-lamp. **(O)**

5 (a) Study the diagram of the candle flame opposite and answer the questions which follow.
(i) Label with an arrow and the letter 'H' the part of the flame which is the hottest. Explain why this part is the hottest.

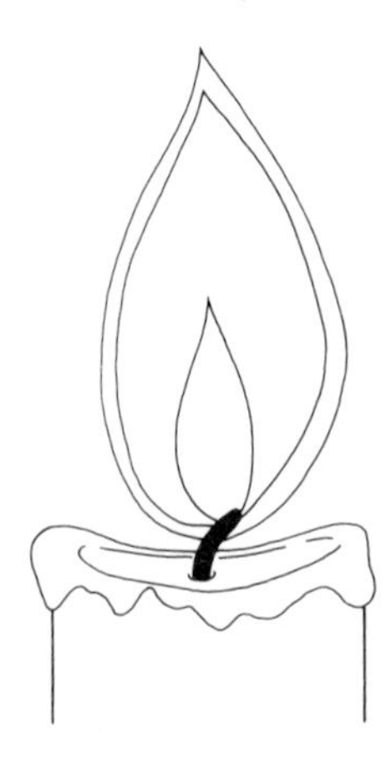

(ii) What would happen to a piece of white porcelain if it was held in the yellow part of the flame? Explain why this happens.

(iii) If a candle flame is blown out and a lighted splint placed *near* the wick, the candle relights immediately, but when a new candle is first lit, there is a long delay before a lighted splint causes the candle to burn. Carefully explain the difference with reference to how a candle actually burns.

(b) Draw a diagram of a bunsen burner flame with the air hole open. Label with an arrow and the letter 'H' the hottest part of the flame. Label with an arrow and the letter 'C' the coldest part of the flame.

(c) (i) Complete the labelling on the fire triangle below which illustrates the three essentials of a fire.

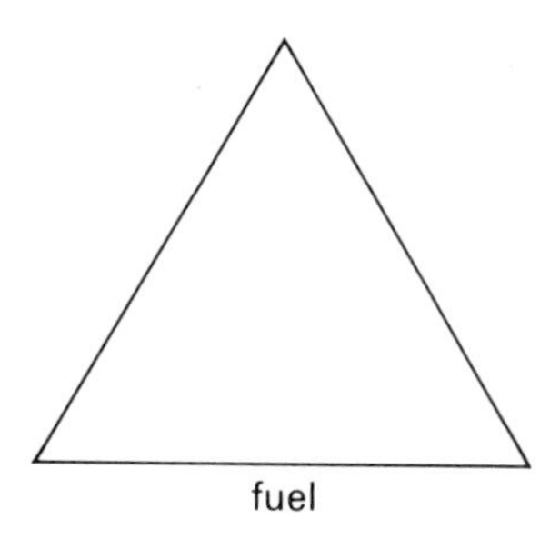

(ii) A fire may be put out by removing any one of the three essentials. For a named type of fire, explain how you would put it out, mentioning which one of the three essentials you are removing in the method you choose.

(d) Explain why a fire in a workshop, where timber had recently been sawn, could lead to an explosion.

(e) The central heating in a house is changed from electric night storage heaters to a gas fired boiler with a gas fire in the lounge. How might this affect the amount of condensation inside the house? Why is this? **(ALSEB)**

6 (a) The pH values of soil samples from various areas are shown below.

Sample	A	B	C	D	E	F	G	H	I	J
Area	Moorland	Moorland	Chalk	Moorland	Chalk	Moorland	Chalk	Loam	Moorland	Loam
pH	4.5	8.1	8.3	4.8	7.9	6.8	8.0	7.2	5.1	7.0

Group together those samples which have a similar pH. (There are three groups.)

Two more soil samples are found; one is a loam, the other is from a chalk area. What would you expect their pH values to be, approximately?

(b) Acidity in the soil may come from pollution in the air. Below is a diagram of one way in which this can happen.

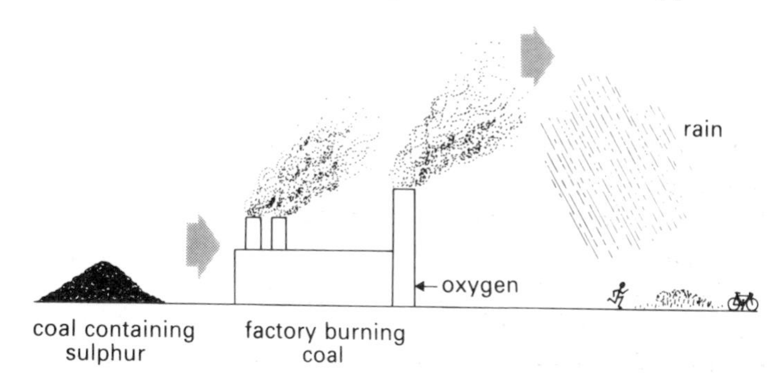

(i) What compound of sulphur is formed when the coal is burnt?

(ii) Write a word equation for the formation of this sulphur compound.

(iii) This compound reacts with water in the air to form an acid called ....

(iv) Write a word equation for the reaction in (iii).

(v) What effects does this pollution have on plants, on things made of steel and on human beings?

(c) Besides being affected by pollution in the air, plants are affected by the presence or absence of small quantities of metals in solution.

Describe an experiment to find out whether iron in solution is important for plant growth. In your description, include (i) any apparatus you would use, (ii) your method of performing the experiment, (iii) the measurements you would take, (iv) the results you might expect (give examples) and (v) how you would use your results to come to a conclusion. **(WMEB)**

7 The experiment below was set up by a pupil in order to determine the conditions under which iron nails will rust.

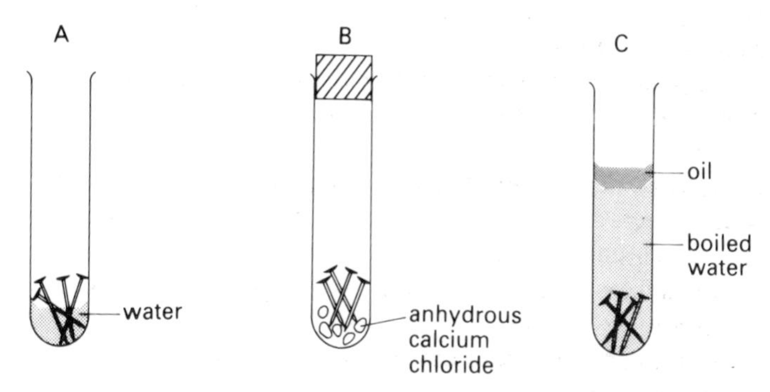

(a) What is the reason for using boiled water in tube C?
(b) What is the reason for having oil on the surface of the boiled water in tube C?
(c) What is the purpose of the anhydrous calcium chloride in tube B?
(d) What is happening to the nails in tube A?
(e) Give *four* methods of preventing iron from rusting.
(f) Is the process of rusting an oxidation or a reduction?
(g) Give a reason for your answer to (f). **(WJEC)**

8 The corrosion of iron was investigated by giving *six* identical iron nails certain treatments. *One* other nail was left untreated. All *seven* nails were then left for several weeks exposed to the atmosphere. The results of the experiments are given in the table below. One of the results contains an error.

Nail	Treatment	Cost of treatment	Mass of nail + coating before exposure to the atmosphere (g)	Mass of nail + coating after exposure to the atmosphere (g)
A	Waxed	Low	5.0	5.3
B	Oiled	Low	5.0	4.1
C	Chromium plated	High	5.0	5.0
D	Painted	Low	5.0	5.4
E	Untreated	—	4.9	6.1
F	Galvanized	Fairly high	5.0	5.1
G	Dipped in salt solution	Low	5.0	6.7

(a) Name *two* naturally occurring substances which must be present for iron to rust.
(b) Which nail was *best protected* against corrosion?
(c) Which nail received a treatment which made the corrosion worse than it would have been had it remained untreated?
(d) Which nail received a treatment which could be used on car bumpers?
(e) Which nail received a treatment which is most suitable for use on iron railings?
(f) In which case was there an obvious mistake in the weighing of the nail and the coating after the experiment?
(g) What treatment not listed in the table would you give to a piece of steel that was to be used for making containers for canning fruits and vegetables?
(h) When iron rusts, iron(III) oxide is one of the products. Write the chemical formula for this oxide. **(SREB)**

THEME C
Water

The water cycle supplies this vital compound for use in the home, in agriculture and in industry

8 Water as a compound

Water is so common that it is easy to forget its importance except, perhaps, during a drought. Almost everyone in Britain is supplied with fresh water and we each use about 160 litres of water per day for washing and drinking. In Britain as a whole, the ten regional water authorities supply about 14 million litres of water per day.

Water is vital for agriculture. On a hot day a fully grown fruit tree may lose up to 5 litres of water per minute by evaporation from its leaves and take up this water from the soil through its roots.

Industry uses water in many ways. The uses of water in the Frasch process and in salt mining are described in chapter 1. Much water is used for cooling. Power stations are built near rivers or the coast so that they can have continuous supplies of water for the cooling towers. Water is also required for washing and as a solvent. Industry needs 250 000 litres to make a tonne of paper, 75 000 litres to make a tonne of steel and 2500 litres to process a barrel of oil. On a scale which is easier to understand, this means that 9 litres of water are needed to make a single newspaper and that about 45 000 litres are involved in the manufacture of a single car.

In industrial societies there is great pressure on water supplies and a continual threat of pollution. Figure 8.1 shows some of the ways in which agriculture and industry can contaminate water supplies.

Water is so abundant and so vital to life that for thousands of years it was thought of as one of the four elements, together with earth, air and fire. Today,

Examples of possible pollution sources
1 Illegal dumping of harmful chemicals. 2 Leakage of chemicals from industrial premises. 3 Illegal oil discharges from launches and ships. 4 Trade effluent discharges containing toxic waste
5 Oil spillages from road tankers. 6 Poor quality sewage works effluent. 7 Chemical spillages on the highway. 8 Liquid farmyard waste. 9 Liquor from farmyard silage stores
10 Heating oil spillages and leaks from storage tanks 11 Illegal disposal of car engine oil into drains.

Figure 8.1

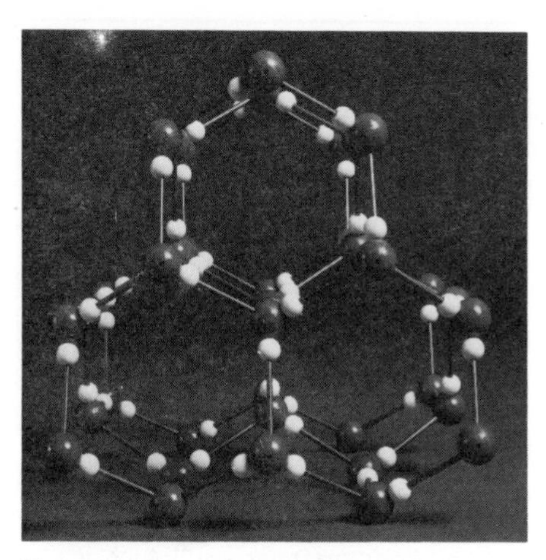

Figure 8.2 The crystal structure of ice

Figure 8.3 A pond skater

chemists describe water as a **compound** of the elements hydrogen and oxygen. Experiment 2c on p. 17 illustrates this.

Water is, in many ways, an unusual compound. Considering its structure, it is surprising that it is a liquid at room temperature (see question **1**). It is one of the few substances which we commonly experience in each of the three physical states: as ice, water and steam. Water is exceptional in that its solid form is less dense than the liquid, so that ice floats on water. Usually the molecules in a solid are packed a little more closely than in a liquid, so that the solid is denser, and sinks if put into its own liquid. In ice, the molecules are arranged in a very open network with empty spaces between (see figure 8.2). The bonds between the atoms in the ice structure are described in chapter 20.

Water molecules seem to attract each other more strongly than other molecules. One result of this is the high **surface tension** of water, which accounts for some surprising observations. Insects such as pond skaters can walk on water (see figure 8.3). With care, a needle can be floated on water. Water, on its own, is not a good wetting agent and will not spread on many surfaces or soak into cloth.

Water is also a remarkably good solvent for a wide range of substances. It is especially good at dissolving compounds made of ions, such as sodium chloride and copper(II) sulphate, which are insoluble in most other solvents. Water is much less good at dissolving substances which consist of molecules, such as iodine, sulphur and naphthalene, all of which are insoluble in it.

Questions

1 Use tables 2 and 3 on pp. 337–40 to compare these substances with water: sulphur dioxide, carbon dioxide, hydrogen sulphide, phosphine, ammonia, butane. What are the formulae of the molecules? Are they solids, liquids or gases at room temperature?
2 Can you see any connection between the shape of a snowflake and the pattern of molecules in ice?

Figure 8.4 Snow crystals

Water can take part in many chemical reactions. The demonstrations described in experiments 8a and b provide further evidence that water is a compound of hydrogen and oxygen.

Experiment 8a
The reactions of magnesium with water and steam

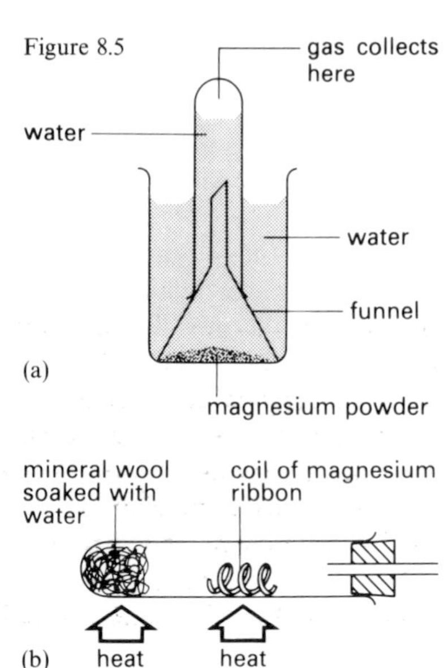

Figure 8.5

The reaction of magnesium with cold water is very slow but it can be demonstrated using the apparatus in figure 8.5(a). Magnesium reacts much faster with steam, in the apparatus shown in figure 8.5(b). The magnesium ribbon is heated strongly, and when it begins to melt, the wet mineral wool is heated to produce steam.

Results
After a day or so the test-tube in figure 8.5(a) fills with a colourless gas which burns with a pop.

When heated in steam, as in figure 8.5(b), the magnesium glows brightly and turns to a white powder. The gas leaving the outlet tube burns.

Discussion
There are only three elements involved in these reactions: magnesium, which is present as the free element at the start, and hydrogen and oxygen, which are combined as water at first. In the reactions, the magnesium takes oxygen from the water and sets free hydrogen, which is the inflammable gas:

$$\text{Magnesium(s)} + \text{Water(l)} \longrightarrow \text{Magnesium hydroxide(s)} + \text{Hydrogen(g)}$$
$$\text{Magnesium(s)} + \text{Water(g)} \longrightarrow \text{Magnesium oxide(s)} + \text{Hydrogen(g)}$$

Follow the procedure given on p. 28 to write full symbol equations for these reactions.

Experiment 8b
The reaction of chlorine with water in sunlight

chlorine

When chlorine gas is bubbled into water, it dissolves to give a pale yellow–green solution. Some of the solution is poured into a long tube and stood in a trough of water (see figure 8.6). The apparatus is placed by a window in sunshine.

Results
After a few hours in bright sunlight, a colourless gas collects at the top of the tube. This gas relights a glowing splint.

Discussion
Which three elements are involved in this reaction? Which element is free at the start? Which element is set free during the reaction? Write a possible word equation for the reaction.

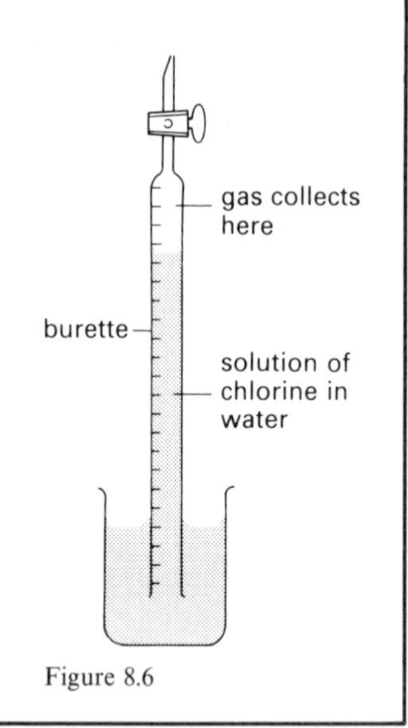

Figure 8.6

Summary

1 With the help of figure 8.1, make a list of ten ways in which water supplies may be polluted. Which of these ways is likely to be a serious source of pollution near where you live?
2 Describe four ways in which water is unusual when compared with other simple compounds which consist of small molecules.
3 How does the synthesis of water show that water is a simple compound of two elements? How do the reactions of water with magnesium and chlorine help to confirm this conclusion?

9 Water as a solvent

9.1 Solutions of solids

Water is the main **solvent** in the home, where it is used for washing, cleaning and cooking. It is the usual solvent in a chemistry laboratory, where most of the reagents are solutions in water. Solutions in water are usually called *aqueous* solutions. This is because the Latin word for water is *aqua*. So we have words such as aquarium, aqualung and aquadrome, which describe things which involve water, or are aquatic. Aqueous solutions are also of great importance in industry.

There is a limit to the amount of a substance which will dissolve in water. A solution is **saturated** when it contains as much of the dissolved substance as possible at a particular temperature. The variation of solubility with temperature is investigated in experiment 9a.

Experiment 9a
Solubility and temperature

potassium chlorate

A weighed sample of crystals is put into a boiling tube and a measured volume of water added. The mixture is heated until all the crystals have dissolved. Then the solution is allowed to cool slowly, while being stirred with a thermometer. The temperature at which crystals first appear is noted. Then more water is added and the new mixture warmed to redissolve the salt. This solution is cooled and crystallized as before. In this way, a series of temperature readings is recorded for increasing volumes of water.

Results
The results in figure 9.1 were obtained with a sample of 2.0 g of potassium chlorate, adding 4 cm^3 of water each time.

Volume of water (cm^3)	Temperature at which crystals appear (°C)	Calculated solubility (grams of solute per 100 g water)
4	92	$\frac{100}{4} \times 2 = 50$
8	63	
12	48	
16	35	
20	27	

Figure 9.1

The solubility of the salt in water, at a given temperature, is the mass of solute which will just saturate 100 g (100 cm^3) of water at that temperature. Calculate the solubility values to complete the table, and then plot a graph showing solubility on the vertical (y) axis and temperature on the horizontal (x) axis.

On the same graph, plot solubility curves for the salts in table 6 on p. 342.

Discussion

Does the solubility of potassium chlorate in water increase or decrease as the temperature rises? Look at table 6 on p. 342 to see whether this is the usual pattern for solids. Which temperature is usually chosen as the standard when comparing the solubilities of different substances? Use the graph from this experiment to work out the temperature at which crystals will start to form when a solution of 30 g potassium chlorate in 100 g water is cooled. What mass of crystals separates out when this solution has cooled to 15 °C?

9.2 Solutions of gases

Water is also a good solvent for many gases. Life depends on this, because respiration in animals and plants depends on oxygen and carbon dioxide being able to dissolve in the water that is present in all living tissues. Fish absorb dissolved oxygen from the water they live in, taking it into their blood through the gills.

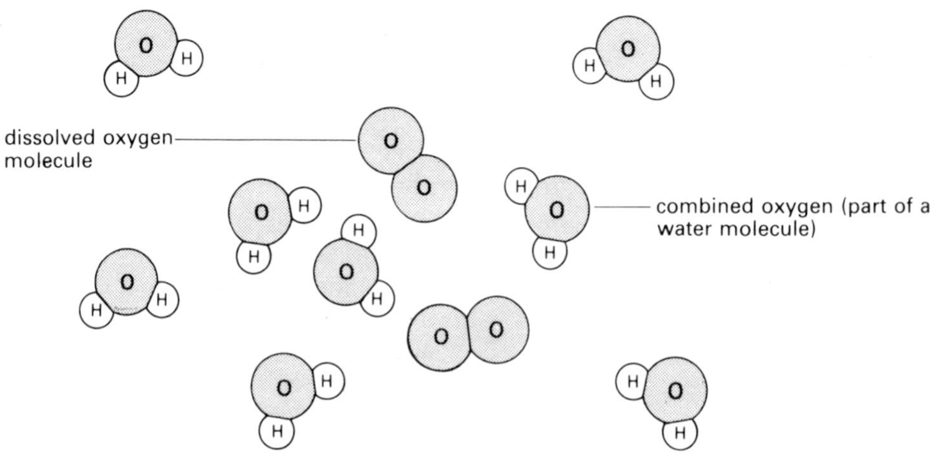

Figure 9.2 The difference between dissolved and combined oxygen in water

Experiment 9b
Measuring the amount of dissolved air in tap water

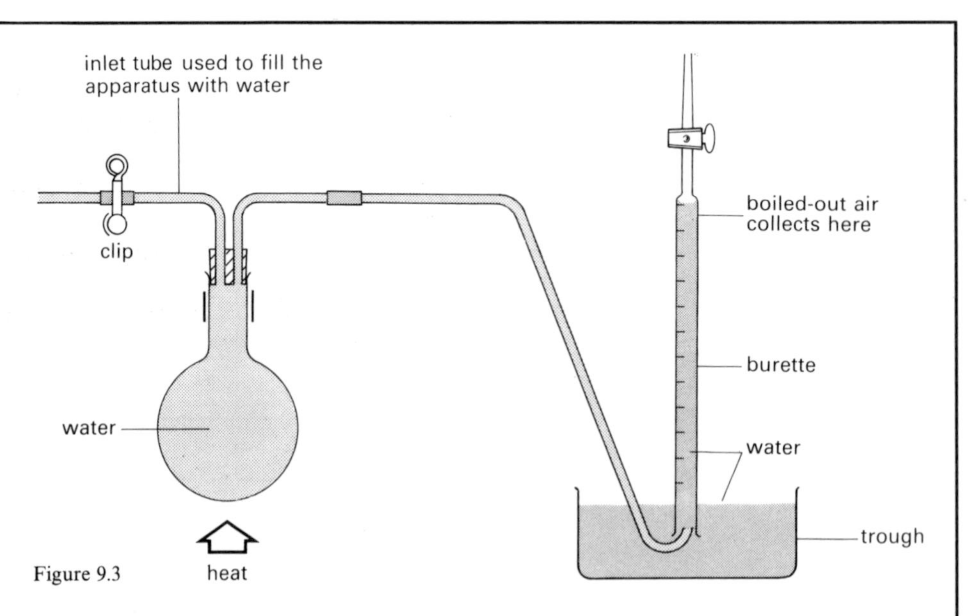

Figure 9.3

The apparatus shown in figure 9.3 is completely filled with tap water, making sure that there are no air bubbles in the top of the flask. The water in the flask is heated until the water boils. The heating is continued until no more gas collects in the upside-down burette.

Results
About 20 cm^3 of gas are collected from a 1 litre sample of water.

Discussion
Do gases get more or less soluble in water as the temperature is raised? Consider the data in table 7 on p. 343. Which is more soluble in water, oxygen or nitrogen? How could you find out, by experiment, the proportion of oxygen dissolved in water?

Of the air dissolved in water, 34 per cent is oxygen and 66 per cent is nitrogen. Fish need this oxygen to live. Polluted water can become oxygen starved. Fish and other water-life are then threatened.

The very high solubility of some gases in water is illustrated by the fountain experiment.

Experiment 9c
The fountain experiment

hydrogen chloride gas

This experiment is carried out with the apparatus shown in figure 9.4. The flask is filled with a gas which is very soluble in water, such as hydrogen chloride.

Discussion
What happens to the pressure inside the flask when the first drops of water enter it? What forces the jet of water up the central tube? Which of the other gases listed in table 7 on p. 343 could be used in the fountain experiment?

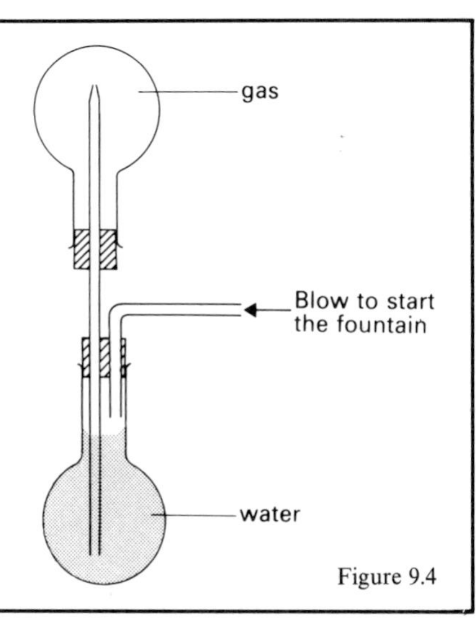

Figure 9.4

Summary

Write a summary of the main ideas in this chapter based on these questions:
1 Why is water the commonest solvent in the home, in laboratories, and in industry?
2 Is it usual for solids to become more or less soluble in water as the temperature is raised?
3 Is it usual for gases to become more or less soluble in water as the temperature is raised?
4 Why is it important to life that gases, particularly oxygen and carbon dioxide, can dissolve in water?

10 Water in the home

10.1 Pure water

The water we use at home comes from rivers, lakes and boreholes. There is a natural water cycle. Water **evaporates** from the sea, lakes and rivers, and from the leaves of plants. The vapour **condenses** to form clouds, and later falls as rain, hail or snow. The rain or melting snow completes the cycle by flowing down streams and rivers to lakes and the sea. Some water seeps through rocks to join natural underground reservoirs. Figure 10.1 shows how the needs of home and industry are met by 'tapping' water from the natural cycle. After use, the water is purified by sewage treatment before being returned to rivers and the sea.

The water from rivers is first stored in reservoirs where the process of purification starts. The water is still, so the larger particles of dirt can settle out. At

Figure 10.1 The water cycle

Figure 10.2 A filter tank in a water treatment works

the surface, oxygen and sunlight break down other impurities and kill some bacteria.

The water from the storage reservoirs is treated in two stages. It is filtered through beds of fine sand to remove suspended solids, then it is treated with chlorine to kill bacteria.

Household tap water is free from harmful bacteria and insoluble dirt, but it is not pure in the chemist's sense because it contains gases and salts in solution. In the home, water is mainly used for drinking and washing. The presence of salts does not matter in water used for drinking, but some salts can affect its use for washing, and damage boilers and hot water pipes.

10.2 Soaps and detergents

Although water is a good solvent for many things, it is not, by itself, good at removing dirt from clothes. Its high surface tension means that drops of water do not readily spread through the cloth. Also, dirt is insoluble in water. This is why **detergents** are needed to help separate it from the cloth and carry it away in the water.

Anything which works with water to get things clean is a detergent. There are two common types of detergent: soaps, which are manufactured from animal fats or vegetable oils, and soapless (synthetic) detergents, which are made from oil. Washing-up liquid is a soapless detergent.

Experiment 10a
Making soap

sodium hydroxide

Soap is made by boiling fat or vegetable oil with a concentrated, aqueous solution of an alkali. The process is called *saponification*. Sodium hydroxide is the alkali usually used. Quick results are achieved using castor oil.

The mixture of oil and alkali is stirred and heated on a steam bath for about ten minutes. Then concentrated salt solution is added. This helps to separate out the soap, which forms a white crust on the surface when the mixture is cooled.

castor oil and concentrated aqueous sodium hydroxide
stirring rod
boiling water
heat

Figure 10.3

Discussion
How pure is the soap made in this experiment? Is it safe to wash with? Suggest other types of fat or oil which are readily available but cheaper than castor oil.

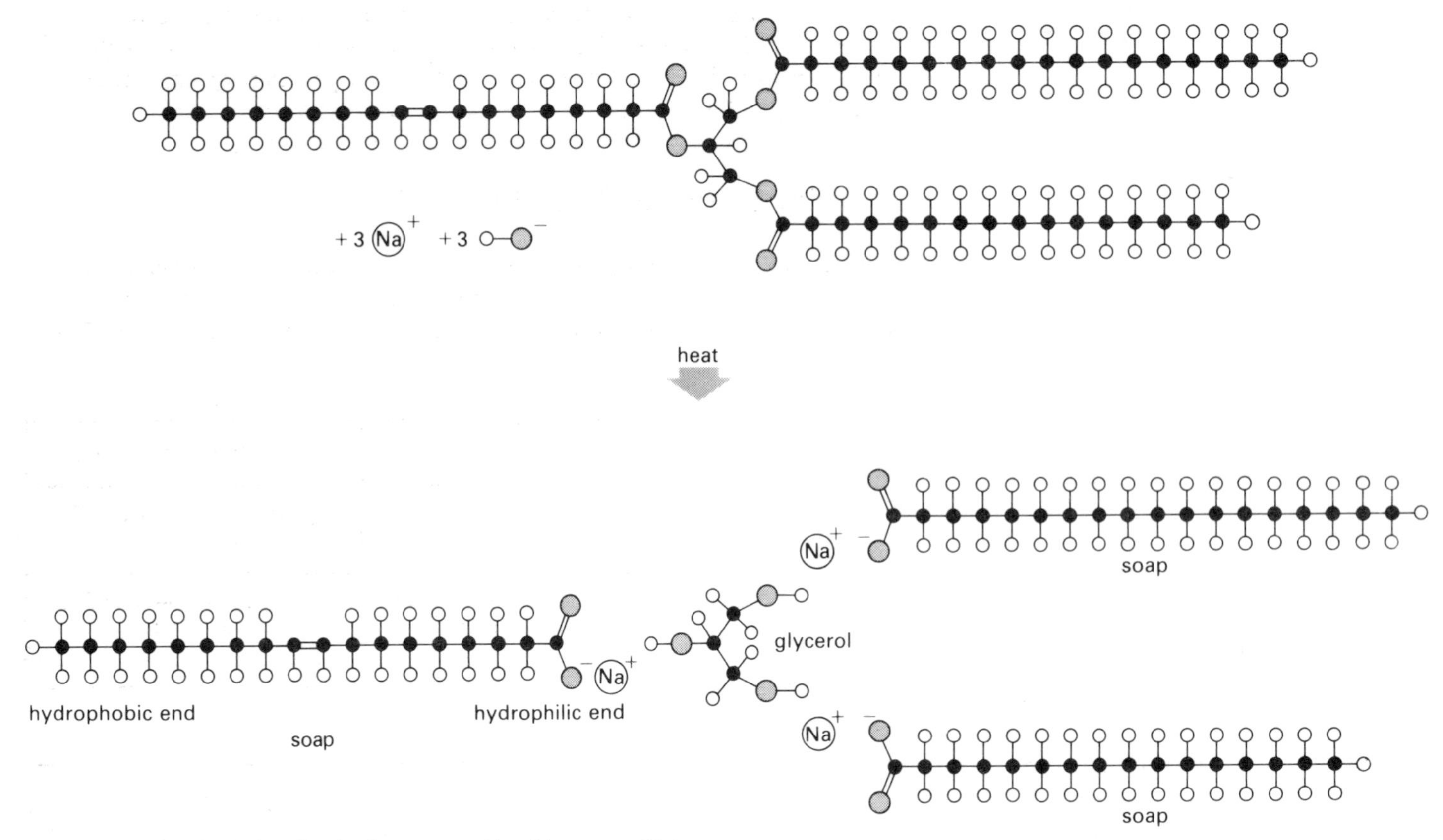

Figure 10.4 Model equation showing how a vegetable oil is converted into soap

> **Questions**
>
> **1** Why is hot water better for washing than cold water? How does raising the temperature affect water and how does it affect dirt?
> **2** Why is it necessary to agitate clothes during washing?

Figure 10.4 shows a molecule of a vegetable oil and how it is converted to soap with alkali. The three long, thin particles with sodium ions are soap. The smaller fragment is a molecule of propane-1,2,3-triol, which is usually called glycerol. Glycerol is a useful by-product of soap making.

Detergents help to get things clean in three ways:

* they reduce the surface tension of the water so that it wets the cloth
* they separate dirt from the fibres of the cloth
* they keep the dirt suspended in the water so that it can be rinsed away.

The third function is particularly important in modern washing machines which spin the dirty water away through the clothes being washed.

The particles of a detergent are 'two-faced'. One end (the *hydrophilic* end) is

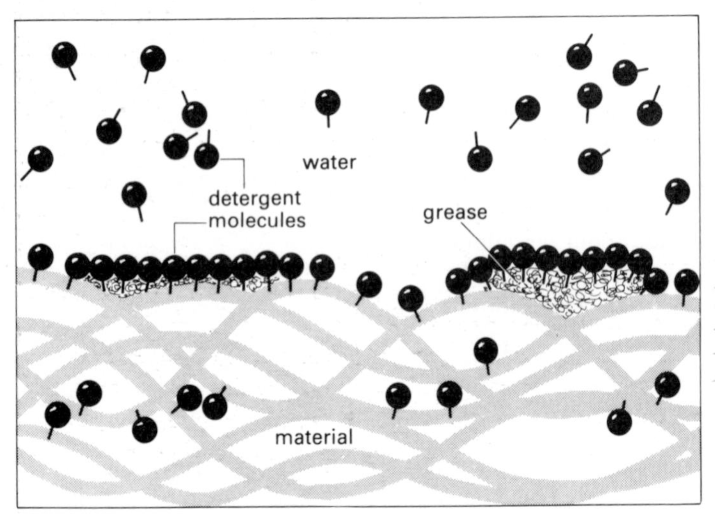

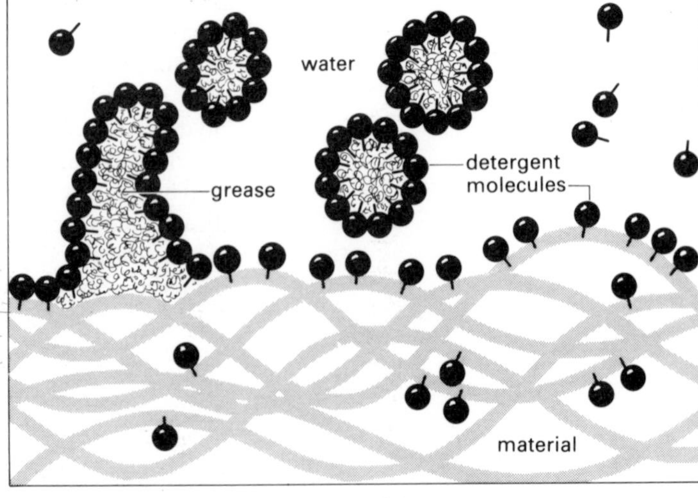

Figure 10.5 Detergent molecules removing grease from cloth fibres

attracted to water and likes to mix with water molecules. The other end (the *hydrophobic* end) is chemically like oil and tends to avoid water and mix with grease. The pictures in figure 10.5 show how detergent particles separate grease from the fibres of cloth and then hold the droplets in suspension.

10.3 Hard and soft water

In some regions it is found that water does not easily form a lather when shaken with soap. Water like this is called *hard* water. The behaviour of soapy and soapless detergents when shaken with hard water is described in figure 10.6. There are different types of hardness.

Type of water	Soapy detergent	Soapless detergent
Soft water	A good lather is formed. The solution stays clear	A good lather is formed. The solution stays clear
Hard water	Very little lather forms. A greasy scum appears and the solution is cloudy	A good lather is formed. The solution stays clear

Figure 10.6

Experiment 10b
Types of hardness

Various types of water are tested by shaking measured samples with soap solution. The amount of soap solution needed to produce a good and permanent lather on shaking is noted. The experiment is repeated with fresh water samples which have been boiled for several minutes and then cooled.

Results
Typical results are shown in figure 10.7.

Sample	Volume of soap solution needed (cm^3)	
	Before boiling	After boiling
Distilled water	0.5	0.5
Water from limestone hills	10.5	1.0
Water from chalk stream	9.0	1.5
Distilled water, after shaking with calcium sulphate and filtering	11.0	11.0

Figure 10.7

Discussion
Why must the same volume of each type of water be used in the tests? What apparatus would you use to add the soap solution to the water samples in a controlled way? Which of the samples tested were hard?

The results for tap water are omitted from the table because they depend on the region where the tests are done. What would be the results for tap water where you live? Which of the samples were changed by boiling? Water is *temporarily hard* if it can be softened by boiling. Boiling does not affect *permanent hardness*.

Experiment 10c
Which ions cause hardness in water?

This experiment involves dissolving a selection of salts in distilled water and then shaking a measured volume of each solution with a little soap solution. The water is soft if a good lather forms. The water is hard if it turns cloudy with scum and there is little, or no, lather.

Results
Some typical results are shown in figure 10.8.

Solution used	Ions present		Height of lather (mm)
	Positive	Negative	
Sodium sulphate	Na^+	SO_4^{2-}	19
Magnesium sulphate	Mg^{2+}		3
Potassium chloride		Cl^-	21
Calcium chloride	Ca^{2+}		1
Sodium nitrate		NO_3^-	20
Magnesium nitrate			2
Calcium nitrate			1

Figure 10.8

Discussion
Copy and complete the table to show the ions present in each salt. Is hardness associated with the positive or the negative ions? Which of the ions make water hard? Which of these ions are present in the minerals chalk, limestone, gypsum and dolomite? Where in Britain can these minerals be found?

Marble, chalk and limestone are all forms of calcium carbonate. Calcium carbonate is insoluble in pure water. If there is carbon dioxide dissolved in the water, it reacts with the calcium carbonate to form calcium hydrogencarbonate, which is soluble. In this way calcium ions are brought into solution and the water becomes hard.

Calcium carbonate(s) + Carbon dioxide(aq) + Water(l) $\rightleftharpoons$ Calcium hydrogencarbonate(aq)

Some carbon dioxide may dissolve in rain-water as it falls. Most of the carbon dioxide is present in water in the soil.

The reaction described by the above word equation can go both ways. It is **reversible**. Boiling decomposes the calcium hydrogencarbonate and drives off the carbon dioxide. As a result, the insoluble calcium carbonate re-forms. This is seen

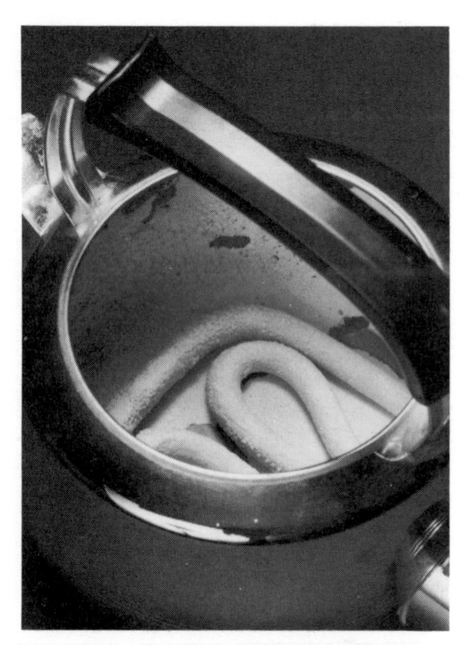

Figure 10.9 A deposit of calcium carbonate

Questions

1 What are stalactites and stalagmites, and where are they found? How are they formed?
2 In what ways is the chemistry of hard water related to the formation of stalactites and stalagmites?

Questions

1 What are the main disadvantages of a hard water supply?
2 In some circumstances hard water may be preferred. What are its advantages over soft water?

as fur in kettles and hot water pipes (see figure 10.9). The water in chalk and limestone regions is temporarily hard because the hardness is removed by boiling.

The chemistry of these reactions is described in more detail in chapter 23.

The mineral gypsum is a form of calcium sulphate. Calcium sulphate is very slightly soluble in water. Enough will dissolve to make the water hard. This hardness is not removed by boiling, and the water is permanently hard.

The most effective method of softening both types of hard water is to use an ion exchange column. The column contains a mass of tiny plastic beads which are chemically treated so that they are negatively charged. These attract and hold on to positive metal ions (see figure 10.10).

Figure 10.10 An ion exchange column

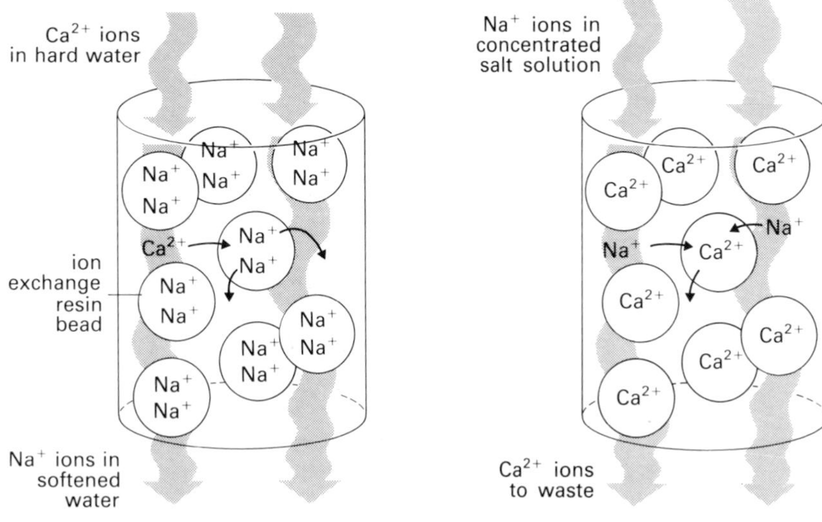

(a) Softening hard water by ion exchange (b) Regenerating the column

In a household water softener the column is prepared for use by running salt solution through the column so that the beads have sodium ions associated with them. As hard water passes through the column the calcium ions are exchanged for sodium ions:

$$\underset{\text{calcium ions in hard water}}{Ca^{2+}} + \underset{\text{sodium ions on beads}}{2Na^{+}} \longrightarrow \underset{\text{calcium ions on beads}}{Ca^{2+}} + \underset{\text{sodium ions in the water}}{2Na^{+}}$$

Sodium ions do not affect soap, so the water leaving the column is soft. Eventually, the beads are completely covered with calcium ions and the column then has to be regenerated. This is done by passing concentrated salt solution through the beads again.

There are other ways of removing hardness by removing calcium ions. These are described in chapter 23.

10.4 Sewage treatment

After use in the home, the water discharged into sewers has to be purified before it can be returned to rivers (see figure 10.1). Sewage treatment is carried out in three stages, but in many places the third stage is omitted.

1 Removal of solids

Metal screens remove rags, pieces of wood and other large objects which would damage the pumps. Then the sewage flows slowly along channels where sand and coarse dirt settle out. Finally, the sewage enters large settling tanks where the

Figure 10.11 Sewage treatment: the filter beds where the organic waste is removed by bacteria

remaining solids slowly sink to form a sludge. Sewage sludge is digested to form methane gas, which can be used as a fuel.

2 Removal of organic waste

Organic waste in the sewage is removed by bacteria. This can be done by sprinkling the sewage from rotating arms so that it trickles through a bed of gravel. The micro-organisms grow on the surface of the gravel where they are exposed to the air and can feed on the sewage. Oxygen is provided to help the bacteria feed and grow.

The water flowing from the gravel beds is pure enough to be fed into a river but a third stage of treatment is desirable.

3 Removal of nitrates and phosphates

The first two stages do not remove nitrates and phosphates. These soluble impurities can cause serious pollution in rivers and lakes because they are plant nutrients and encourage the growth of algae. If the supply of nutrients is too great the plant growth is excessive. The river or lake becomes choked with rotting plants. The animal life is starved of oxygen and dies. The third stage of sewage treatment is designed to remove nitrates and phosphates.

Washing powders include phosphates to help give a whiter wash but this causes an increase in the amount of phosphates in sewage.

Summary

Write a summary of the main ideas in this chapter based on the following questions.

1 (a) What are the two main stages of water treatment before it is used in the home?
(b) What are the three main stages of water treatment after it has been used?

2 What does the term detergent mean? What are the two main types of detergent? In what three ways do detergents help to get things clean?

3 Which ions cause hardness in water? What is the difference between temporary and permanent hardness of water? How does water become hard naturally and how may it be softened?

Review questions

1 (a) A group of pupils, concerned with pollution in a small river close to their school, carried out an investigation. Various measurements were taken at five sites along the river bank. The sites are shown on the map and the corresponding measurements are shown in the table.

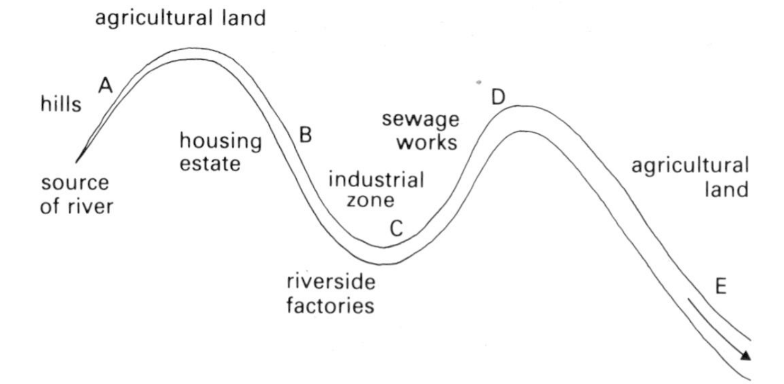

Samples of water	Site A	Site B	Site C	Site D	Site E
Temperature (°C)	6	7	12	11	11
pH value	7	8.5	5	6	6
Dissolved oxygen (parts per million)	14	12	10	3	5
Existence of fish life	Yes	Yes	Little	No	No

Study the map and table and then suggest reasons for the following:

(i) the large variation in temperature between B and C,

(ii) the large drop in pH value between B and C,

(iii) the large drop in dissolved oxygen content between C and D.

(iv) Which sample of water do you think would be the most pure?

(v) Why do you think this sample is the most pure?

(b) At E, the river is flowing very slowly through agricultural land. It has become choked with water plants, especially chickweed, at the surface.

(i) Name a pollutant that may have caused this excessive growth.

(ii) How could the pollutant have entered the river?

(iii) Name the element which is most responsible for the excessive plant growth.

(iv) There are no fish at D or E. Suggest different reasons for this.

(c) The pupils at E needed some distilled water. They had with them a camping stove, a kettle and some bottles.

(i) Explain how they could have produced some distilled water. Draw a sketch if this helps to explain your method.

(ii) What would be the pH value of the distilled water produced? **(WMEB)**

2 (a) How does the gaseous mixture which may be boiled out of tap-water differ in its oxygen–nitrogen ratio from air, and why?

(b) Calculate the solubility of potassium chloride in water at 15 °C from the following data:

Mass of evaporating basin = 23.62 g
Mass of basin + saturated solution = 53.41 g
Mass of basin + potassium chloride = 30.91 g

(c) Name a substance which causes permanent hardness in water and explain, briefly, why the addition of sodium carbonate softens this water.

(d) Explain how temporarily hard water is formed when water containing dissolved carbon dioxide runs through limestone.

(e) State what happens when carbon dioxide is bubbled into limewater for a long time, and the solution is then boiled. **(SUJB)**

3 Describe, with diagrams where necessary, how to carry out experiments to measure

(a) the solubility of potassium chloride in water at temperatures between 25 °C and 75 °C,

(b) the volume of air dissolved in 1 litre of tap water and the percentage of oxygen in this dissolved air. **(L)**

4 What is the biological importance of the fact that gases can dissolve in water?

Describe how you would demonstrate that gases become less soluble in water as the temperature is raised.

Carbon dioxide is dissolved in drinks to make them ‘fizzy’. What happens when a bottle, or can, of a ‘fizzy’ drink is opened and what does this show about the effect of pressure on the solubility of the gas? **(L)**

5 (a) Describe and explain how you would convert a sample of fat to soap. The formula of a typical fat is

$$\begin{array}{l} CH_2O.CO(CH_2)_{16}CH_3 \\ | \\ CHO.CO(CH_2)_{16}CH_3 \\ | \\ CH_2O.CO(CH_2)_{16}CH_3 \end{array}$$

(b) Explain how soap serves to remove dirt and account for the fact that it is less successful in hard water districts.
(c) *Either* explain in terms of its structure how a typical detergent differs from soap *or* give a short explanation of the widespread popularity of detergents over soap powders. **(O)**

6 (a) (i) A soap molecule and a detergent molecule have a similar chemical structure. Describe this structure. (You may draw a labelled diagram if you wish.)
(ii) Explain how these molecules remove grease during their washing action. (Diagrams may be used.)
(b) Name the raw materials used to make a detergent.
(c) What is a biodegradable detergent?
(d) A needle is carefully floated on the surface of a bowl of water and then a few drops of washing-up liquid are added to the water. The needle immediately sinks to the bottom of the bowl. Explain why the needle behaves in this way in both parts of this experiment.
(e) Some people living in Derbyshire have hard water piped into their homes. Describe and explain how this would affect (i) the inside of their kettles and (ii) the bath when the water is emptied.
(f) (i) Shops in Derbyshire sell an ion exchange resin canister which may be fitted onto the cold water tap. How does this soften the water?
(ii) How may this resin be regenerated when it ceases to soften the tap water? **(ALSEB)**

7 What is meant by 'hard water'? Describe experiments you have carried out which show which ions cause hardness in water, and explain how you could remove these ions (a) by precipitation and (b) by ion exchange. **(L)**

8 A class of pupils was provided with a sample of hard water and asked to use any method they wished to soften the water. 25 cm^3 of the original sample without any treatment required 30 cm^3 of the soap solution before a lather lasting two minutes was observed. At the end of their experiments, each pupil took 25 cm^3 of product and ran soap solution in until a two minute lather was obtained on shaking. The methods used and results obtained are summarized as follows:

Method used	Volume of soap solution required for 25 cm^3 of product
Filtering	30 cm^3
Distilling	2 cm^3
Boiling	30 cm^3
Adding washing soda	2 cm^3
Ion exchange column	2 cm^3

(a) Which methods were successful in softening the water?
(b) (i) Was the original sample provided temporarily or permanently hard? Give a reason for your answer.
(ii) Name a chemical which could have been causing the hardness.
(c) Explain why distillation and filtration produced such different results. **(NWREB)**

9 Describe, with examples, some of the ways in which pollution is a danger to water supplies. Give an account of some of the methods which are being considered for obtaining an increasing volume of fresh water in the future. **(L)**

10 Describe the treatment, by the water authorities, of water between entering the reservoir and being used in the home. Explain what happens to the water after we have used it and before it is returned to the rivers. **(L)**

THEME D

The metals

Metals in action on a grand scale. How many different uses of metals are pictured here?

11 The physical properties and uses of metals

Most of the elements are metals – over three-quarters of them. Some metals have given their names to periods of history, for example, the Bronze Age and the Iron Age. Some metals, such as gold and silver, are highly valued for their rarity and appearance. Other metals, such as titanium and zirconium, have become essential to high technology.

The uses of important metals illustrate their typical properties. Some metals are used in the pure state, but often metals are mixed by being melted together to make *alloys.*

The link between the properties of metals and their structure is described in chapter 18.

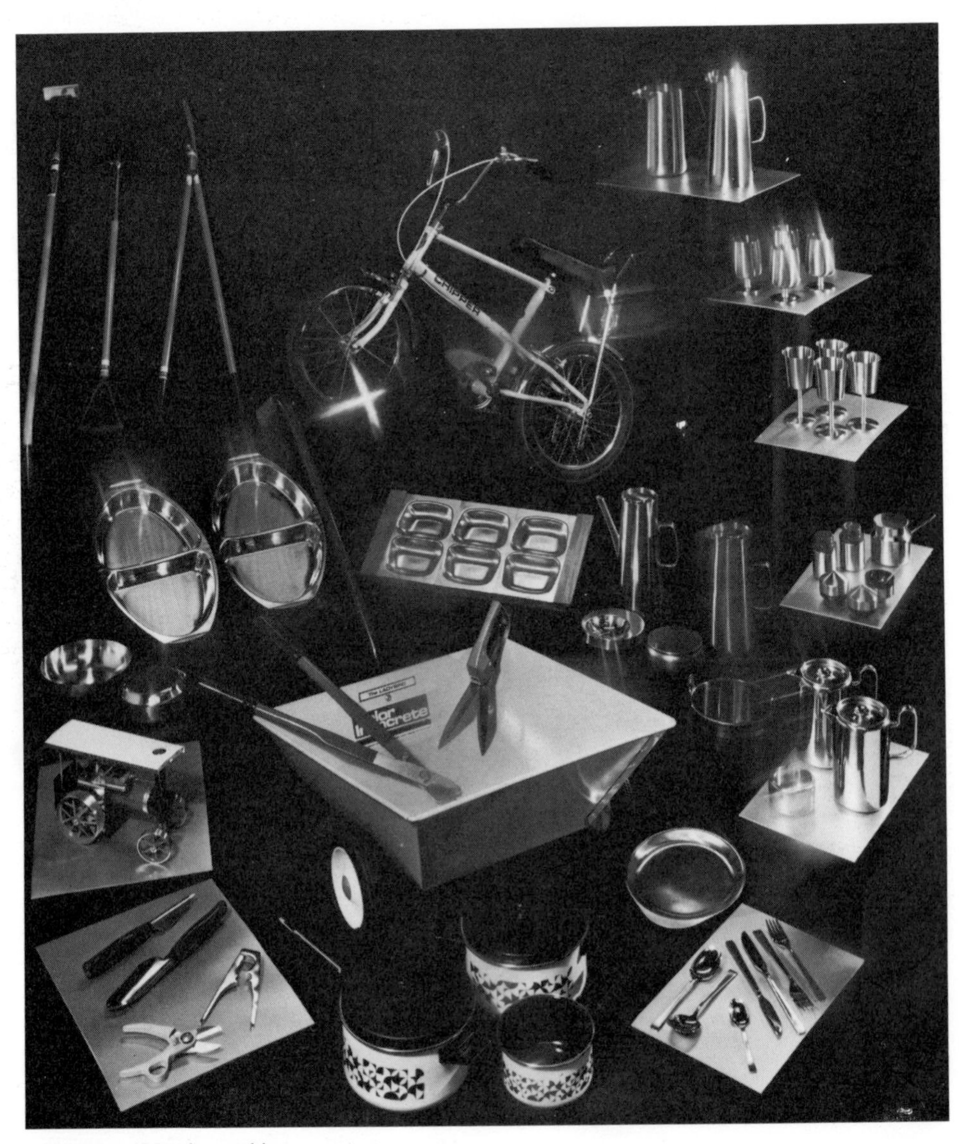

Figure 11.1 Metals are shiny

Metals are shiny

All metals are shiny when freshly cut and polished, but many of them lose their shine in the air. The value of platinum and gold in jewellery is that they stay shiny.

The elements copper, silver and gold are sometimes called the coinage metals because they are used for making money. Today, only copper is cheap enough for making everyday coins. 'Silver' coins are made from an alloy of copper and nickel.

Mirrors are made by coating glass with silver. Silver quickly loses its shine in polluted air or when exposed to certain foods, such as egg. It is also very expensive. This is why stainless steel, which is an alloy of iron with chromium and nickel, is often used for making tableware.

Metals conduct electricity

Pure copper is an excellent conductor of electricity and is used for wire in transformers and motors. Aluminium is not such a good conductor, but it is much less dense and so it is preferred for making overhead cables.

Metals bend and stretch

Metals can be hammered or rolled into shape – they are malleable. Metals can be stretched into wire – they are ductile. The ductility of metals is exploited in the manufacture of lightweight drink cans. Pure aluminium can be rolled into foil (often called 'silver' paper) which is used to wrap sweets, to cover milk bottles and to make food containers.

Lead is also relatively soft and bendable. It is used to protect underground cables and for roofs and pipes. Lead is suitable for outdoor use because it resists corrosion.

The hot and cold rolling processes in steelmaking depend on steel's ductility and malleability. They are important methods for shaping this alloy and for changing its properties.

Figure 11.2 Metals are malleable

(a) A large forging press

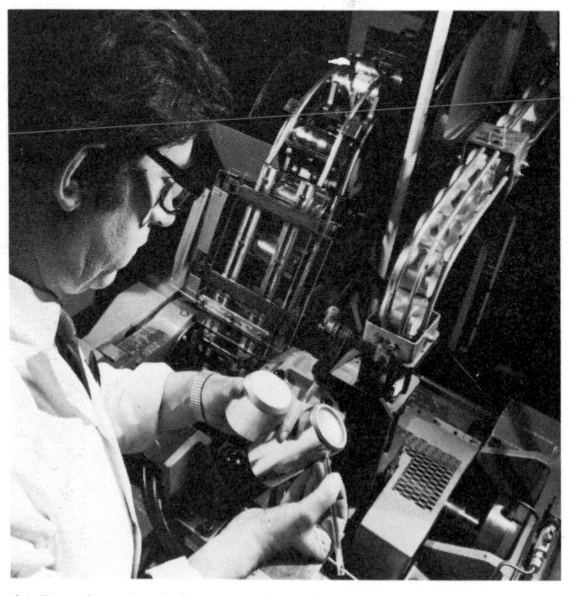

(b) Pressing aluminium cans into shape

Metals conduct heat

Metals feel cold when you touch them because they quickly conduct heat away from your hand. Aluminium and steel are used to make kettles and saucepans. They are able to conduct heat away from the source to the food being cooked.

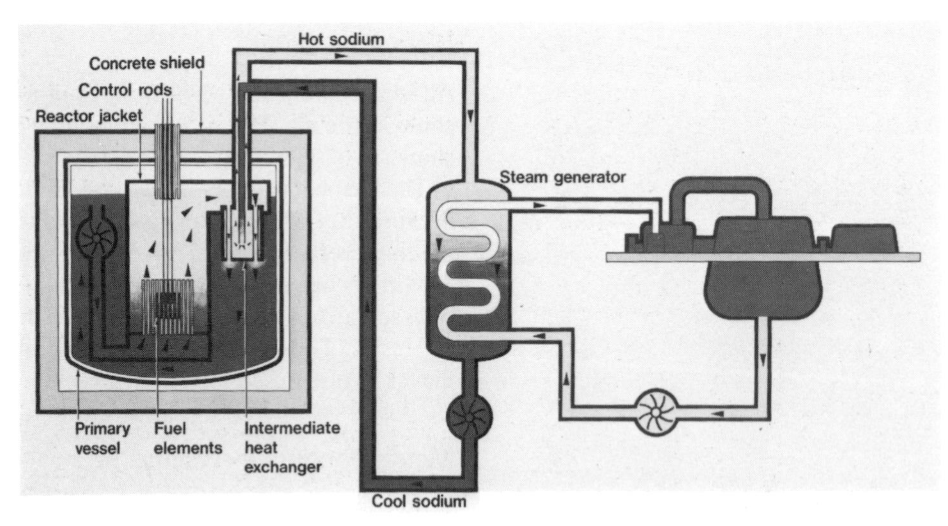

Figure 11.3 Metals conduct heat: liquid sodium carries the heat away from the core of a nuclear reactor. The heat is used to produce steam, which drives turbines and generates electricity

Liquid sodium is used to carry the heat from the core of some types of nuclear reactor to the heat exchangers where steam is raised.

Metals have high melting points

Not all metals melt at high temperatures. Mercury is a liquid at room temperature, and sodium and potassium melt below 100 °C. Solder is an alloy of tin and lead which can be melted with a soldering iron and used to join wires in electric circuits. But in general, most metals do melt at high temperatures – much higher temperatures than most non-metals.

Tungsten is used to make the filaments of light bulbs because of its very high melting point. Alloys of chromium and nickel are used to make the heating elements of electric fires. Metals are exposed to high temperatures inside the engines of cars and aeroplanes. Titanium is used to make light, strong alloys for supersonic aircraft. These alloys must remain strong at high temperatures because they are heated by friction as the aircraft speeds through the air.

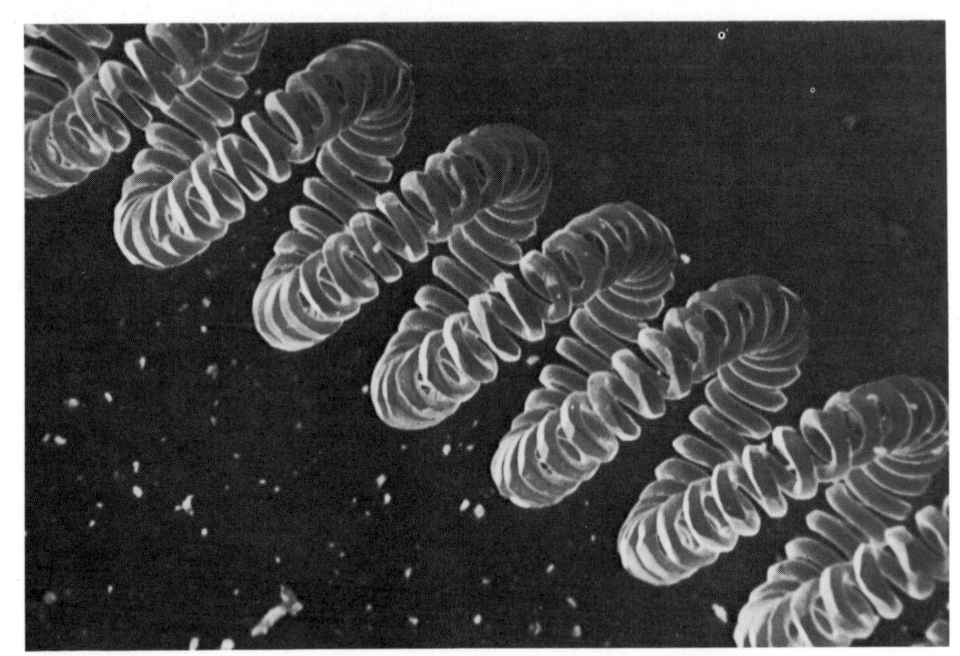

Figure 11.4 Metals have high melting points: tungsten filaments in electric light bulbs get very hot without melting

Question

Use table 1 on p. 336 to make a comparison of the melting points of a selection of common metals and non-metals. How many metals melt below 100 °C? How many non-metals melt above 100 °C?

Questions

1 Make a list of metals which are used for their strength. Describe their uses. Include examples of tensile strength, hardness and toughness.
2 Not all metals are strong. Make a list of metals which are mechanically weak.
3 Do metals have high or low densities? Use table 1 on p. 336 to list metals in each of these ranges of density: 0–1.0 g/cm^3, 1.0–10.0 g/cm^3, 10.0–20.0 g/cm^3, above 20 g/cm^3.
4 Which metals and alloys can be magnetized? How is magnetism used?
5 Which metals and alloys make a ringing sound when struck? What use is made of this property of metals?

Metals are strong

The term strength has different meanings. Perhaps the most obvious type of strength is tensile strength. This is the ability to support a heavy load. Steel has a high tensile strength and is used to make girders, hawsers and chains. For some purposes, hardness is a more important property. Steel can be hardened by heat treatment or alloying. Saw and chisel blades must be hard so they are not easily blunted. Tungsten steel cutting tools stay sharp at high speeds and high temperatures. The head of a hammer must be not only hard but tough – it must not shatter when it hits a nail. Toughness and brittleness are opposites. Cast iron, which is used for manhole covers, is brittle.

Figure 11.5 Metals are strong: the galvanized steel suspension cables in the Severn Bridge have very high tensile strength

Summary

Metals are useful because they

* are shiny
* conduct electricity
* bend and stretch
* conduct heat
* have high melting points
* are strong.

1 Draw pictures or line diagrams to illustrate the properties and uses of metals described in this chapter. Name the metals and alloys included in your pictures. You may be able to find some suitable pictures in magazines which you can cut out and stick into your notes.
2 With the help of this chapter, other reference books and your general knowledge, make a list of the main uses of each of the following metals: iron, copper, lead, aluminium, zinc.

12 The chemical properties of metals

12.1 Reactions with air, water and acids

Why is it so difficult to stop cars going rusty? Why do oven cleaning agents damage aluminium? Why are lead water pipes dangerous if the water supply is soft? How can it be safe to use liquid sodium as the coolant in a nuclear reactor? To answer these and many other questions, it is necessary to know something about the chemical behaviour of metals.

Two general properties of metals are mentioned in chapters 3 and 5:

* metals react with oxygen to form basic oxides, which are alkalis if they are soluble in water
* metals are deposited at the cathode during electrolysis because metal atoms turn into positive ions when they react.

Metals are thus the chemical opposites of non-metals, which form acidic oxides and negative ions.

Metals in the air

The corrosion of metals such as iron and copper is slow in moist air at room temperature. The details of the chemistry are complicated. Electrical and chemical changes are involved. It is easier to understand what happens when metals are heated in the air. The bright flare produced when magnesium burns is familiar in the laboratory. Iron filings sparkle if sprinkled into a flame. Copper foil loses its colour and is covered with a grey–black coating when heated. In these reactions the metals combine with oxygen to form oxides. The metals are **oxidized**:

$$\text{Metal} + \text{Oxygen} \longrightarrow \text{Metal oxide}$$

Symbol equations can be written, following the procedure outlined on p. 28:

$$\text{Magnesium} + \text{Oxygen} \longrightarrow \text{Magnesium oxide}$$
$$2Mg(s) + O_2(g) \longrightarrow 2MgO(s)$$

Question

Write balanced symbol equations for the reactions of zinc, aluminium, iron and copper with oxygen. Iron forms iron(III) oxide and copper forms copper(II) oxide.

Metals in water and steam

The reactions of magnesium with water and steam are described in chapter 8. Only the more reactive metals, such as potassium, sodium, lithium and calcium will react with cold water. These reactions are described in experiment 12a.

Experiment 12a
Metals and cold water

lithium
sodium
potassium

Lithium, sodium, potassium and calcium are all metals which give an immediate and obvious reaction with cold water. The reactions of sodium and potassium can be so violent that they are dangerous. They must be carried out in an open dish behind a safety screen.

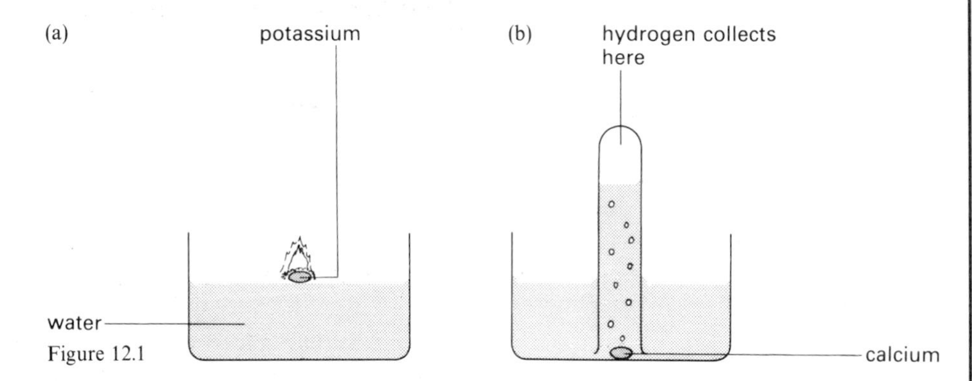

Figure 12.1

Results
Typical results are shown in figure 12.2.

Metal	Apparatus used	Observations
Potassium	(a)	The metal melts and forms a silvery bead. It skates about rapidly on the water surface. A gas is given off which catches fire. The flame is coloured mauve. The potassium quickly disappears. The solution is alkaline.
Sodium	(a)	It behaves like potassium but the reaction is less violent. The gas may catch fire. The flame is coloured yellow.
Lithium	(b)	The metal floats on water. A steady stream of bubbles forms. The gas collected burns with a 'pop'. After the metal dissolves, the solution is alkaline.
Calcium	(b)	The metal sinks. A steady stream of bubbles forms. The gas collected burns with a 'pop'. The water becomes cloudy. The solution is alkaline.

Figure 12.2

Discussion
The pattern is the same in all these reactions:

Reactive metal + Water $\longrightarrow$ Alkaline solution + Inflammable gas

The alkaline solution contains the hydroxide of the metal. Hydroxide ions make water alkaline. The hydroxides of lithium, sodium and potassium are soluble in water. Calcium hydroxide is only slightly soluble, so most of the hydroxide formed separates as a white insoluble solid. The inflammable gas is hydrogen. Thus the word equation becomes:

Reactive metal + Water $\longrightarrow$ Metal hydroxide + Hydrogen

Write symbol equations for the reactions of lithium, sodium, potassium and calcium with water. Use table 8 on p. 343 to work out the formulae of the hydroxides.

Questions

1 Design and draw an apparatus to demonstrate the reaction of iron with steam. The apparatus should make it possible to collect and test the hydrogen formed.
2 Write symbol equations for the reactions of zinc and iron with steam. Iron forms an oxide with the formula Fe_3O_4.

Metals such as magnesium, zinc and iron do not react readily with cold water, but do react if heated in steam. The apparatus for demonstrating the reaction of magnesium with steam is shown in figure 8.5 on page 62. The reaction is:

Metal + Steam ⟶ Metal oxide + Hydrogen

Some metals are so unreactive that they will not react with water or steam. Examples are lead, copper and silver.

Metals and acids

Many, but not all, metals react with acids. Hydrogen gas is formed, and the metal usually dissolves in the acid to form a **salt**:

Metal + Acid ⟶ Salt + Hydrogen

Experiment 12b
The laboratory preparation of hydrogen

hydrogen

dilute sulphuric acid

Hydrogen forms explosive mixtures with air. In the laboratory it is usually safer to use hydrogen from a gas cylinder. If no cylinder is available, the gas can be made by reacting zinc with sulphuric acid, as shown in figure 12.3.

The hydrogen can be collected over water. If dry hydrogen is needed, it can be passed through a U-tube containing silica gel (a drying agent) and collected by upward delivery. If the reaction is slow it can be speeded up by adding a few drops of copper(II) sulphate solution.

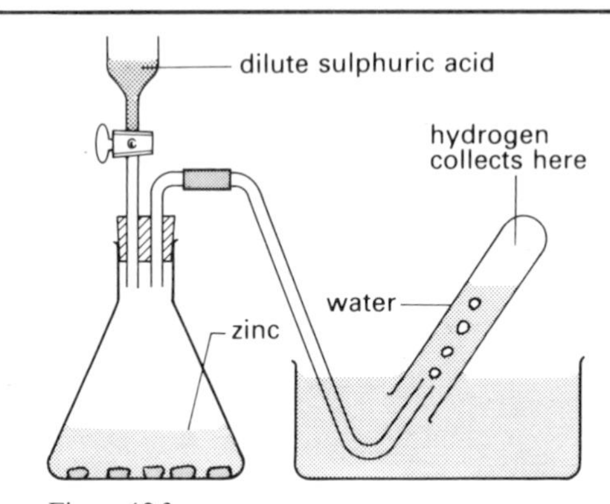

Figure 12.3

Discussion
Suggest reasons for using zinc as the metal for making hydrogen. Why are sodium and magnesium not chosen? Why are lead and copper not used? Write word and symbol equations for the reaction of zinc with sulphuric acid.

The method of collecting the gas illustrates two properties of hydrogen. Is hydrogen soluble in water? Is hydrogen more or less dense than air? Which chemical test is used to identify hydrogen? Why is it not safe to ignite the hydrogen at the end of the delivery tube in figure 12.3?

The properties of hydrogen are summarized in figure 12.4.

Hydrogen
is a colourless gas
has no smell
is much less dense than air
is insoluble in water
is neutral
is very inflammable and explodes when ignited in air

Figure 12.4 The properties of hydrogen

Question

Write word and symbol equations for the reaction of
(a) magnesium with hydrochloric acid.
(b) aluminium with hydrochloric acid.
(c) iron with sulphuric acid to form iron(II) sulphate.

12.2 The activity series

There are obvious differences in the chemical activities of metals. Sodium and potassium are so reactive that they have to be stored under oil to protect them from the air and water vapour. Iron is less reactive but rusts in moist air. Gold is so unreactive that it keeps its shine.

Metals can be listed in order of their chemical activity. The list is called the *activity series.* Metals high in the list are very reactive. Figure 12.5 shows the activity series and includes a summary of the reactions described in the previous section.

<table>
<tr><th>Metal</th><th>Reaction with oxygen in air</th><th>Reaction with cold water</th><th>Reaction with steam</th><th>Reaction with dilute hydrochloric acid</th><th></th></tr>
<tr><td>Potassium</td><td rowspan="5">Burn in air on heating to form the metal oxide</td><td rowspan="4">React with cold water to form hydrogen and the metal hydroxide</td><td rowspan="6">React with steam to form hydrogen and the metal oxide</td><td rowspan="2">Violent reaction to give hydrogen</td><td>K</td></tr>
<tr><td>Sodium</td><td>Na</td></tr>
<tr><td>Calcium</td><td rowspan="4">React to form hydrogen and the chloride of the metal</td><td>Ca</td></tr>
<tr><td>Magnesium</td><td>Mg</td></tr>
<tr><td>Zinc</td><td rowspan="4">Do not react with cold water</td><td>Zn</td></tr>
<tr><td>Iron</td><td rowspan="3">Do not burn but react to form metal oxide on heating</td><td>Fe</td></tr>
<tr><td>Lead</td><td rowspan="2">Do not react with steam</td><td rowspan="2">Do not react</td><td>Pb</td></tr>
<tr><td>Copper</td><td>Cu</td></tr>
</table>

Figure 12.5

Experiment 12c
Competition for oxygen

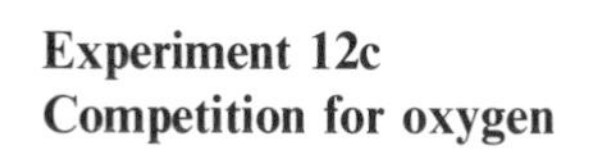

aluminium powder barium peroxide

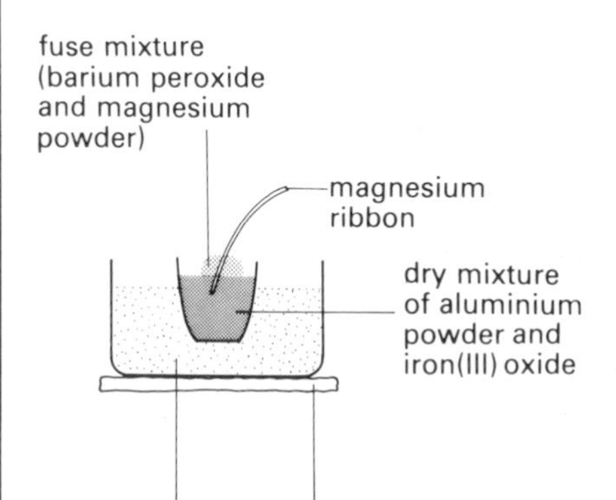

Figure 12.6

Figure 12.6 shows how to set up a chemical competition for oxygen between aluminium and iron. At the start of the experiment, the iron is in possession of the oxygen, as iron(III) oxide. The reaction is started by the magnesium ribbon, which ignites a fuse mixture of magnesium powder and barium peroxide.

Results
The reaction is very violent. There is a flash of light as the fuse starts the reaction, then a shower of sparks. At the end a glowing pool of molten iron is left at the bottom of the crucible.

Discussion
Aluminium is a more reactive metal than iron. It combines with oxygen more strongly, so it can **reduce** iron oxide. This is an example of the *thermit* process. Iron from a thermit reaction is used to weld railway lines together. The thermit process is also used to extract chromium and manganese from their oxide ores.

Figure 12.7 Welding railway lines together using the thermit process: casting the iron into a mould at 2000 °C

Questions

1 Write word and symbol equations for the thermit reaction of aluminium with iron(III) oxide.
2 Use the activity series to predict what will happen when the following mixtures are heated: (a) magnesium powder and calcium oxide, (b) iron filings and copper oxide, (c) iron filings and zinc oxide. If you predict that there will be a reaction, write an equation to show what will happen.
3 Chromium is below zinc but above iron in the activity series. Predict what will happen when chromium is (a) heated in air, (b) put into cold water and (c) added to dilute hydrochloric acid.

The thermit reaction is an example of a more reactive metal displacing a less reactive metal from one of its compounds. Similar changes can happen in solution.

Experiment 12d
Displacement reactions

lead(II) nitrate solution

silver nitrate solution

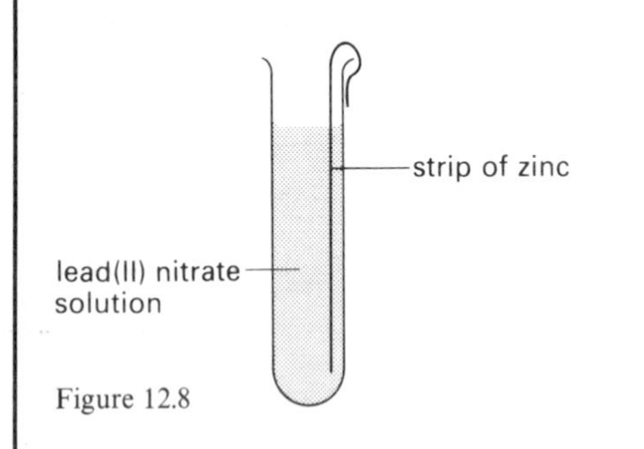

Figure 12.8

A series of test-tubes is used in this experiment. In each tube, a strip of one metal is dipped into a solution of a compound of another metal. Figure 12.8 shows a typical arrangement.

Results
Figure 12.9 shows what happens in the tube containing zinc dipping into a solution of lead(II) nitrate. Crystals of lead have grown out from the surface of the zinc. Similarly, copper displaces silver from silver nitrate solution. Copper does not displace magnesium from magnesium sulphate. If the solution is dilute and the reaction is slow then the displaced metal appears as attractive crystals.

Figure 12.9

Discussion
The activity series is also a displacement series for metals. A metal high in the series will displace any metal lower in the series from solutions of its compounds.

Question

Use the activity series to predict what will happen when
(a) iron filings are added to copper(II) sulphate solution.
(b) copper powder is added to zinc sulphate solution.
(c) magnesium is added to lead(II) nitrate solution.

Predictions based on the activity series are usually, but not always, accurate. The behaviour of aluminium is puzzling. Aluminium is used to make window frames and aircraft. The statue of Eros in Piccadilly Circus is also made of aluminium. In these applications the aluminium is apparently unprotected. It is exposed to the air and to rain, yet it stays shiny. The metal seems to be quite unreactive. But the thermit reaction shows that aluminium is above iron in the activity series. Aluminium becomes much more reactive towards air and water if its surface is treated with mercury(II) chloride. If a piece of aluminium foil which has been rubbed with cotton wool soaked in mercury(II) chloride solution is exposed to the air, it becomes very hot, and whiskery white growths appear. The treated metal reacts vigorously with dilute acids to produce hydrogen.

The thermit reaction shows that aluminium combines strongly with oxygen. In the air the surface of the metal is rapidly covered by a thin but continuous layer of oxide, which protects it from further attack. Aluminium is above mercury in the activity series. When aluminium is dipped into mercury(II) chloride solution, it displaces mercury. Aluminium, like other metals, will dissolve in mercury to form an amalgam. So the mercury breaks down the oxide film on the aluminium, and this allows attack by air, water and acids.

The activity series can also be used to predict the properties of metal compounds. The compounds of reactive metals are much more difficult to decompose than the compounds of less reactive metals. The commonest way of decomposing compounds is to heat them. Reactive metals form compounds which are difficult to decompose by heating.

Experiment 12e
The effect of heat on metal carbonates

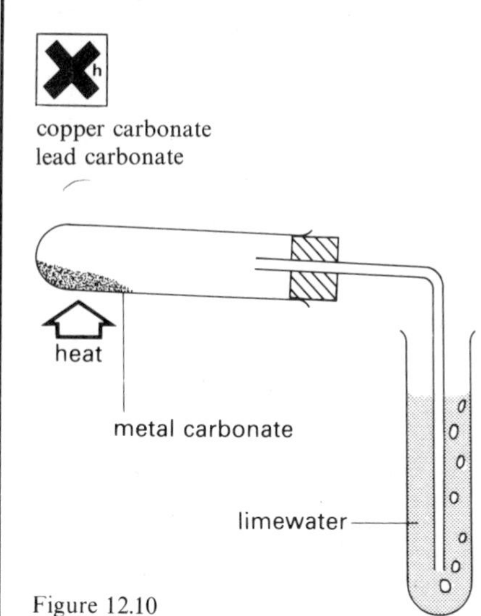

copper carbonate
lead carbonate

Figure 12.10

Separate samples of metal carbonates are heated in the apparatus shown in figure 12.10.

Results
If the carbonate decomposes there is a steady stream of bubbles through the limewater, which turns milky. Most carbonates decompose when heated in this way. Calcium carbonate needs very strong heating. The carbonates of sodium and potassium do not decompose.

Sometimes there is a colour change in the heated tube. Copper(II) carbonate changes from green to black. Lead carbonate changes from white to yellow. Zinc carbonate is white: it decomposes to give a solid which is yellow when hot but turns white again on cooling.

Discussion
The pattern for these changes is:

Metal carbonate ⟶ Metal oxide + Carbon dioxide

What are the colours of the oxides of copper and lead? What colour is zinc oxide when hot and when cold? Work out symbol equations for the action of heat on the carbonates of lead, zinc and copper.

Experiment 12f
The effect of heat on lead(II) nitrate

lead(II) nitrate

nitrogen dioxide

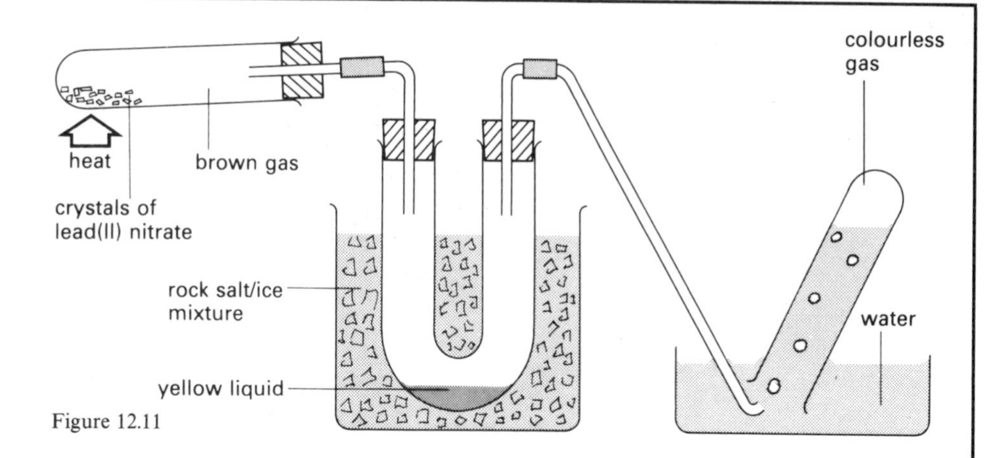

Figure 12.11

This experiment illustrates the effect of heat on the nitrates of most metals.

Results
When heated, the solid in the test-tube spits and crackles. It turns yellow as the tube fills with a brown gas. When the gas passes into the U-tube, part of it condenses to a yellow liquid; the rest is colourless and can be collected over water. The colourless gas relights a glowing splint. The liquid in the U-tube turns back to a brown gas if warmed to room temperature.

Discussion
The yellow solid residue is lead(II) oxide. The brown gas is nitrogen dioxide and the colourless gas is oxygen.

Lead(II) nitrate(s) ⟶ Lead(II) oxide(s) + Nitrogen dioxide(g) + Oxygen(g)

This is the usual pattern for most metal nitrates. The nitrates of sodium and potassium are the only ones which do not behave in this way. They give off only oxygen on strong heating, leaving the nitrite as the residue:

Potassium nitrate(s) ⟶ Potassium nitrite(s) + Oxygen(g)

The effect of heat on carbonates and nitrates can be summarized alongside the activity series as shown in figure 12.12, which also shows the effect of heat on metal hydroxides.

Metal	Effect of heat on carbonate	Effect of heat on nitrate	Effect of heat on hydroxide	
Potassium Sodium	Do not decompose	Decompose to nitrite and oxygen	Do not decompose	K Na
Calcium Magnesium Zinc Iron Lead Copper	Decompose to oxide and carbon dioxide – decomposition gets easier down the series	Decompose to oxide, nitrogen dioxide and oxygen – decomposition gets easier down the series	Decompose to oxide and water – decomposition gets easier down the series	Ca Mg Zn Fe Pb Cu

Figure 12.12

Questions

1 What colour changes would you expect to see on heating (a) copper(II) nitrate, (b) zinc nitrate and (c) calcium nitrate?
2 Write word and symbol equations for the action of heat on the nitrates of copper, zinc, and sodium.

Summary

Copy and complete the diagram in figure 12.13 to summarize the chemistry of zinc. Put the names and formulae of the missing compounds in the empty boxes. Draw similar diagrams to show the chemistry of sodium and copper.

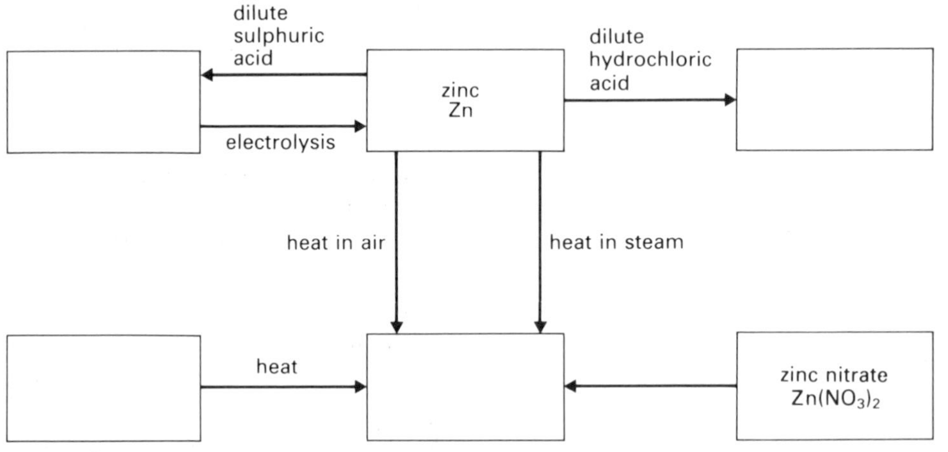

Figure 12.13

13 The extraction of metals from their ores

Question

In the Goldschmidt process, sodium reacts with titanium(IV) chloride to produce titanium metal. Write word and symbol equations for the reaction.

13.1 Extraction methods

Metals low in the activity series, such as silver and gold, may be found free in nature, but most metals occur as compounds – oxides, carbonates, sulphides and chlorides. These compounds are found in an impure state in rocks. After mining, the rocks are processed in various ways to separate the valuable ore (see chapter 1).

There are two main methods for extracting metals from their purified and concentrated ores:

* reduction of the metal oxide by carbon (coke)
* electrolysis of the molten chloride or oxide.

Wherever possible, reduction with coke is used because it is cheaper than electrolysis. The oxides and sulphides of some metals near the bottom of the activity series are so unstable that they can be decomposed by the action of heat alone.

In special cases, a more reactive metal may be used to extract a less reactive one. The use of aluminium to extract chromium and manganese in the thermit process is described in chapter 12. Titanium is obtained by the action of sodium on titanium(IV) chloride.

Figure 13.1 shows how the method of extraction is related to the position of the metal in the activity series.

Metal	Method of extraction
Sodium Calcium Magnesium Aluminium	Electrolysis of molten chloride or oxide
Zinc Iron Lead	Reduction of the oxide by coke in a blast furnace
Copper	Thermal decomposition of the sulphide in the presence of oxygen

Figure 13.1

13.2 Extraction of metals by electrolysis

Extraction of sodium

Sodium is obtained by the electrolysis of molten sodium chloride. Pure salt cannot be used as the electrolyte because it melts at 808 °C. At this temperature, sodium is near to its boiling point (900 °C), and tends to evaporate. Also at this temperature, sodium dissolves in molten sodium chloride. The solution to these problems was found in the 1920s, when the Down's cell was developed. The design of the cell is

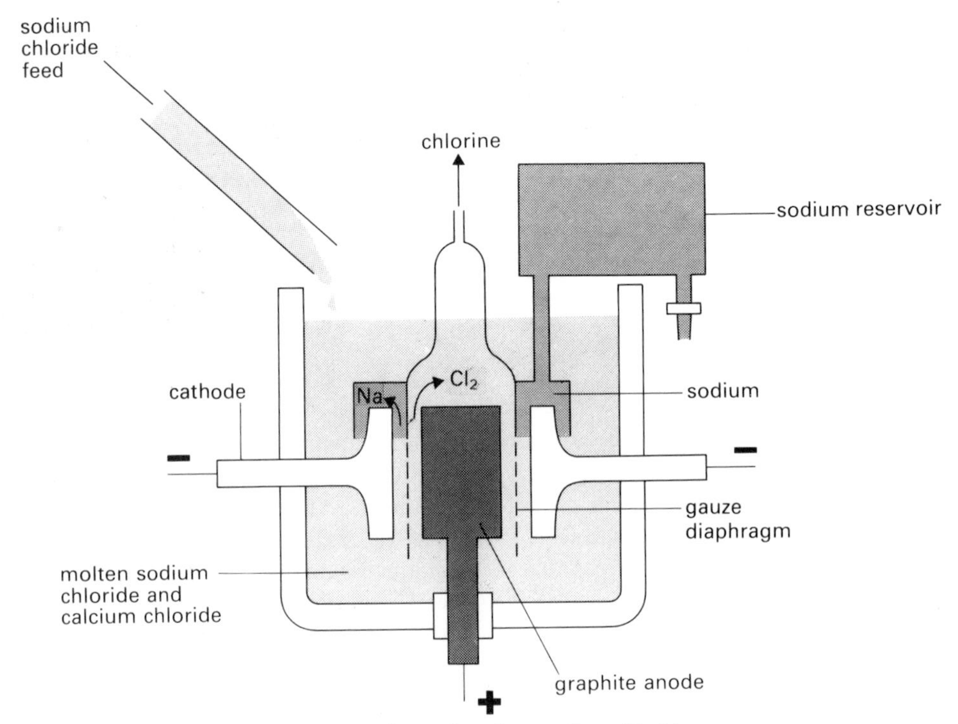

Figure 13.2 The Down's cell for producing sodium from sodium chloride

shown in figure 13.2. The electrolyte is a mixture of sodium chloride (40 per cent) and calcium chloride (60 per cent) which melts below 600 °C.

At the *cathode*:

Sodium ions + Electrons (from the cathode) ⟶ Sodium atoms

$$Na^+ + e^- \longrightarrow Na$$

Symbol equations for electrode reactions are explained in chapter 40.

More energy is needed to discharge calcium ions than sodium ions, so little calcium is formed at the cathode. Any calcium that is formed dissolves in the sodium, but crystallizes out when the sodium is removed from the cell and cooled. The liquid sodium is less dense than the electrolyte. It rises to the surface and is piped off into a receiver.

At the *anode:*

Chloride ions ⟶ Chlorine molecules + Electrons (taken by the anode)

$$2Cl^- \longrightarrow Cl_2 + 2e^-$$

The chlorine must be kept separate from the sodium, to prevent the two elements recombining to form salt. The elements are separated by a cylindrical steel gauze between the cathode and the anode.

Extraction of aluminium

The ore of aluminium is bauxite, which is treated to obtain pure aluminium oxide. Aluminium oxide melts at 2015°C, so large-scale extraction by electrolysis was not possible until it was discovered that the oxide will dissolve in molten cryolite (Na_3AlF_6) at a much lower temperature. Electrolysis takes place in carbon-lined steel tanks called 'pots'. The carbon lining forms the cathode. The electrolyte is a molten solution of aluminium oxide in cryolite, aluminium fluoride (AlF_3) and calcium fluoride (CaF_2). The anodes are blocks of carbon.

At the *cathode*:

Aluminium ions + Electrons ⟶ Aluminium atoms

$$Al^{3+} + 3e^- \longrightarrow Al$$

Question

Aluminium smelters are often built in areas where hydroelectric power is available. Suggest a reason for this.

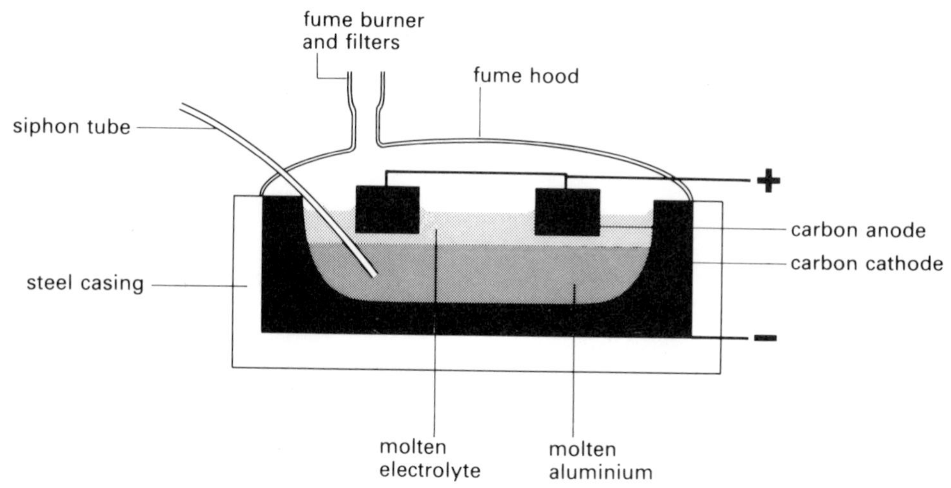

Figure 13.3 An electrolytic cell for producing aluminium from aluminium oxide

The aluminium is liquid at the temperature of the molten electrolyte and it collects at the bottom of the pot. The molten metal is siphoned off once every 24 hours.

At the *anode*:

Oxide ions ⟶ Oxygen molecules + Electrons

$$2O^{2-} \longrightarrow O_2 + 4e^-$$

The oxygen reacts with the carbon in the anodes to form carbon dioxide. The anodes burn away. A new set of anodes is installed once every 24 hours.

The waste gases from the cell contain fluorides and have to be thoroughly cleaned to avoid pollution of the air and the surrounding countryside.

13.3 Extraction of metals using carbon

Extraction of iron

Iron is extracted in a *blast furnace*. The furnace gets its name from the blast of hot air blown in at the bottom. A cross-section of a blast furnace is shown in figure 13.4. The furnace is built of steel lined with heat-resistant bricks. Once started, the furnace runs continuously for two years or more, until the lining begins to fail. Ore, coke and limestone are fed in at the top. Iron and slag are tapped from the bottom.

The main ore of iron is haematite, Fe_2O_3. High grade ore contains up to 60 per cent of iron. The impurities are sand and clay. Most of the ore now used in Britain is imported from Scandinavia, Australia, North and South America, and Africa. It is transported in ships carrying up to 300 000 tonnes.

The coke has three jobs to do in the furnace. It must be strong enough to support the mixture of ore and limestone yet be porous enough to allow the gases to rise to the top. Also, it is the burning of the coke in the air blast which heats the furnace.

$$C(s) + O_2(g) \longrightarrow CO_2(g)$$

Finally, the carbon dioxide rising up the furnace reacts with more coke to form carbon monoxide. Carbon monoxide is the main **reducing** agent for removing the oxygen from the iron ore.

$$CO_2(g) + C(s) \longrightarrow 2CO(g)$$

$$Fe_2O_3(s) + 3CO(g) \longrightarrow 3Fe(s) + 3CO_2(g)$$

As the iron moves down the furnace it melts, and runs down to collect in a pool at the bottom.

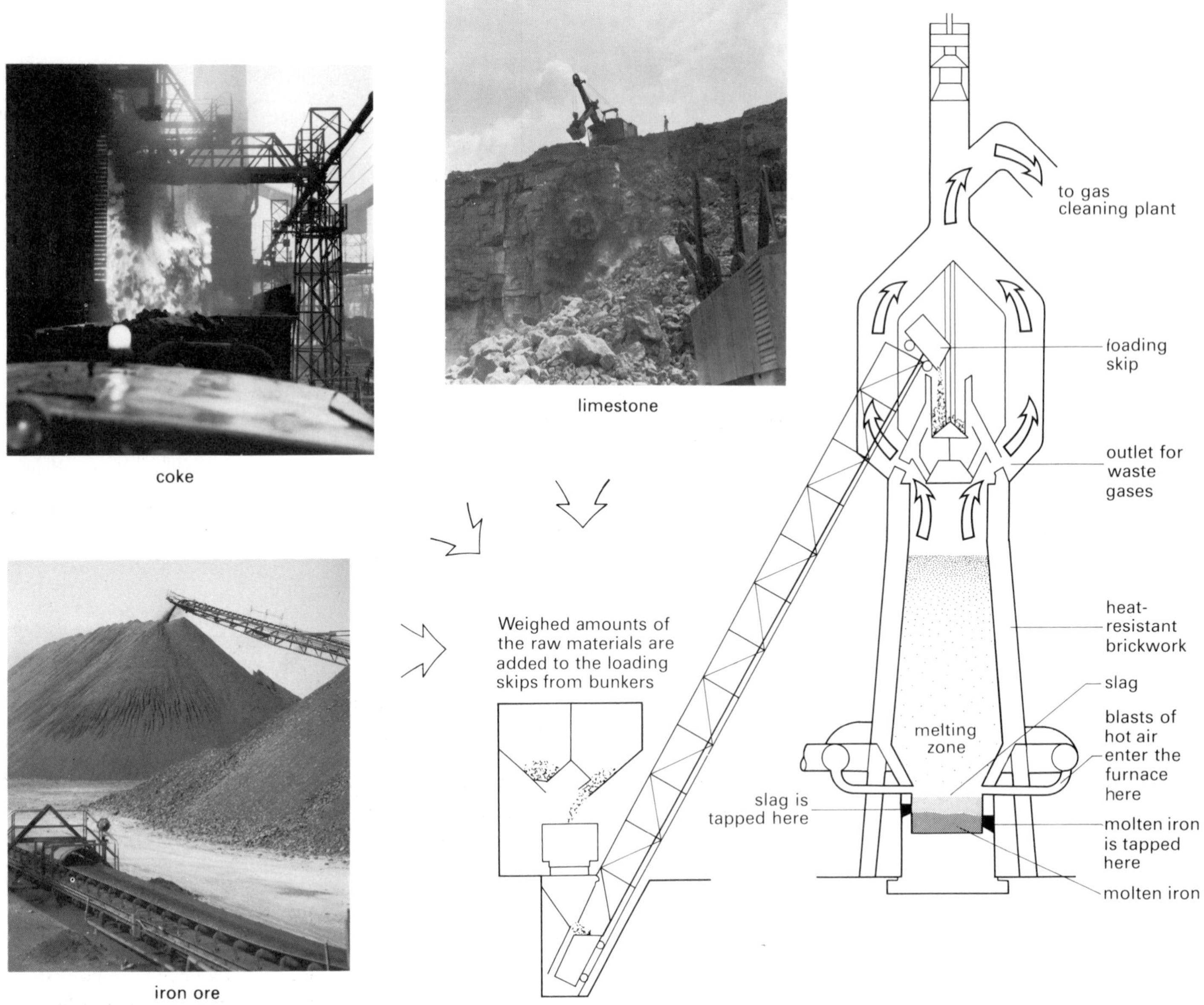

Figure 13.4 The blast furnace

The purpose of the limestone is to act as a flux. The furnace runs continuously and so the impurities, such as sand, must be removed so that they do not choke up the furnace. Sand (SiO_2) will not melt at the temperatures in the furnace, but it will react with calcium oxide to form a liquid slag. This runs down to the bottom and floats on top of the iron. The slag is tapped off from time to time. The calcium oxide is formed when the limestone decomposes.

$$CaCO_3(s) \longrightarrow CaO(s) + CO_2(g)$$

$$\underset{\text{basic metal oxide}}{CaO(s)} + \underset{\text{acidic non-metal oxide}}{SiO_2(s)} \longrightarrow \underset{\text{slag}}{CaSiO_3(l)}$$

Steelmaking

The iron tapped from a blast furnace is about 90–95 per cent pure. The main impurities are carbon, sulphur, phosphorus, manganese and silicon. Some of the hot metal is cast as pig iron but most of it is transferred to a furnace to be converted into steel. The importance of steel is described in chapter 18.

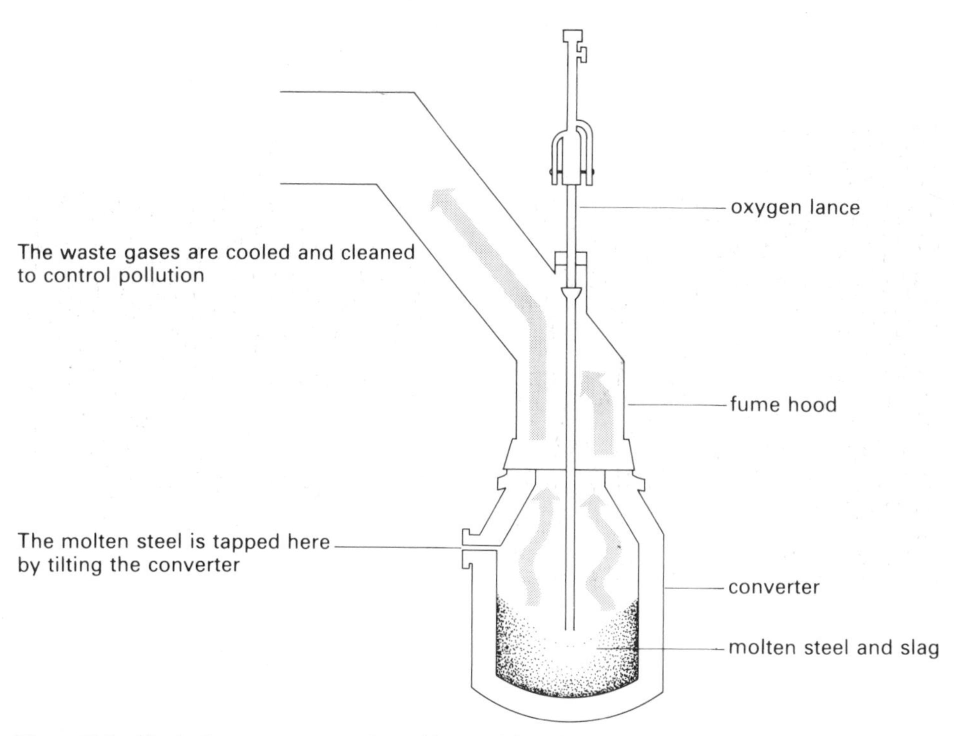

Figure 13.5 The basic oxygen process for making steel from iron

Steel is now made by the basic oxygen process, which removes most of the impurities from the iron. The furnace is charged with hot metal from the blast furnace together with about 30 per cent scrap metal. A high speed blast of pure oxygen is blown onto the metal through a water-cooled lance. This converts the impurities to their oxides. The oxides of carbon and sulphur are gases and are blown off. The oxides of the other elements combine with added limestone to form a slag, which is poured away after the steel has been tapped off. To avoid pollution, the gases from the furnace are collected, cooled and cleaned before being discharged to the atmosphere.

Question

Work out word and symbol equations to describe the reactions involved in the roasting and smelting of the sulphides of zinc and lead.

Extraction of zinc and lead

The ores of lead and zinc consist mainly of sulphides. Zinc sulphide is found as zinc blende and lead sulphide occurs as the mineral galena. These ores are often found together and are processed together.

The sulphides are roasted in air to convert them to oxides. They are then mixed with coke and limestone and fed into a smaller version of the blast furnace used to extract iron. The furnace is heated by burning coke. The carbon monoxide produced reduces the oxides to metals. Lead and slag are tapped from the bottom of the furnace. Zinc boils at 913 °C and evaporates in the furnace. It rises up and escapes with the waste gases. To prevent the zinc recombining with oxygen in the air, it is rapidly cooled and condensed in a spray of molten lead. The mixture of lead and zinc is collected. As it cools, the zinc separates and floats on the surface and is run off. The lead, which is still saturated with zinc, is used again to condense more zinc vapour.

Figure 13.6 Removing the slag from the bottom of a zinc–lead blast furnace

13.4 Extraction of copper

Copper is so low in the activity series that it can be obtained from its ores without using a reducing agent. Copper pyrites, $CuFeS_2$, is the main ore. It is mined and concentrated before reduction. The ore is mixed with limestone and heated in a furnace to remove impurities as a slag. After removal of the slag, the furnace is tapped to obtain a mixture of copper and iron sulphides called *matte*.

Question

Suggest which method of extracting metals might be used to obtain
(a) tin from tinstone, SnO_2.
(b) mercury from cinnabar, HgS.
(c) manganese from pyrolusite, MnO_2.
(d) magnesium from magnesium chloride, $MgCl_2$.

Figure 13.7 Charging a converter with molten matte

In a second furnace, called a converter, air is passed through the red-hot molten matte. The iron is oxidized and forms a slag which is removed. Sulphur is removed as sulphur dioxide gas.

The copper obtained from the converter is 98–99.5 per cent pure. It must be further purified before it is suitable for its modern applications. The impurities affect its mechanical properties and its electrical conductivity. This purification is described in chapter 40.

Summary

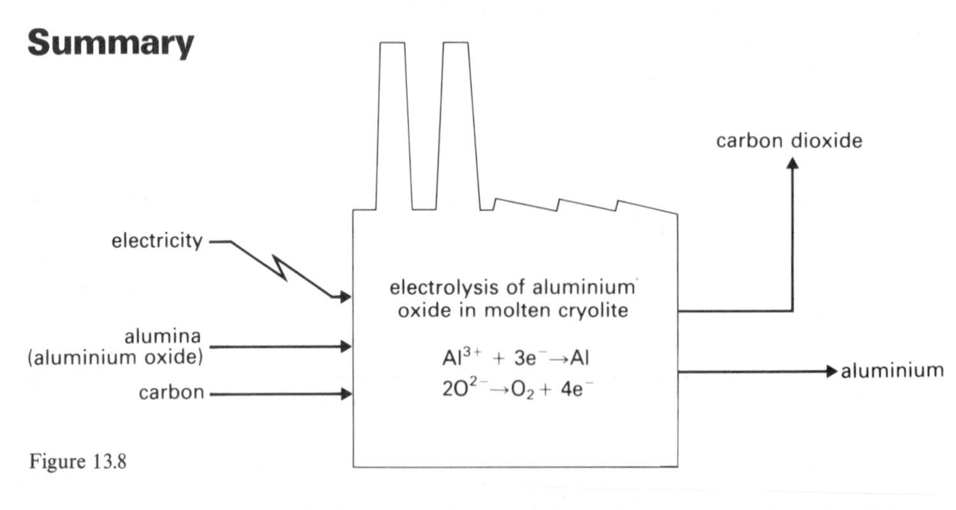

Figure 13.8

Figure 13.8 shows the main inputs and outputs for a factory extracting aluminium from its ore. Draw similar diagrams to summarize the manufacture of the metals iron, sodium, copper, zinc and lead.

Review questions

1 (a) The following is a list of some metals in *decreasing* order of reactivity: potassium, sodium, calcium, magnesium, aluminium, zinc, iron, lead, copper.

Name from the above metals only

(i) a metal used for protecting iron from rust.

(ii) a metal used in aircraft construction because of its low density.

(iii) a metal that can only be produced from its compounds by electrolysis of one of its molten salts.

(iv) a metal than can be produced easily by heating its oxide with carbon. (Write a word equation for this reaction.)

(v) a metal that is unaffected by water or dilute sulphuric acid.

(vi) a metal that produces a flame when placed on the surface of cold water.

(b) No change occurred when powdered copper was stirred into zinc sulphate solution. When powdered zinc was stirred into copper(II) sulphate solution, its blue colour disappeared and a red–brown solid was left.

(i) What was the red–brown solid?

(ii) What type of reaction was taking place?

(iii) Why did the solution lose its blue colour?

(c) Complete this table concerning the names, composition and uses of common alloys:

Name of alloy	Major constituents	Example of use
	Iron, small amounts of carbon	Girders
Duralumin	Mostly aluminium, some copper	
Solder		Joining electrical wires together
Stainless steel	Mostly iron, some chromium, nickel	
	Copper, zinc	Electrical connections

(d) Carbon when heated with the oxide of a metal X will remove the oxygen from it. Similarly, when carbon is heated with the oxide of a metal Y the oxygen is removed.

No reaction takes place between carbon and the oxide of a metal Z.

Metal Y when heated with the oxide of metal X will remove its oxygen.

(i) On the basis of this evidence, place the metals in order of reactivity, the most reactive first.

(ii) If the formula of the oxide of metal X is XO, write a symbol equation for the reaction between the oxide and carbon. **(WMEB)**

2 What properties would you consider to be important for a metal chosen for manufacturing

(i) a central-heating radiator?

(ii) a car wing?

(iii) an electric fuse-wire?

What metals might be used for these applications and why are they preferred?

How would you investigate experimentally the differences in properties between a zinc casting used for several years as a flame-spreader on a gas stove and a similar brand-new casting? **(L)**

3 A well-known manufacturer of stainless steels discovered recently that some of their steels used to make cutlery were prone to corrosion when in contact with solutions containing dissolved chlorides. Describe how you would set up laboratory tests to investigate this problem, and suggest any steps that could be taken to minimize or eliminate the problem. **(EMREB)**

4 The following is a series of metallic elements in order of their chemical reactivity: sodium (most reactive), X, calcium, magnesium, Y, zinc, iron, Z, lead, copper, silver, gold (least reactive). X, Y and Z are unknown metals in this list.

(a) (i) Select a metal from the series which will *not* react with cold water, but will when heated in steam, and draw a labelled diagram of a laboratory apparatus used to demonstrate this. Make sure that you show how to generate the steam and how to collect any gas produced.

(ii) Name the gas which is collected.

(iii) Write a chemical equation for the reaction you have chosen.

(b) (i) In what special way might element X be stored?

(ii) Explain the reason for this storage.

(c) (i) Under what conditions would you expect metal Y to react with hydrochloric acid?

(ii) Name the products.

(d) (i) Name a compound of metal Z which is likely to be found in the ground.
(ii) Give a reason for your answer.
(e) Name *two* metals from this series which would probably be extracted by electrolysis of their fused chlorides.
(f) Name a metal from the series which is found native (chemically uncombined) in the ground.
(g) (i) What would you expect to happen if dry hydrogen gas was passed over the heated oxide of metal X?
(ii) Explain your answer.
(h) (i) Some powdered magnesium was thoroughly mixed with lead(II) oxide (PbO) and the mixture was heated on a piece of ceramic wool paper. Describe what you would *see* happen.
(ii) Explain the reaction.
(i) Zinc will prevent iron from rusting, even if the zinc coating is scratched to expose the iron. Explain the reason for this.
(j) (i) What would you expect to see if a piece of metal Z was put into copper(II) sulphate solution?
(ii) Write a word equation for the reaction which occurs. **(ALSEB)**

5 (a) Outline the extraction of iron from a named ore by the blast furnace process, explaining the chemical reactions involved.
(b) State and explain *three chemical* properties of iron or its compounds that are typical of a metal.
(c) State *two physical* properties of iron that make it an important industrial material, giving one example of how the property determines the use in each case. **(WJEC)**

6 Most industrial extraction processes involve a sequence of events which can be represented by the diagram shown below.
(a) Illustrate the scheme below by referring to the extraction of iron metal from any *one* of its more common ores. Write down balanced equations to summarize any chemical reactions which you include in your account.
(b) To what extent does the process which you have described in (a) add to the world's pollution problems? Explain your answer. **(O)**

7 (a) Study the following passage carefully, and then answer the questions related to the metal, aluminium.

For some sixty years, aluminium has been manufactured from bauxite, an ore containing aluminium oxide, found in various parts of the world.

The ore, after extensive purification, is dissolved in molten cryolite and electrolysed in a steel cell lined with carbon, which forms the cathode and uses carbon blocks as the anodes. The process involves a large consumption of electrical power and the frequent replacement of the graphite anodes. Hence, aluminium is an expensive metal. However, its low density and good heat and electrical conducting properties make it a superior metal to steel for many purposes. Coupled with this is its resistance to atmospheric corrosion which is often enhanced by anodizing. This is a process in which the aluminium article is made the anode of an electrolytic cell and oxygen is liberated at the metal surface from an aqueous electrolyte.

(i) Why is aluminium extracted from its oxide by electrolysis rather than by using a chemical reducing agent?
(ii) Explain, with equations, the processes occurring at the anode which make necessary 'the frequent replacement of the graphite anodes'.
(iii) Give a reason, from information in the passage, for the location of an aluminium extraction plant in the Highlands of Scotland.
(iv) Why is aluminium resistant to atmospheric corrosion?
(v) Suggest how 'anodizing' increases the resistance of aluminium to corrosion.
(b) Give *two* different uses of aluminium, which depend upon
(i) its low density and good electrical conduction,
(ii) its low density and resistance to atmospheric corrosion. **(JMB)**

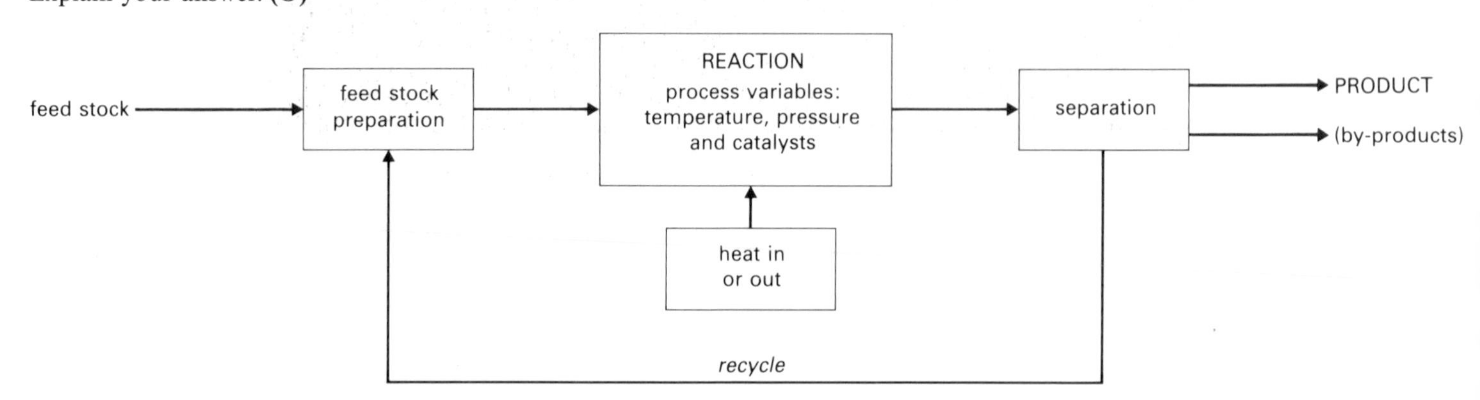

THEME E
The periodic table

Dmitri Mendeléev, 1834–1907, the father of the periodic table of the elements

14 Looking for patterns

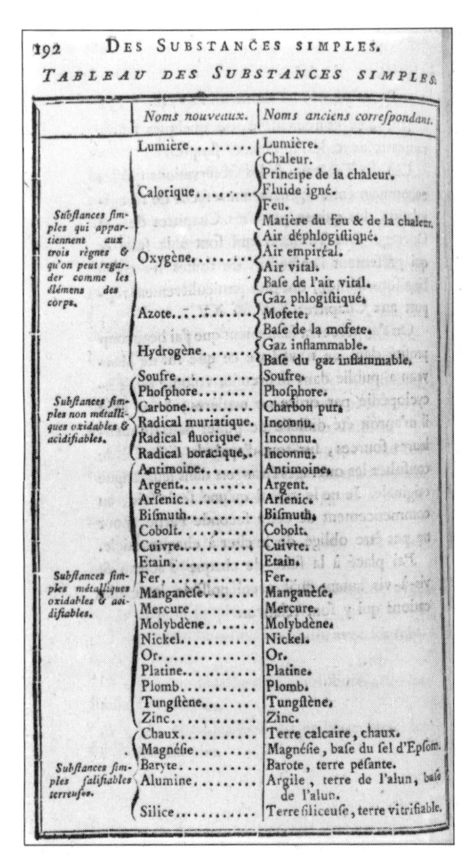

192 DES SUBSTANCES SIMPLES.

TABLEAU DES SUBSTANCES SIMPLES.

	Noms nouveaux.	*Noms anciens correſpondans.*
Subſtances ſimples qui appartiennent aux trois règnes & qu'on peut regarder comme les élémens des corps.	Lumière	Lumière.
	Calorique	Chaleur. Principe de la chaleur. Fluide igné. Feu. Matière du feu & de la chaleur.
	Oxygène	Air déphlogiſtiqué. Air empiréal. Air vital. Baſe de l'air vital.
	Azote	Gaz phlogiſtiqué. Mofete. Baſe de la mofete.
	Hydrogène	Gaz inflammable. Baſe du gaz inflammable.
Subſtances ſimples non métalliques oxidables & acidifiables.	Soufre	Soufre.
	Phoſphore	Phoſphore.
	Carbone	Charbon pur.
	Radical muriatique.	Inconnu.
	Radical fluorique	Inconnu.
	Radical boracique	Inconnu.
Subſtances ſimples métalliques oxidables & acidifiables.	Antimoine	Antimoine.
	Argent	Argent.
	Arſenic	Arſenic.
	Biſmuth	Biſmuth.
	Cobolt	Cobolt.
	Cuivre	Cuivre.
	Etain	Etain.
	Fer	Fer.
	Manganèſe	Manganèſe.
	Mercure	Mercure.
	Molybdène	Molybdène.
	Nickel	Nickel.
	Or	Or.
	Platine	Platine.
	Plomb	Plomb.
	Tungſtène	Tungſtène.
	Zinc	Zinc.
Subſtances ſimples ſalifiables terreuſes.	Chaux	Terre calcaire, chaux.
	Magnéſie	Magnéſie, baſe du ſel d'Epſom.
	Baryte	Barote, terre péſante.
	Alumine	Argile, terre de l'alun, baſe de l'alun.
	Silice	Terre ſiliceuſe, terre vitrifiable.

Figure 14.1 A page from Lavoisier's *Traité Elémentaire de Chimie*, showing his list of the chemical elements

Antoine Lavoisier is famous for his theories of burning and for bringing system and method to the study of chemistry. John Dalton is remembered because of his atomic theory. The work of both these scientists was important in the lead-up to the discovery of the *periodic table*.

In 1789, Lavoisier made a list of the substances which he thought to be elements and gave them new names to fit in with his theories (see figure 14.1).

Question

Substances included in Lavoisier's list of elements were light, heat, oxygen, nitrogen, hydrogen, phosphorus, antimony, cobalt, platinum, magnesia, barytes, alumina and silica. Which of these do we still class as elements? Which of them are compounds? What are the chemical names for the compounds?

An important idea suggested by Dalton was the thought that it might be useful to find out the relative masses of the particles of different gases. In time, Dalton arrived at his atomic theory and went on to do many experiments in chemistry to confirm this theory.

According to Dalton's theory each element is made up of its own kind of **atom**. There are as many kinds of atom as there are elements. Atoms cannot be split and are unchanged during chemical reactions. Compounds consist of the atoms of two or more elements joined together.

Carbon is an element and consists of carbon atoms. In Dalton's theory all the carbon atoms are the same and have the same mass. Oxygen is another element. All oxygen atoms are the same as each other, but they differ in mass from carbon atoms. When carbon burns in oxygen it forms a compound. The particles of this compound are made up of carbon atoms joined to oxygen atoms.

Questions

1 How does Dalton's theory explain the fact that compounds can be decomposed but elements cannot?
2 Which parts of Dalton's theory are no longer thought to be true?

The discovery of **electrolysis** was very important in the history of chemistry because it made it possible to decompose substances which previously could not be split up. The famous English chemist, Humphry Davy (1778–1829), used electrolysis to discover potassium, sodium, barium, calcium and magnesium in the two years 1807 and 1808. Davy also isolated boron and established that chlorine and iodine were also elements.

Following the discovery of many new elements by Davy and others, chemists began to look for patterns in the properties of the elements. One idea, which seemed crazy at first, was to look for a connection between the chemistry of

Question

Which of these groups of three elements behave like Döbereiner's triads: (a) lithium, sodium and potassium;
(b) chlorine, bromine and iodine;
(c) copper, silver and gold?
Relative atomic masses are given in table 1 on p. 336.

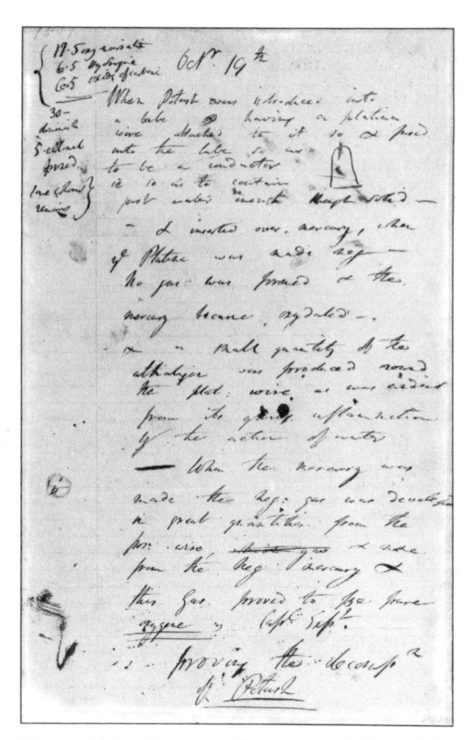

Figure 14.2 Humphry Davy's description of the experiment that led to the discovery of potassium

elements and the relative masses of their atoms. For example, Johann Döbereiner noticed that there were several examples of groups of three elements (triads) with similar properties (such as calcium, strontium and barium), such that the **relative atomic mass** of the middle element was the mean of the relative atomic masses of the other two.

Other early attempts to find a connection between chemical properties and atomic mass were greeted with ridicule. However, in the end, Dmitri Mendeléev (1834–1907) and Lothar Meyer (1830–1895) showed that this was not a hare-brained scheme but a pattern of real meaning. Mendeléev's inspiration was to realize that not all of the elements had yet been discovered. He arranged the elements in order of the mass of their atoms, leaving gaps for undiscovered elements when this was necessary to produce a sensible pattern.

When Mendeléev and others arranged the elements in order of atomic mass they saw a repeating pattern. At intervals along the line of elements they noticed elements with similar properties. Nowadays, we see that the third, eleventh and nineteenth elements (lithium, sodium and potassium) are very similar.

A pattern which repeats is called a *periodic pattern.* Examples of periodic patterns are waves on water, prints on wallpaper and the swing of a pendulum. Most periodic patterns keep repeating as time passes. The pattern on wallpaper repeats at regular intervals along the roll of paper. The pattern of the chemical properties of the elements repeats at intervals if the elements are listed in order of atomic mass. The table of elements showing the elements in this order is called the *periodic table* (see figure 14.3). Each element has a number, its **atomic number**.

In the periodic table, the elements are arranged in horizontal rows, one above the other. Each row is called a **period**. Elements with similar properties come

The periodic table

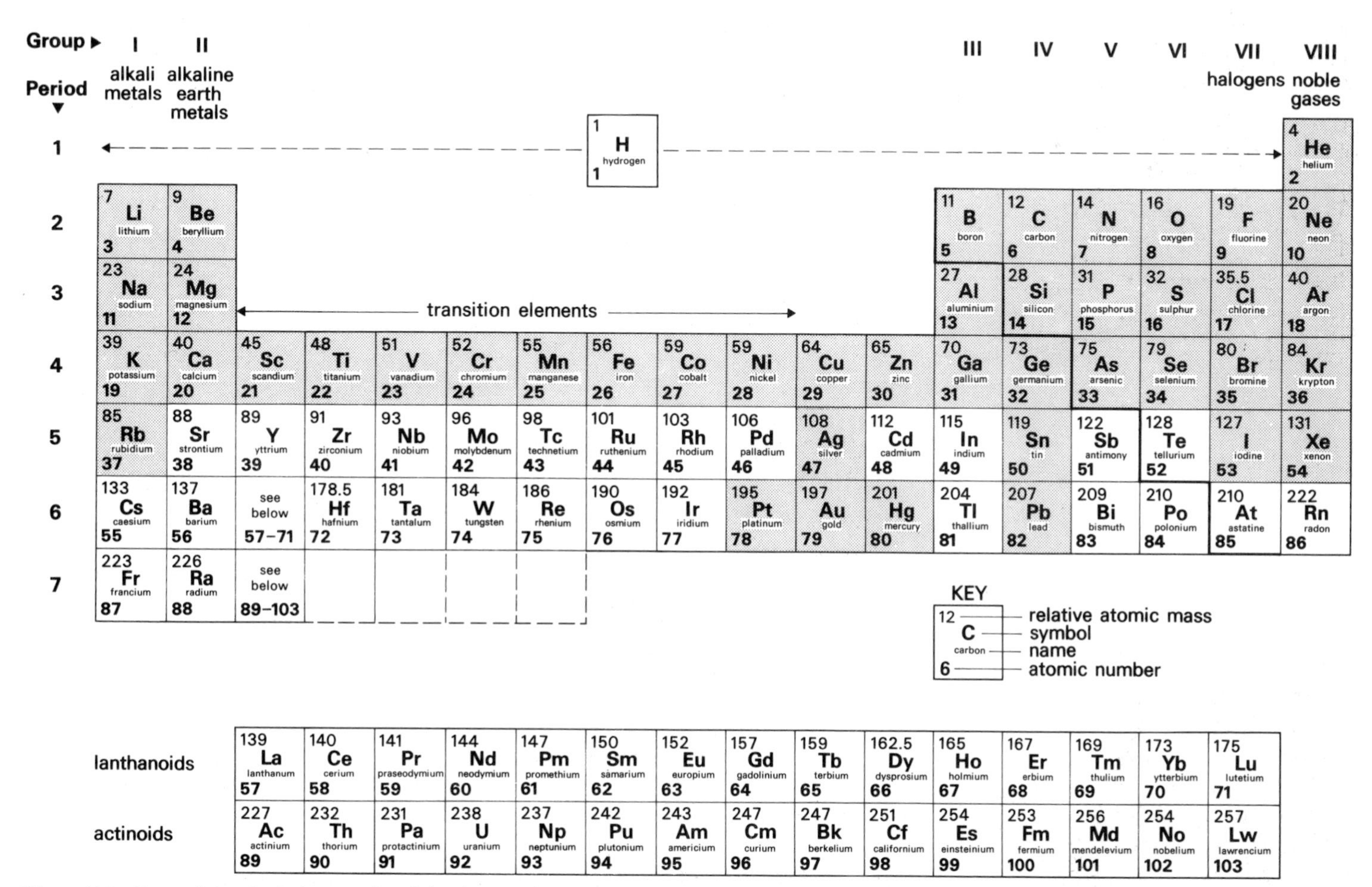

Period ▼ / Group ▶	I alkali metals	II alkaline earth metals											III	IV	V	VI	VII halogens	VIII noble gases
1								1 H hydrogen 1										4 He helium 2
2	7 Li lithium 3	9 Be beryllium 4											11 B boron 5	12 C carbon 6	14 N nitrogen 7	16 O oxygen 8	19 F fluorine 9	20 Ne neon 10
3	23 Na sodium 11	24 Mg magnesium 12	← transition elements →										27 Al aluminium 13	28 Si silicon 14	31 P phosphorus 15	32 S sulphur 16	35.5 Cl chlorine 17	40 Ar argon 18
4	39 K potassium 19	40 Ca calcium 20	45 Sc scandium 21	48 Ti titanium 22	51 V vanadium 23	52 Cr chromium 24	55 Mn manganese 25	56 Fe iron 26	59 Co cobalt 27	59 Ni nickel 28	64 Cu copper 29	65 Zn zinc 30	70 Ga gallium 31	73 Ge germanium 32	75 As arsenic 33	79 Se selenium 34	80 Br bromine 35	84 Kr krypton 36
5	85 Rb rubidium 37	88 Sr strontium 38	89 Y yttrium 39	91 Zr zirconium 40	93 Nb niobium 41	96 Mo molybdenum 42	98 Tc technetium 43	101 Ru ruthenium 44	103 Rh rhodium 45	106 Pd palladium 46	108 Ag silver 47	112 Cd cadmium 48	115 In indium 49	119 Sn tin 50	122 Sb antimony 51	128 Te tellurium 52	127 I iodine 53	131 Xe xenon 54
6	133 Cs caesium 55	137 Ba barium 56	see below 57–71	178.5 Hf hafnium 72	181 Ta tantalum 73	184 W tungsten 74	186 Re rhenium 75	190 Os osmium 76	192 Ir iridium 77	195 Pt platinum 78	197 Au gold 79	201 Hg mercury 80	204 Tl thallium 81	207 Pb lead 82	209 Bi bismuth 83	210 Po polonium 84	210 At astatine 85	222 Rn radon 86
7	223 Fr francium 87	226 Ra radium 88	see below 89–103															

KEY: 12 — relative atomic mass; C — symbol; carbon — name; 6 — atomic number

lanthanoids	139 La lanthanum 57	140 Ce cerium 58	141 Pr praseodymium 59	144 Nd neodymium 60	147 Pm promethium 61	150 Sm samarium 62	152 Eu europium 63	157 Gd gadolinium 64	159 Tb terbium 65	162.5 Dy dysprosium 66	165 Ho holmium 67	167 Er erbium 68	169 Tm thulium 69	173 Yb ytterbium 70	175 Lu lutetium 71
actinoids	227 Ac actinium 89	232 Th thorium 90	231 Pa protactinium 91	238 U uranium 92	237 Np neptunium 93	242 Pu plutonium 94	243 Am americium 95	247 Cm curium 96	247 Bk berkelium 97	251 Cf californium 98	254 Es einsteinium 99	253 Fm fermium 100	256 Md mendelevium 101	254 No nobelium 102	257 Lw lawrencium 103

Figure 14.3 Some of the physical properties of the elements in the shaded boxes in this periodic table are listed in table 1 on p. 336

above each other in a vertical column. The columns of elements form families. They are called **groups**.

Perhaps the most obvious repeating pattern in the table is the change from metals on the left of each period to non-metals on the right. Lothar Meyer discovered periodic patterns by plotting graphs of **physical properties** (such as melting point, boiling point and density) against the relative atomic masses of the elements. The formulae of simple compounds and the charges on ions also show periodic patterns when written as in the periodic table (see figures 14.4 and 14.5).

I	II		III	IV	V	VI	VII	VIII
LiCl	$BeCl_2$		BCl_3	CCl_4	NCl_3	OCl_2	FCl	no chloride formed
NaCl	$MgCl_2$		$AlCl_3$	$SiCl_4$	PCl_3	SCl_2	—	
KCl	$CaCl_2$		$GaCl_3$	$GeCl_4$	$AsCl_3$	$SeCl_2$	BrCl	

Figure 14.4 The formulae of these chlorides show a repeating pattern in the periodic table

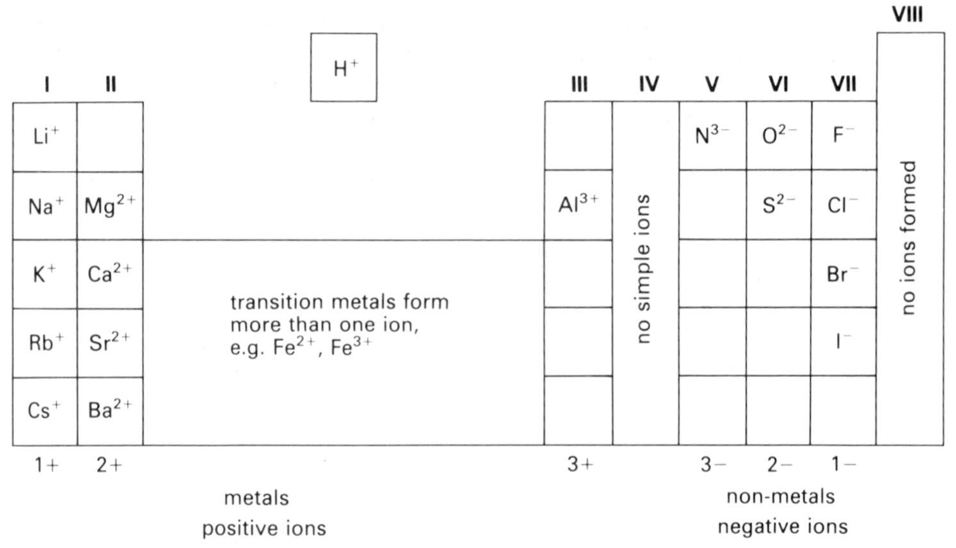

Figure 14.5 The charges on some simple ions

Summary

1 Write a sentence or two to state the main contributions made by the following scientists to the development of the modern form of the periodic table: Lavoisier, Dalton, Davy, Döbereiner, Mendeléev, Meyer.
2 Explain briefly the meanings of the following terms: periodic pattern, period, group, atomic number.

Questions

1 With the help of table 1 on p. 336, plot bar charts to show how (a) the boiling points and (b) the densities of elements 1–20 vary with relative atomic mass (or atomic number).
2 With the help of table 2 on p. 337, make tables to show the formulae of (a) the oxides and (b) the hydrides of elements 3–18. Compare your tables with figure 14.4.

15 Metals in the periodic table

15.1 Family names

Figure 15.1 shows some of the commoner metals in the periodic table. Some of the groups and blocks of elements have family names. Groups I and II in the table include the very reactive metals at the top of the activity series. The elements in Group I are called the **alkali metals** and the elements in Group II are called the **alkaline earth metals**.

Tin and lead at the bottom of Group IV are also metals. The change from non-metals (carbon and silicon) to metals (tin and lead) down Group IV illustrates a general trend in the table: down any group there is a tendency for the elements to become more like reactive metals and less like non-metals.

The metals in the block of elements in the middle of the periodic table are called **transition metals**. This block is divided into groups. One of these groups consists of the coinage metals (copper, silver and gold). They were first used to make coins because of their scarcity and resistance to corrosion. But the transition metals have many general properties, so they are often studied as a block.

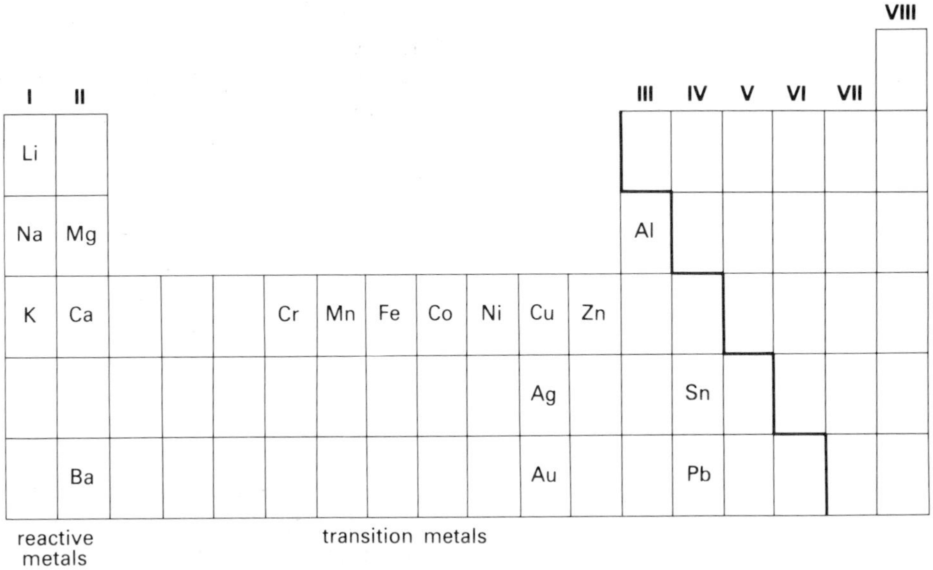

Question

Attempt to list the metals shown in figure 15.1 in the order of the activity series. Beside each metal, write its family name (alkali metal, alkaline earth metal or transition metal).

I	II											III	IV	V	VI	VII	VIII
Li																	
Na	Mg											Al					
K	Ca				Cr	Mn	Fe	Co	Ni	Cu	Zn						
										Ag			Sn				
	Ba									Au			Pb				

reactive metals — transition metals

Figure 15.1 Some metals in the periodic table

15.2 The alkali metals

Lithium, sodium and potassium are called the alkali metals because they react with water to form alkaline solutions. They are very reactive and have to be stored under oil to protect them from air and moisture. They are very soft and can be cut with a knife. The freshly-cut surface is shiny but tarnishes quickly. The metals are less dense than water, which means that they float on water as they react with it. The properties of the metals are summarized in figure 15.2.

All the elements in Group I are metals. The trend towards more metallic

Questions

1 With the help of table 1 on p. 336, plot graphs of (a) melting point and (b) density against relative atomic mass for the first four alkali metals.
Use the graphs to predict the melting point and density of caesium, Cs. (The relative atomic mass of caesium is 133.)

2 Predict the colours, solubilities in water and formulae of the oxide, the hydroxide, the chloride and the sulphate of francium, Fr.

Element	Reaction with water	Reaction with chlorine	Flame colour	Symbol of ion	Formula of simplest oxide	Salts (e.g. nitrates, sulphates, carbonates)
Li	Reacts with cold water to give hydrogen and an alkali, LiOH(aq)	Reacts on heating to give white LiCl(s)	Crimson red	Li^+	Li_2O	White crystalline solids which are soluble in water
Na	Reacts vigorously with cold water to give hydrogen and an alkali, NaOH(aq)	Vigorous reaction on heating to give white NaCl(s)	Bright yellow	Na^+	Na_2O	White crystalline solids which are soluble in water
K	Reacts violently with cold water to give hydrogen and an alkali, KOH(aq)	Very vigorous reaction to give white KCl(s)	Mauve	K^+	K_2O	White crystalline solids which are soluble in water

Figure 15.2

behaviour down the group is shown by the fact that the metals become increasingly reactive as metals from lithium to potassium.

15.3 The alkaline earth metals

The hydroxides of the metals in Group II are also alkaline. These alkalis are much less soluble in water. They are often used as suspensions of insoluble particles in water. The suspension of magnesium hydroxide is called milk of magnesia and is used to treat stomach acidity. The suspension of calcium hydroxide in water is called milk of lime (because it is made from limestone) and is used in industry as a cheap alkali. Limewater is a **saturated solution** of calcium hydroxide in water.

The Group II elements are called *earth* metals because many of their compounds are found as minerals in rocks. This is possible because the compounds are insoluble in water, so they are not washed away by rain. Chalk, marble and limestone are forms of calcium carbonate. Dolomite consists of a mixture of calcium and magnesium carbonates. Fluorspar, which is found in

Element	Reaction with cold water	Reaction with dilute acids	Symbol of ion	Flame colour	Formula of oxide	Formula, colour and solubility of salts	
						Chloride	Sulphate
Mg	Very slow reaction to give hydrogen and insoluble $Mg(OH)_2$	Hydrogen evolved rapidly, and magnesium salt formed	Mg^{2+}	White	MgO	$MgCl_2$ white soluble	$MgSO_4$ white soluble
Ca	Rapid reaction to give hydrogen and slightly soluble $Ca(OH)_2$	Very rapid formation of hydrogen and calcium salt	Ca^{2+}	Brick red	CaO	$CaCl_2$ white soluble	$CaSO_4$ white insoluble
Ba	Very rapid reaction to give hydrogen and soluble $Ba(OH)_2$	Violent reaction to give hydrogen and barium salt	Ba^{2+}	Lime green	BaO	$BaCl_2$ white soluble	$BaSO_4$ white insoluble

Figure 15.3

Derbyshire as Blue John, is a form of calcium fluoride. Heavy spar (barytes) is made of barium sulphate. Calcium sulphate occurs naturally as gypsum and as alabaster. Some of these are shown in figure 18.1 on p. 118. The properties of the Group II elements are summarized in the table in figure 15.3.

Question

Strontium comes between calcium and barium in Group II. Use table 1 on p. 336 to predict the melting and boiling points of strontium. How would you expect strontium to react with air, water and acids? Predict the formulae of the oxide, hydroxide and chloride of strontium.

15.4 The transition metals

The transition metals are hard, strong metals used in engineering. They are much denser than the alkali metals. Most of the metals form two, or more, series of compounds by forming ions with different charges. The compounds are coloured, and this distinguishes them from the compounds of the Group I and Group II metals, which are usually white. Many important **catalysts** are based on the transition metals or their compounds. Catalysts are described in chapter 26.

These general properties are illustrated by the chemistry of iron and copper.

Iron

Iron is below zinc but above lead in the activity series. It reacts with steam and dilute acids as described in chapter 12. When iron reacts with dilute hydrochloric acid, the products are hydrogen and a solution of iron(II) chloride, $FeCl_2$. Anhydrous iron(II) chloride is formed when iron is heated in a stream of hydrogen chloride gas:

$$Fe(s) + 2HCl(g) \longrightarrow FeCl_2(s) + H_2(g)$$

The crystals of iron(II) chloride are yellow. But the commonest compound containing Fe^{2+} ions is iron(II) sulphate, which is pale green.

Solutions of iron(II) compounds are converted to iron(III) compounds by oxygen in the air. The commonest iron(III) compound is iron(III) oxide which is the main ore of the metal. In the laboratory, iron(III) chloride is frequently used to make solutions containing $Fe^{3+}(aq)$ ions. Solutions of iron(III) compounds are yellow. Anhydrous iron(III) chloride can be made by heating iron in a stream of chlorine gas:

$$2Fe(s) + 3Cl_2(g) \longrightarrow 2FeCl_3(s)$$

Solutions containing iron(II) and iron(III) ions can be distinguished by using sodium hydroxide solution. This is described in chapter 23.

The most important use of iron as a catalyst is in the Haber process for making ammonia, which is described in chapter 34.

Questions

1 Draw and label a diagram of the apparatus you would use to make anhydrous iron(III) chloride. Iron(III) chloride sublimes on heating and is collected as a solid.
2 Hydrogen chloride gas is formed when concentrated sulphuric acid is added to sodium chloride (salt). It is sometimes called 'salt gas'. Explain how the reaction of salt gas with iron could be used as evidence that one of the elements in salt gas is hydrogen.
3 Write down the formulae of copper(I) oxide, copper(I) chloride, copper(II) oxide, copper(II) chloride and copper(II) sulphate.
4 What are the colours of solutions containing the following transition metal ions in the states shown: manganese(II), manganate(VII), chromium(III), chromate(VI), nickel(II)?
5 Give examples of the use of the following as catalysts: manganese(IV) oxide, vanadium(V) oxide, platinum, nickel.

Copper

Copper is low in the activity series. It is so low that it does not react with water or dilute hydrochloric acid. The metal is so unreactive that it can sometimes be found free in nature. Copper forms copper(I) and copper(II) compounds, but the copper(II) compounds are more common.

On heating in air, copper combines with oxygen to form black copper(II) oxide. Copper(II) oxide dissolves in dilute sulphuric acid to form copper(II) sulphate, which is blue. Benedict's and Fehling's solutions are blue because they contain copper(II) ions. They are used to test for glucose and other similar sugars. When they are warmed with glucose, the blue colour goes and an orange or yellow precipitate of copper(I) oxide forms. Most copper(I) compounds are insoluble in water. The only other common example is copper(I) chloride, which is a white solid when freshly made. It turns green in the air as it begins to be converted to copper(II) chloride by oxygen in the air.

The use of copper(II) sulphate to speed up the reaction of zinc with sulphuric acid is an example of a copper compound being used as a catalyst (see experiment 12b on p. 82).

Summary

1 Copy and complete the table in figure 15.4 to compare the properties of a typical Group I element and a typical Group II element.

Property	Sodium	Magnesium
The element Behaviour in moist air Reaction with cold water		
The hydroxide Formula Solubility in water		
The chloride Formula State at room temperature Solubility in water		
The carbonate Formula Solubility in water		
Ion Symbol for the common ion		

Figure 15.4

2 Copy and complete the table in figure 15.5 to show that iron and copper are typical transition metals.

Property	Iron	Copper
The element Melting point Density Behaviour in moist air Reaction with dilute acid		
Ions Symbols for common ions		
Colours Colours of compounds		
Catalysts Examples of the use of the element or its compounds as catalysts		

Figure 15.5

16 Non-metals in the periodic table

	III	IV	V	VI	VII	VIII
H						He
	B	C	N	O	F	Ne
		Si	P	S	Cl	Ar
					Br	Kr
					I	Xe

Figure 16.1 Some non-metals in the periodic table

16.1 Family names

Figure 16.1 shows the non-metals in the periodic table. Two of the groups have family names. The elements in Group VII are called the **halogens** and the elements in Group VIII are called the noble gases.

The discovery and uses of the noble gases are described in chapter 4. They are very unreactive. It was a long time before they were discovered because they do so little chemistry that nobody noticed them. The reason for the inactivity of these elements is given in chapter 17.

Questions

Use table 1 on p. 336 to help you answer these questions:

1 Draw up a table to show the similarities of the noble gases. Include the following headings: colour, state at room temperature, boiling point, chemical reactions, structure.

2 Plot bar charts to show how the melting and boiling points of the noble gases vary with the relative atomic masses of the elements.

Group IV includes two very important non-metals: carbon and silicon. The chemistry of living things is based on the complexity and variety of ways in which carbon atoms can join together. Carbon chemistry is so important that it makes up a large part of several of the themes in this book, including themes I ('Chemicals from oil and gas') and J ('Chemistry and food').

The crust of the earth consists mainly of two non-metals, silicon and oxygen. The feldspars and mica in granite (see figure 1.1 on p. 2) are silicate minerals. Quartz, sand and sandstone consist mainly of silica (silicon dioxide, SiO_2). The structures of some of these minerals are illustrated in chapter 19.

The similarities between carbon and silicon are not so obvious as those between sodium and potassium, or chlorine and bromine. But a close look at their chemistry shows that there are good reasons for their being in the same group. The table in figure 16.2 shows a few of the similarities between the chemistries of these elements. Some differences are described in chapter 19.

Element	Structure	Electrical conductivity	Properties of the common oxide	Chloride	Simplest hydride
C	Two allotropes: diamond and graphite	Graphite conducts; diamond does not	CO_2(g), an acidic oxide; dissolves in water to give an acid solution	Colourless liquid, CCl_4(l)	Colourless gas, CH_4(g); burns in air when ignited
Si	Has the same structure as diamond	Semiconductor when pure	SiO_2(s), an acidic oxide; insoluble in water	Colourless liquid, $SiCl_4$(l)	Colourless gas, SiH_4(g); catches fire in air at room temperature

Figure 16.2

16.2 The halogens

Fluorine, chlorine, bromine and iodine are all very reactive non-metals. They are interesting because they do much vigorous chemistry. They are dangerous because they are reactive. For the same reason, they are not found free in nature – they are found combined with metals. The compounds are salts. Hence the family name 'halo-gen', meaning 'salt-former'.

Calcium fluoride is mined as the mineral fluorspar. Chlorine is available on a very large scale from sodium chloride (salt), as described in chapter 22. Bromine can be extracted from the sea, in which it occurs as metal bromides. Iodine comes mainly from Chile, where it is found as sodium iodate in a mineral called caliche, which consists mainly of potassium nitrate. Experiment 1d shows how iodine can be extracted from seaweed.

In a school laboratory, out of the four halogens, most use is made of chlorine. This can be made from concentrated hydrochloric acid.

Experiment 16a
The laboratory preparation of chlorine

concentrated hydrochloric acid

potassium manganate(VII)

potassium manganate(VII)
manganese(IV) oxide

chlorine

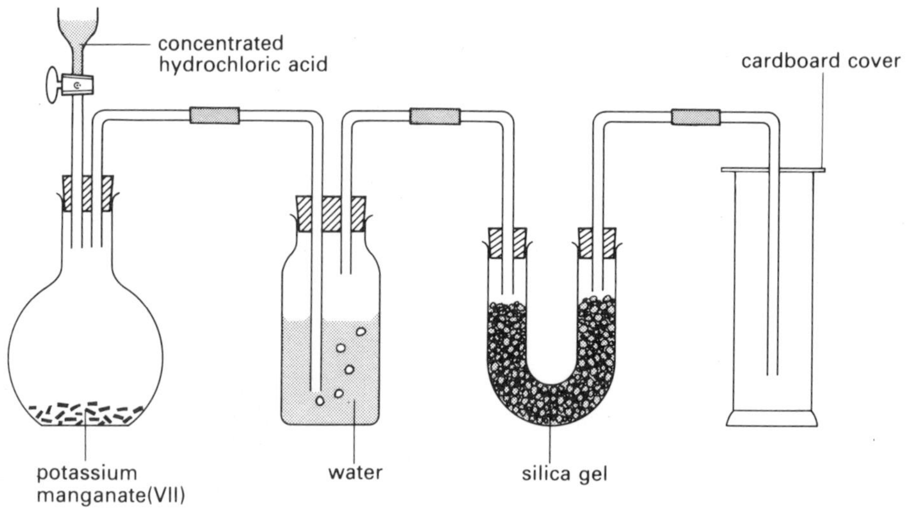

Figure 16.3

Concentrated hydrochloric acid can be converted to chlorine using either manganese(IV) oxide or potassium manganate(VII). Heat is required if manganese(IV) oxide is used. With potassium manganate(VII), the reaction goes at room temperature. A suitable apparatus is shown in figure 16.3.

This experiment must be done in a fume cupboard because chlorine is very poisonous.

Results
Chlorine is a dense, green gas, and it can be seen to fill the gas jar from the bottom, displacing the air upwards. It is easy to see when the jar is full. A strong, choking smell can be detected unless the experiment is carried out in a very efficient fume cupboard.

Discussion
What is the purpose of the wash bottle and the U-tube in the apparatus in figure 16.3? Using manganese(IV) oxide, the products are chlorine, manganese(II) chloride and water. Use this information to work out word and symbol equations for the reaction.

The properties of chlorine are shown in figure 16.4. The great usefulness of the halogens and their compounds is illustrated by figures 16.5–7. The table in figure 16.8 shows some of the ways in which the halogens are similar.

Figure 16.4 The properties of chlorine

(a)

(b)

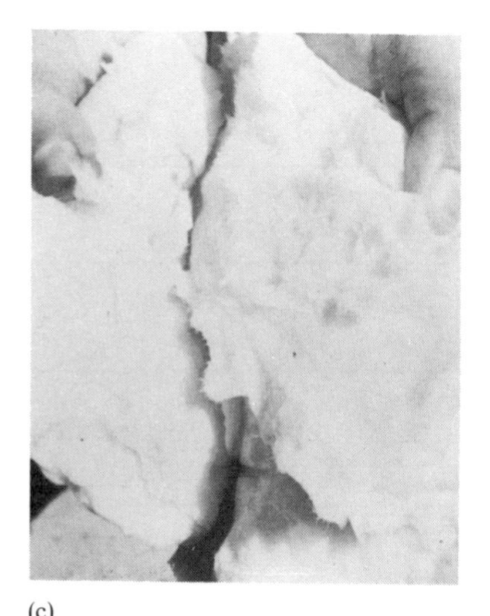

(c)

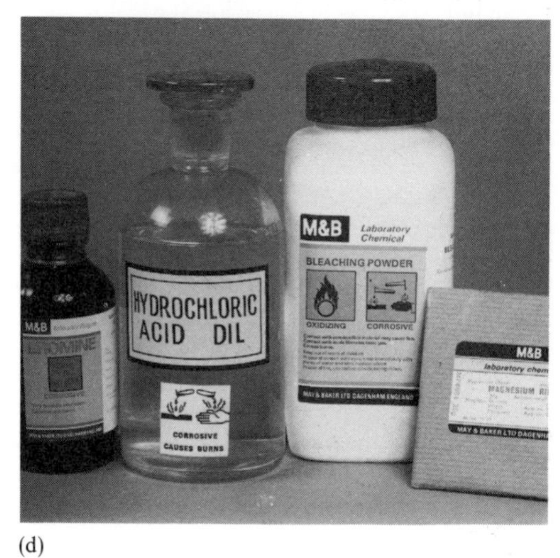

(d)

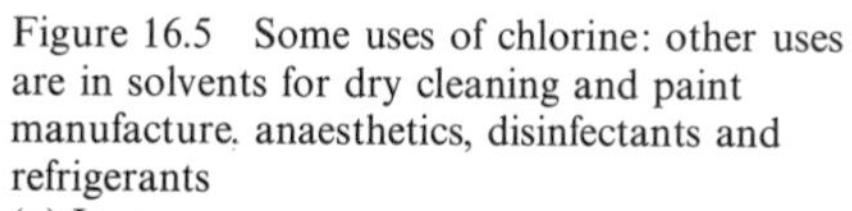

Figure 16.5 Some uses of chlorine: other uses are in solvents for dry cleaning and paint manufacture. anaesthetics, disinfectants and refrigerants
(a) In pvc
(b) In aerosol propellants
(c) To bleach wood pulp
(d) For making chemicals

(a)

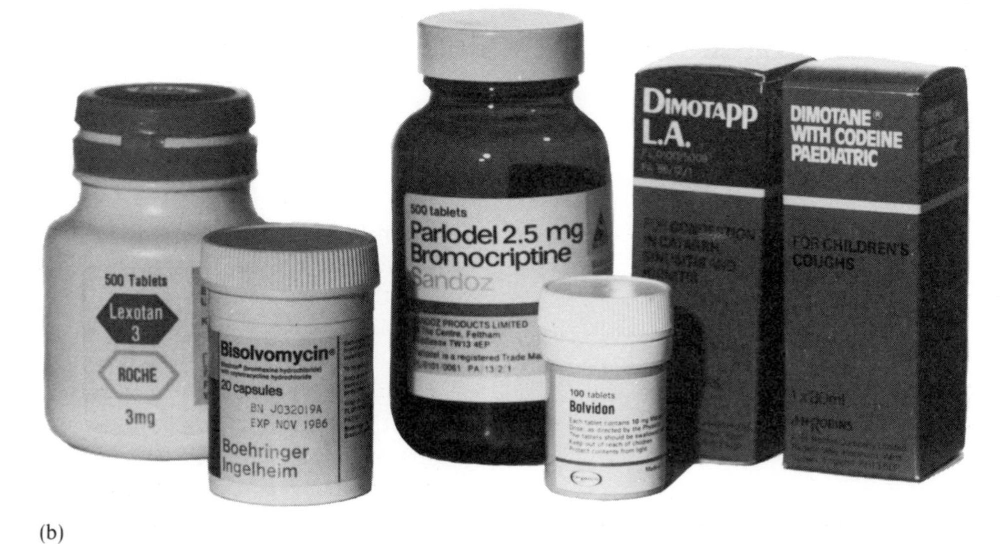

(b)

Figure 16.6 Two uses of bromine: it is also used to make anaesthetics, dyestuffs, fire extinguishers and additives for petrol
(a) In silver bromide, which is used in photographic paper and film
(b) For making drugs

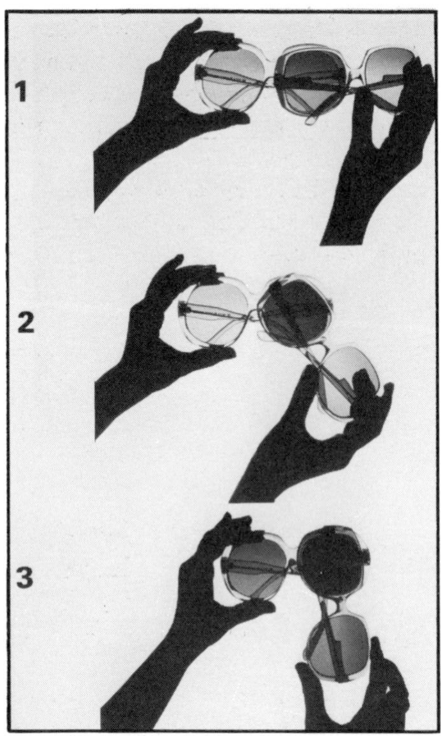

Figure 16.7 Two uses of iodine: it is also used to make photographic materials and as an antiseptic
(a) In polarizing filters that are used in some sunglasses to cut out glaring light
(b) Small amounts of iodine are essential in the diets of all animals, including humans

(a) (b)

Element	Colour and state at room temperature	Colour of vapour	Structure	Hydride	Symbol of ion	Potassium salt
Cl	Green gas	Green	Molecules $Cl_2(g)$	Fuming, colourless, acid gas, $HCl(g)$	Cl^-	White, crystalline solid, $KCl(s)$; soluble in water
Br	Dark red liquid	Orange–brown	Molecules $Br_2(l)$	Fuming, colourless, acid gas, $HBr(g)$	Br^-	White, crystalline solid, $KBr(s)$; soluble in water
I	Black solid with slight shine	Purple	Molecules $I_2(s)$	Fuming, colourless, acid gas, $HI(g)$	I^-	White, crystalline solid, $KI(s)$; soluble in water

Figure 16.8

Questions

1 Plot bar charts showing how the boiling points of the halogens vary with atomic number.
2 With the help of chapter 29, draw the structures of the following molecules:
(a) chloroethane, which is used to make the plastic pvc,
(b) 1,2-dichloroethane, which is used as a solvent and as an anaesthetic,
(c) 1,1,2,2-tetrachloroethane, which is used for dry-cleaning.
3 With the help of the information in chapter 1, draw up a flow chart to show the main stages by which salt underground in Cheshire is converted into chlorine gas for use in the laboratory.

An element and its compounds are quite different. The element chlorine is dangerously reactive – a choking, poisonous gas. It is like an unfed lion, savage in its hunt for food. Chlorine gas molecules, Cl_2, are chemically 'hungry' for electrons.

After catching its prey, a lion relaxes and sleeps peacefully. After combining with a metal, chlorine becomes safe and harmless. The chloride ions in a compound such as sodium chloride are chemically 'peaceful'. So when thinking about halogen chemistry, it is very important to distinguish between the hungry molecules ('ine') and the peaceful ions ('ide').

Down the group, from chlorine to iodine, there is a clear trend in the reactivity of the elements. This is illustrated by the reactions of the elements with iron and with hydrogen.

Experiment 16b
The reactions of the halogens with iron

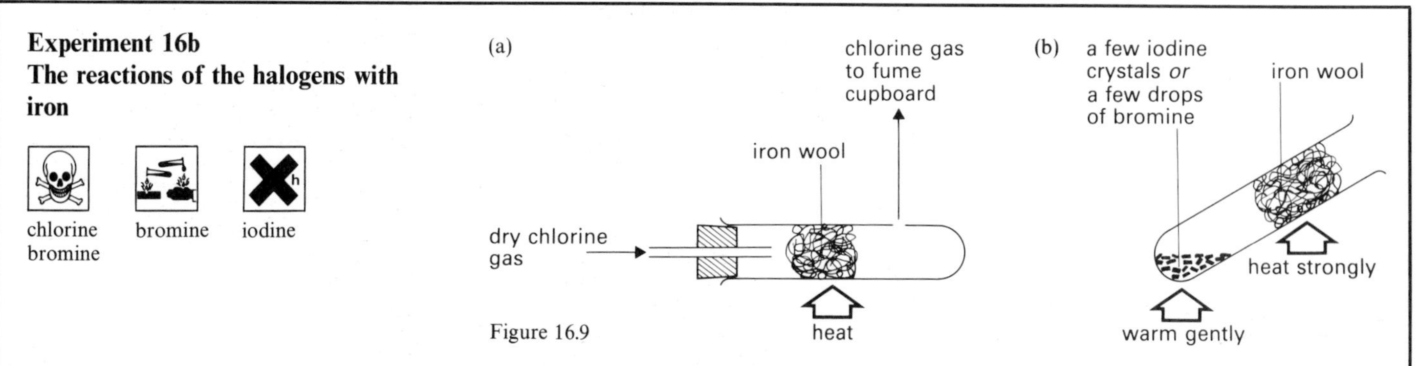

Figure 16.9

Figure 16.9 shows how these reactions may be carried out on a small scale.

The chlorine gas is generated as described in experiment 16a. The vapours of bromine and iodine are produced by warming small amounts of the elements in the apparatus.

Results

The iron glows brightly when heated in chlorine. The product sublimes and appears as a rust-brown solid along the tube. The solid turns yellow as it meets moisture in the air. The iron also glows when heated in bromine vapour, but less brightly. The reaction is similar but less vigorous. There is even less sign of change when the reaction is repeated with iodine.

Discussion

Further investigation shows that chlorine forms iron(III) chloride, $FeCl_3$, with iron, and bromine forms iron(III) bromide, $FeBr_3$. Iodine only reacts to the extent of forming iron(II) iodide, FeI_2.

Write balanced symbol equations for these reactions. What tests could be used to find out whether the product is an iron(II) or an iron(III) compound (see chapter 23)? What is the trend in reactivity down the group from chlorine to iodine?

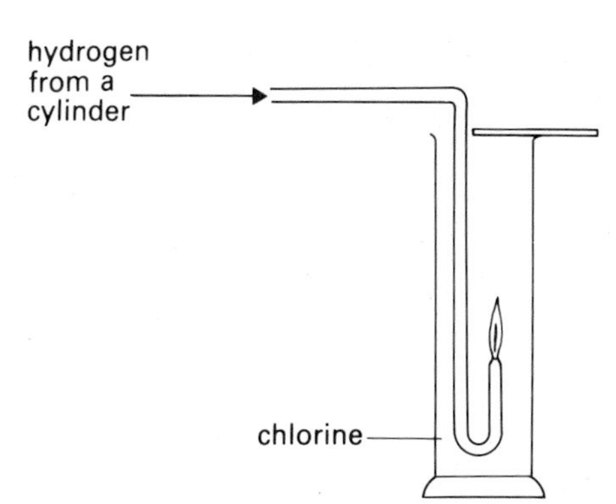

Figure 16.10 Burning hydrogen in chlorine

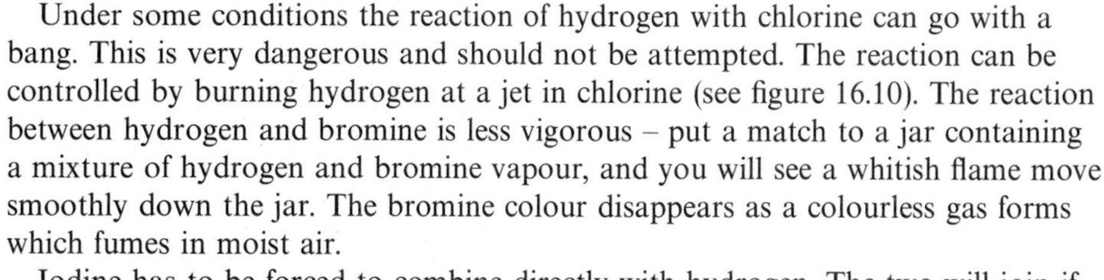

Under some conditions the reaction of hydrogen with chlorine can go with a bang. This is very dangerous and should not be attempted. The reaction can be controlled by burning hydrogen at a jet in chlorine (see figure 16.10). The reaction between hydrogen and bromine is less vigorous – put a match to a jar containing a mixture of hydrogen and bromine vapour, and you will see a whitish flame move smoothly down the jar. The bromine colour disappears as a colourless gas forms which fumes in moist air.

Iodine has to be forced to combine directly with hydrogen. The two will join if mixed in a sealed tube and heated. However, the hydrogen iodide formed is easily decomposed. It splits up into its elements if a hot wire is lowered into a jar of the gas.

Hydrogen chloride, hydrogen bromide and hydrogen iodide are similar in many ways (see figure 16.11). They illustrate very clearly the idea that the compounds of elements in the same group in the periodic table have similar formulae and properties.

Compound	Formula and state of pure compound	Colour and behaviour in moist air	Effect on blue litmus paper	Solubility in water	Reaction with ammonia gas
Hydrogen chloride	HCl(g)	Colourless; fumes in air	Turns red	Very soluble; forms hydrochloric acid	Forms white smoke of solid ammonium chloride
Hydrogen bromide	HBr(g)	Colourless; fumes in air	Turns red	Very soluble; forms hydrobromic acid	Forms white smoke of solid ammonium bromide
Hydrogen iodide	HI(g)	Colourless; fumes in air	Turns red	Very soluble; forms hydriodic acid	Forms white smoke of solid ammonium iodide

Figure 16.11

Questions

1 Draw a gas jar like the one in figure 16.4 to summarize the properties of hydrogen chloride.
2 What would you expect to happen when
(a) hydrogen bromide gas is mixed with ammonia gas?
(b) blue litmus paper is dipped into aqueous hydrogen bromide?
(c) magnesium ribbon is added to a solution of hydrogen iodide in water?
3 Write equations, including state symbols, for the reactions of (a) hydrogen with bromine, (b) hydrogen iodide with ammonia and for (c) the decomposition of hydrogen iodide.

It is possible to set up a displacement series for non-metals, just as it is for metals. The elements higher in the list are more reactive and will displace elements lower in the list from their compounds. Experiment 16c shows how this can be done for the halogens.

Experiment 16c
A displacement series for the halogens

1,1,1-trichloroethane

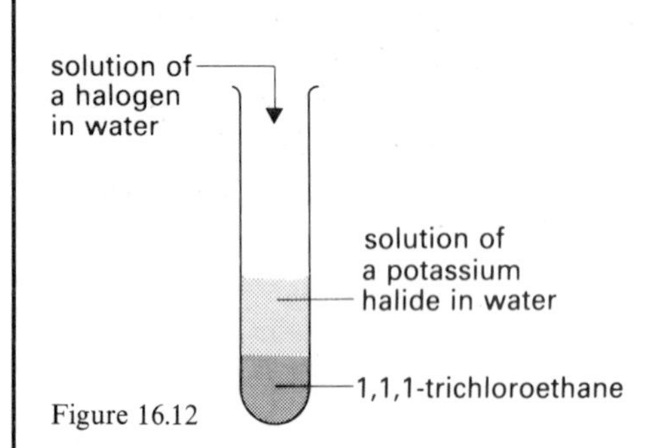

Figure 16.12

The reactions are carried out by mixing solutions of the halogens in water with solutions of the potassium salts. The mixtures are shaken in the presence of a little 1,1,1-trichloroethane. Only free halogen molecules will dissolve in the 1,1,1-trichloroethane layer. The salts stay in the water. Thus the colour of the lower layer shows which halogen is present in the free state at the end.

Results
The results are summarized in the table in figure 16.13.

Halogen solution added	Potassium salt in solution	Colour of the lower layer after shaking
Chlorine	Potassium bromide	Orange
Chlorine	Potassium iodide	Red
Bromine	Potassium chloride	Orange
Bromine	Potassium iodide	Red
Iodine	Potassium chloride	Red
Iodine	Potassium bromide	Red

Figure 16.13

Discussion
From the table in figure 16.13, it can be seen that chlorine will displace bromine from potassium bromide:

Chlorine(aq) + Potassium bromide(aq) ⟶
Potassium chloride(aq) + Bromine(aq)

Which other displacement reactions will happen? Will chlorine displace iodine? What will bromine and iodine displace? Write word equations for the reactions which go. What is the displacement series?

Halide ions in solution can be identified using two reagents: silver nitrate solution and ammonia solution. A **precipitate** forms when aqueous silver nitrate is added to a solution containing chloride, bromide or iodide ions. The colour of the precipitate depends on which halide ion is present, and so does the solubility of the precipitate in ammonia solution. The results are shown in the table in figure 16.14.

Questions

1 Use table 8 on p. 343 to work out the formulae of silver chloride, silver bromide and silver iodide.
2 Write word and symbol equations for the reactions of silver nitrate solution with solutions of (a) sodium bromide and (b) potassium iodide.

Halide ion	Colour of precipitate formed when silver nitrate solution is added	Effect of adding ammonia solution to the precipitate
Chloride	White	Dissolves in dilute ammonia solution
Bromide	Pale yellow	Dissolves in concentrated ammonia solution
Iodide	Yellow	Insoluble in ammonia solution

Figure 16.14

Summary

Chlorine is such an important element that it is worthwhile bringing together all the information about its chemistry in a single summary. Use the headings below to guide you. (The information is in this chapter unless other references are given.)

- *Sources* (a) in industry (see p. 157), (b) in the laboratory.
- *Physical properties* (a) state at room temperature, (b) colour, (c) density relative to air, (d) solubility in water (see table 7 on p. 343).
- *Chemical properties* its reactions with (a) metals such as sodium (see p. 140) and iron, (b) non-metals, such as hydrogen, (c) water (see p. 62), (d) hydrocarbons (see p. 220), (e) iron(II) ions (see p. 322), (f) sulphide ions (see p. 325), and the test used to identify the gas (see p. 155).
- *Uses.*

17 Atomic structure

1804
Dalton's solid atom

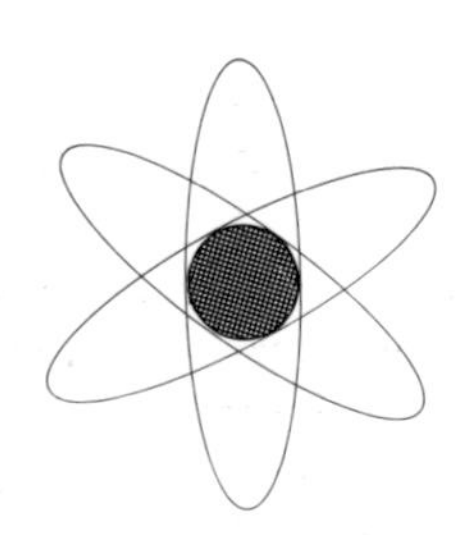

1913
The Bohr-Rutherford 'solar system' atom, in which electron 'planets' orbit round a nuclear 'sun'

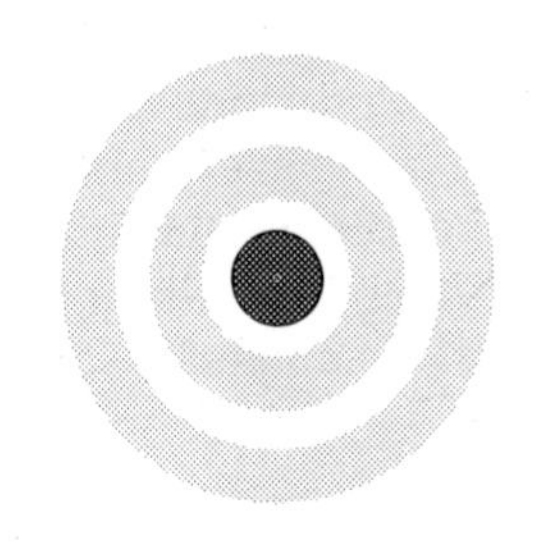

1924
The de Broglie atom in which the electron is no longer treated as a particle

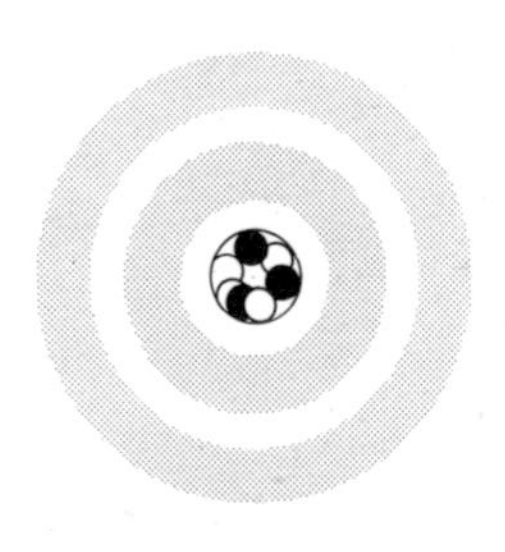

1932
The atom in which the nucleus is built up from neutrons as well as protons

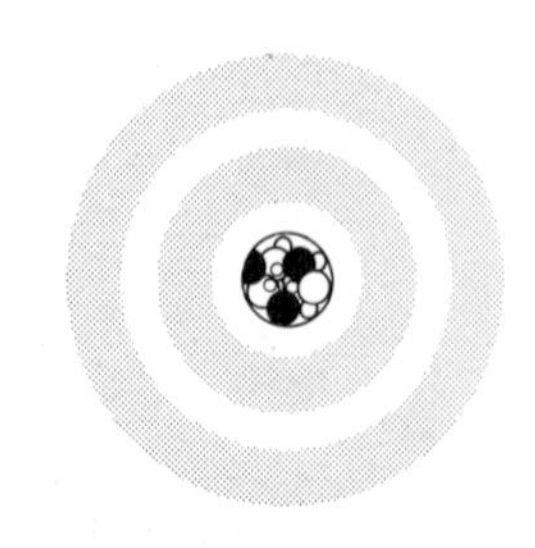

1984
The present-day atom in which the nucleus is built up from many kinds of particles

Figure 17.1 Some models of atomic structure

Each element has its own kind of atom. To explain why one element is different from another, it is necessary to have a picture of what atoms are like. A picture, or model, of an atom can be used to understand how atoms join together to form compounds and how they regroup during chemical reactions. Theories of atomic structure are fundamental to modern chemistry.

Different models are used to solve different problems. This seems odd. Surely there should be an atomic theory which is 'true' or 'correct', so that no other theory is needed? But it does not work like this. A simple model can be used to solve easy problems. More complicated models are used for more difficult problems. It is like using maps to travel through London. The usual Underground map (see figure 17.2) is a very useful guide for getting from one tube station to another. It helps us to plan a route. It is 'true' in that it shows how the lines and stations connect. But it cannot be used to solve all problems. It does not show when the trains run. It does not show which lines are overground and which are underground. It does not indicate which trains will be crowded, nor does it show how the tube lines relate to roads and buildings on the surface. Another map is

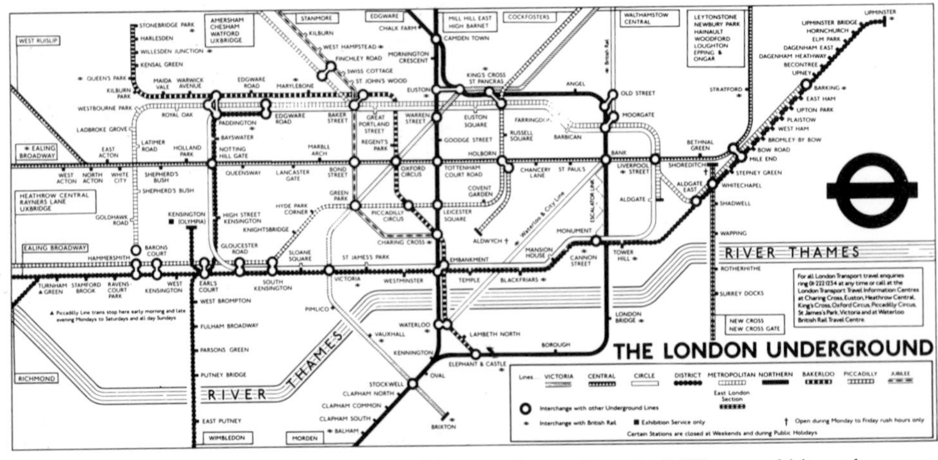

Figure 17.2 When would you be able to use this map of part of London? When would it *not* be useful?

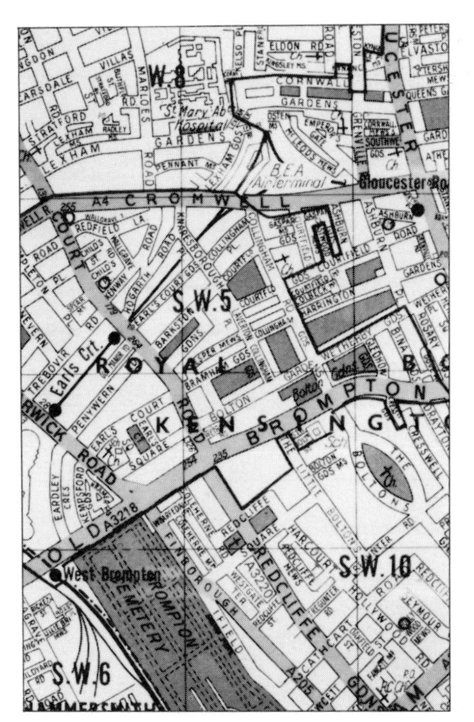

Figure 17.3 When would this map of part of London be more useful than the one in figure 17.2?

needed to find out how to get to the station at the start of a journey (see figure 17.3). Travellers use different maps to solve different aspects of planning a journey. Chemists use different models of the atom to explain different aspects of the behaviour of the elements and their compounds.

Dalton's atomic theory was introduced in chapter 3. The main ideas in his theory are still important and useful. Each element does have its own kind of atom and the atoms of different elements differ in mass. Dalton used his theory to find the formulae of compounds by experiment. This is still done today, as described in chapter 25. The idea that chemical equations must balance comes from Dalton's theory. Atoms are not created or destroyed during reactions. They are rearranged so that the number of atoms of each type stays the same.

But Dalton's theory is limited, like the map in figure 17.2. It cannot explain the pattern of elements in the periodic table. It cannot explain why metals form positive ions and non-metals form negative ions. It does not explain why sodium forms ions with a 1+ charge while magnesium forms ions with a 2+ charge. Nor can the theory explain how atoms join together in elements and compounds.

The history of how famous scientists, including J. J. Thomson, Marie and Pierre Curie, Rutherford, Moseley, Chadwick and many others, discovered more about the 'insides' of atoms is a fascinating story. Today millions of pounds are being spent to build particle accelerators. In these instruments, higher and higher energies are being used to discover more about the internal structure of atoms.

The model of the atom described in this chapter is useful but limited. It dates from 1932, when Chadwick discovered the neutron. The use of this model is justified by the chemistry that it can be used to explain.

The mass of the atom is concentrated in the central **nucleus**, which is very small. The particles in the nucleus are **protons** and **neutrons**. The protons are positively charged and the neutrons are uncharged. Protons and neutrons have the same mass. Around the nucleus are the **electrons**. The electrons are negatively charged. The mass of the electron is so small that it can often be ignored. In an atom, the number of electrons equals the number of protons in the nucleus. Thus, the total negative charge equals the total positive charge, and overall the atom is uncharged.

	Relative mass	Charge
Protons	1	+1
Neutrons	1	0
Electrons	$\frac{1}{1870}$	−1

Figure 17.4

Electrons are arranged in a series of shells around the nucleus. Each shell can only contain a limited number of electrons. The shell nearest to the nucleus fills first. When it is full, the electrons go into the next shell, and so on. Figure 17.5 shows how this model pictures the three simplest atoms. The first shell can only hold two electrons, so in the lithium atom the third electron goes into the second shell. A hydrogen atom has only one proton and no neutrons. It is the simplest and lightest atom, with a relative mass of 1. Helium, with two protons and two neutrons, has a relative mass of 4.

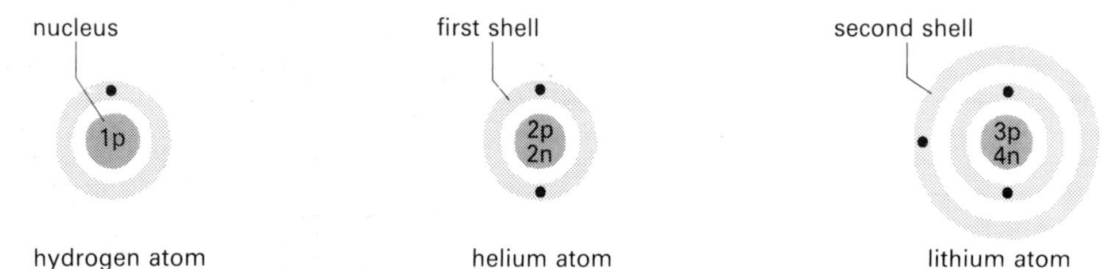

Figure 17.5 Models of hydrogen, helium and lithium atoms: protons, neutrons and electrons are shown by p, n and •

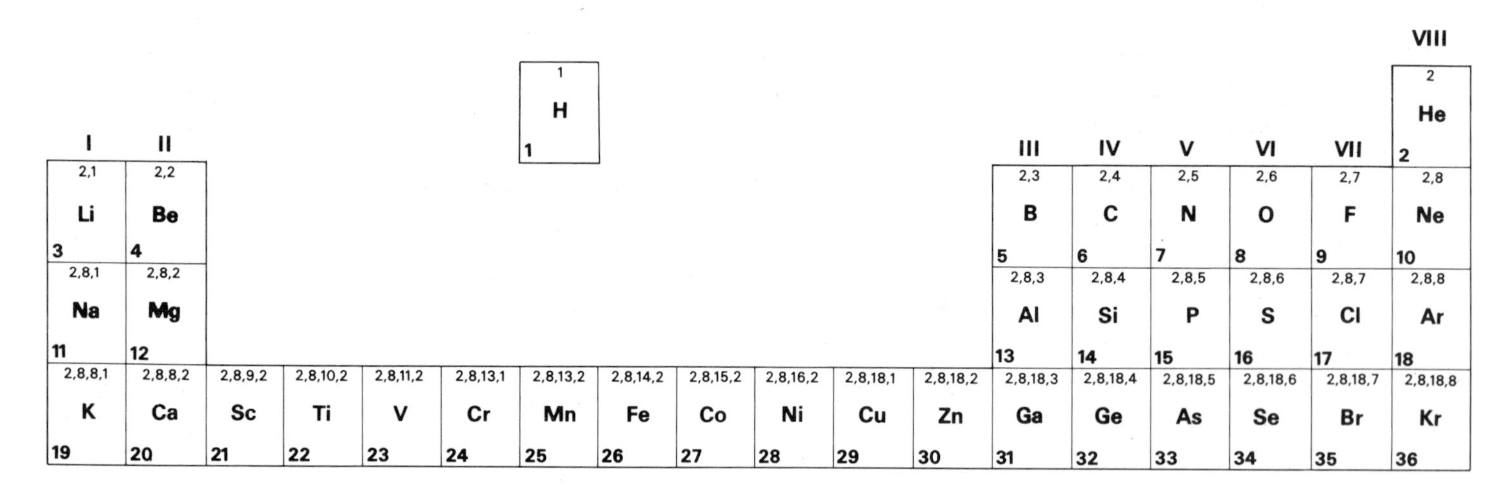

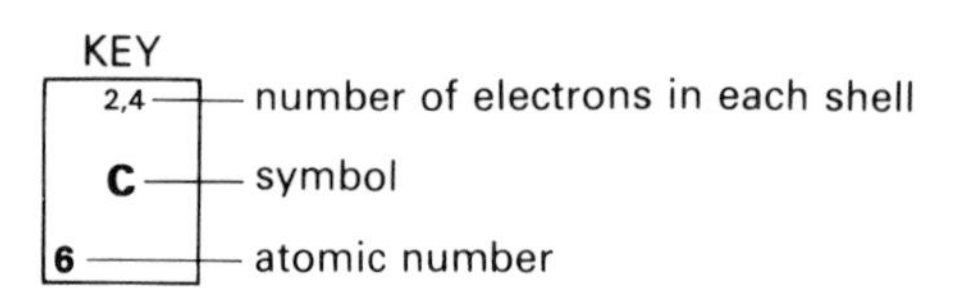

Figure 17.6 The electron structures of the first 36 elements

All atoms are built up of these three basic particles. They differ in mass because of the differing numbers of protons and neutrons in the nucleus. Figure 17.6 shows how this model of the atom relates to the periodic table, for the first 36 elements.

The position of the element in the table is given by its **atomic number**. The atomic number is the number of protons in the nucleus. This is the same as the number of electrons in the atoms of that element. The relative masses of the elements increase across each period, but the rise is uneven and cannot be predicted in a simple way, because the number of neutrons can vary. The neutrons are uncharged and do not affect the electrical balance of an atom. They just add to the atomic mass.

Each horizontal row (**period**) in the table corresponds to the filling of an electron shell. The first shell can hold two electrons, so there are two elements in the first period. The second shell holds eight electrons, so there are eight elements in the second period. There are eight elements in the third period too, and then, with potassium and calcium, the fourth shell starts to fill. In fact, the third shell can hold up to 18 electrons. This shell is completed from scandium to zinc before the fourth shell continues to fill, from gallium to xenon. Why this happens cannot be explained by the simple theory described here.

When atoms bump into each other and react, it is the electrons on the outsides of the atoms which are important. The atoms react by exchanging electrons or by sharing electrons. Elements in the same **group** in the periodic table have similar properties because they have the same number of electrons in their outer shells. The number of electrons in the outer shell is the same as the group number in the periodic table. Thus, all the alkali metals in Group I have one electron in the outer shell. Down a group there is a gradual change in properties, with the increasing number of inner, full shells. The outer electrons get further from the nucleus.

The gases in group VIII have eight electrons in the outer shells of their atoms. They are now called the *noble gases*. The term 'noble' has been used for hundreds of years to describe metals such as gold and silver. These elements were inert to most of the reagents used by the alchemists and early chemists. The metals were seen to be valuable and to have admirable qualities. They are special elements which do not get involved in everyday chemistry.

Other elements, when they react, behave as if they want to become like a noble gas. The atoms gain, lose or share electrons in such a way that they end up with the electron arrangement of the nearest noble gas in the periodic table.

Questions

1 Draw diagrams to show the number and arrangement of electrons in a magnesium atom and in a magnesium ion. What is the charge on the magnesium ion?

2 Draw diagrams to show the electron arrangements in a sulphur atom and in a sulphide ion. What is the charge on the sulphide ion?

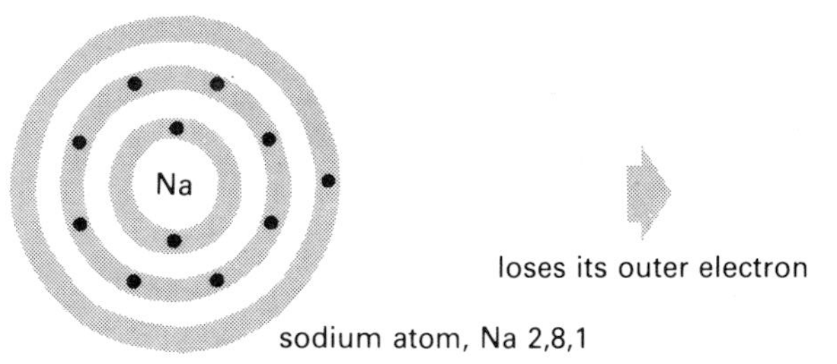

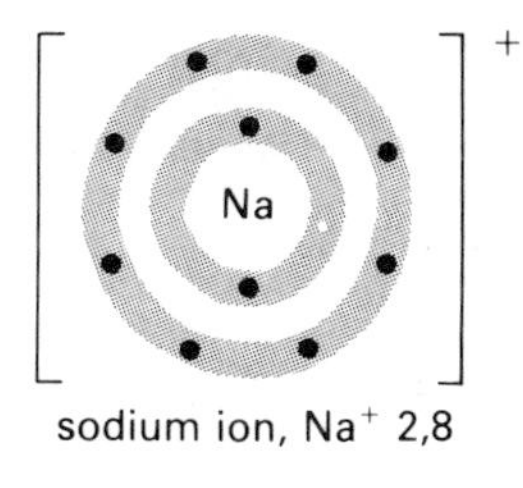

Figure 17.7 Formation of an ion of sodium (a metal)

Cl

gains an electron

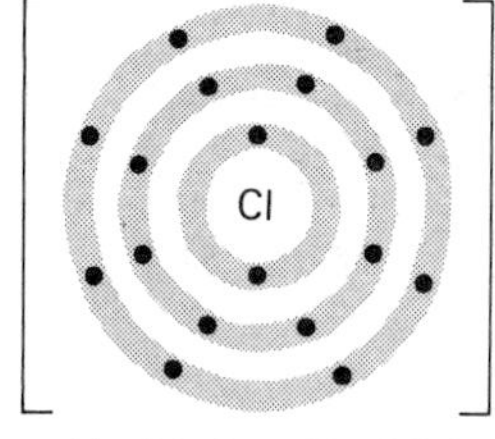

chlorine atom, Cl 2,8,7

chloride ion, Cl^- 2,8,8

Figure 17.8 Formation of an ion of chlorine (a non-metal)

The metals on the left-hand side of the table form ions by losing the few electrons in the outer shell. This means that in the ions there are more protons than electrons and so the ions are positively charged. The non-metals on the right-hand side of the table form ions by gaining electrons to fill the outer shell. In these ions there are more electrons than protons, so they are negatively charged.

This does not explain *why* the electron arrangement in a noble gas is so special. Nor is it easy to explain why the transition metals can form ions with different charges.

The way in which this theory of atomic structure can start to describe the different types of chemical bonding is explained in theme F.

Summary

Copy the table in figure 17.9 and extend it to include the first 20 elements. The terms 'atomic number' and 'mass number' are defined in the glossary on p. 346.

Element	**Atomic number**	**Mass number**	**Number of particles in nucleus**		**Number of electrons in each shell**			
			Protons	**Neutrons**	**First shell**	**Second shell**	**Third shell**	**Fourth shell**
Hydrogen	1	1	1	0	1			
Helium	2	4	2	2	2			
Lithium	3	7	3	4	2	1		

Figure 17.9

Review questions

1 The table shows part of the periodic table, with only a few symbols for elements included.

group / period	I	II											III	IV	V	VI	VII	0
1																		He
2																	F	
3	Na												Al	Si		S		
4		Ca				Cr											Br	
5																		
6																		

(a) Using only the elements in the table, write down the *symbol* for each of the following:
(i) a metal stored in oil,
(ii) a noble or inert gas,
(iii) a transition metal,
(iv) a halogen which is a gas at room temperature and pressure,
(v) the element with the largest number of protons in the nucleus of each atom,
(vi) the element which would contain most atoms in a 100 g sample,
(vii) an element extracted from sand when a mixture of dry sand and magnesium powder is heated.
(b) When Mendeléev devised the periodic table, many of the elements that are now known had not been discovered. The following account refers to such an element. It is represented by the symbol X but this is not the usual symbol for this element.

X is an element with a melting point of 30 °C and a boiling point of 2440 °C. It conducts electricity at room temperature.
It burns in oxygen to form an oxide which has a pH of 7. The oxide is a colourless solid with a formula X_2O_3.
X forms a compound with fluorine. This compound has a high melting point and it conducts electricity when molten. The similar compound with bromine has a formula XBr_3 but is a low melting point solid.
The approximate relative atomic mass of X is 70.

(i) Is X a solid, liquid or gas at room temperature (20 °C)?
(ii) Give a reason why X is regarded as a metal.
(iii) Predict the formula for the fluoride of X.
(iv) What are the products of the electrolysis of the molten fluoride of X?
(v) Give the symbol for the element in the table which most closely resembles X.
(vi) In which group and period of the periodic table will X be placed? **(EAEB)**

2 Use the shortened form of the periodic table opposite to answer the questions set out below.

(a) Give the symbol for the most reactive metal to be found in the table and include a reason for your choice.
(b) Give the symbol of one element from this table which is metallic and which is liquid at 100 °C, supporting your answer by an experimental observation.
(c) Give the symbol of an element which does not react with hydrogen, oxygen or chlorine.
(d) The atomic masses of sulphur (S), tellurium (Te) and bromine (Br) are 32, 128 and 80 respectively. What do you predict that the approximate atomic mass of selenium (Se) will be?
(e) What will be the formula for the compound of selenium (Se) with hydrogen?
(f) From your knowledge of the periodic table, predict *three* properties of the metal cobalt (Co) or of its compounds.
(g) Will astatine (At) be a solid, a liquid or a gas, at 20 °C? Explain how you have reached your conclusion.
(h) What type of structure would you expect astatine to have? **(L)**

						H	He										
Li	Be											B	C	N	O	F	Ne
Na	Mg											Al	Si	P	S	Cl	Ar
K	Ca	Sc	Ti	V	Cr	Mn	Fe	Co	Ni	Cu	Zn	Ga	Ge	As	Se	Br	Kr
Rb	Sr														Te	I	Xe
Cs	Ba														Po	At	Rn

3 Describe the experiments you would carry out to illustrate the following patterns found in the periodic table. State what you would do and what you would expect to see. Name the chemicals you would use and, if possible, write equations for any reactions which take place.
(a) Metals on the left of the table form basic oxides, non-metals on the right of the table form acidic oxides, and there are elements in between, such as aluminium, which form oxides with both basic and acidic properties.
(b) In the series of halogens (chlorine, bromine and iodine), chlorine is the most reactive element and iodine is the least reactive.
(c) In the series $MgSO_4$, $CaSO_4$ and $BaSO_4$, magnesium sulphate is the most soluble in water and barium sulphate is the least soluble. **(L)**

4 In the periodic table, elements in vertical groups show a similarity but a gradation in properties. Explain how this statement can be justified by the following comparisons of pairs of elements and their compounds.
(a) *Sodium and potassium*
(i) the outer electron structure,
(ii) the reaction of the elements with water,
(iii) the properties of the oxides,
(iv) the action of heat on the hydrogencarbonates (bicarbonates).
(b) *Chlorine and iodine*
(i) the outer electron structure,
(ii) the colours of the elements in their vapour state,
(iii) the reaction of the elements with hydrogen,
(iv) the reaction of chloride and iodide ions with aqueous silver nitrate,
(v) the reaction of the sodium salts with bromine water.
(WJEC)

5 Make use of the following information about silicon (Si) and its compounds to answer the questions below.

Silicon, atomic number 14, is the element immediately below carbon in group IV of the periodic table. It does not react with water nor with dilute acids. It can be obtained by heating sand with an excess of magnesium. Sand is an oxide of silicon.

When sand is heated with carbon at a high temperature, carbon monoxide and carborundum are formed. Carborundum is a compound of silicon and carbon only and it is a very hard substance.

(a) State the characteristic valency of silicon and hence write down the chemical formulae for (i) sand and (ii) sodium silicate.
(b) Write the equation for the reaction between sand and magnesium.
(c) Describe how you would obtain pure *dry* silicon from the products of reaction (b).
(d) How are the electrons arranged in the silicon atom?
(e) Give (i) *two* physical differences and (ii) *one* chemical similarity, between carbon dioxide and sand.
(f) Suggest a formula for carborundum.
(g) Write the equation for the reaction of carbon with sand.
(h) Name another substance that you would expect to have the same crystal structure as carborundum. **(CLES)**

6 (a) The diagram below summarizes the results of a series of chemical reactions.

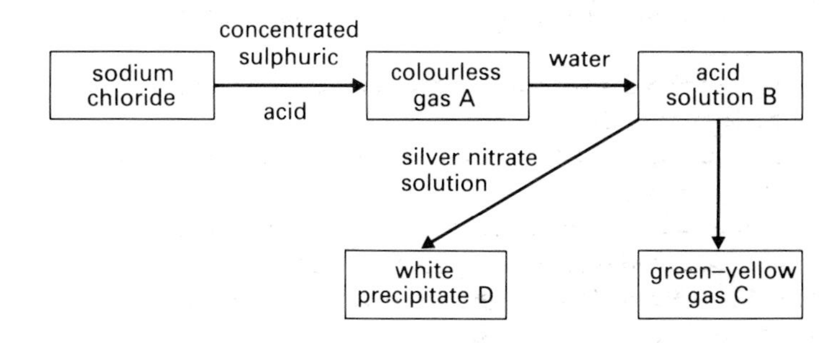

(i) Name A.
(ii) Name an indicator. What is the effect of B on this indicator?
(iii) Name B.
(iv) Name C.
(v) What reagent would you use to convert B into C?
(vi) What is the effect of C on the indicator you named in (ii)?
(vii) Name D.
(b) Chlorine is used as a bleaching agent.
(i) What other substance must be present before chlorine will bleach?
(ii) When cotton sheets are bleached with chlorine, what precaution must be taken after bleaching and why?
(iii) Give *one* other use of chlorine.

(c) When chlorine is passed over heated iron, the iron glows brightly, even when the source of heat is removed, and the product of the reaction is reddish–brown in colour. When bromine vapour is passed over heated iron, the iron glows even when the source of heat is removed, and the product is reddish–brown in colour. When iodine vapour is passed over heated iron the only evidence that a reaction has taken place is the reddish–brown product.

(i) What evidence is there in the above paragraph that the family of elements chlorine, bromine and iodine are chemically similar?
(ii) What can you deduce from the above paragraph about the relative reactivity of the elements chlorine, bromine and iodine? Explain your deduction.
(iii) State what you would *see* when chlorine gas is bubbled through a solution of potassium bromide. Explain your observation and give the equation for the reaction which takes place.
(iv) Explain why you would not expect iodine to react with potassium bromide solution. **(SEREB)**

7 The element astatine (symbol At) is a radioactive element which follows iodine in group VII of the periodic table in the sequence chlorine–bromine–iodine–astatine.
(a) Would you expect the compounds which astatine should form with (i) sodium, (ii) silver and (iii) lead, to be very soluble, sparingly soluble, or almost insoluble in water?
(b) Would you expect the element itself to be a solid, liquid or gas at atmospheric temperature and pressure?
(c) How would you expect the stability to heat and the acidity of the compound of astatine with hydrogen to compare with these properties for hydrogen iodide?
Give briefly the reasoning behind your answers in (a), (b) and (c). **(O & C)**

8 Arrange the following elements in three pairs so that the elements in each pair have similar chemical properties: bromine, calcium, chlorine, magnesium, potassium, sodium. For *each* pair describe fully *two* tests which you would carry out to show this similarity, explaining what happens in each case. **(L)**

9 In answering (a) to (d), you *must* select elements chosen *only* from the following table:

Element	Atomic number	Relative atomic mass
Carbon	6	12
Oxygen	8	16
Neon	10	20
Sodium	11	23
Magnesium	12	24
Aluminium	13	27
Sulphur	16	32
Chlorine	17	35.5
Argon	18	40

(a) Name *two* of these elements which are
(i) in the same group of the periodic table.
(ii) in the same period of the periodic table.
(iii) metals, giving a large scale use of *one* of them.
(iv) able to form acidic oxides.
(v) not readily able to form compounds.
(b) Give the names and formulae of *two* compounds, each of which contains *three* of the elements.
(c) Give the formulae of *three* ions which have the same electronic configuration as neon.
(d) If a 1 g sample of each element were available, which one would contain the greatest number of atoms? Show your reasoning. **(CLES)**

10 (a) Complete the following table.

	Proton	Neutron	Electron
Electrical charge			
Relative mass			

(b) An element has the atomic number 3 and a mass number of 7. Draw the structure of *one* atom of the element to show clearly the number and positions of protons, neutrons and electrons in the atom.
(c) Another element has the electronic structure 2. 8. 1.
(i) What is the valency of this element?
(ii) State whether the element will be a metal or a non-metal and give a reason for your answer.
(iii) This element reacts vigorously with cold water. Write the equation using words *or* symbols, for the reaction that takes place. **(WJEC)**

THEME F
Structure and bonding

Halite is sodium chloride. The ions are arranged in a cubic giant structure

18 Investigating structure

To understand how a car engine works, we need to know how it is put together. The same is true of chemical substances. Some of their properties can only be explained if we know how the atoms, ions or molecules are arranged (the structure of the substance) and what holds them together (the bonding).

Why is polythene easily bent whereas cast iron is not? Why is graphite soft and flaky whereas diamond is the hardest substance known? Why does copper conduct electricity when sulphur does not? Why is steel very strong whereas wrought iron is much weaker? We need to know about structure and bonding to answer these questions.

The problem in studying the structure of substances is that atoms are very small and cannot be seen easily.

18.1 Looking for evidence of structure

Crystals can be grown in the laboratory from **saturated solutions**. Figure 18.1 shows some minerals. The crystalline shapes were formed in the earth's crust as it cooled. The regular shapes of the crystals suggest that they consist of particles arranged in a regular way. The shape of the halite (sodium chloride) crystal on p. 117 makes sense when compared with the cubic pattern of ions in the structure shown in figure 20.4 on p. 141.

Figure 18.1 Some naturally-occurring crystalline minerals

(a) Fluorite (calcium fluoride)

(b) Galena (lead(II) sulphide)

(c) Barytes (barium sulphate)

(d) Calcite (calcium carbonate), showing double refraction

(e) Sulphur

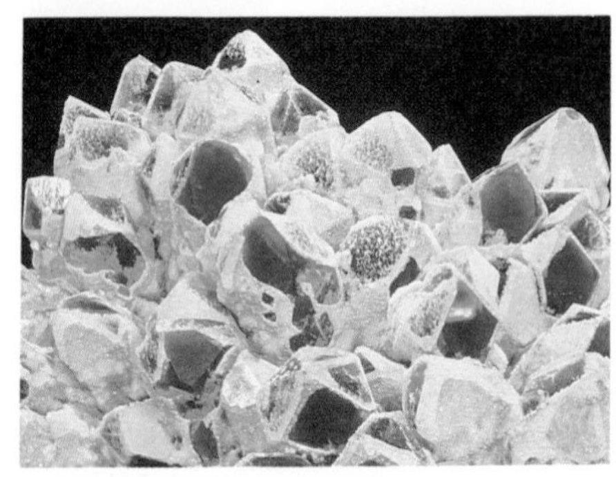

(f) Amethyst – a variety of quartz (silicon dioxide)

(a) A simplified diagram of his apparatus

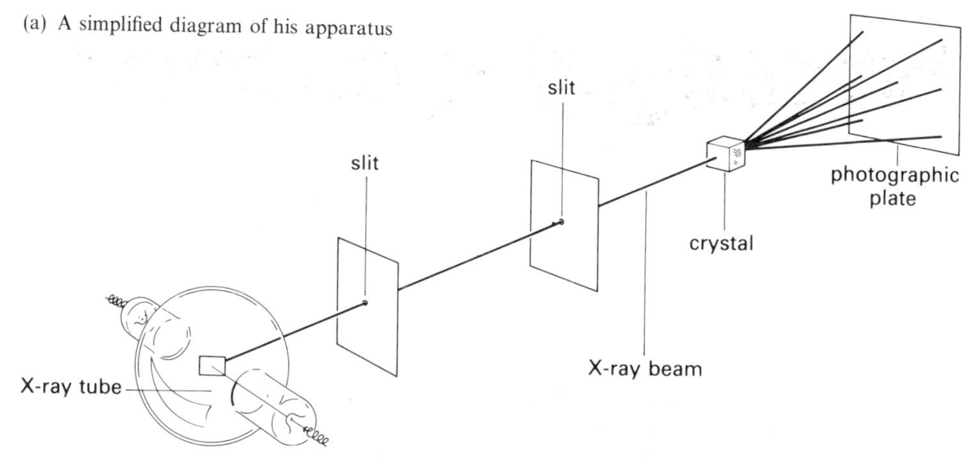

(b) A pattern of dots obtained from a sample of zinc blende

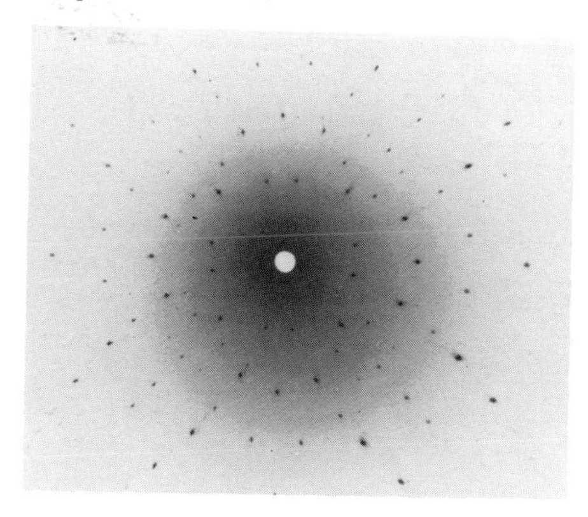

Figure 18.2 Von Laue's X-ray diffraction experiment

In 1912 a German scientist, Max von Laue, found that when a beam of X-rays was passed through a form of zinc sulphide called zinc blende, a pattern of dots formed on a photographic plate. A simplified diagram of his apparatus is shown in figure 18.2. On a larger scale, a similar effect is seen if a finely-woven handkerchief is held between a light bulb and the eye (see figure 18.3). A pattern of bright dots is seen which corresponds to the pattern of the weave. If the handkerchief is rotated, the observed pattern also rotates. The wavelength of visible light is about 500 nm. The wavelength of X-rays can be as short as 0.1 nm. X-rays can be used to investigate structure because their wavelength is about the same as the distance between atoms in a crystal.

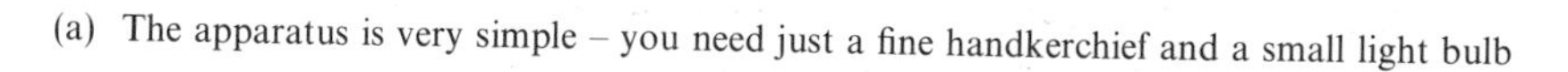

(a) The apparatus is very simple – you need just a fine handkerchief and a small light bulb

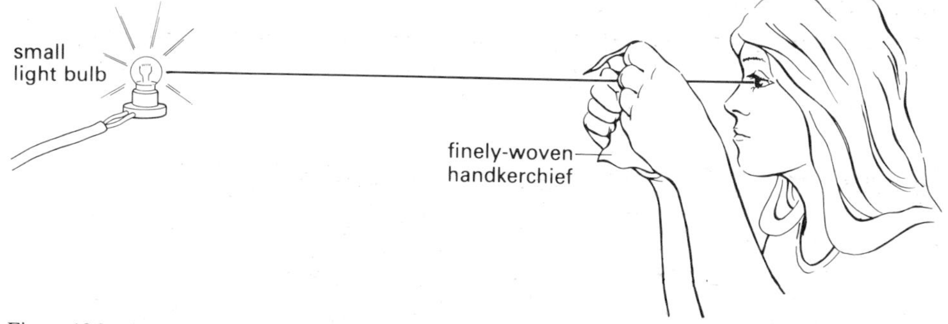

(b) When you look through the handkerchief at the light, you should see a pattern of bright dots

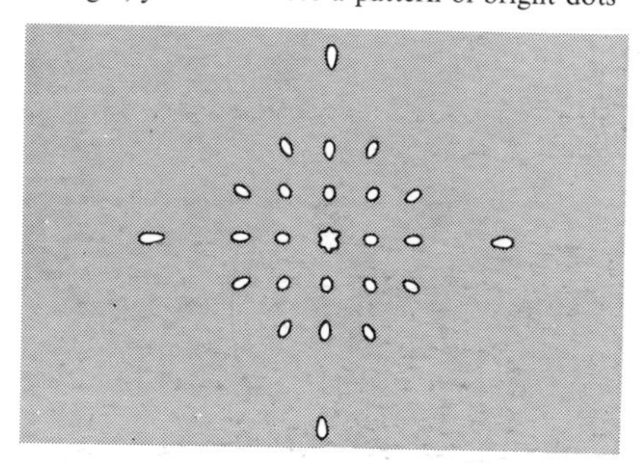

Figure 18.3 An optical analogue of von Laue's experiment

Sir Lawrence Bragg and his father, Sir William Bragg, showed that the pattern and spacing of the dots on a photographic plate could give clues to the arrangement of particles in the crystal. Figure 18.4 shows a model of the structure of zinc blende which they worked out, alongside a crystal of the mineral.

Question

The experiment shown in figure 18.3 is an analogue of von Laue's experiment. In his experiment, what corresponded to (a) the handkerchief, (b) the eye, (c) the light bulb and (d) the visible light beam?

Figure 18.4 (a) Crystals of zinc blende (zinc sulphide) (b) Their crystal structure

Figure 18.6 Zinc crystals on a lamp post made of galvanized iron

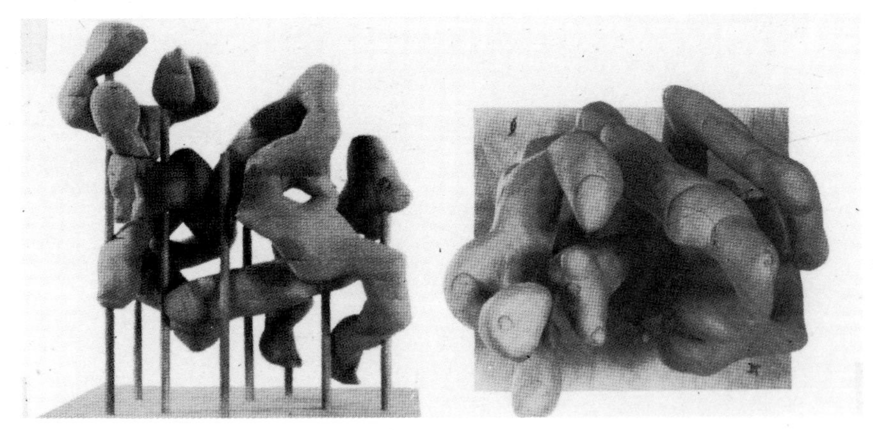

Figure 18.5 Model showing the shape of a lysozyme molecule
(a) Front view
(b) View from above

This technique, called X-ray diffraction, has since been used to solve many crystal structures, including some very complicated ones like those of proteins (see figure 18.5).

18.2 The structures of metals

Although it is not always obvious, metals are crystalline. Look at the surface of galvanized iron railings or inside a new dustbin or fire bucket. Irregularly shaped zinc crystals can be seen (see figure 18.6). Metal crystals can be seen in the laboratory by displacing metals from their salts or by rapidly cooling molten metals.

Experiment 12d on p. 84 shows how displacement reactions can be used to grow metal crystals from solution. When a strip of zinc is dipped into a solution of a lead salt, crystals of lead form quickly on the surface of the zinc. They sparkle when light is shone onto the tube.

Experiment 18a
How do crystals form from molten lead?

Figure 18.7 shows the steps in the experiment. Sometimes a scum forms on the surface of the lead. This is removed with a spatula before pouring the molten lead.

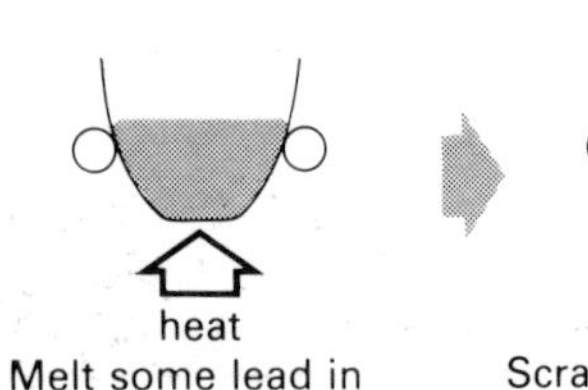

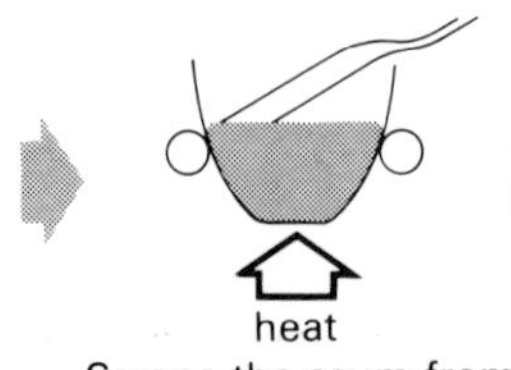

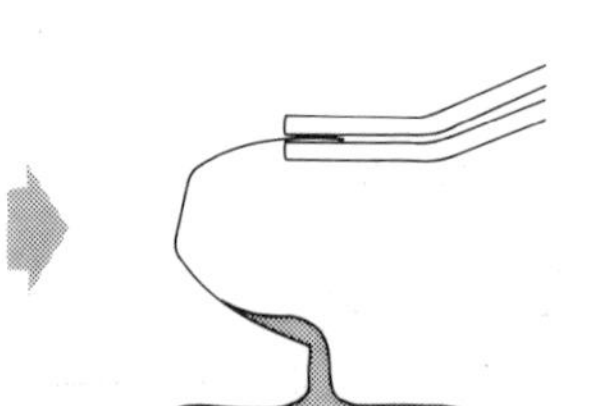

Melt some lead in a crucible

Scrape the scum from the surface

Pour the molten lead on to a plate of smooth metal

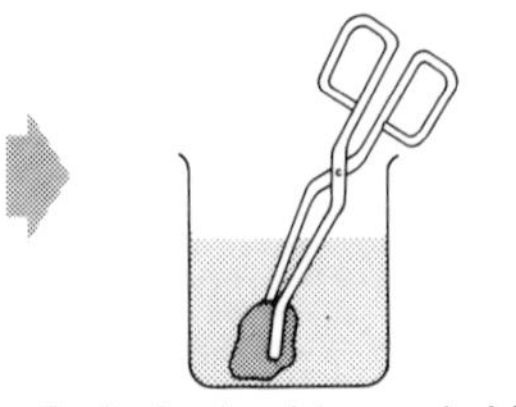

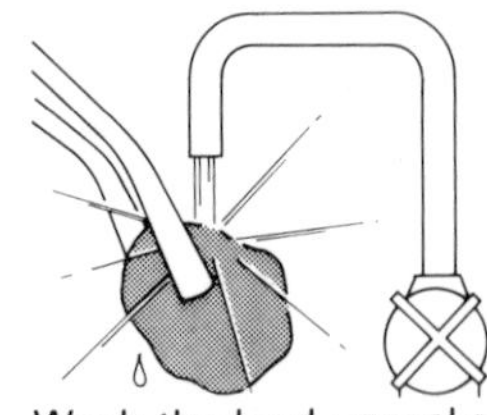

Wash the lead pancake with water and leave to dry

Figure 18.7

Figure 18.8 A lead pancake

Results
The lead solidifies quickly when it is poured onto the sheet. What is the purpose of the metal sheet? Little can be seen at first, but when the lead 'pancake' is dipped into dilute nitric acid for about one minute, crystalline areas begin to sparkle. Some of them are large and long. Others are much smaller. Can you suggest a reason why?

Discussion
The crystalline areas are called *grains*. They do not have straight sides and regular shapes like copper sulphate crystals, but the internal structure of the grains

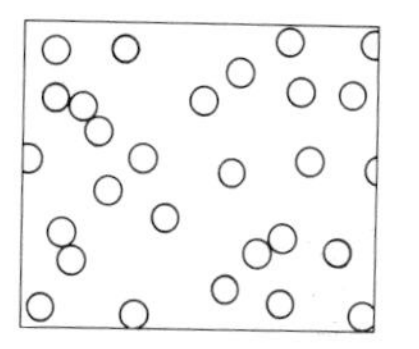

The moving atoms are in random positions in the molten metal

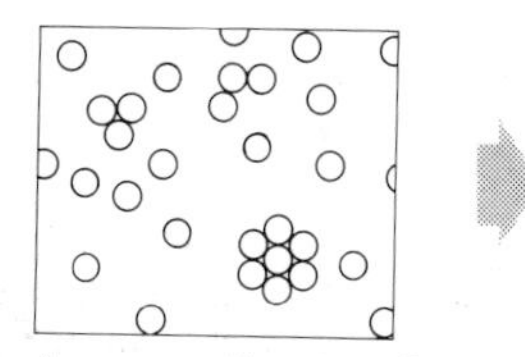

A group of atoms forms a small cluster with a regular pattern

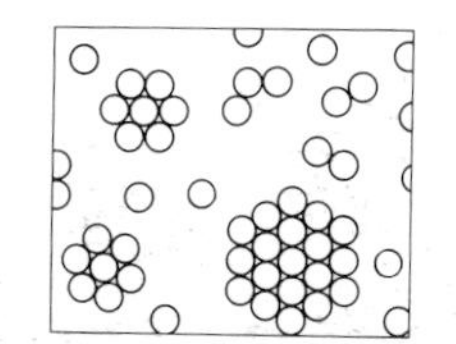

As the metal cools, more atoms are added to the cluster from the molten metal, and other clusters form

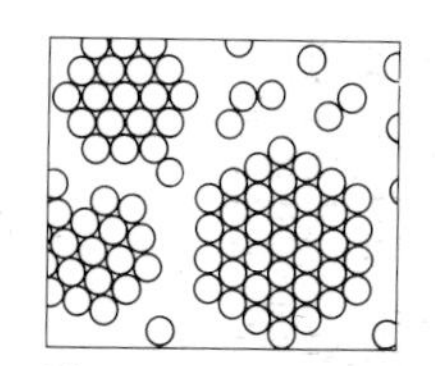

The clusters continue to grow, forming grains

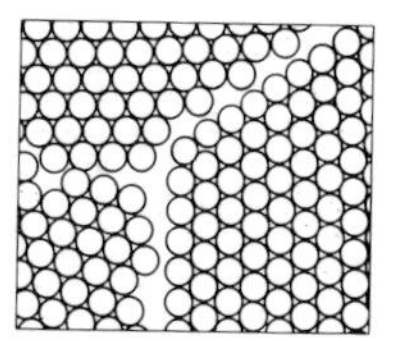

Eventually, the grains meet. The patterns of atoms do not line up along the boundaries between the grains

Figure 18.9 Formation of grains as a metal cools

is regular. When the molten metal cools rapidly, moving atoms slow down and begin to group together in a regular way. Many small groups – the crystal nuclei – form and grow into larger grain areas. Eventually, growing grains meet. A grain boundary separates one grain from the next. The process is shown in figure 18.9. The grains only become obvious when the surface has been etched by the acid. It attacks the grain boundaries first, making them clearer.

X-ray diffraction experiments confirm that metal crystals have a regular three-dimensional structure. They are built up from layers of atoms. There are two types of layers. Models of these are shown in figure 18.10(a) and (b). One type, (a), is described as close packed because the spheres are packed as close together as possible. Type (b) layers are more open.

In three dimensions, the layers fit together. Close packed layers can stack in two possible ways, shown in figure 18.10(c) and (d). In type (c), called *hexagonal* close packing, the third layer is directly over the first layer. In type (d), called *face-centred cubic* close packing, it is different. The open layers, type (b), fit together to give an open structure called *body-centred cubic.* This is shown in figure 18.10(e).

Most metal structures fit into one of these three types. The low density metals,

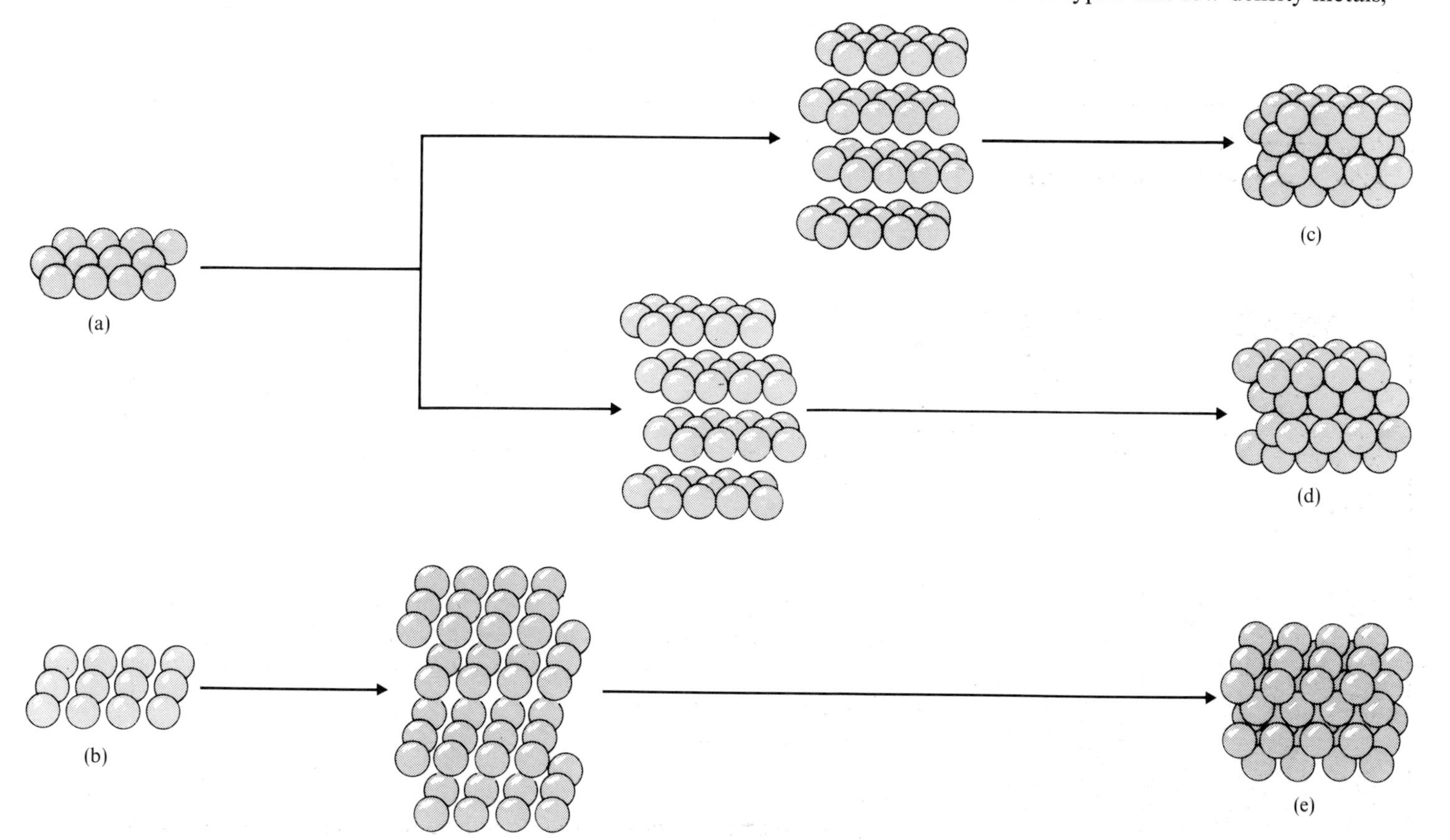

Figure 18.10 Three common metal structures that are built up by packing spheres together in different ways

Question

Choosing an inner sphere in each case, how many spheres surround and touch it
(a) in close packed layers, as in figure 18.10(a)?
(b) in the layers shown in figure 18.10(b)?
(c) in a hexagonal close packed structure, as in figure 18.10(c)?
(d) in a face-centred cubic structure, as in figure 18.10(d)?
(e) in a body-centred cubic structure, as in figure 18.10(e)?

Figure 18.11 Copper

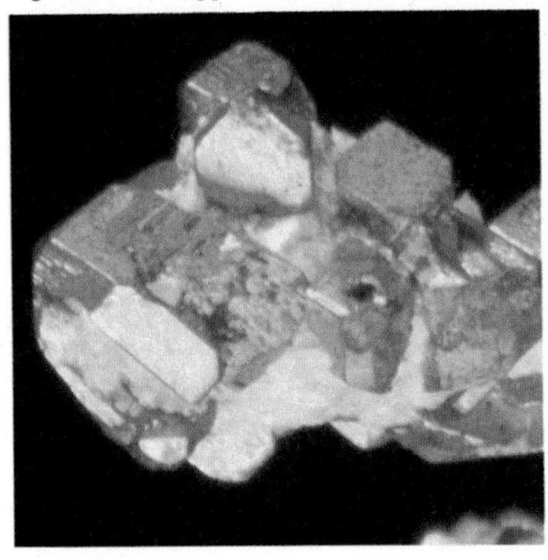

(a) Some naturally-occurring crystals

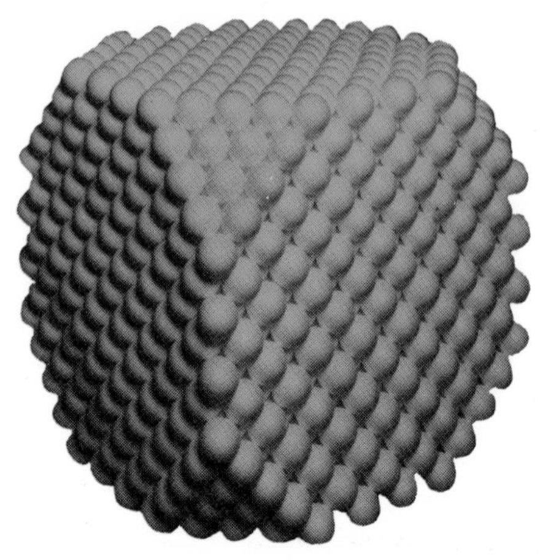

(b) The crystal structure (face-centred cubic)

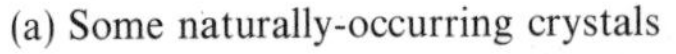

sodium and potassium, have the open body-centred cubic structure shown in figure 18.10(e). Atoms in copper crystals are close packed with the face-centred cubic structure. Figure 18.11 shows crystals of native copper alongside a model of the structure. Can you see the similarities? Is copper a high density metal? Look at table 1 on p. 336.

In a single metal crystal the size of a pin-head, millions upon millions of atoms are packed together. The bigger the crystal, the more atoms are packed together. For this reason, we say that metals have a **giant structure** of atoms.

18.3 Explaining the properties of metals

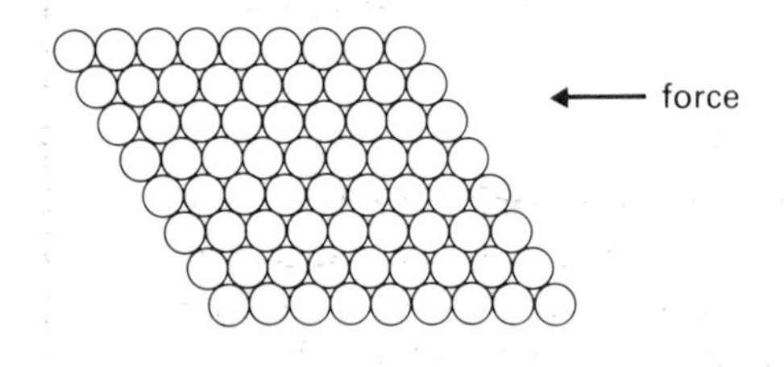
force

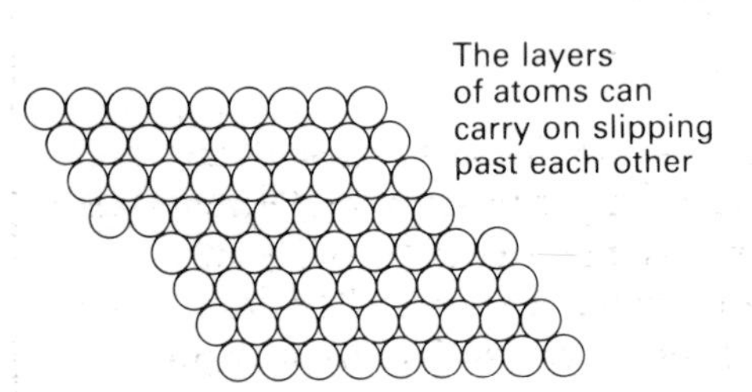
The layers
of atoms can
carry on slipping
past each other

Figure 18.12 When force is applied to a metal, the layers of atoms can slip over each other

The **physical properties** of metals are described in chapter 11:

* they usually have high densities
* they have high melting points and boiling points
* they conduct heat and electricity
* they are malleable and ductile.

Knowing about the structure helps to explain some of these properties.

Metals have high densities because the atoms are packed closely together in giant structures. The most dense metals usually have close packed structures. Can you suggest other factors which might affect the density?

Metals can be flattened and pulled into wires because layers of atoms in grains can slip over each other. Forces applied to the metal make this happen (see figure 18.12). Slip stops at grain boundaries.

Metals have high melting points, and the energy needed to melt metals is high, because the atoms are packed closely and the bonds holding them together are very strong. In a simple model of the bonding in metals, the outer electrons of the atoms are pictured as being free to move in the structure, leaving positive metal ions behind. The electrons attract all nearby positive ions and so pull them

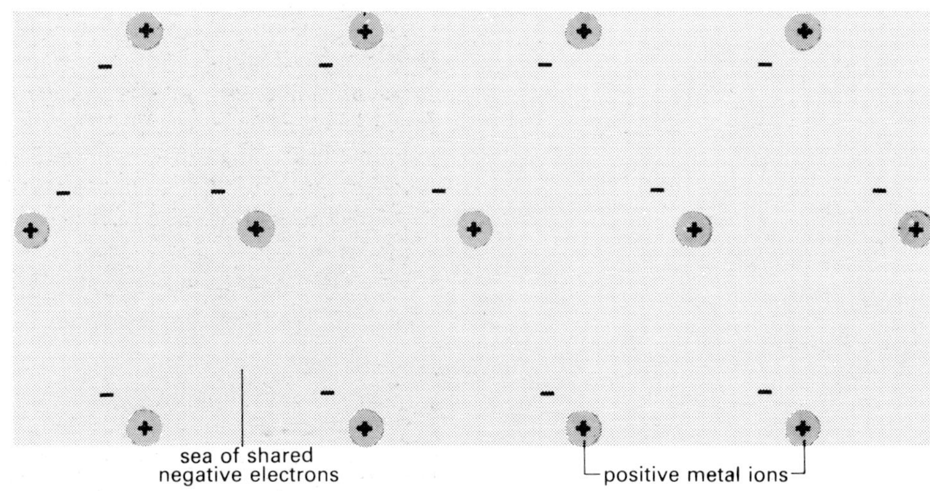

Figure 18.13 Metallic bonding

together. A metal crystal consists of positive metal ions held together by a sea of negative electrons (see figure 18.13).

Metals conduct electricity because the electrons can move about inside the structure. An electric current passing through a metal wire is a flow of electrons.

Metal structures are not perfect! A bubble raft is a good two-dimensional model of a metal and shows how some faults can arise in metal structures.

Experiment 18b
A bubble raft model of a metal

Figure 18.14

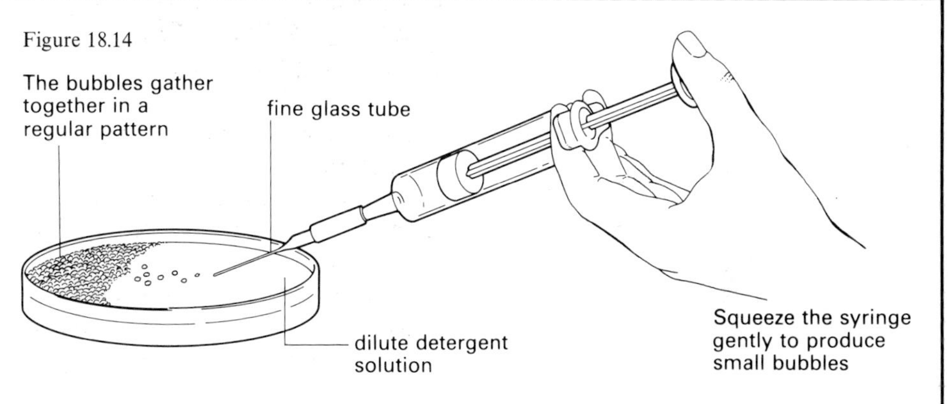

Figure 18.14 shows a bubble raft being made by bubbling air through a capillary tube into dilute detergent solution.

Results

A completed bubble raft is shown in figure 18.15. The bubbles represent atoms in a layer. Can you find grain areas where the bubbles are packed together in an orderly way? As in real metals, the grain areas are of different sizes. Can you see any grain boundaries, where lines of bubbles do not fit together properly, separating one grain area from another? Can you see rows of bubbles which seem to stop suddenly or are pushed out of line, disrupting the regular arrangement? What seems to start these *dislocations*? Can you see areas where there is no regular pattern? What causes these areas?

Figure 18.15 A bubble raft

Discussion

In real metal structures, dislocations occur that are similar to those in the bubble raft. Holes where atoms are missing, or impurity atoms of different sizes, can disrupt the regular arrangement and reduce the sizes of the grain areas. There may be millions of dislocations in each cubic centimetre. They are important because they affect the properties of a metal such as its strength, hardness and malleability.

Knowing about the structure of a metal is obviously important to engineers. They may need metals with particular properties for specific jobs. By controlling the grain size, metals can be modified to give the properties which are needed. One simple relationship is that structures in which grains are small, are stronger, harder and more brittle than those with larger grains.

Questions

1 Design and describe experiments which could be carried out to investigate the strength, hardness and malleability of some metals.
2 In terms of strength, hardness, malleability and ductility, what characteristics should metals have for making the following: (a) bridges, (b) hammer heads, (c) ornamental vases, (d) electrical wires, (e) engine blocks, (f) intricately-designed gates?

Making alloys with other metals is one of the most common ways of changing the properties of metals. Alloys are formed by mixing molten metals together and allowing them to cool.

Alloying often results in a metal which is stronger than any of its components. Brass and bronze are two familiar alloys. Both contain copper, but in brass there is also 40 per cent zinc, whilst bronze contains about 10 per cent tin. They are both stronger and harder than the pure metals and are used for making gears, ships' propellers and rudders, bearings, and statues. Aluminium is a low density metal which is not very strong. In an alloy with 4 per cent copper and smaller amounts of other elements, it gives a metal which combines strength and lightness and is ideal for aircraft building.

Figure 18.16 Some uses of alloys
(a) A large propeller made of a special bronze
(b) Different coloured alloys are used to make coins
(c) Lightweight aluminium alloys are used in aircraft

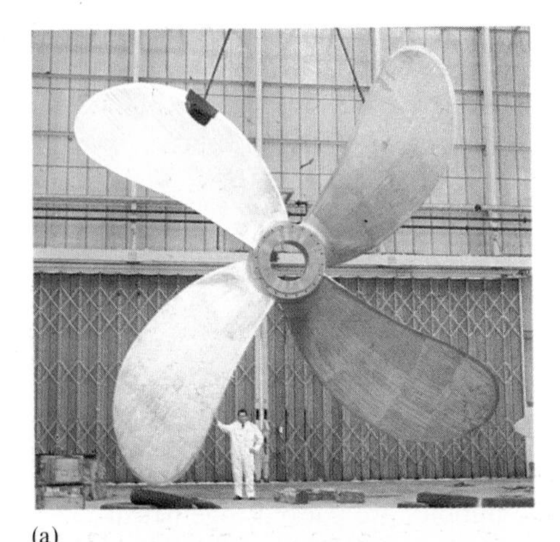

(a)

(b)

(c)

Questions

1 What alloys are used in coinage metals? What are the properties of these metals?
2 Solder is an alloy. What metals does it contain? What are the special properties of solder?

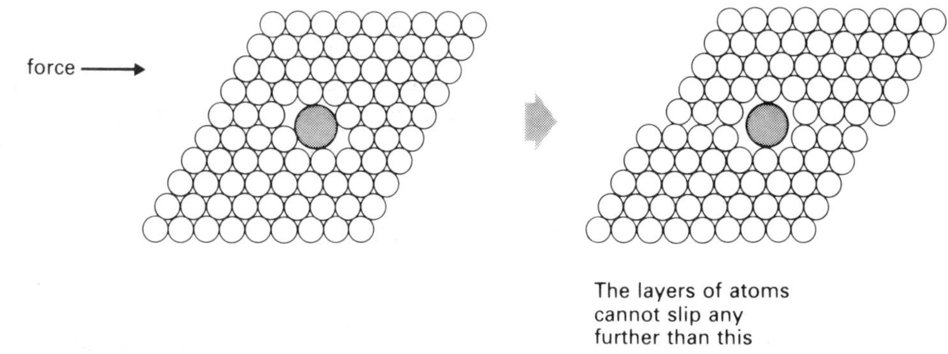
force
The layers of atoms cannot slip any further than this

Figure 18.17 When force is applied to an alloy, the impurity atoms stop the layers from slipping very far

The 'impurity' atoms reduce slip between layers by 'keying' them together as shown in figure 18.17, and also help in reducing grain size. The strength increases because slip cannot occur across grain boundaries. As strength is increased, the malleability usually decreases. In practice, the most suitable proportions of metals, to give the best compromise of properties, are worked out by experiment.

The most widely used alloy is steel. It is called an alloy even though in many steels the 'impurity' is carbon. Cast iron produced in the blast furnace contains 2–4 per cent carbon. It is very hard and brittle because flakes of graphite form in the structure and act like tiny cracks. Pure iron is softer and has little strength. It is suitable for decorative work. In between are the steels with carefully controlled carbon contents, with uses ranging from nails and screws to car bodies and bridges. The production of steel is described in chapter 13. Mild steel contains about 0.3 per cent carbon, in the form of iron carbide. Crystals of iron carbide in the structure make this steel strong, and yet it is still malleable. It is used mainly for car bodies and bridges. As the carbon content is increased, the steel becomes stronger and harder, and is suitable for railway lines.

Special alloy steels contain other metals in addition to carbon. High-speed steel used for drills and cutting tools contains tungsten. It remains hard even when it is very hot. Stainless steel for cutlery and surgical instruments contains nickel and chromium. Steel for spanners and torque wrenches contains chromium and vanadium. A small amount of molybdenum in steel prevents brittleness. In fact, by

Figure 18.18

(a) Cast iron bridge at Ironbridge, Salop

(b) Wrought iron gateway at Littlecote House, Wiltshire

Figure 18.19 Some uses of steel
(a) Most 'tin' cans are actually made of steel
(b) Medical instruments are made of stainless steel
(c) Construction of an oil rig, using steel girders

(a)

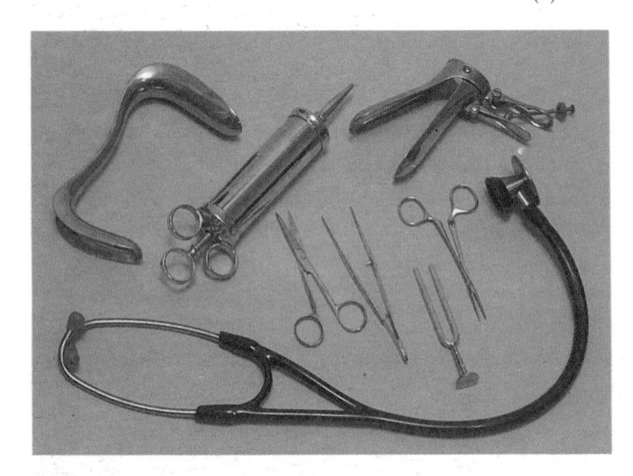

(b)

(c)

using varying amounts of about ten possible alloy metals, steels can be 'tailor-made' for almost any purpose (see figure 18.19).

The structure and properties of metals can also be changed by 'working' them – by beating or rolling – and by heat treatment. Try bending a piece of copper or lead backwards and forwards. Does it become harder and stronger before it finally breaks? What effect do you think the working has had on the metal structure? Work hardening can be carried out when the metal is cold or hot.

Experiment 18c shows how the properties of a metal can change with heat treatment.

Experiment 18c
The effect of heat treatment on steel

Three simple experiments are carried out using steel needles:
(a) A needle is heated until red-hot and then cooled slowly above the flame.
(b) A needle is heated to red heat and then plunged into cold water.
(c) A needle is treated as in (b), and then reheated until a blue colour appears. It is then slowly cooled.

Results
Needle (a) becomes soft and very easy to bend. Needle (b) becomes harder to bend than the original. It is brittle and breaks. Needle (c) has become springy like the original needle.

Discussion
The effects of heat treating steel have been known for many centuries but the explanations are relatively recent and quite complicated. The structure of iron crystals is different above and below 900 °C. The change is normally reversible, but when hot steel is cooled rapidly (quenched), the structure is 'frozen' in the high temperature form. This form is hard and brittle. When the steel is tempered (reheated and cooled slowly) the structure returns to its low temperature form. The changes in properties are due to structure changes.

Questions

1 What three processes have been mentioned in this chapter which will increase the strength of a metal?

2 List some items you have come across which are made of steel. What do you think are the main characteristics of the steel for each use?

Figure 18.20 Quenching: hot steel is sprayed with cold water as it emerges from the rolling mills

Figure 18.21 Cutting some hot-rolled steel to length

Hot steel is sprayed with water to quench it (see figure 18.20). Very hard steels used for railway lines are produced in this way. Hot rolling to give steel plates and girders combines work hardening and slow cooling from high temperatures. What will be the main characteristics of this steel? What might it be used for?

Summary

The following terms are not included in the glossary at the end of this book: X-ray diffraction, etching, close packing, face-centred cubic structure, body-centred cubic structure, dislocation, malleable, quenching. Write your own definitions for the terms in a form suitable for inclusion in a glossary.

19 Covalent molecules and giant structures

19.1 Molecular or giant structure?

Melting points and boiling points are important clues to the structure of substances.

Most non-metal elements and compounds have low melting points and boiling points. They may be gases, liquids or solids at room temperature. If they are solid, they are usually easily vaporized on gentle heating. They have molecular structures. Molecules are particles which usually consist of a few atoms joined tightly together. The molecules are easily separated from each other though, because the attractions between them are small. Bromine vaporizes easily, but the vapour still consists of bromine molecules, not atoms (see figure 19.1). Molecular compounds behave rather like nuts and bolts. You can think of a nut and bolt screwed together as a molecule. A box of these 'molecules' is easy to separate, but the nuts and bolts stay screwed together. More effort is needed to separate the nuts from the bolts.

Questions

1 List the melting and boiling points of the elements, oxygen, bromine, helium, phosphorus, carbon and sulphur.
(a) Show whether they are gases, liquids or solids at room temperature.
(b) Show which have molecular and which have giant structures.

2 List the melting and boiling points of hydrogen chloride, silicon tetrachloride, silicon dioxide, carbon dioxide and hexane.
(a) Show whether they are gases, liquids or solids at room temperature.
(b) Show which have molecular and which have giant structures.

Figure 19.1 Representations of the arrangements of the molecules in bromine liquid and vapour

A few non-metals and non-metal compounds have very high melting and boiling points. These have **giant structures**. The atoms are tightly joined together, but in a massive network. It is difficult to separate particles which are held tightly from all sides.

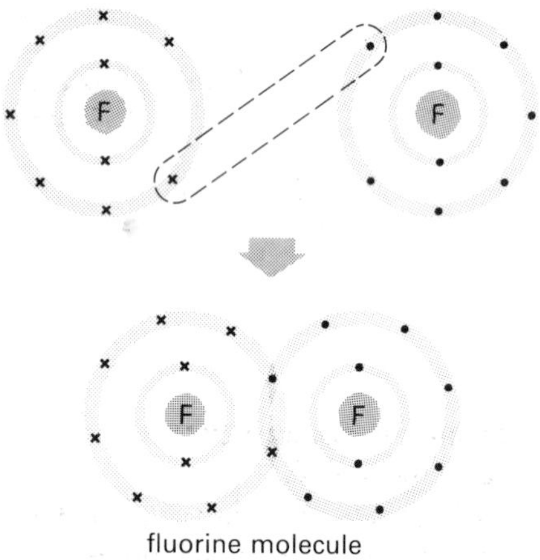

19.2 Covalent bonding

Non-metal atoms are joined together by covalent bonds. Outer electrons of the atoms are shared.

A fluorine atom has an incomplete shell of seven electrons (see chapter 17). When two fluorine atoms combine to form a molecule, they share two electrons. Each atom then has a complete shell of eight electrons. The electron structure of each atom is now like that of neon, the nearest noble gas. The hungry lions have been satisfied!

fluorine atoms

fluorine molecule

Figure 19.2 Formation of a fluorine molecule: the electrons are shown as dots and crosses

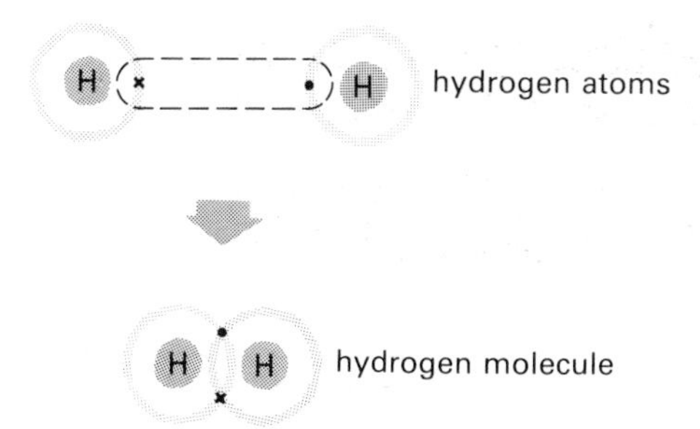

Figure 19.3 Formation of a hydrogen molecule

A shared pair of electrons, called a covalent bond, is represented by '—'. A fluorine molecule can be written

F—F

This is the structural formula. It shows how the atoms in the molecule are bonded together. The molecular formula of fluorine is F_2. It shows how many of each type of atom there are in the molecule.

Figure 19.3 shows the electrons in a hydrogen molecule. After sharing electrons, both hydrogen atoms have complete outer electron shells. What are the structural and molecular formulae of hydrogen?

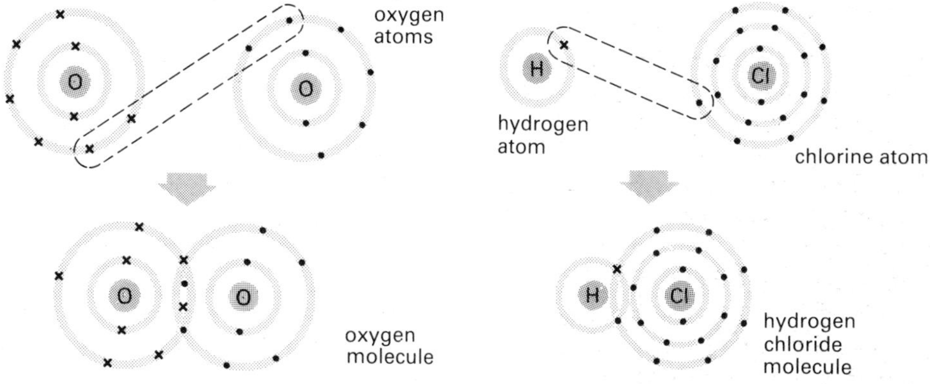

Figure 19.4 Formation of an oxygen molecule

Figure 19.5 Formation of a molecule of hydrogen chloride

In an oxygen molecule, two pairs of electrons must be shared, because each oxygen atom needs two extra electrons to have a complete shell. A 'double bond' is formed when two pairs of electrons are shared. It is represented by '='. The structural formula of oxygen is

O=O

The molecular formula is O_2.

In non-metal compounds, covalent bonds hold different atoms together.

A molecule of hydrogen chloride has one atom of hydrogen covalently bonded to an atom of chlorine as shown in figure 19.5 In the molecule both atoms have complete outer shells. Figure 19.6 shows the covalent bonds in methane. What are its structural and molecular formulae?

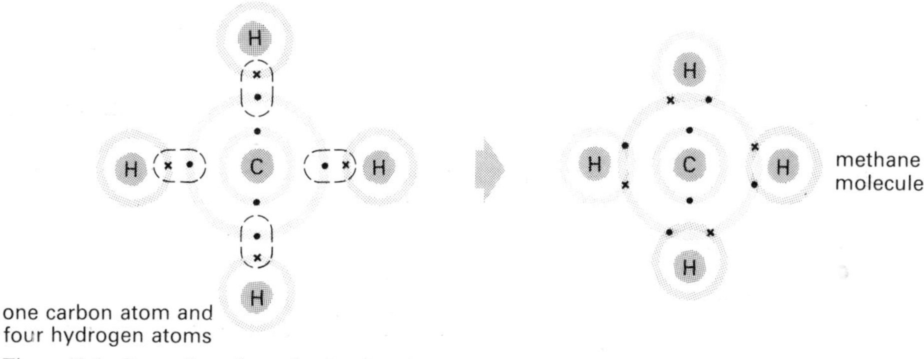

Figure 19.6 Formation of a molecule of methane

Atoms in molecules are held tightly together because the negative electrons between the atoms attract the positive nuclei and pull them together.

Figure 3.15 on p. 25 shows the 'combining powers' (**valencies**) of some non-metal atoms. You should now see that the combining power is the same as the number of covalent bonds formed by an atom. This is the same as the number of electrons which it needs to gain to have a complete shell.

Questions

1 Draw electron diagrams and structural formulae for each of the following molecules: (a) Cl_2, (b) N_2 (this needs three shared pairs of electrons), (c) H_2O, (d) NH_3, (e) CCl_4.

2 Draw electron diagrams and molecular formulae for each of the following:

(a) O=C=O

```
        H   H
        |   |
(b) H—C—C—H
        |   |
        H   H

        H   H
        |   |
(c) H—C—C—Cl
        |   |
        H   H

    H   H
    |   |
(d) C=C
    |   |
    H   H
```

Can you name these compounds?

3 Why do the noble gases not form molecules, such as Ne_2 and He_2?

19.3 Covalent molecules

Models of some simple molecules are shown in figure 19.7. Can you work out the colour code? Ball-and-spring models show which atoms are bonded together and the shape of the molecule, but they give little information about the space occupied by the atoms. Space-filling models help us to appreciate this.

Covalent molecular substances can be recognized by their properties. We have already seen that they have low melting and boiling points. In experiment 19a the properties of iodine (a non-metal element) and naphthalene (a non-metal compound) are studied.

<table>
<tr><th>Name</th><th>Molecular formula</th><th>Structural formula</th><th>Ball and spring model</th><th>Space-filling model</th></tr>
<tr><td>Hydrogen</td><td>H_2</td><td>H—H</td><td> 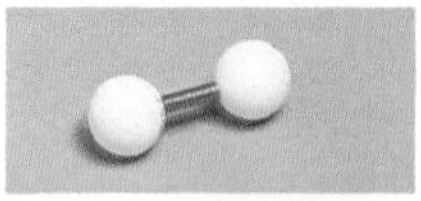</td><td> </td></tr>
<tr><td>Hydrogen chloride</td><td>HCl</td><td>H—Cl</td><td> </td><td> </td></tr>
<tr><td>Water</td><td>H_2O</td><td>O
/ \
H H</td><td></td><td></td></tr>
<tr><td>Methane</td><td>CH_4</td><td>H
|
H—C—H
|
H</td><td></td><td></td></tr>
<tr><td>Ethane</td><td>C_2H_6</td><td>H H
| |
H—C—C—H
| |
H H</td><td></td><td></td></tr>
<tr><td>Propane</td><td>C_3H_8</td><td>H H H
| | |
H—C—C—C—H
| | |
H H H</td><td></td><td></td></tr>
<tr><td>Butane</td><td>C_4H_{10}</td><td>H H H H
| | | |
H—C—C—C—C—H
| | | |
H H H H</td><td></td><td></td></tr>
<tr><td>2-methylpropane</td><td>C_4H_{10}</td><td>H H H
| | |
H—C—C—C—H
| | |
H | H
C
H / | \ H
H</td><td></td><td></td></tr>
<tr><td>Ethanol</td><td>C_2H_5OH</td><td>H H
| |
H—C—C—O—H
| |
H H</td><td></td><td></td></tr>
</table>

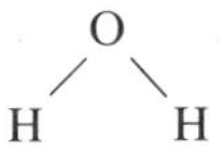

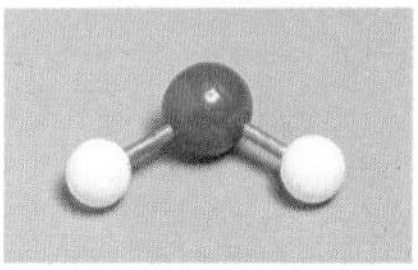

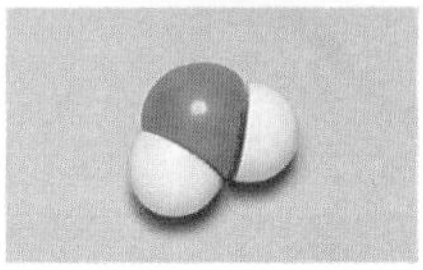

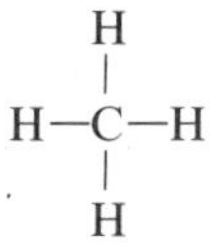

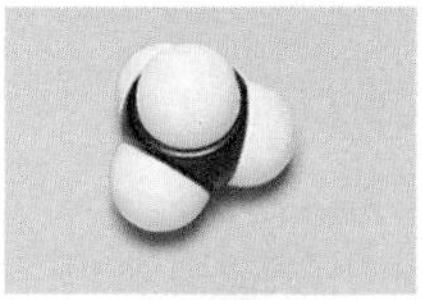

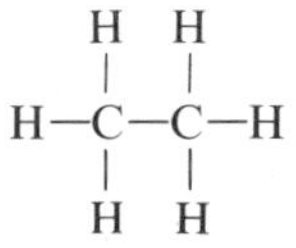

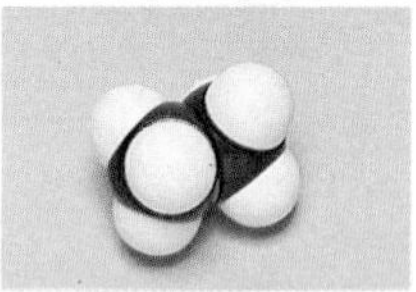

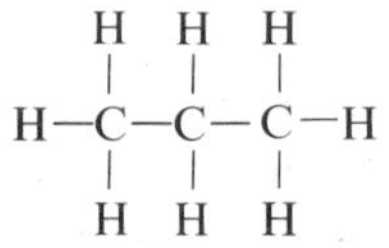

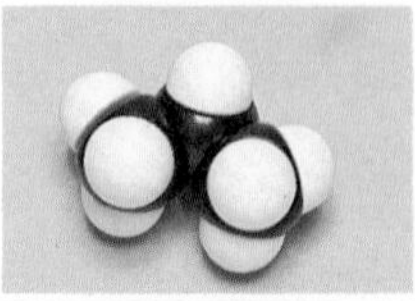

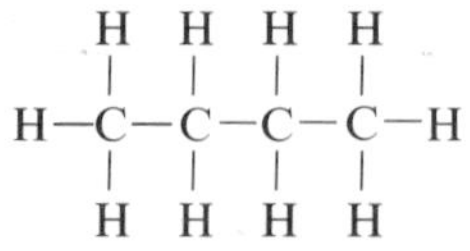

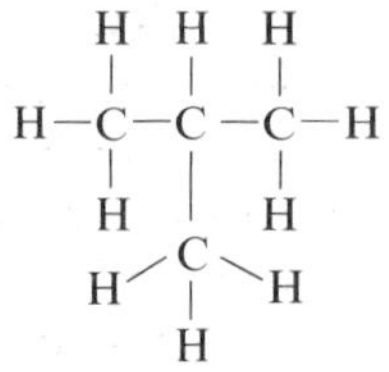

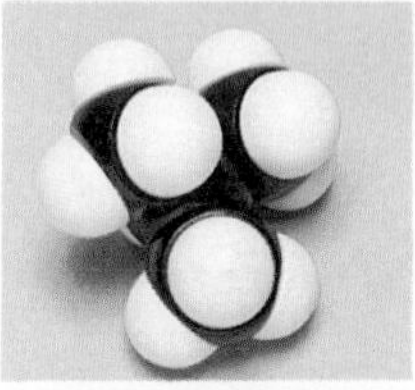

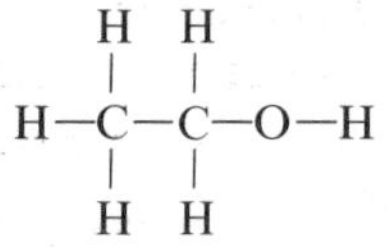

Figure 19.7 Different ways of representing some simple molecules

Experiment 19a
Investigating the properties of iodine and naphthalene

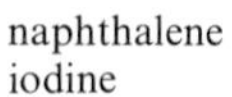

naphthalene iodine

cyclohexane

The appearance and smell of both substances are studied, along with the effect of heat, their solubilities in water and in cyclohexane (a petrol-like solvent), and the conductivities of the substances and of their solutions.

Results
Typical results are shown in figure 19.8.

Property	**Iodine**	**Naphthalene**
Appearance	Flaky grey crystals	White crystals
Smell	Slight disinfectant smell	Smell of mothballs
Effect of gentle heat	Vaporizes easily, giving a purple vapour	Vaporizes easily, subliming onto cooler part of tube
Does the pure substance conduct electricity?	No	No
Does it dissolve in water?	Slightly	No
Does it dissolve in cyclohexane?	Yes, giving a purple solution	Yes, giving a colourless solution
Does the solution in cyclohexane conduct electricity?	No	No

Figure 19.8

Discussion
Iodine and naphthalene are examples of solids which are easily vaporized. Many covalent molecular compounds have a smell – think of ethanol (alcohol), trichloroethane (a dry-cleaning solvent) and hexane (in petrol). Why can you smell them?

The two substances are not conductors of electricity because covalent molecules are neutral and the electrons are in fixed positions in the molecules. They are not free to move as they are in metals.

Cyclohexane is a covalent molecular solvent. Solutes like iodine and naphthalene are soluble in it because they have similar structures.

The main properties of covalent molecular substances are:

- they have low melting points and boiling points
- the energy needed to separate the molecules is low
- they often have a smell
- they do not conduct electricity in any state when pure
- they are more soluble in solvents like cyclohexane than in water, and the solutions do not conduct electricity.

Of course, rules are made to be broken. Some covalent molecular compounds are very soluble in water. For example, sugar dissolves to give a solution which does not conduct electricity. Hydrogen chloride, sulphur dioxide and silicon tetrachloride are molecular, and are very soluble. They react with water to produce acidic solutions which conduct electricity. The reaction of hydrogen chloride with water is described in chapter 24.

19.4 Some non-metal elements

The molecular formulae of some non-metals are shown in figure 3.12 on p. 24. You may recall that most of the common non-metals have molecules containing two atoms. They are *diatomic*.

The halogens

The chemistry of the group of elements which includes chlorine, bromine and iodine is described in chapter 16. Chlorine is a gas at room temperature; bromine is a dark red, heavy liquid which vaporizes easily; iodine is a grey, shiny, flaky solid. Figure 19.1 shows bromine and the arrangement of molecules in the liquid and vapour. Does figure 19.9 help to explain why iodine is a flaky solid? Draw a similar picture to represent chlorine at room temperature.

Figure 19.9 Representation of the arrangement of the molecules in an iodine crystal

Figure 19.10 A molecule of sulphur, S_8. (In this set of models, atoms of oxygen and sulphur are shown in red because both elements are in Group VI; in many other sets, sulphur atoms are shown in yellow.)

Sulphur

Sulphur is a yellow crystalline solid which melts at 115 °C. A sulphur molecule consists of eight sulphur atoms in a ring. Figure 19.10 shows a space-filling model of it. Sulphur exists in two crystalline forms. One type is called orthorhombic sulphur and the other is monoclinic sulphur. Different crystalline forms of the same element are called **allotropes**. Crystals of the two allotropes of sulphur can be made in the laboratory.

Experiment 19b
Making crystals of sulphur

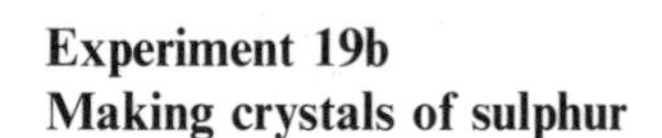

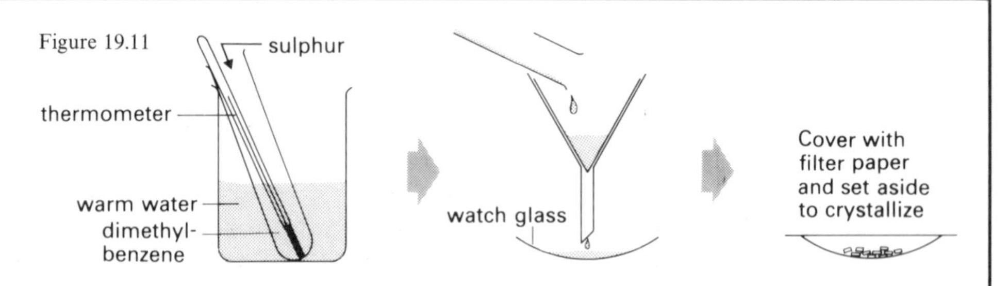

Figure 19.11

(a) Figure 19.11 shows the steps in making orthorhombic sulphur.
(b) A method for making monoclinic sulphur is shown in figure 19.12.

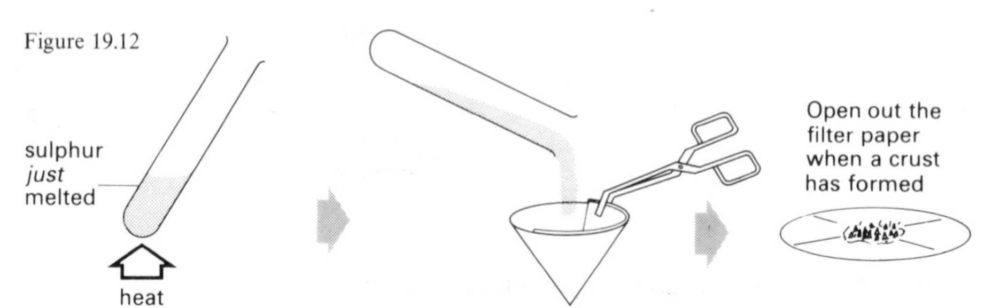

Figure 19.12

Results

The crystals from experiment (a) are usually small, but rhombic shaped crystals can be seen through a magnifying glass. A large orthorhombic crystal is shown in figure 19.13(a). In experiment (b), long needle-like crystals, like those in figure 19.13(b), are obtained.

Discussion

What is the main difference in conditions which gives rise to orthorhombic sulphur in experiment (a) but monoclinic sulphur in (b)? What can you say about the nature of the solvent, dimethylbenzene, which is used to dissolve the sulphur?

Question

The extraction of sulphur from the ground is described in chapter 1.
(a) Who developed the process?
(b) At what temperature does the sulphur come out of the ground?
(c) When the sulphur crystallizes, which allotrope will be formed?

Figure 19.13 (a) Orthorhombic sulphur (b) Monoclinic sulphur

Below 96 °C, orthorhombic sulphur is the stable allotrope. Above 96 °C, monoclinic sulphur is more stable. The allotropes are identical chemically but the crystals have different shapes because the molecules are packed in different ways. A sample of monoclinic sulphur below 96 °C will change slowly to the orthorhombic form.

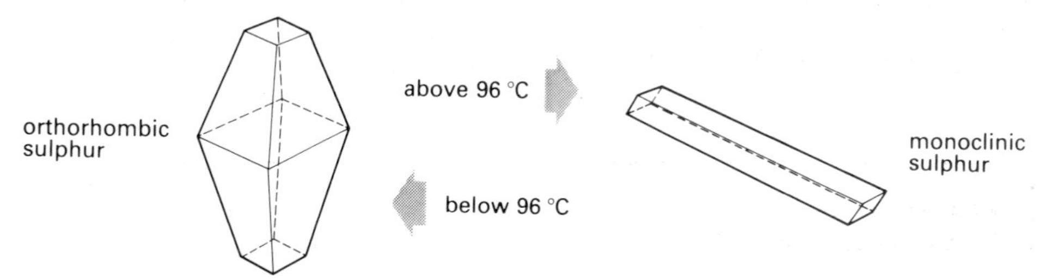

Figure 19.14 Orthorhombic sulphur can be turned into monoclinic sulphur

Interesting changes occur when sulphur is heated. When sulphur first melts, a free-flowing, amber liquid is formed. It contains S_8 rings sliding past each other. After more heating, the liquid becomes viscous and red (see figure 19.15). Bonds in

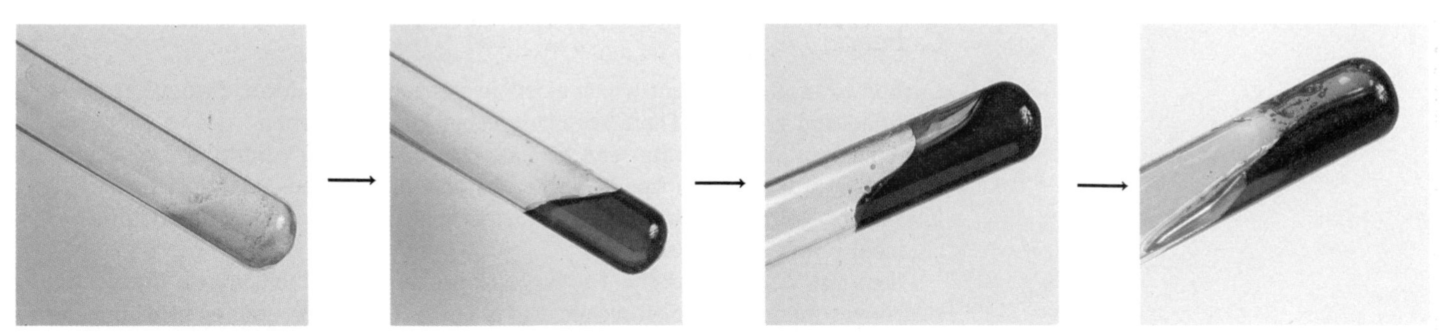

Figure 19.15 The sequence of changes that occurs when sulphur is heated

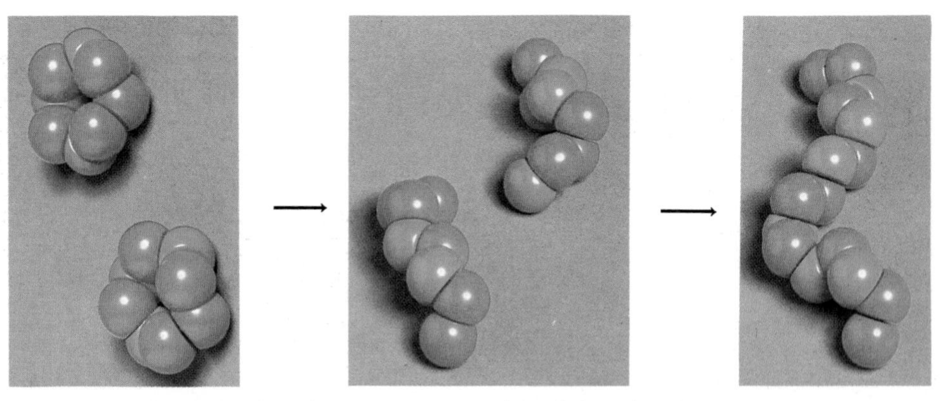

Figure 19.16 The changes in molecular structure as sulphur is heated

the S_8 rings are beginning to break, and chains of atoms are formed. They link up with other chains. Eventually, long chains of sulphur atoms, containing millions of atoms, are present in the liquid. They become tangled and make the liquid very **viscous**.

Near its boiling point, liquid sulphur is very dark, but quite runny again. The long chains are still present. If the liquid is poured rapidly into cold water, the atoms do not have time to change back to rings, and the liquid solidifies with them still in long chains tangled together. Plastic sulphur is formed (see figure 19.17). This is another kind of sulphur, but it is not crystalline. Because there is no order in it, it is described as **amorphous**. It is rubber-like and can be stretched. On standing, it hardens and becomes brittle again. Can you suggest why?

(a)

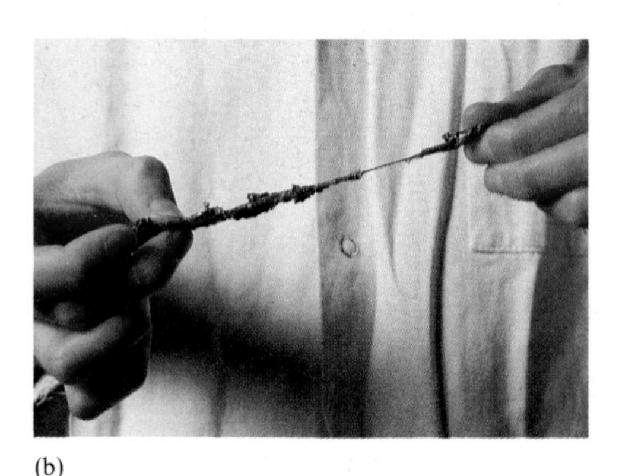

(b)

Figure 19.17 Plastic sulphur
(a) Pouring the molten sulphur into water
(b) Plastic sulphur can be stretched like rubber

Carbon

Carbon is another element which exists in more than one form. Two allotropes are diamond and graphite. There are many other kinds of carbon, such as soot and charcoal. Any of these forms can be shown to be carbon by burning it in oxygen: only carbon dioxide is formed – rather a waste of diamonds, perhaps! In the laboratory, you may have carried out experiments to show that graphite and charcoal both contain carbon.

Diamonds occur naturally. The largest one found was the Cullinan diamond, which was found in 1905. It weighed over 600 g. Diamonds used as gemstones are cut from the rather unattractive crude stones. The largest cut stone is over 100 g. It

(a)

(b)

Figure 19.18
(a) A natural diamond
(b) Some cut diamonds: a rare naturally-occurring black diamond, and two that have been coloured by irradiation

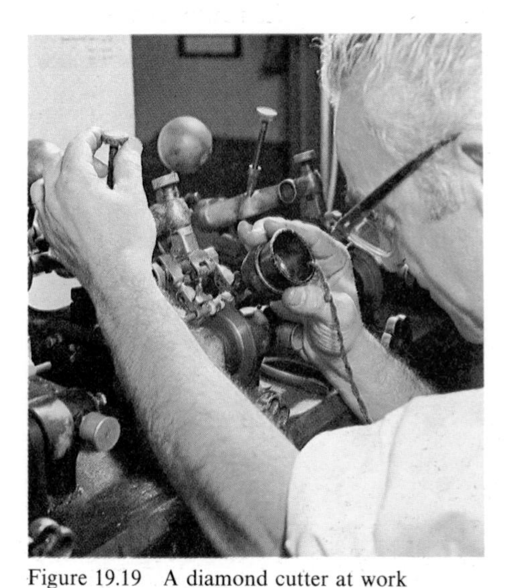

Figure 19.19 A diamond cutter at work

is now in the royal sceptre. Cutting diamonds is technically a very skilled job, and can only be done by a diamond cutter who knows the structure of the diamond.

Graphite also occurs naturally. It is much cheaper than diamond, but has important uses. By heating diamond in the absence of oxygen, it can be changed to a worthless black powder. There are interesting stories about some early attempts to reverse the process, and to change graphite and charcoal into diamond. This can now be done at extremely high pressures and temperatures. The synthetic stones have industrial uses only: they are not acceptable as gem stones.

Diamond and graphite are very different. Diamond is the hardest substance known. It is 70 times harder than corundum, which is used to make emery paper. It is used for glass cutting and engraving. Cutting tools and drills used for cutting stone are edged with small diamonds. A diamond-tipped saw will cut through anything! Some of these uses are shown in figure 19.20. Graphite, on the other hand, is very soft and flaky. It is used in pencil 'leads' and as a lubricant in 'graphited' penetrating oils. Because of its unusual property of conducting electricity, it is used in dry cells and, more important, as electrodes in industrial electrolysis processes such as the extraction of aluminium. Some of these uses are shown in figure 19.21.

Figure 19.20 Some uses of diamonds

(a) Around the edge of a cutting wheel for slicing rock specimens

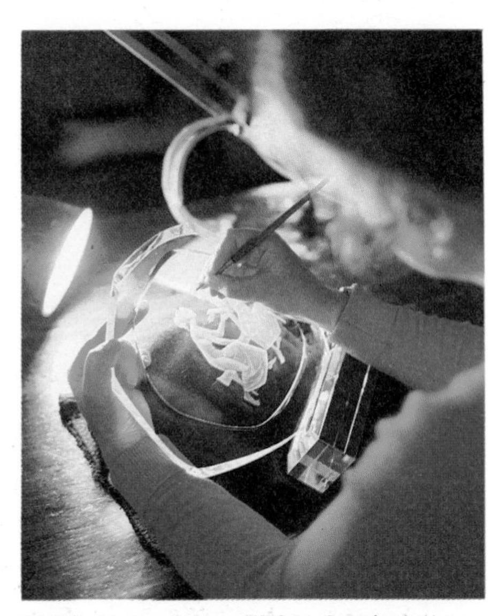

(b) On the tip of a 'pencil' for engraving glass

(c) In a drill bit for cutting through rock

Figure 19.21 Some uses of graphite

(a) In pencil 'leads'

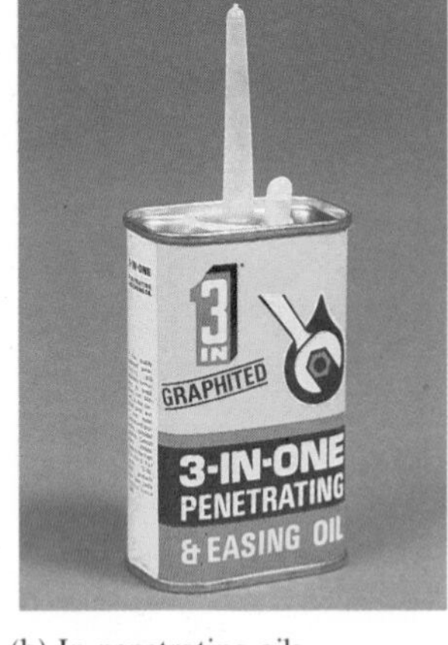

(b) In penetrating oils

An obvious difference between carbon (diamond, graphite or any other form) and other non-metals is the very high melting point. Find the melting points of diamond and graphite in table 1 on p. 336. Diamond and graphite have giant structures but the atoms are covalently bonded together.

Figure 19.22 shows a ball-and-spring model of part of the structure of diamond. Each carbon atom has bonds to four other carbon atoms arranged tetrahedrally around it. A diamond crystal the size of a pin-head will contain millions upon millions of atoms, all firmly held together by strong covalent bonds. You can appreciate why diamond is so hard and has a high melting point. Figure 19.23

Figure 19.22 The crystal structure of diamond

Questions

1 With the help of table 1 on p. 336, draw up a table comparing the physical and chemical properties of diamond with those of graphite.
2 Compare the structures of graphite and iodine.
(a) In what ways are they similar?
(b) What similar properties do graphite and iodine have?
3 Pencil 'leads' are a mixture of graphite and clay.
(a) Pencil leads got their name because graphite was confused with lead metal. How are graphite and lead similar? How do they differ?
(b) How is it possible to get pencil leads ranging from very soft to very hard?

Figure 19.23 A giant structure!

Figure 19.24 The layered structure of graphite

shows an unusual view of a model of the diamond structure. You can picture the giant structure. Can you see the tetrahedral arrangement of atoms?

The graphite structure (see figure 19.24), as you would expect, is different. Carbon atoms are held tightly together in layers. Each carbon atom has three covalent bonds to other carbon atoms. Hexagons of carbon atoms are formed. Weak bonds hold every layer to the one above and the one below. The forces between layers are weak, so they can slide past each other. This explains why graphite is soft and flaky. The electrons between the layers are free to move, as in metals, allowing graphite to conduct electricity in the direction of the layers. The layers themselves are very strong. Graphite fibres with the layers arranged along the fibre are stronger than steel. They are used as reinforcement in metals. Graphite fibres have even been used to reinforce broken bones in animals.

19.5 Some covalent compounds based on carbon and silicon

Carbon and silicon are both in group IV of the periodic table. They have four electrons in the outer shells of their atoms and form four covalent bonds with other atoms. Silicon is a brown–grey powder and has a structure similar to diamond. Find its melting point in table 1 on p. 336. Does its melting point suggest that it has a giant structure?

Most molecules are simple, containing just a few atoms. Carbon forms many simple molecules, such as methane, CH_4, and carbon dioxide, CO_2, but it can also form some very large molecules containing thousands of atoms. Carbon atoms can link together to form an amazing variety of chains and rings. These molecules are important in materials like plastics, and in living cells. Substances like polythene, starch, rubber and insulin have very large covalent molecules. Figure 19.25 shows a model of a molecule of polythene which might contain up to 50 000 atoms. It is a **polymer** – a giant molecule built up of repeating units. Although the molecule is large, the melting point range is still fairly low. There is no exact melting point because a piece of polythene contains molecules of all different lengths.

Polythene is easily stretched. Is this evidence for a tangled arrangement of molecules? What do you think happens to the molecules when polythene is stretched? Some plastics are difficult to stretch. Can you suggest why? Bakelite is a rigid material used for electrical fittings like plugs and switches.

Plastics are described in more detail in chapter 32, while starch, an important naturally-occurring large molecule, is discussed in chapter 33.

Carbon dioxide is a simple molecule. It is a colourless gas. We might expect silicon dioxide to be similar – but chemistry is not always that predictable! Sand,

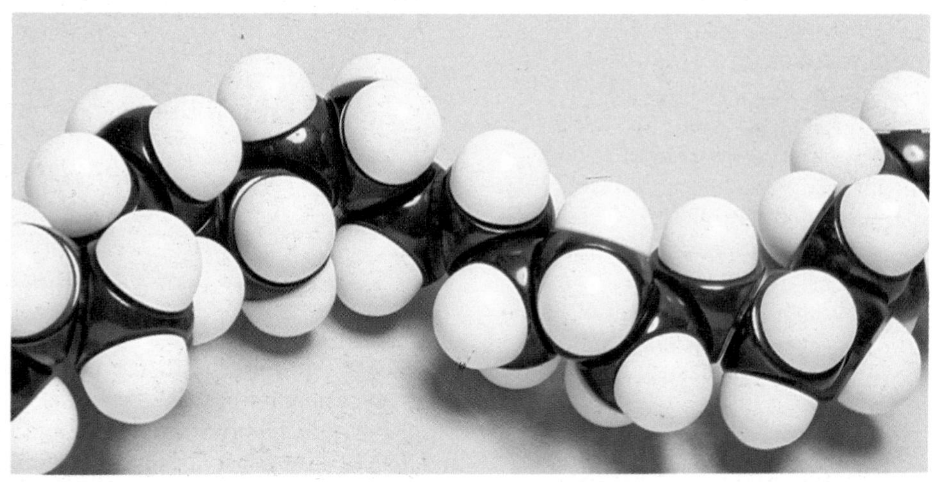

Figure 19.25 A very long molecule!

Figure 19.26 Quartz

(a) Some naturally-occurring crystals

(b) The crystal structure

flint and quartz (see figure 19.26) are common forms of silicon dioxide found in rocks. They are hard substances. Flint was discovered by Stone Age people. It was a very good material for spear heads.

Look at the melting point of quartz in table 2 on p. 337. Quartz must have a giant structure. Part of it is shown in figure 19.26. Are there any ways in which it resembles diamond?

Figure 19.27 Asbestos

(a) Fibres of asbestos

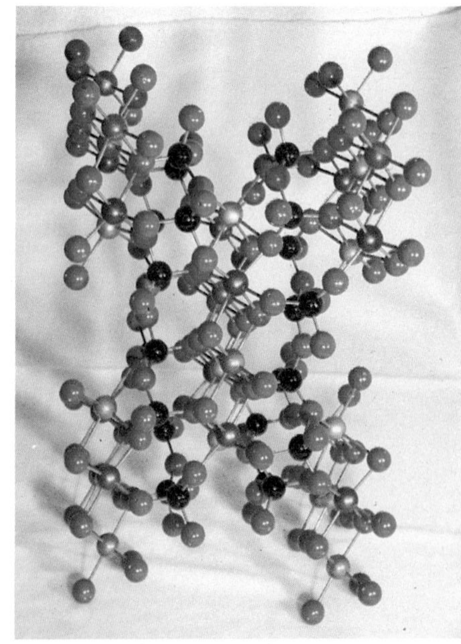

(b) The crystal structure

Silicon and oxygen are also combined in silicate minerals. Asbestos and mica are two examples of these. Asbestos is a fibrous material (see figure 19.27). It used to be used in fireproof insulation and brake linings. Mica (see figure 19.28) is like slate and cleaves very easily to give thin films. Can you see a link between this property and the structure shown?

Questions

1 Using table 2 on p. 337, draw up a table comparing the formulae, properties and structures of carbon dioxide and silicon dioxide.

2 Tetrachloromethane, CCl_4, and silicon tetrachloride, $SiCl_4$, are colourless liquids. CCl_4 does not mix with water but $SiCl_4$ reacts violently, giving a steamy acidic gas and leaving a white powder.

(a) Which is more typical of covalent molecular compounds?

(b) If the white solid is silicon dioxide, what must the gas be?

(c) Write an equation for the reaction of $SiCl_4$ with water.

Figure 19.28 Mica

(a) A crystal of mica separating into thin flakes

(b) The crystal structure

Summary

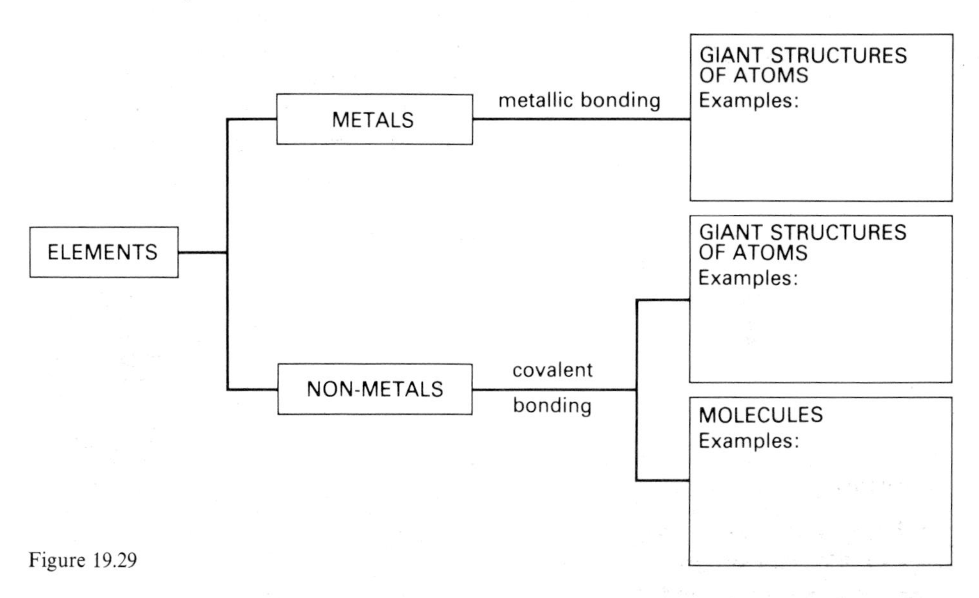

Figure 19.29

Figure 19.29 is a summary of the bonding and structure of elements. Copy the diagram and complete it by adding examples. Among your examples include carbon, chlorine, copper, lead, magnesium, nitrogen, oxygen and zinc.

20 Ions and ionic crystals

Figure 20.1 Hot sodium being lowered into chlorine

Sodium chloride, lead(II) bromide, copper(II) sulphate and potassium nitrate are well-known compounds. They consist of metals combined with non-metals. Look up their melting points in table 2 on p. 337.

Compounds like this are all solids at room temperature and they have high melting and boiling points. Magnesium oxide and calcium oxide are used to line furnaces because they have such high melting points. They all have **giant structures**. They conduct electricity when molten or when dissolved in water because they contain charged particles called **ions**.

Most compounds of metals with non-metals have giant structures in which ions are held tightly together by ionic bonds. Ions are formed when atoms lose or gain electrons. Metals, on the left-hand side of the periodic table, have just a few electrons in their outer shells. Non-metal atoms have almost complete shells (see chapter 17). When compounds form between metals and non-metals, the metal atoms lose their outer electrons to the non-metals. In this way, the electron structure of each atom becomes like that of the nearest noble gas.

Sodium burns in chlorine with a bright yellow flame (see figure 20.1). The elements combine to give sodium chloride:

$$2Na(s) + Cl_2(g) \longrightarrow 2NaCl$$

In the reaction, each sodium atom loses the electron from its outer shell. The electron is taken by a chlorine atom. Sodium atoms become sodium ions, while chlorine atoms become chloride ions. Notice that in this reaction the bond in the chlorine molecule must be broken.

A sodium ion has a charge of 1+ because, after losing an electron from a neutral atom, there is one more positive proton than there are electrons. The chlorine atom gains an electron, giving an ion with a charge of 1−. The reaction can be shown in ionic equations:

$$2Na \longrightarrow 2Na^+ + 2e^-$$

$$Cl_2 + 2e^- \longrightarrow 2Cl^-$$

sodium atom, Na

Na

Cl

chlorine atom, Cl

sodium ion, Na^+

Na

Cl

chloride ion, Cl^-

Figure 20.2 Formation of an ion pair of sodium chloride

When calcium fluoride is formed, each calcium atom loses both its outer-shell electrons, becoming a calcium ion, Ca^{2+}. Two fluorine atoms can each gain one electron, becoming fluoride ions, F^-. For each calcium ion there are two fluoride

Questions

1 Write equations to show the formation of ions from the following atoms: (a) potassium, (b) magnesium, (c) aluminium, (d) oxygen, (e) bromine, (f) sulphur.

2 Draw electron diagrams to show the formation of:
(a) lithium chloride,
(b) magnesium chloride,
(c) calcium oxide and
(d) magnesium oxide.

3 Look up the melting points of magnesium oxide, calcium oxide, sodium chloride and potassium chloride in table 2 on p. 337. Suggest why the melting points of magnesium oxide and calcium oxide are very much higher than those of sodium chloride and potassium chloride.

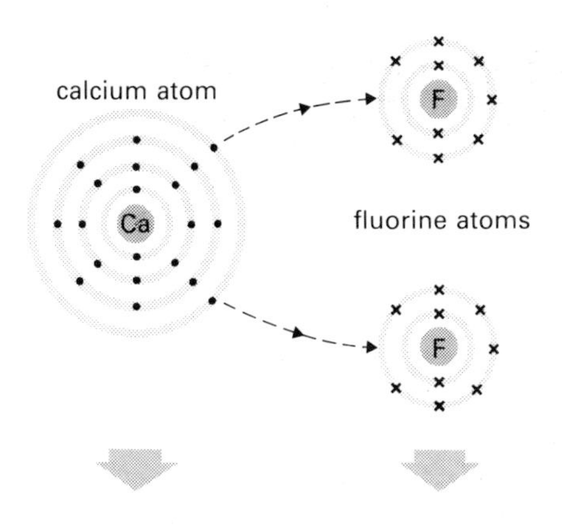

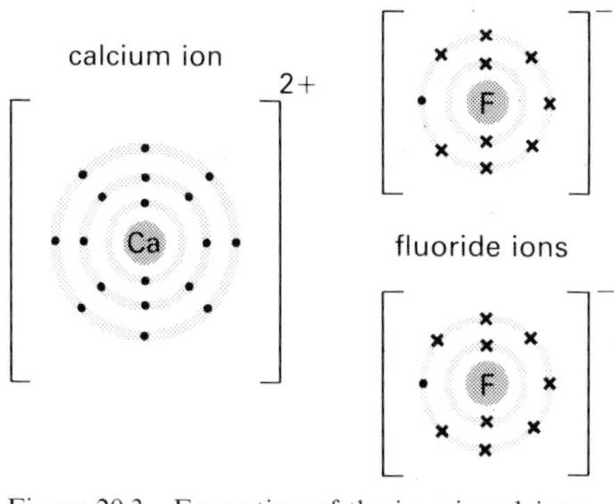

Figure 20.3 Formation of the ions in calcium fluoride

ions. The formula of calcium fluoride is CaF_2. Equations to describe the changes are:

$$Ca \longrightarrow Ca^{2+} + 2e^-$$

$$F_2 + 2e^- \longrightarrow 2F^-$$

The charges of some common ions are given in table 8 on p. 343. These can be used to work out formulae of compounds as shown on p. 26.

In an ionic structure, the oppositely charged ions attract each other. The ionic bond is an electrostatic attraction. It is very strong.

The main properties of compounds with a giant ionic structure are:

* they have high melting points and boiling points
* the energy needed to separate the particles is high
* they are usually crystalline, and the crystals can be cleaved
* many are soluble in water
* they do not conduct electricity when solid, but are good conductors when molten or in solution.

Crystal shape depends on how the ions are arranged in the giant structure. The photograph on p. 117 shows a crystal of sodium chloride. Compare the shape of the crystal with the structure, shown in figure 20.4. The space-filling model gives an idea of the relative volumes occupied by each ion. In this simple cubic pattern, the ions closest to each other have opposite charges. The ball-and-stick model helps to show how the layers of ions fit together.

Figure 20.4 Models of the crystal structure of sodium chloride. (The green spheres represent chloride ions in both models.)

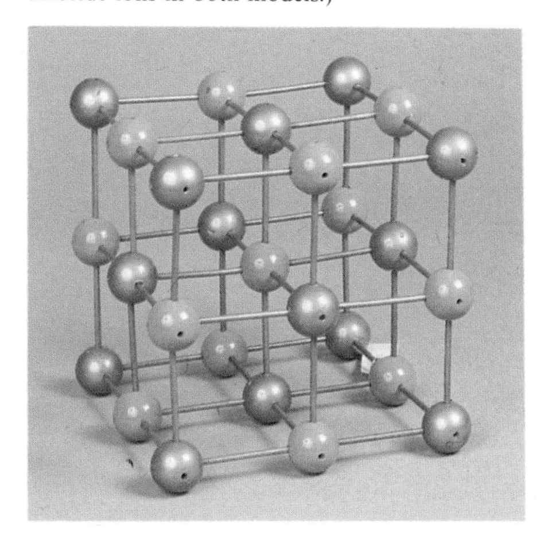

(a) Ball-and-stick

(b) Space-filling

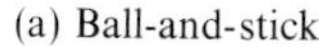

A sharp tap with a hammer on a blade held parallel to one of the faces of a crystal of salt or calcite causes the crystal to split cleanly (see figure 20.5). This is

Figure 20.5 Cleaving a calcite crystal

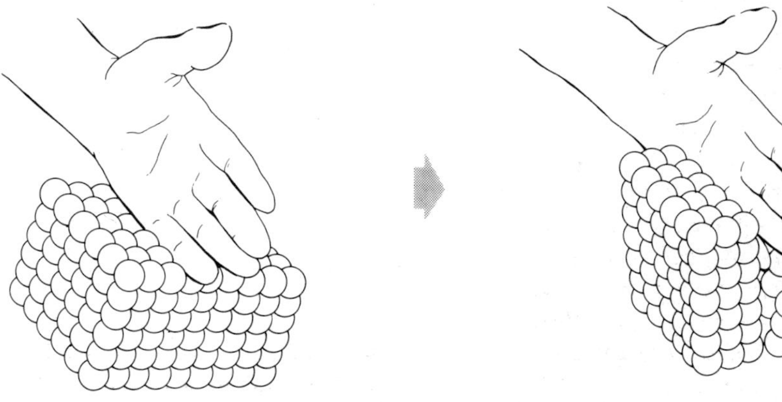

Figure 20.6 A model of what happens when a crystal is cleaved

Question

Graphite and mica are easily cleaved. Would you expect diamond to cleave easily? The models in chapter 19 will help you decide.

called *cleaving*. The fragments have the same shape as the parent crystal. If the blade is wrongly placed, the crystal may shatter. The way in which a crystal cleaves can give useful information about the structure, because cleavage can only occur in planes between layers of particles.

At high enough temperatures, ionic compounds can be melted. The ions are then free to move. Molten sodium chloride conducts electricity and electrolysis takes place (see figure 20.8). Positive sodium ions move to the cathode and the negative chloride ions move to the anode. The positive ions (**cations**) and negative ions (**anions**) act as 'carriers' of the electric current. The processes occurring at the electrodes are the reverse of what happens when ionic compounds are formed from atoms. The electrolysis of molten sodium chloride in the Down's cell for the extraction of sodium is described on p. 87. Evidence for the movement of ions during the electrolysis of solutions is described on p. 310 in chapter 40.

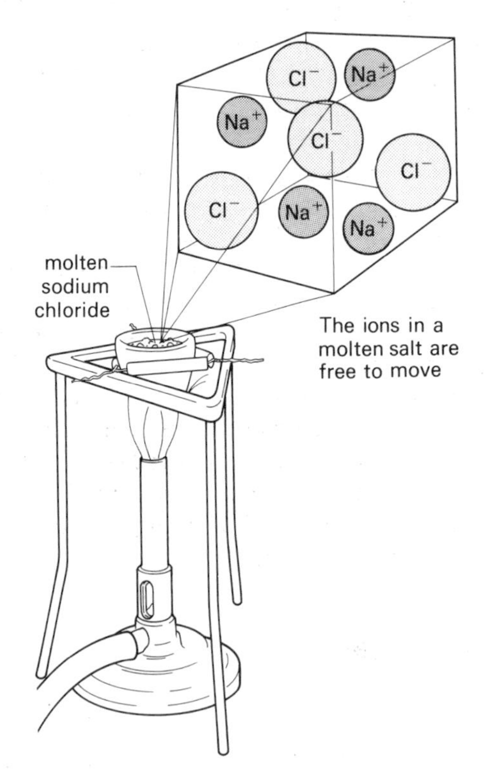

Figure 20.7 The behaviour of the ions in molten sodium chloride

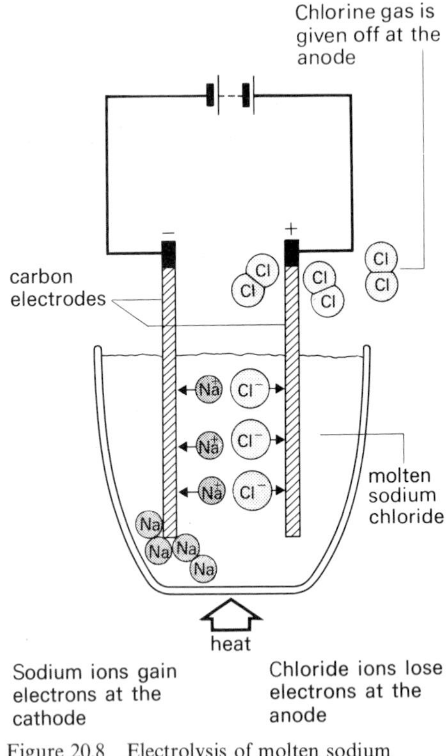

Figure 20.8 Electrolysis of molten sodium chloride

Question

Consider the experiment shown in figure 20.8.
(a) How would you show that molten sodium chloride conducts electricity?
(b) What will happen when the molten salt is allowed to cool?
(c) How could you prevent the chlorine formed at the anode from mixing freely in the molten salt and attacking the cathode product?

Many ionic substances are soluble in water. Figure 20.9 shows the familiar model of a water molecule. The covalent bond between each pair of hydrogen and oxygen atoms contains a shared pair of electrons. But these are not shared equally by oxygen and hydrogen. Oxygen has a greater pull on the shared electrons because of its greater nuclear positive charge. So the oxygen end of the molecule is slightly more negative than the hydrogen end. This is shown in figure 20.9. The Greek sign 'δ' (delta) means 'slightly'. Electrically, the water molecule has a positive end (pole) and a negative end (pole). It is said to be *polar*.

Polar water molecules are attracted to ions in a crystal structure and can surround them. The attraction can be so great that the forces holding ions together are overcome. The ions become free to move in solution, surrounded by water molecules. The negative ends of the water molecules are attracted to the positive ions, and the positive ends to the negative ions (see figure 20.10).

Solutions of ions conduct electricity because the ions are free to move towards the electrodes. The electrolysis of aqueous solutions is described in chapter 40.

The polarity of water molecules can be used to explain some of the other unusual properties of water described in chapter 8. List some of these properties.

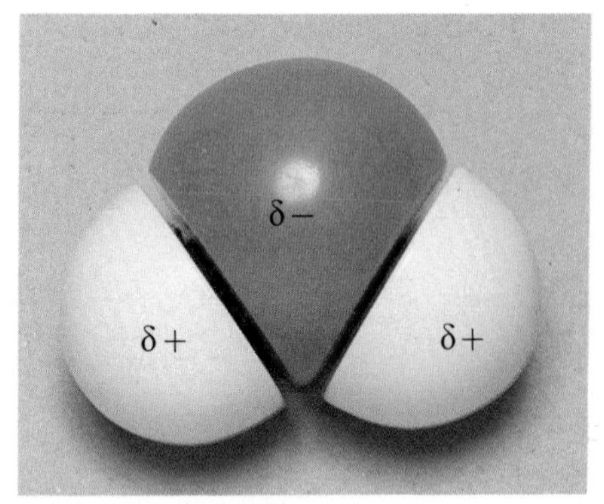

Figure 20.9 The positive and negative poles in a water molecule

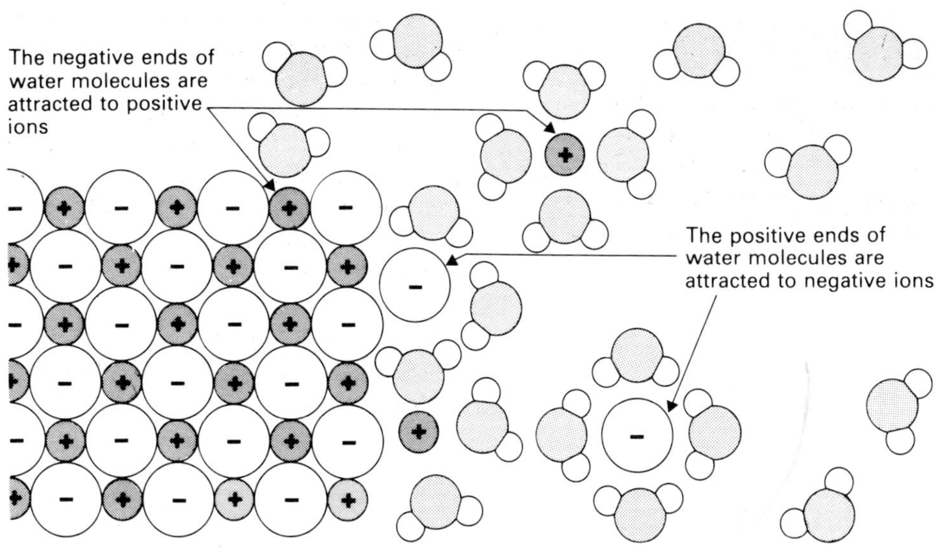

Figure 20.10 Dissolving an ionic crystal

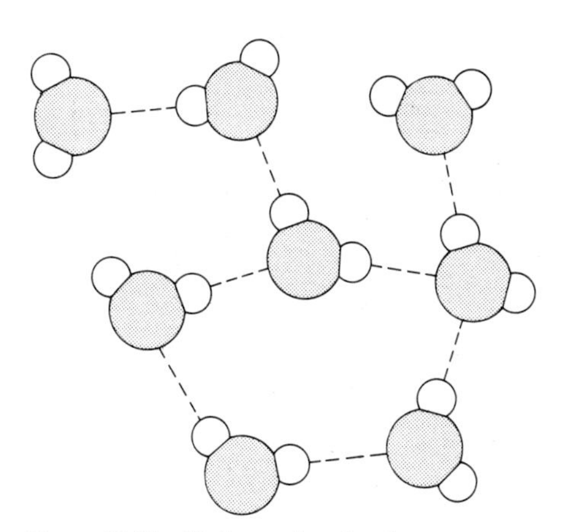

Figure 20.11 Hydrogen bonding in water

Water molecules attract each other, and weak bonds form between molecules. $\delta-$ oxygen atoms are attracted to $\delta+$ hydrogen atoms. Although the bonds are weak compared with the bonds within the molecules, they are stronger than is normal between molecules. The weak bonds are called *hydrogen bonds.* Hydrogen bonding between water molecules is shown in figure 20.11. The attraction of water molecules for each other accounts for the relatively high melting and boiling points of water.

In water at room temperature, hydrogen bonds are constantly made and broken. As water cools, more bonds are made than broken until, at 0 °C, the open structure of ice is formed, with hydrogen bonds holding the water molecules apart in a rigid structure. This is shown in figure 8.2 on p. 61. In the ice structure, can you see the hexagonal symmetry which is found in the shape of many ice and snow crystals (see figure 8.4 on p. 61)?

Summary

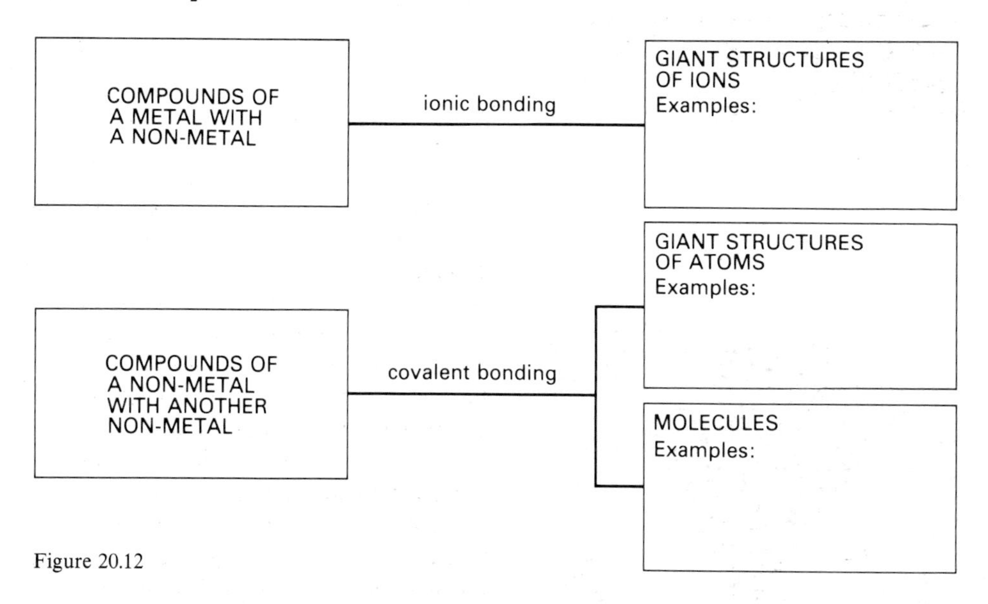

Figure 20.12

Figure 20.12 is a summary of the bonding and structure of compounds. Copy the diagram and complete it by adding examples. Among your examples include carbon dioxide, hydrogen chloride, lead(II) bromide, potassium iodide, silicon dioxide, sodium chloride, water and zinc oxide.

Review questions

1 Hydrogen 1, Helium 2, Lithium 3, Carbon 6, Nitrogen 7, Oxygen 8, Fluorine 9, Neon 10, Sodium 11, Magnesium 12, Chlorine 17

Hydrogen	1	Fluorine	9
Helium	2	Neon	10
Lithium	3	Sodium	11
Carbon	6	Magnesium	12
Nitrogen	7	Chlorine	17
Oxygen	8		

(a) From the above list of elements and their atomic numbers, name

(i) a gas which is greenish-yellow in colour,

(ii) the element which has an allotrope called diamond,

(iii) two elements whose atoms each have a complete outer shell of electrons,

(iv) two elements whose atoms each have *one* electron on their outer shells,

(v) two elements which form compounds by gaining or losing of electrons from their atoms.

(b) X, Y and Z represent *three of the elements in the list.* They each combine with hydrogen to form the compounds XH_4, YH_3, and HZ. Identify X, Y and Z. **(EMREB)**

2 A motor car provides examples of the uses of many different metals and alloys, all of which enable the car to be safe and efficient. Describe the production from its ore of one of the metals used in the construction of a motor car.

Explain why an alloy (such as brass which is an alloy of copper and zinc) is sometimes used in preference to either of its constituent metals. **(L)**

3 Make a list of *six* metals in common use in the home or in industry, and for each one, state its use and the particular property which makes it more suitable for that use than any of the other metals in your list.

Copper is said to be twice as hard as aluminium. Describe an experiment you would carry out to see if this is so. **(L)**

4 It is possible to make crystals of some metals by displacement from a solution of one of their salts.

(a) Name a metal which can be made in the form of crystals by this method.

(b) Name a metal which you could use to displace it.

(c) Why would this metal be suitable to use?

(d) Draw a diagram to show how you would do the experiment.

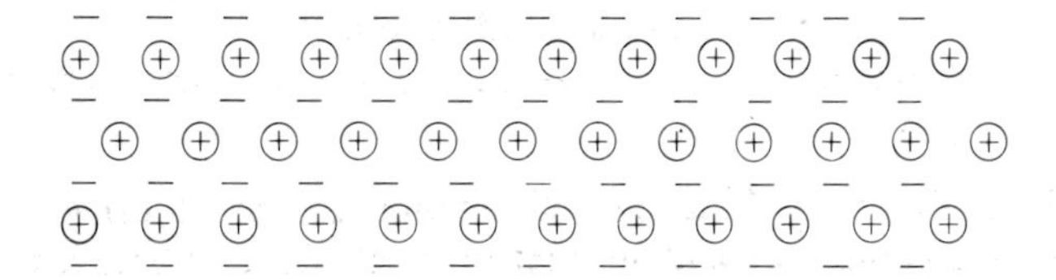

The diagram shows the structure of a metal. The signs represent electrons and the metal ions each have a positive charge.

(e) Write down one method which is used to find how the atoms of a metal are arranged.

(f) By consulting the diagram, explain why a metal with a structure like that is able to conduct electricity.

(g) Crystals of ionic compounds, such as sodium chloride, NaCl, may be made by another method, not involving displacement. Write down in the correct order the words which would describe how you would make crystals of sodium chloride from rock salt using this method. (The best answer will have six words, using one word twice.)

Saturated	Drying	Neutralization
Melting	Distillation	Cooling
Filtering	Dissolving	

(h) Crystals of sodium chloride have a different kind of structure from a metal. It is an ionic structure. Using Na^+ and Cl^- for the ions, draw a simple diagram to show the pattern of the sodium chloride structure.

(i) Write down one property of sodium chloride which shows it is an ionic compound. **(EMREB)**

5 The following is a list of substances: butane, diamond, ethane, ethanol, ethene, graphite, iodine, 2-methylpropane, propane, sodium chloride. Explain the meaning of each of the following terms, choosing suitable examples from the above list of substances to illustrate your answer: (a) allotropes, (b) structural isomers, (c) the difference between empirical and molecular formulae, (d) molecular crystals. Each substance may be used once, more than once, or not at all. **(JMB)**

6 (a) Element X has atomic number 9 and element Y has atomic number 11.

(i) Give the arrangement of electrons in one atom of X.

(ii) Give the arrangement of electrons in one atom of Y.

(iii) Draw a diagram to show the arrangement of the electrons in the outer shells when two atoms of X join to form a molecule.

(iv) Draw a diagram to show the arrangement of the electrons in the outer shells when one atom of X and one atom of Y form a compound.

(v) Name the type of bonding shown with the molecule X_2.

(b) The table below gives information about four different substances A, B, C and D. Study the table and answer the questions printed below it.

Substance	Melting point and boiling point	Electrical conductivity when solid	Electrical conductivity when molten
A	Above 200 °C	Poor	Good
B	Above 200 °C	Good	Good
C	Above 200 °C	Poor	Poor
D	Below 200 °C	Poor	Poor

State the substance most likely to be
(i) silver.
(ii) diamond.
(iii) bromine.
(c) Explain why sodium chloride conducts electricity when molten but not when solid. **(YHREB)**

7 (a) Write down the electronic structures of (i) a lithium atom and (ii) the lithium ion. On the basis of your answers in (i) and (ii), explain why lithium is classified as a metal.
(b) Write down the electronic structures of (i) a fluorine atom and (ii) the fluoride ion. On the basis of your answers in (i) and (ii), explain why fluorine is classified as a non-metal.
(c) (i) What type of bonding is present in lithium fluoride?
(ii) Would you expect this compound to have a low or a high melting point?
(iii) Explain what happens when this compound melts.
(iv) What change (if any) happens to the electrical conductivity of lithium fluoride when it melts? **(O & C)**

8 (a) Give an account of the bonding between the atoms in a molecule of water. Include in your answer a description of the shape of a water molecule.
(b) Many elements and compounds are changed when water is added to them. Choose *four* substances which are changed by water, each in a *different* way, and describe what happens during these changes. **(L)**

9 Atoms can combine to form ionic crystals, simple molecules, giant molecular structures and metallic crystals. Choose *one* example of each of these structures and describe the bonding present in the substances *you have chosen.*

Compare the physical properties of the substances you have chosen and explain the differences or similarities in these properties. **(L)**

10 The element phosphorus has an atomic number of 15 and a mass number of 31. Phosphorus occurs as two allotropes: a red form and a white form. Only the white form is soluble in the solvent carbon disulphide. The element can combine with chlorine to form two chlorides, of formulae PCl_3 and PCl_5.

(a) State the numbers of protons and neutrons present in the phosphorus nucleus and give a simple diagram to show the arrangement of the electrons in an atom of phosphorus.
(b) Explain what is meant by *allotropes* and name *two* other elements that exist as allotropes.
(c) How could a pure sample of red phosphorus be obtained from a mixture of the red and white allotropes?
(d) Predict the formulae of the compounds you would expect phosphorus to form with (i) bromine and (ii) oxygen.
(e) The chloride PCl_3 reacts with water to form the acids H_3PO_3 and HCl, and the chloride PCl_5 reacts to form the acids H_3PO_4 and HCl. Write equations for these two reactions. **(CLES)**

11 The figure below shows the bonding within a molecule of an oxide of the element phosphorus.

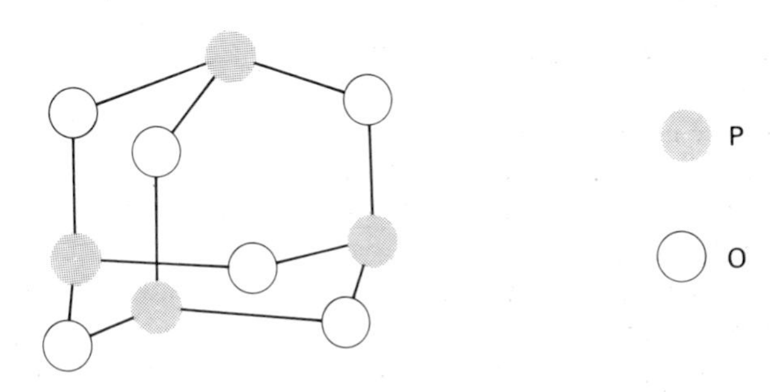

This oxide reacts with cold water to give the colourless, soluble acid H_3PO_3.
(a) Write down (i) the molecular formula and (ii) the empirical formula, of this *oxide.*
(b) Name (i) *one* other compound which is made up of small molecules and (ii) *one* element which is solid at room temperature and is made up of small molecules.
(c) Magnesium oxide has the same geometric form of crystal lattice as sodium chloride. Draw a diagram showing the arrangement and type of particles present in a crystal of magnesium oxide.
(d) Give *two* characteristic physical properties in *each* case of (i) compounds made of simple molecules and
(ii) compounds with the same *type* of structure as sodium chloride and magnesium oxide.
(e) A suspension of magnesium oxide in water is gradually added, until present in excess, to the solution formed by adding the phosphorus oxide (drawn above) to water containing litmus. Suggest how the colour of the solution changes during this experiment. Construct equations for the reactions that occur. **(CLES)**

12 (a) Draw a labelled diagram to show the industrial apparatus used to electrolyse molten sodium chloride (e.g. the Downs cell). Explain why calcium chloride is almost always added as a secondary electrolyte.
(b) Write down 'half equations' to show the principal reactions taking place at the anode and cathode.
(c) Give one large-scale use for each of the products formed in (b). **(O)**

13 (a) For *each* of the crystalline solids, sodium chloride and diamond, state and account for (i) the type of bonding

present and (ii) the structure of the crystals, explaining how the physical properties are related to their structures.
(b) Describe how the following processes may be used in the separation of the components of a mixture, using an appropriate example in *each* case: (i) filtration, (ii) chromatography.
(c) Explain why the colour of bromine vapour spreads to fill the whole space when some liquid bromine is put in a gas jar of air and the lid closed. **(WJEC)**

14 (a) Potassium ions and calcium ions have identical electronic structures, yet the chemistry of these two elements shows marked differences. Illustrate these differences by comparing, for potassium and calcium, (i) the reactivity of the metals towards water, (ii) the solubility of their hydroxides and (iii) the stability to heat of their hydroxides and carbonates.
(b) Describe, and explain briefly, one large-scale use of *two* of the following compounds: calcium oxide; calcium sulphate; calcium hydroxide.
(c) Although potassium chloride and calcium oxide have exactly the same type of crystal structure, the melting point of potassium chloride (770 °C) is much lower than that of calcium oxide (2580 °C). Suggest a reason for this. **(O & C)**

15 Using your knowledge of the structure of the water molecule, and with the help of diagrams, write an account of the structure of ice.

How can the fact that ice floats on water be explained?

By what means can water molecules attach themselves to ions when salts dissolve in water? **(L)**

THEME G
Acids, bases and salts

Natural acid–base reactions produced the stalactites and stalagmites in this cave in Wales

21 Acids

21.1 What do acids do?

The word **acid** sounds dangerous! Nitric, sulphuric and hydrochloric acids are very dangerous when they are concentrated. They have to be handled with great care. But when these acids are diluted with water they are less of a hazard. Dilute hydrochloric acid will not hurt your skin if washed away quickly, but it will sting in a cut and will slowly rot clothing. Safety spectacles should always be worn when using dilute or concentrated acids.

Not all acids are dangerous to life. Many acids are part of life itself. They are found in living things. Citric acid gives orange and lemon juices their sharp taste. Acetic acid, which chemists call ethanoic acid, is the main ingredient of vinegar.

The easiest way to recognize an acid is to test a solution with an **indicator**. Litmus turns red in any acid solution. Many indicators, like litmus, are coloured extracts from plants. Full range indicator is a synthetic indicator. It turns to shades of orange and red, depending on the strength and concentration of the acid.

The pH scale is a number scale which shows the acidity or alkalinity of a solution in water. Most laboratory solutions have pH values in the range 1–14.

The effect of acids on indicators is one of their best-known properties. There are other important reactions which are shown by all acids.

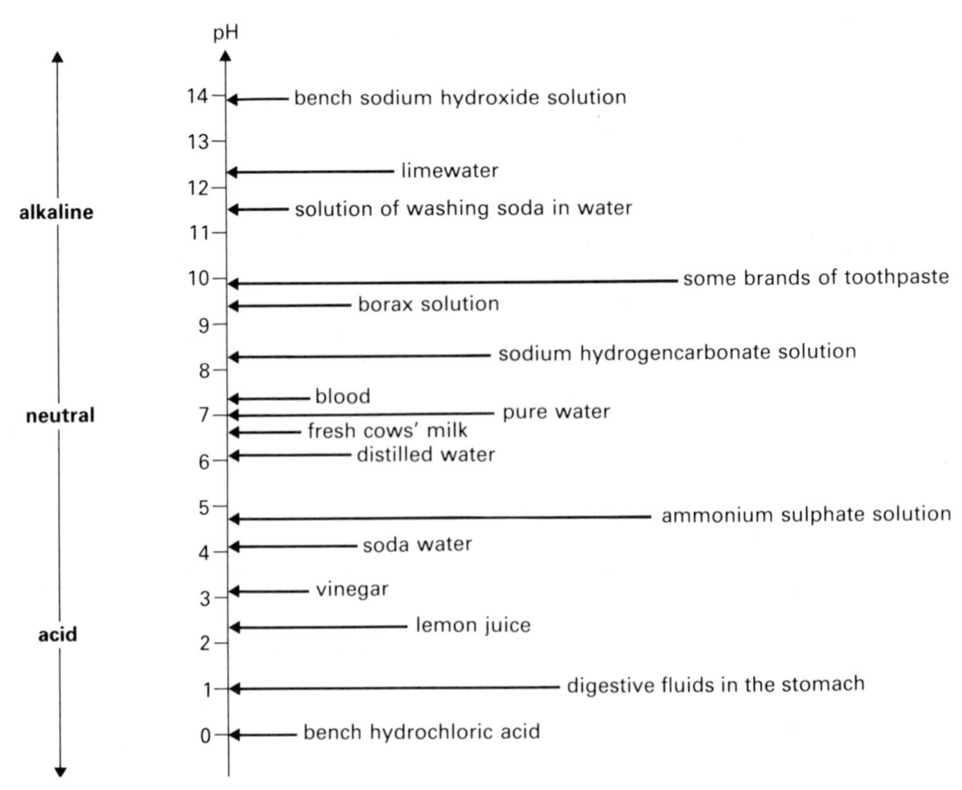

Figure 21.1 The pH scale

Experiment 21a
Investigating the reactions of acids

dilute hydrochloric acid
dilute sulphuric acid

copper(II) oxide
copper(II) carbonate

Various acids are tested with indicators, metals, metal oxides and carbonates. All the reactions can be carried out in test-tubes. To make a fair comparison, the dilute acids are chosen to have the same concentration.

Results
The results for hydrochloric, sulphuric and ethanoic (acetic) acids are shown in the table in figure 21.2.

Test	Dilute hydrochloric acid	Dilute sulphuric acid	Dilute ethanoic acid
Indicators			
Litmus	Turns red	Turns red	Turns red
Full range (universal) indicator	Dark red, pH below 1	Dark red, pH below 1	Orange, pH about 3
Metals			
Magnesium	Rapid fizzing, and gas evolved which burns with a pop	Rapid fizzing, and gas evolved which burns with a pop	Slow fizzing, and gas evolved which burns with a pop
Zinc	Gas formed which burns with a pop	Gas formed which burns with a pop	Very slow formation of gas
Copper	No reaction	No reaction	No reaction
Metal oxides			
Copper(II) oxide	Black solid dissolves on warming to give a green solution	Black solid dissolves on warming to give a blue solution	Black solid dissolves on warming to give a blue solution
Carbonates			
Sodium carbonate	Gas formed which turns limewater milky	Gas formed which turns limewater milky	Gas formed which turns limewater milky – slow reaction
Calcium carbonate (marble chip)	Gas formed which turns limewater milky	A few gas bubbles, then reaction stops	Gas formed slowly which turns limewater milky

Figure 21.2

Discussion
The results show that these acids have similar properties. One of them seems less reactive than the other two: which one? For reasons which are explained on p. 171, sulphuric and hydrochloric acids are called strong acids and ethanoic acid is called a weak acid.

What is the colour of litmus in an acid solution? How high must a metal be in the activity series for it to react with an acid? Which gas is evolved when an acid reacts with a carbonate? What signs of reaction are there when an acid reacts with a metal oxide?

Indicators

Indicators change colour to show whether a solution is acidic or alkaline. Litmus turns red in acidic solutions. Full range indicator is probably more useful: it shows that ethanoic acid solution is less acidic than solutions of sulphuric and hydrochloric acids at the same concentration.

> **Question**
>
> Write a word equation for the reaction of
> (a) sodium carbonate with hydrochloric acid.
> (b) copper(II) oxide with sulphuric acid.
> (c) zinc with hydrochloric acid.

Metals

Hydrogen gas is formed when metals react with acids. Only metals high in the activity series react. There is no reaction with metals below lead in the series. Even with lead, it is difficult to detect any change in a short time.

Metal + Acid ⟶ Salt + Hydrogen

The salts of hydrochloric acid are called chlorides. The salts of sulphuric acid are called sulphates.

Carbonates

All carbonates give off carbon dioxide when they react with acids. The normal laboratory preparation of this gas is based on the reaction of hydrochloric acid with calcium carbonate (in the form of marble chips). The reaction between sulphuric acid and marble chips soon stops. Can you explain this, given that calcium chloride is soluble in water, whereas calcium sulphate is insoluble?

Metal carbonate + Acid ⟶ Salt + Carbon dioxide + Water

Metal oxides and hydroxides

Most metal oxides and hydroxides are insoluble in water. They react with acids and dissolve. No gas is evolved. The acids are neutralized.

Metal oxide + Acid ⟶ Salt + Water

The oxides and hydroxides which behave in this way are called **bases**.

21.2 The structures and formulae of acids

Pure, anhydrous acids may be solids (e.g. citric and tartaric acids), liquids (e.g. sulphuric, nitric and ethanoic acids), or gases (e.g. hydrogen chloride, which becomes hydrochloric acid when dissolved in water). All of these acids have molecular structures. The particles of the pure acids are molecules. Figure 21.3 shows the structures and formulae of some common acids.

Figure 21.3 Some common acids

H—Cl

(a) Hydrogen chloride (a gas)

$H-C(H)(H)-C(=O)-O-H$

(b) Ethanoic acid (a liquid)

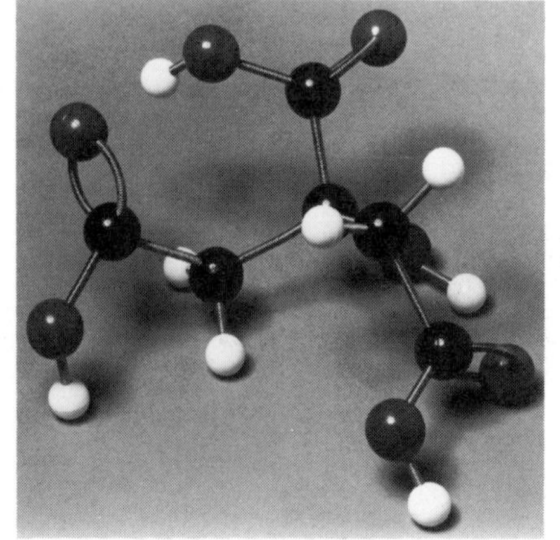

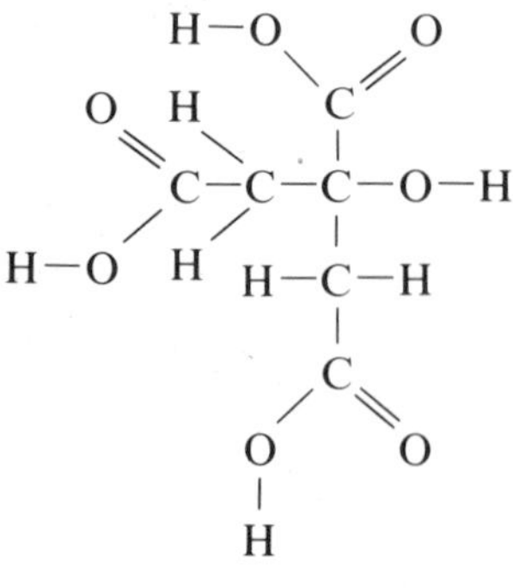

(c) Citric acid (a solid)

> **Question**
>
> Work out the molecular formula of each of the acids in figure 21.3.

All acids have similar properties. Why should this be? Lavoisier thought that oxygen was the acid-forming element. Hence the name he gave to the gas – 'oxygen', meaning 'acid-former'. Look at figure 21.3 again. Do all the acids have oxygen in their structures? Is there any one element present in the formulae of all the acids?

21.3 The mineral acids

Sulphuric acid, hydrochloric acid and nitric acid are sometimes called the mineral acids because they come from mineral rather than organic sources.

Hydrochloric acid used to be made directly from salt. The oldest way of making sulphuric acid was the distillation of hydrated iron(II) sulphate. The old name for the acid was 'oil of vitriol', so the green crystals of iron(II) sulphate were called 'green vitriol'. Nitric acid was made from potassium nitrate (saltpetre).

Now these acids are made from simpler starting materials. Very pure hydrogen chloride is made by direct combination of the two elements. The most important source of hydrogen chloride is as a by-product of the reaction of chlorine with hydrocarbons (see chapter 29). Hydrogen chloride is dissolved in water to make hydrochloric acid. Concentrated hydrochloric acid is a solution containing 27.66 per cent by mass of hydrogen chloride in water. Hydrochloric acid has many important industrial uses, and some of these are illustrated in figure 21.4.

Figure 21.4 Two uses of hydrochloric acid: other uses are in the dyestuffs, textiles, food and leather industries

(a) Removing the oxide film from steel pipes before they are galvanized

(b) Making printed circuit boards

In the laboratory, hydrochloric acid can be made by generating hydrogen chloride, and then dissolving the gas in water. The gas is made by the action of concentrated sulphuric acid on salt.

Experiment 21b
Making hydrochloric acid from salt

concentrated sulphuric acid

dilute hydrochloric acid

The apparatus is shown in figure 21.5. The inverted funnel is necessary to prevent 'suck-back' of water.

Results
The main observations are frothing in the reaction flask and swirling currents in the beaker of water as the gas dissolves to form a dense solution.

Discussion
Why are lumps of rock salt often used in preference to fine crystals? Which property of hydrogen chloride can lead to 'suck-back' in this preparation? How does an inverted funnel prevent 'suck-back'?

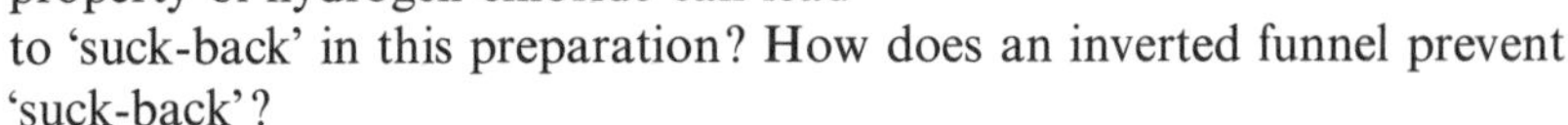

concentrated sulphuric acid
funnel
rock salt
water

Figure 21.5

The equation for the reaction between sulphuric acid and sodium chloride is:

$$NaCl(s) + H_2SO_4(l) \rightleftharpoons NaHSO_4(s) + HCl(g)$$

The reaction can go both ways (see chapter 27). In this experiment, the hydrogen chloride escapes as a gas, so the reverse reaction cannot happen.

Questions

1 How would you modify the apparatus in figure 21.5 to collect hydrogen chloride gas? (Can hydrogen chloride gas be collected over water? Is hydrogen chloride more or less dense than air?)
2 Nitric acid can be made in a similar way, by the action of concentrated sulphuric acid on potassium nitrate. The nitric acid formed is a liquid at room temperature, so it has to be distilled from the reaction mixture. Draw and label an apparatus suitable for making the acid in this way.

The industrial processes used to manufacture sulphuric and nitric acids are based on the fact that the oxides of non-metals form acids when they react with water. Sulphuric acid is made from sulphur, oxygen and water in the contact process. This is described in chapter 28. The final step in the process involves the combination of sulphur trioxide with water. Sulphuric acid has enormous commercial importance and its main uses are shown in figure 28.1 on p. 211.

Nitric acid is manufactured from nitrogen, hydrogen, oxygen and water, as described in chapter 35. First, ammonia is made, and then this is converted to

Figure 21.6 Some uses of nitric acid

(a) For making fertilizers

(b) For making explosives

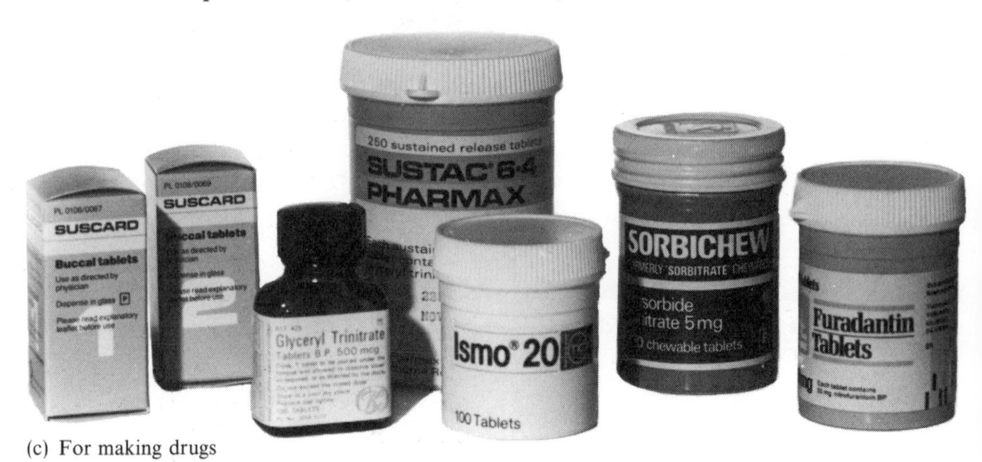

(c) For making drugs

nitrogen dioxide. Finally, nitrogen dioxide is dissolved in water to form nitric acid. Figure 21.6 illustrates the importance of this acid.

In school laboratories, it is hydrochloric and sulphuric acids that are used most often. Many of the reactions used to make gases in the laboratory use hydrochloric or sulphuric acids. Some are described in the next section. The chemical reactions of sulphuric acid are shown in figure 21.7.

Concentrated Sulphuric Acid is...

an oxidizing agent

Hot concentrated sulphuric acid will oxidize metals and non-metals. The acid is reduced to sulphur dioxide and water
e.g.
$Cu + 2H_2SO_4 \rightarrow CuSO_4 + SO_2 + 2H_2O$
$C + 2H_2SO_4 \rightarrow CO_2 + 2H_2O + 2SO_2$

a strong involatile acid

It is used in laboratory preparations of hydrogen chloride and nitric acid. Sulphuric acid does not evaporate and it drives off the more volatile products

$H_2SO_4(l) + NaCl(s) \longrightarrow NaHSO_4(s) + HCl(g)$

← involatile → given off as gas or vapour

$H_2SO_4(l) + NaNO_3(s) \xrightarrow[\text{heat}]{} NaHSO_4(s) + HNO_3(g)$

this is condensed

a dehydrating agent

Blue copper sulphate turns white in the acid, as it is dehydrated

$CuSO_4.5H_2O \xrightarrow{-5H_2O} CuSO_4$

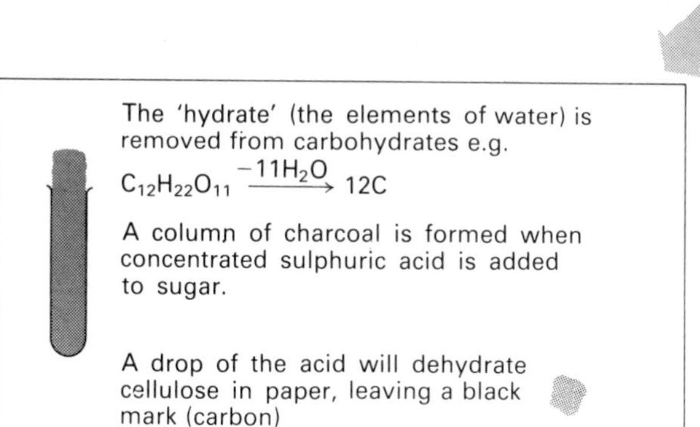
The 'hydrate' (the elements of water) is removed from carbohydrates e.g.
$C_{12}H_{22}O_{11} \xrightarrow{-11H_2O} 12C$
A column of charcoal is formed when concentrated sulphuric acid is added to sugar.
A drop of the acid will dehydrate cellulose in paper, leaving a black mark (carbon)

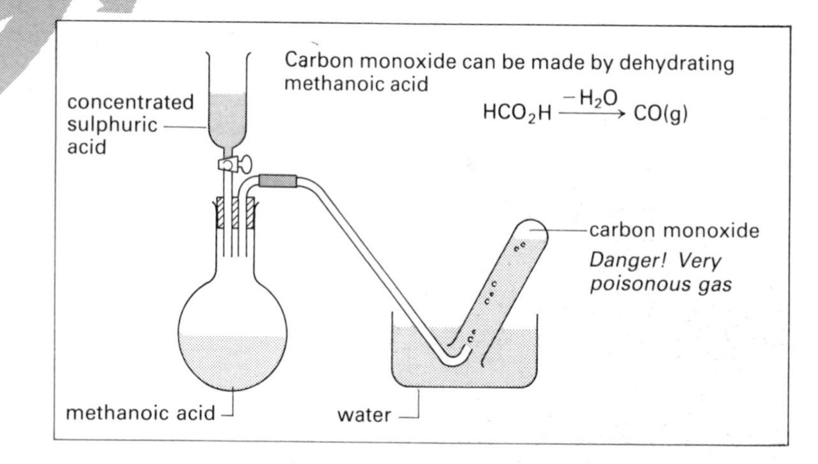
Carbon monoxide can be made by dehydrating methanoic acid
$HCO_2H \xrightarrow{-H_2O} CO(g)$
concentrated sulphuric acid
carbon monoxide
Danger! Very poisonous gas
methanoic acid
water

Figure 21.7 The chemistry of concentrated sulphuric acid

21.4 Making gases

Nowadays, most common gases are available in small cylinders, and this is the safest and most convenient way of obtaining them. However, in school laboratories, it is generally cheaper to make the gases by chemical reactions.

In any chemical preparation there are four important stages:

* carrying out the reaction under controlled conditions in a suitable container
* separating the product from the reaction mixture
* purifying the product
* collecting and identifying the product.

These stages have already been described for the preparation of oxygen (see p. 38) and chlorine (see p. 104). The preparation of ammonia is described in chapter 35 on p. 259.

1 Carrying out the reaction

The preparations of gases in this section all involve the addition of dilute hydrochloric or sulphuric acid to a suitable solid. Sometimes the mixture needs to be warmed to drive off the gas, but otherwise no special conditions are needed.

2 Separating the product

There is no problem when the product is a gas, so long as only one gas is formed. The gas escapes from the flask, leaving everything else behind.

3 Purifying the product

If hydrochloric acid is used, some hydrogen chloride may escape from the solution with the gas formed. If the gas being prepared is insoluble, the hydrogen chloride can be removed by bubbling the mixture through water.

The main impurity is usually water vapour. There is no point in removing this if the gas is going to be collected over water. If a dry sample of gas is needed, a drying agent must be found which will absorb the water but not react with the gas. For most common gases, silica gel can be used. However, it is not suitable for ammonia, which can be dried with soda lime. Calcium chloride is another drying agent.

4 Collecting and identifying the product

The commonest method for collecting gases is to use a gas jar, or test-tube, full of water and turned upside-down in a trough. This method cannot be used if the gas is very soluble in water. It cannot be used if a dry sample is needed. Small samples of dry gas can be collected in a gas syringe.

Upward, or downward, delivery can be used if the gas is much less, or much more, dense than air.

Figure 21.8 shows how these techniques can be used to prepare carbon dioxide, sulphur dioxide and hydrogen.

Figure 21.8 Preparation of gases

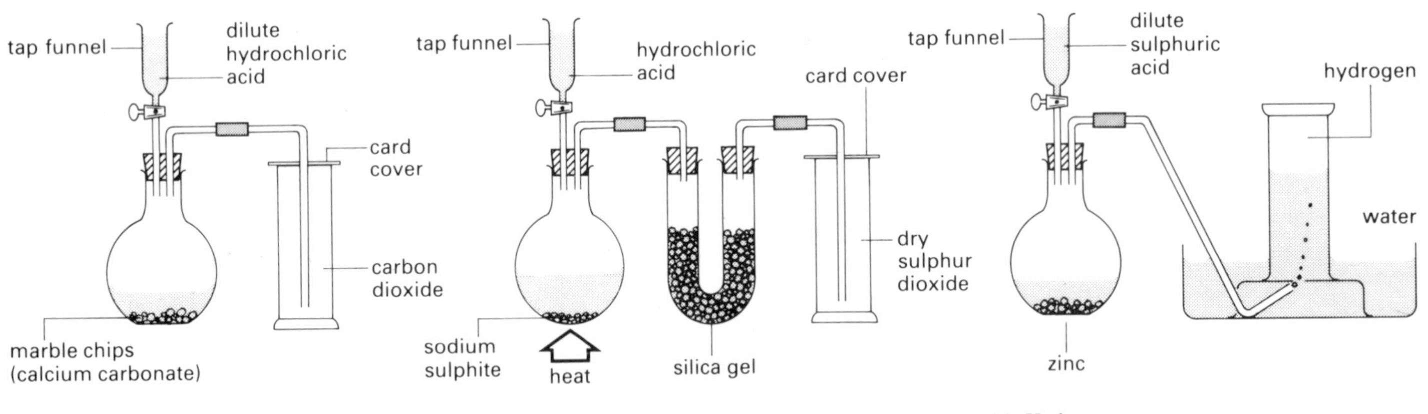

(a) Carbon dioxide

(b) Sulphur dioxide

(c) Hydrogen

Question

Draw and label the apparatus you would use to make and collect a sample of hydrogen sulphide, by the action of dilute hydrochloric acid on lumps of iron(II) sulphide. The properties of hydrogen sulphide are shown in figure 21.9. (Note that hydrogen sulphide smells foul and is extremely poisonous.)

Simple tests are used to check the identity of the gas prepared. The tests used for common gases are summarized in the table in figure 21.10.

Hydrogen
Sulphide
is a colourless gas
smells of bad eggs and is poisonous
has about the same density as air
is slightly soluble in water
is weakly acidic
burns with a blue flame
turns lead ethanoate paper black

Figure 21.9 The properties of hydrogen sulphide

Gas	Tests	Results
Hydrogen	Burning splint	Burns with a pop
Oxygen	Glowing splint	Relights
Carbon dioxide	Limewater	Turns milky white
Hydrogen chloride	Smell Blue litmus Ammonia stopper	Pungent Turns red White smoke
Chlorine	Colour Blue litmus Moist starch/iodide paper	Yellow–green Bleached (may turn red first) Turns blue–black
Sulphur dioxide	Smell Blue litmus Dichromate paper	Pungent Turns red Turns green
Hydrogen sulphide	Smell Burning splint Lead(II) ethanoate paper	'Bad eggs' Gas burns – sulphur deposits Turns brown–black
Ammonia	Smell Red litmus	Pungent Turns blue
Nitrogen dioxide	Colour Blue litmus	Orange–brown Turns red
Water vapour	Appearance Cobalt chloride paper	'Steams' in the air Turns from blue to pink

Figure 21.10

Summary

Give examples to illustrate each of the following typical properties of acids:

- acids change the colour of indicators
- acids react with the more reactive metals
- acids react with carbonates
- acids are neutralized by metal oxides and hydroxides (bases).

Where possible write word or symbol equations for the reactions. Use as many acids as you can in choosing your examples.

22 Bases

22.1 What do bases do?

The term **base** is a technical word in chemistry. The word has a different meaning in everyday speech. The word **alkali** is better known. On medicine labels a familiar term is antacid. 'Ant-acid' means 'against-acid'. Bases are antacids. They are the chemical opposites of acids. A base will neutralize an acid, and when this happens a **salt** is formed. Metal oxides and hydroxides are bases, and most of them are insoluble in water.

Bases which dissolve in water form alkaline solutions. The solutions turn litmus blue. They have a pH above 7. An alkali is a base which is soluble in water. Common alkalis are sodium hydroxide, potassium hydroxide, calcium hydroxide, ammonia and sodium carbonate.

Alkalis are used in the home for two purposes: to neutralize acids and to remove grease. Toothpaste is mildly alkaline. It neutralizes the acids which attack teeth. These acids are formed when bacteria in saliva act on the sugars in food. Milk of magnesia (a suspension of magnesium hydroxide in water) is used to cure stomach upsets. It neutralizes some of the acid in the stomach if there is too much. Ammonia is commonly added to mild household cleaners to remove oily, and greasy, dirt. More powerful cleaners contain sodium hydroxide. Tough deposits of grease in an oven may be removed with special cleaners based on potassium hydroxide.

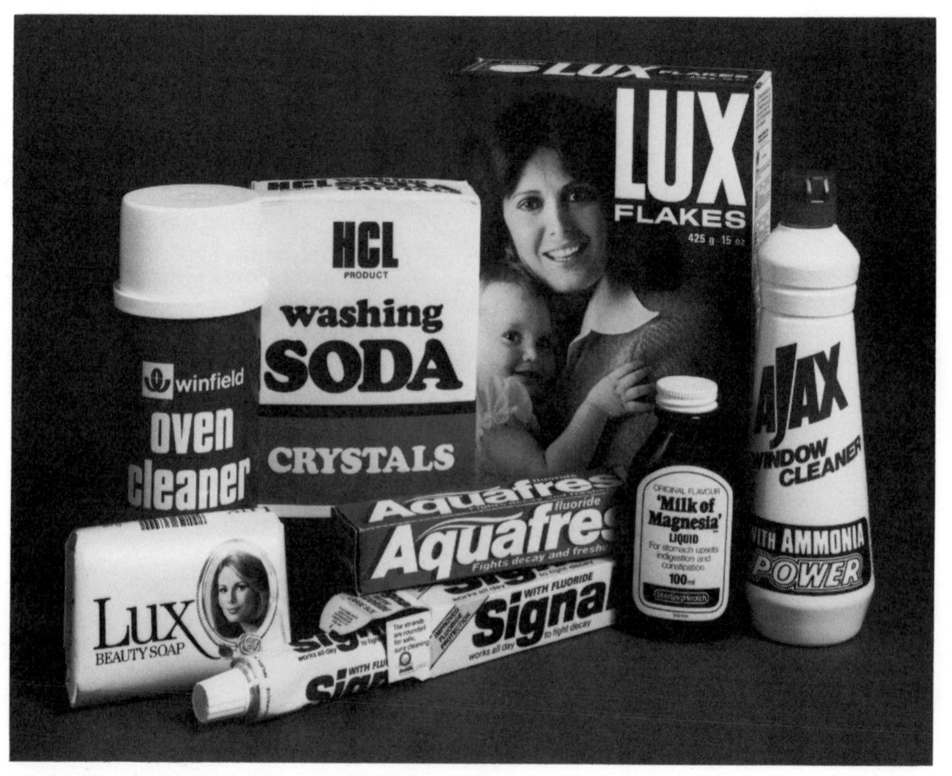

Figure 22.1 Alkalis are used in these domestic products

Question

Write word and symbol equations for the reaction of
(a) copper(II) oxide with hydrochloric acid.
(b) calcium hydroxide with nitric acid.
(c) zinc oxide with sulphuric acid.

Sodium and potassium hydroxides attack skin. They are *caustic*. Even dilute solutions can be hazardous, especially to the eyes. If you rub a drop of dilute sodium hydroxide solution between the tips of two fingers, they will soon begin to feel soapy as the alkali attacks the grease on the skin. In industry, sodium and potassium hydroxides are still better known by their older names – caustic soda and caustic potash.

Bases can be recognized by what they do:

* A base will neutralize an acid to form a salt as the main product. For metal oxides and hydroxides, the word equation is

  Acid + Base ⟶ Salt + Water

* Most bases are insoluble in water. Bases which do dissolve form alkaline solutions. The pH is above 7. Litmus turns blue in alkali.

22.2 The manufacture of alkalis

The manufacture of alkalis in Britain is based on the salt industry in Cheshire. The **electrolysis** of salt solution produces sodium hydroxide and chlorine, both of which are of great importance in the chemical industry. Hydrogen is also produced as a by-product.

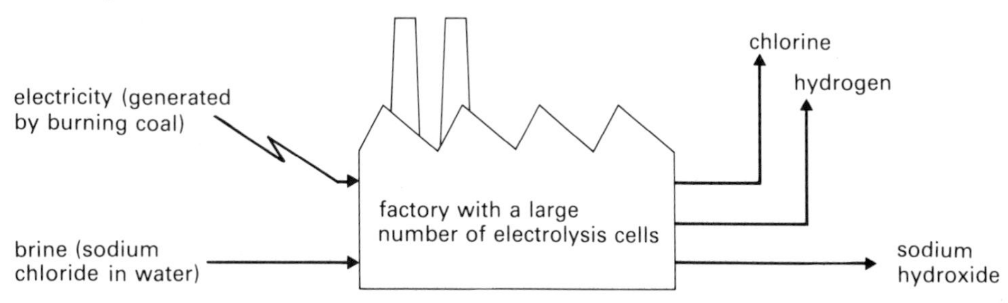

Figure 22.2 The electrolysis of brine (salt solution)

A solution of sodium chloride in water contains four types of ion: sodium, hydrogen, chloride and hydroxide ions. The hydrogen and hydroxide ions come from the water. During electrolysis, hydrogen ions turn into hydrogen at the cathode, and chloride ions turn into chlorine at the anode. This leaves a solution containing sodium and hydroxide ions: a solution of sodium hydroxide. The changes which take place during the electrolysis of salt solution (brine) are illustrated in figure 22.3.

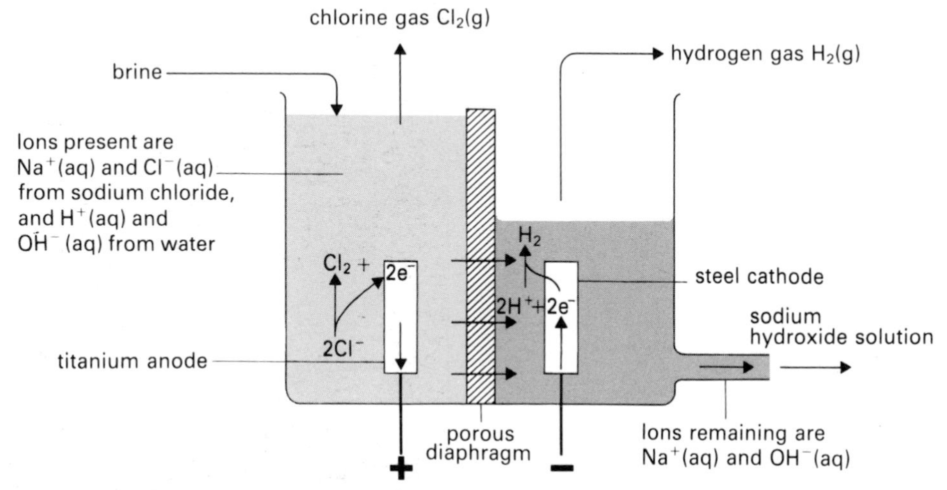

Figure 22.3 As the brine passes through the cell, the chloride ions are discharged at the anode and the hydrogen ions are discharged at the cathode, leaving sodium ions and hydroxide ions in solution

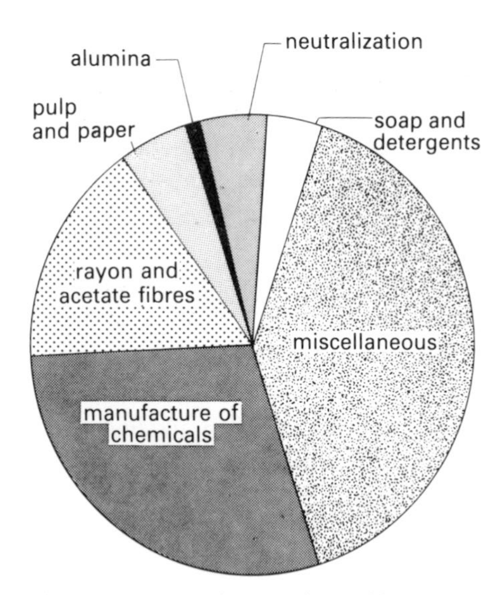

Figure 22.4 Uses of sodium hydroxide

Figure 22.5 Some hydrogen is used for filling meteorological balloons, but its main use is in the chemicals industry

One problem is that chlorine gas reacts with sodium hydroxide solution. The electrolysis cell has to be designed so that they are kept apart. One solution to the problem is the diaphragm cell shown in figure 22.3. The porous diaphragm allows the ions to pass through it but keeps the chlorine gas away from the alkali.

Figures 22.4 and 22.5 illustrate some uses of two of the products of the electrolysis of brine. The uses of chlorine are shown in figure 16.5 on p. 105.

The manufacture of chlorine and sodium hydroxide requires brine, fuel to generate electricity, and good transport facilities to distribute the products. In Britain, the industry is based in and around Runcorn, close to the salt deposits of Cheshire and the coal fields of Lancashire. Runcorn has good road, rail and sea links with the rest of Britain and the world. The main ICI works in Runcorn opened in 1897. At that time the system of inland waterways was still an important part of the transport system.

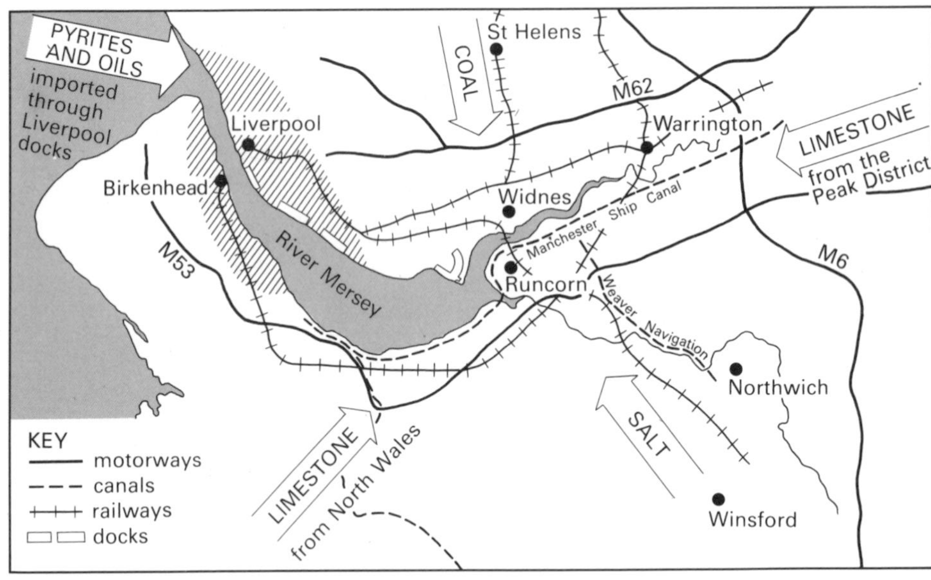

Figure 22.6 Map showing why a chemical industry grew up around Runcorn

The manufacture of sodium carbonate

Sodium carbonate is made on a very large scale. The main ICI works in Cheshire produces about 5000 tonnes of sodium carbonate each day. The plant is close to salt deposits. There is a good rail link to a large limestone quarry in Derbyshire. Coke comes from northern coke ovens. There are good road, rail and river systems to distribute the product. Major users of sodium carbonate, which include the glass and textile industries, have factories nearby.

Sodium carbonate is made from salt by the ammonia–soda process, developed in Belgium by the Solvay family in 1865. At first sight the process seems simple. It is described by this overall equation:

$$2NaCl + CaCO_3 \longrightarrow Na_2CO_3 + CaCl_2$$

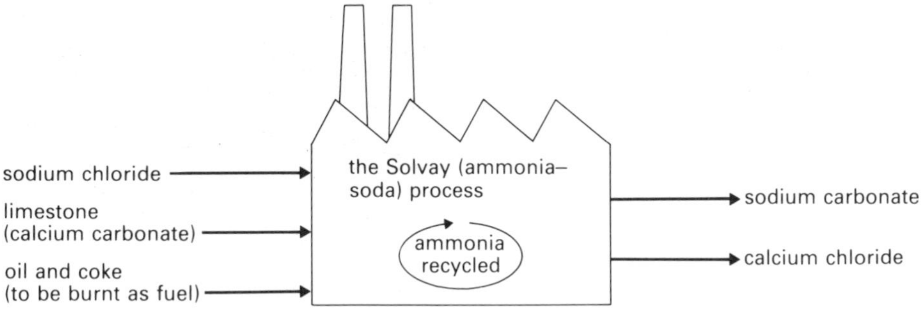

sodium chloride →
limestone (calcium carbonate) →
oil and coke (to be burnt as fuel) →
the Solvay (ammonia–soda) process
ammonia recycled
→ sodium carbonate
→ calcium chloride

Figure 22.7 Summary of the Solvay process

Unfortunately, this reaction will not go. The reverse reaction happens much more easily. When solutions of sodium carbonate and calcium chloride are mixed, they immediately form a precipitate of calcium carbonate and a solution of sodium chloride.

The Solvay process uses a complex sequence of reactions to form the required products. A simplified version of the process is shown in figure 22.8.

Questions

1 Calcium hydroxide is an industrial alkali which is not made from salt. With the help of chapter 2, draw a diagram similar to figure 22.7 to show the main inputs and outputs for the manufacture of calcium hydroxide.

2 Name another industrial alkali which is not made from salt (see chapter 34). What is it made from and what is it used for?

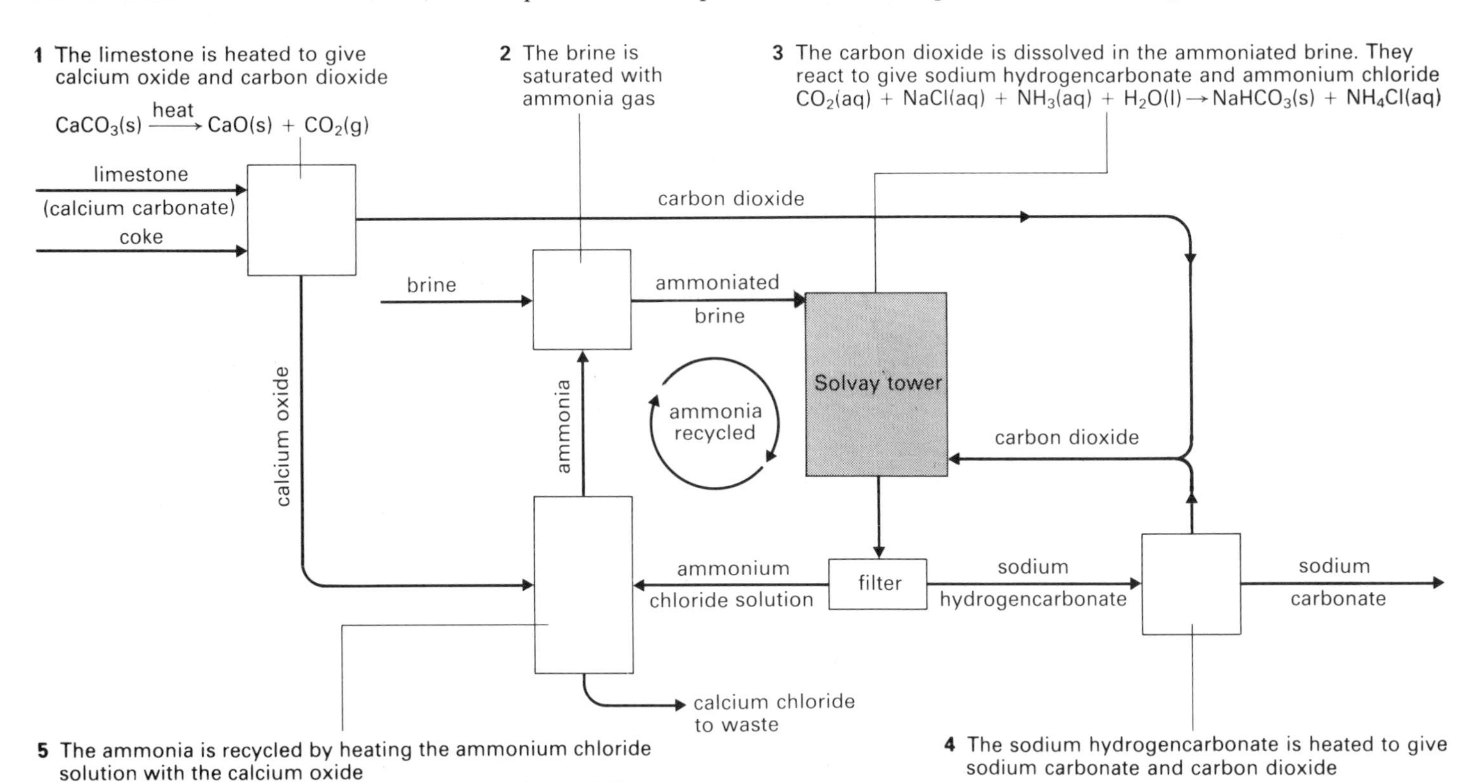
1 The limestone is heated to give calcium oxide and carbon dioxide
$CaCO_3(s) \xrightarrow{heat} CaO(s) + CO_2(g)$
2 The brine is saturated with ammonia gas
3 The carbon dioxide is dissolved in the ammoniated brine. They react to give sodium hydrogencarbonate and ammonium chloride
$CO_2(aq) + NaCl(aq) + NH_3(aq) + H_2O(l) \rightarrow NaHCO_3(s) + NH_4Cl(aq)$
limestone
(calcium carbonate)
coke
carbon dioxide
brine
ammoniated brine
Solvay tower
calcium oxide
ammonia
ammonia recycled
carbon dioxide
ammonium chloride solution
filter
sodium hydrogencarbonate
sodium carbonate
calcium chloride to waste
5 The ammonia is recycled by heating the ammonium chloride solution with the calcium oxide
$2NH_4Cl(aq) + CaO(aq) \rightarrow 2NH_3(g) + CaCl_2(aq) + H_2O(l)$
4 The sodium hydrogencarbonate is heated to give sodium carbonate and carbon dioxide
$2NaHCO_3(s) \xrightarrow{heat} Na_2CO_3(s) + CO_2(g) + H_2O(l)$

Figure 22.8 Flow diagram of the Solvay process for making sodium carbonate

Figure 22.9 A Solvay plant

A large chemical plant, such as is used in the Solvay process, can have a very damaging effect on the environment. It may be noisy, ugly and cause pollution of air and water. The biggest problem in the Solvay process is the disposal of the large volumes of cooling water and process water. The process water is very salty. It is a solution of calcium and sodium chlorides. Suspended in it are all the impurities from the limestone.

In industry, sodium carbonate is still known by its traditional name – soda ash. It is produced in two grades; light ash and heavy ash. The two grades are chemically identical but they differ in the size of their particles. Their uses are shown in figure 22.10.

Figure 22.10 Uses of sodium carbonate

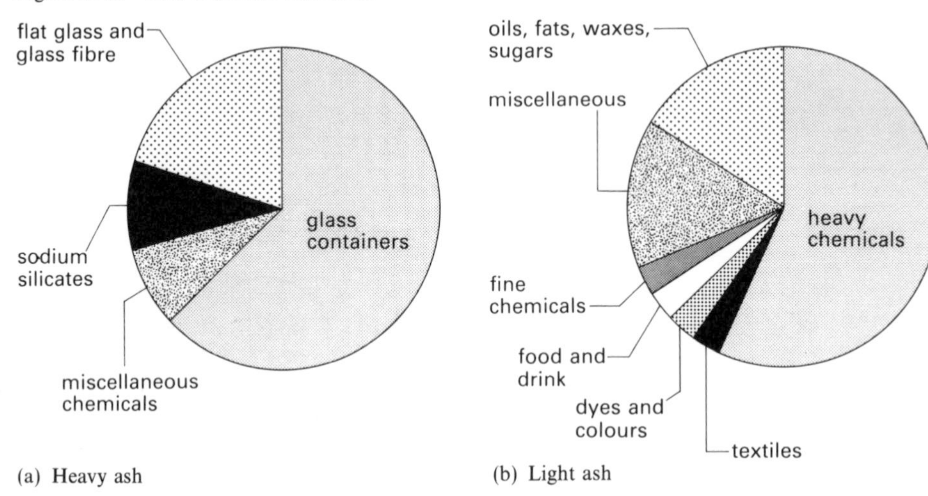

(a) Heavy ash (b) Light ash

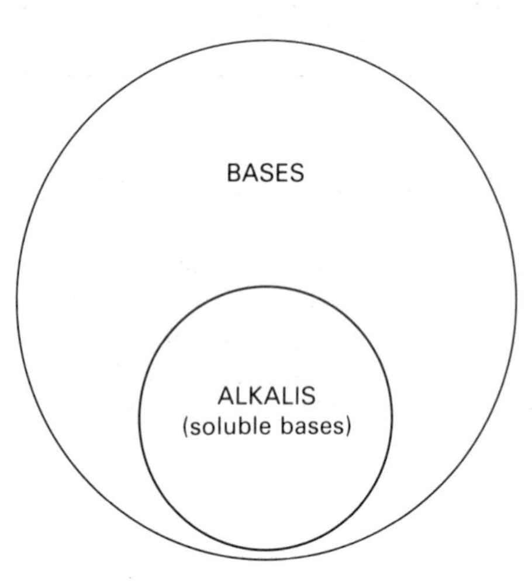

Figure 22.11

Summary

Copy the Venn diagram in figure 22.11 and complete it by writing the names and formulae of common examples in the correct areas.

23 Salts

23.1 What are salts?

A **salt** is formed when an **acid** is neutralized by a **base**. In this way, each salt may be thought of as having two parents. Salts are related to a parent acid (see figure 23.1) and to a parent base (see figure 23.2).

Acid	Salts
Hydrochloric acid, HCl	Sodium chloride, $NaCl$ Magnesium chloride, $MgCl_2$ Copper(II) chloride, $CuCl_2$ Aluminium chloride, $AlCl_3$
Sulphuric acid, H_2SO_4	Sodium sulphate, Na_2SO_4 Magnesium sulphate, $MgSO_4$ Copper(II) sulphate, $CuSO_4$ Aluminium sulphate, $Al_2(SO_4)_3$

Figure 23.1

Base	Salts
Sodium hydroxide, $NaOH$	Sodium chloride, $NaCl$ Sodium sulphate, Na_2SO_4 Sodium nitrate, $NaNO_3$
Magnesium oxide, MgO	Magnesium chloride, $MgCl_2$ Magnesium sulphate, $MgSO_4$ Magnesium nitrate, $Mg(NO_3)_2$

Figure 23.2

Most salts contain metal ions, which are positive, combined with non-metal ions, which are negative.

Salt crystals are often formed by **crystallization** from aqueous solutions. When this happens, water molecules may become part of the crystal structure. This water is called **water of crystallization**, and is included in the formula of the salt. Examples are hydrated copper(II) sulphate, $CuSO_4.5H_2O$, and hydrated sodium carbonate, $Na_2CO_3.10H_2O$. These are called copper(II) sulphate-5-water and sodium carbonate-10-water (see chapter 2).

Question

With the help of table 8 on p. 343, work out the formula of
(a) iron(II) sulphate-5-water.
(b) calcium sulphate-2-water.
(c) calcium chloride-6-water.

23.2 Salts as minerals

Many minerals are composed of single salts. It is not too difficult to find recognizable specimens of minerals such as fluorite (calcium fluoride), galena (lead(II) sulphide), gypsum (calcium sulphate) and calcite (calcium carbonate) in

Figure 23.3 Uses of two forms of calcium sulphate

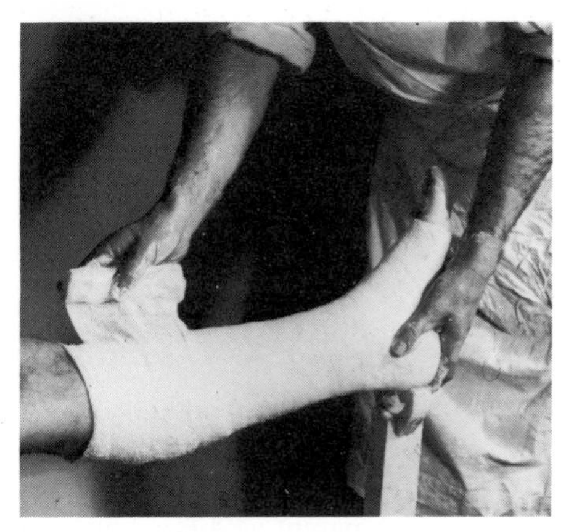

(a) Setting a broken bone using plaster of Paris

(b) An alabaster ornament

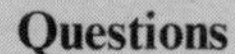

Questions

1 Find out which salts are present in the following, and give their chemical names and formulae: Epsom salts, washing soda, chalk, plaster of Paris, saltpetre, calamine lotion, baking powder, Bordeaux mixture.
2 Find out about the uses of these salts: barium sulphate, copper(II) sulphate, potassium chlorate, iron(II) sulphate, ammonium sulphate, sodium hydrogensulphate.

areas such as the Peak District, North Wales and the Lake District. Colour pictures of a selection of minerals are included in chapters 18 and 19.

Two uses of different forms of calcium sulphate are illustrated in figure 23.3.

23.3 Soluble or insoluble?

If you are going to use a salt or make a salt, you need to know whether or not it is soluble in water. At the same time, it is useful to have some idea of the general patterns of solubility for acids and bases.

All common acids are soluble in water. The solubilities of bases and salts are summarized in the table in figure 23.4.

	Soluble		**Insoluble**
Bases	The **alkalis**: sodium and potassium hydroxides; calcium hydroxide, which is slightly soluble; ammonia		All other metal **oxides** and **hydroxides**
Salts	All **nitrates**		
	All **chlorides**	*except*	Silver chloride and lead chloride (lead chloride is soluble in hot water)
	All **sulphates**	*except*	Barium sulphate and lead sulphate; calcium sulphate is slightly soluble
	Sodium and potassium carbonates		All other **carbonates**

Figure 23.4

23.4 Making soluble salts

Three general reactions of acids were described in chapter 21:

Reactive metal + Acid ⟶ Salt + Hydrogen

Metal carbonate + Acid ⟶ Salt + Carbon dioxide + Water

Base + Acid ⟶ Salt + Water

Each of these reactions can be used to make soluble salts.

Experiment 23a
Method 1 for making soluble salts

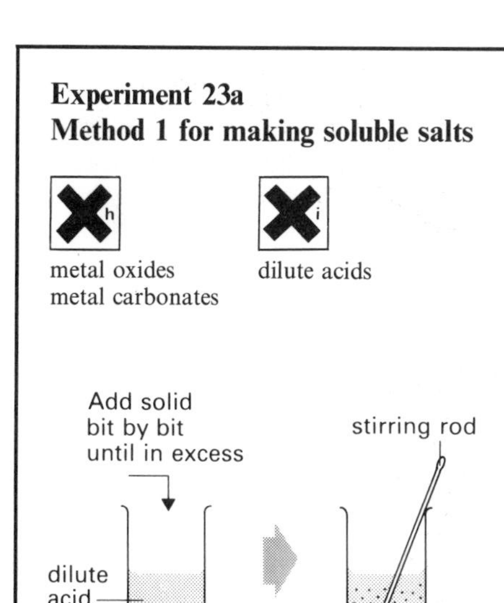
metal oxides
metal carbonates
dilute acids
Add solid
bit by bit
until in excess
dilute
acid
gentle heat

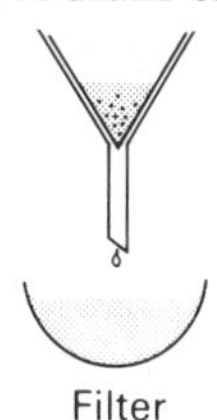
stirring rod
The excess solid
will not dissolve

This method is only suitable if the metal, carbonate or base to be used does not react with water or dissolve in it. The procedure is shown in the series of diagrams in figure 23.5.

Discussion

How does this method make sure that the salt crystals are not contaminated with any of the acid used to start with? How does the method ensure that the salt is not contaminated with any of the metal, carbonate or base added? Why is it essential that the metal, carbonate or base does not react with, or dissolve in, water? Why is this method only suitable for making *soluble* salts?

Which of the following preparations could be done successfully by this method: copper(II) sulphate from copper; zinc nitrate from zinc carbonate; sodium chloride from sodium hydroxide?

Filter
Heat gently to
evaporate off *some*
of the water, until
crystals start to
form round the
edge of the solution
gentle heat

Set aside to
cool and
crystallize

Figure 23.5

In method 1 an excess of solid is added to make sure that all the acid is used up. The excess is then removed by **filtration**. This method cannot be used to make sodium or potassium salts because the hydroxides and carbonates of these metals are soluble in water. If added in excess they will dissolve in the salt solution and contaminate the product. The way round this problem is to mix exactly the right amount of acid with the soluble base or carbonate to give a neutral solution of the salt required. An **indicator** can be used to show when the solution is neutral. The procedure described in experiment 23b is called a **titration**.

Experiment 23b
Method 2 for making soluble salts

dilute sulphuric acid

dilute sodium hydroxide

The method is described in figure 23.6.

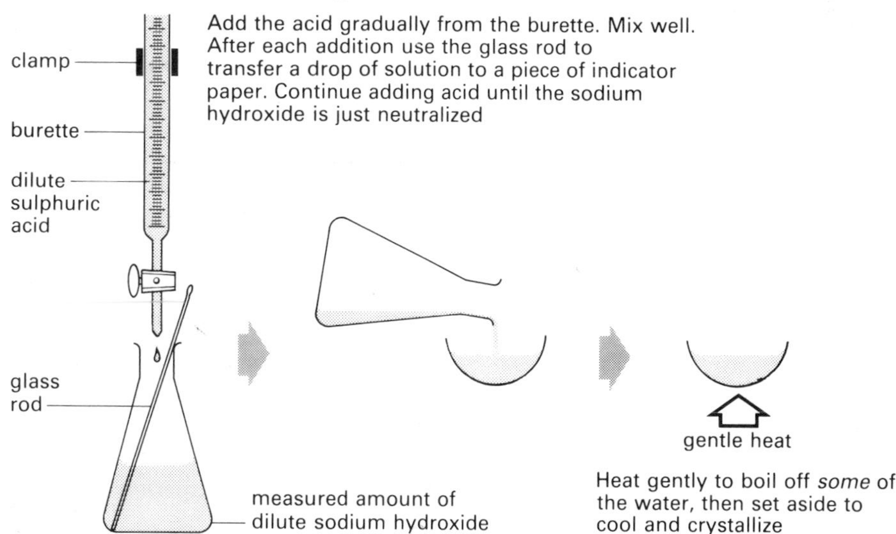
Add the acid gradually from the burette. Mix well.
After each addition use the glass rod to
transfer a drop of solution to a piece of indicator
paper. Continue adding acid until the sodium
hydroxide is just neutralized
clamp
burette
dilute
sulphuric
acid
glass
rod
measured amount of
dilute sodium hydroxide
gentle heat
Heat gently to boil off *some* of
the water, then set aside to
cool and crystallize

Figure 23.6

Discussion

Which salt is being made in figure 23.6? Why is it better to use a glass rod to test the solution than to put the indicator in the flask? Name two other soluble salts which could be made by this method and give word and symbol equations for the reactions.

23.5 Precipitation reactions

Sometimes, when two solutions of soluble salts are mixed, a solid forms and drops out of solution. The solid is called a **precipitate**, and the reaction a precipitation reaction. Precipitation reactions are used to make insoluble salts.

You see a precipitate form every time you wash your hands in hard water. The whitish scum described in chapter 10 is a precipitate of calcium stearate, which is insoluble. This is wasteful because the soap used to form the scum is not available for cleaning.

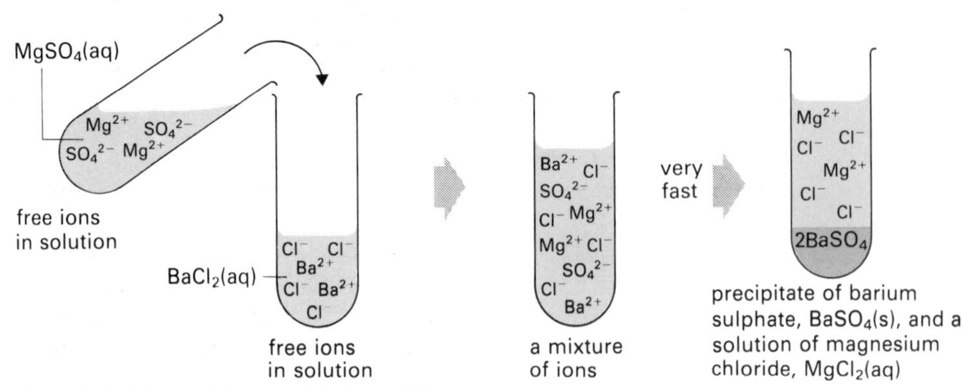

Figure 23.7 The precipitation of barium sulphate

Figure 23.7 shows what is happening during a precipitation reaction. Magnesium sulphate and barium chloride are both soluble salts. When solutions of the two salts are mixed, two new combinations of ions are possible – magnesium ions with chloride ions, and barium ions with sulphate ions. Magnesium chloride is soluble but barium sulphate is not. A precipitate of barium sulphate forms. This can be summarized in the form of an equation:

$$\underbrace{Mg^{2+}(aq) + SO_4^{2-}(aq)}_{\text{magnesium sulphate solution}} + \underbrace{Ba^{2+}(aq) + 2Cl^-(aq)}_{\text{barium chloride solution}} \longrightarrow \underbrace{BaSO_4(s)}_{\text{barium sulphate precipitate}} + \underbrace{Mg^{2+}(aq) + 2Cl^-(aq)}_{\text{magnesium chloride solution}}$$

The magnesium ions and chloride ions are in solution at the end, just as they were at the beginning. They have, as it were, stood on the sidelines and looked on. They are sometimes called 'spectator ions'. Leaving out the spectator ions gives a much simpler equation which shows the essential precipitation process:

$$Ba^{2+}(aq) + SO_4^{2-}(aq) \longrightarrow BaSO_4(s)$$

This is an example of an ionic equation.

Experiment 23c
Making predictions and testing them by experiment

solutions of lead, barium and copper salts

silver nitrate solution

In this experiment, the solubility rules given in figure 23.4 are used to predict whether or not a precipitate will form when two solutions are mixed. The mixtures tested are shown in figure 23.8. The first two examples have been completed. Predict which of the other mixtures will form precipitates, then test your predictions by experiment.

Discussion
A precipitate forms when an ion from one soluble salt joins with an ion from another soluble salt to produce an insoluble salt. Six of the mixtures listed in figure 23.8 produce precipitates. Do you agree? The ionic equation for the reaction of lead(II) nitrate with potassium chloride is:

$$Pb^{2+}(aq) + 2Cl^-(aq) \longrightarrow PbCl_2(s)$$

Write ionic equations for the five other reactions which produce precipitates.

Solution 1			Solution 2			Formula of any precipitate	Colour of any precipitate
Soluble salt	Ions present		Soluble salt	Ions present			
Lead(II) nitrate	Pb^{2+}	NO_3^-	Potassium chloride	K^+	Cl^-	$PbCl_2(s)$	White
Sodium chloride	Na^+	Cl^-	Potassium nitrate	K^+	NO_3^-	None	—
Silver nitrate			Calcium chloride				
Zinc sulphate			Barium chloride				
Sodium chloride			Potassium sulphate				
Potassium nitrate			Sodium chloride				
Sodium carbonate			Calcium chloride				
Potassium nitrate			Copper(II) sulphate				
Zinc sulphate			Sodium carbonate				
Lead(II) nitrate			Barium chloride				

Figure 23.8

The reactions investigated in experiment 23c can be used to make insoluble salts if they are carried out on a larger scale. After mixing the two solutions, the precipitate is separated by filtration, or in a centrifuge. The spectator ions are washed away, leaving a product which can be dried to give a pure solid.

Experiment 23d
Making insoluble salts

lead(II) nitrate solution

The procedure is described by the series of diagrams in figure 23.9.

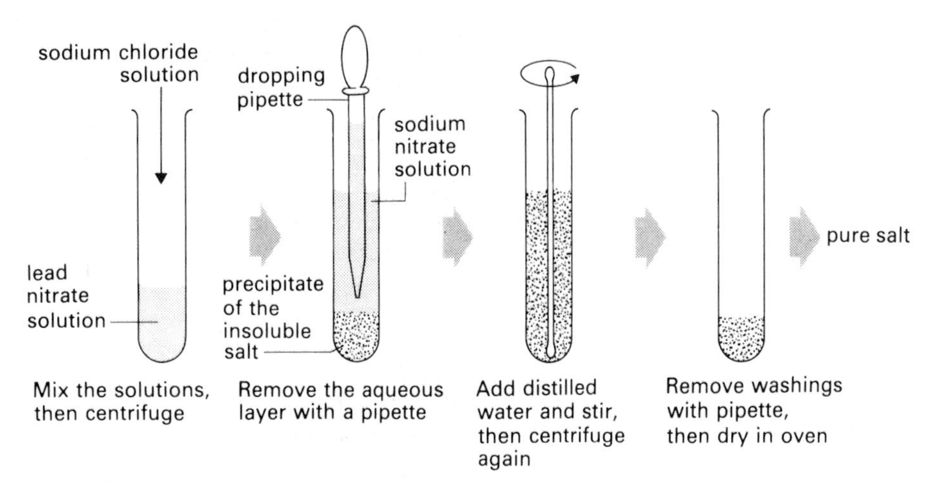

Figure 23.9

Discussion

Which salt is being prepared in figure 23.9? Which spectator ions are being washed away in the purification stages? Write an ionic equation for the reaction. Suggest pairs of soluble salts which could be used to make the following insoluble salts: silver chloride, lead(II) sulphate, lead(II) iodide.

Question

Which method would you use to make the following salts:

(a) sodium nitrate from sodium hydroxide,
(b) magnesium sulphate from magnesium oxide,
(c) zinc chloride from zinc,
(d) lead(II) nitrate from lead(II) carbonate,
(e) potassium sulphate from potassium carbonate?

Questions

1 Write an equation to show the effect of boiling temporarily hard water.
2 The cloudiness in milky limewater is a precipitate of calcium carbonate. Why does the limewater go clear again when more carbon dioxide is bubbled through it?
3 Some kettle descalers use methanoic acid (formic acid), HCO_2H, to remove the scale. Calcium methanoate is soluble in water. Explain the action of a descaler, giving word and symbol equations.

To decide which method to use to make a salt, ask yourself two questions:
1 Is the salt soluble in water?
If the answer is 'No', use the method of experiment 23d.
If the answer is 'Yes', go on to question 2.
2 Does the metal, base or carbonate to be used react with, or dissolve in, water?
If the answer is 'No', use the method of experiment 23a.
If the answer is 'Yes', use the method of experiment 23b.

23.6 Hard water and soap

Experiments to investigate hard water are described in chapter 10. The chemistry of hard water and washing illustrates the reactions of acids, bases and salts.

The formation of hard water

Hard water is formed in chalk and limestone regions where the rocks consist of calcium carbonate. Experiment 23a is done on a grand scale in nature. The acid is carbonic acid, which is formed when carbon dioxide dissolves in water:

$$H_2O(l) + CO_2(aq) \longrightarrow H_2CO_3(aq)$$

The soluble salt formed is calcium hydrogencarbonate:

$$CaCO_3(s) + H_2CO_3(aq) \longrightarrow Ca(HCO_3)_2(aq)$$

The calcium ions in the water make it hard. Hard water formed in this way is *temporarily hard*. Calcium hydrogencarbonate is decomposed by heat. On boiling, it splits up into calcium carbonate, carbon dioxide and water. The calcium carbonate appears as 'fur' in kettles and hot water pipes.

Washing in hard water

Washing with soap in hard water is like doing experiment 23d, and then washing the product down the drain. Soap is the sodium salt of a long-chain organic acid. It is a soluble salt. The other soluble salt is the calcium hydrogencarbonate in the hard water. On mixing the two soluble salts, a precipitate of an insoluble calcium salt of the soap-acid forms:

$$\underset{\text{soap}}{\text{Sodium stearate(aq)}} + \text{Calcium hydrogencarbonate(aq)} \longrightarrow$$
$$\underset{\text{scum}}{\text{Calcium stearate(s)}} + \text{Sodium hydrogencarbonate(aq)}$$

Water softening

Water softening by ion exchange is described in chapter 10. Bath salts are used to soften water and save soap. They are coloured and scented forms of sodium sesquicarbonate, $Na_2CO_3.NaHCO_3.2H_2O$. They work by precipitating calcium ions as calcium carbonate:

$$\underset{\text{in hard water}}{Ca^{2+}(aq)} + \underset{\text{from bath salts}}{CO_3^{\ 2-}(aq)} \longrightarrow CaCO_3(s)$$

Questions

Magnesium ions, like calcium ions, make water hard.
1 What will happen to soap in hard water containing magnesium ions?
2 What will happen to magnesium ions in hard water in an ion exchange resin (see chapter 10)?
3 What will happen when bath salts are added to hard water containing magnesium ions?

23.7 Analysis of salts

The reactions of acids, bases and salts are useful in analysis. The job of an analyst is to find out what things are made of. Modern analysis is often based on instrumental methods not available in schools.

A reaction can be used in a test-tube test if something can be seen to happen.

Figure 23.10 A modern analytical laboratory. These instruments are used to separate and identify the different compounds in mixtures

Maybe a gas is given off, or a precipitate forms, or there is a colour change. Changes such as these can be used in a scheme to identify simple salts.

Analysis starts with a preliminary examination of the unknown salt, for example, by heating it. Examples of the type of observation which may be made and the conclusions that can be drawn are shown in figure 23.11. The colour of the salt can also give useful clues.

Observations on heating	**Conclusion**
Water vapour/steam evolved, turning cobalt chloride paper pink	Crystals contain **water** of crystallization, or the solid is a **hydroxide** which decomposes
Colourless gas evolved which relights a glowing splint	Oxygen from a **nitrate** of potassium or sodium
Brown gas evolved and a glowing splint relights	Nitrogen dioxide and oxygen from the decomposition of a **nitrate**
Gas given off which turns limewater cloudy	Carbon dioxide from the decomposition of a **carbonate**
Pungent gas evolved which turns acid dichromate paper from orange to green	Sulphur dioxide from the decomposition of a **sulphate**
Sublimate forms on a cool part of the tube	Likely to be an **ammonium** salt. (Ammonia may also be detected with moist red litmus: it turns it blue.)
Residue turns yellow when hot and then white again when cold	Zinc oxide, which may have been formed by the decomposition of another **zinc compound**
Residue which is red when hot and yellow when cold	Lead(II) oxide, which may have been formed by the decomposition of another **lead compound**

Figure 23.11

A scheme for identifying the positive ions (cations) in a salt is shown in figure 23.12. The main test is based on the properties of the hydroxides of metals. Most hydroxides are insoluble in water. This means that a precipitate will usually form when sodium hydroxide is added to a solution of a metal salt. The hydroxides of transition metals can be recognized by their colours. Further tests are needed to identify the other hydroxides.

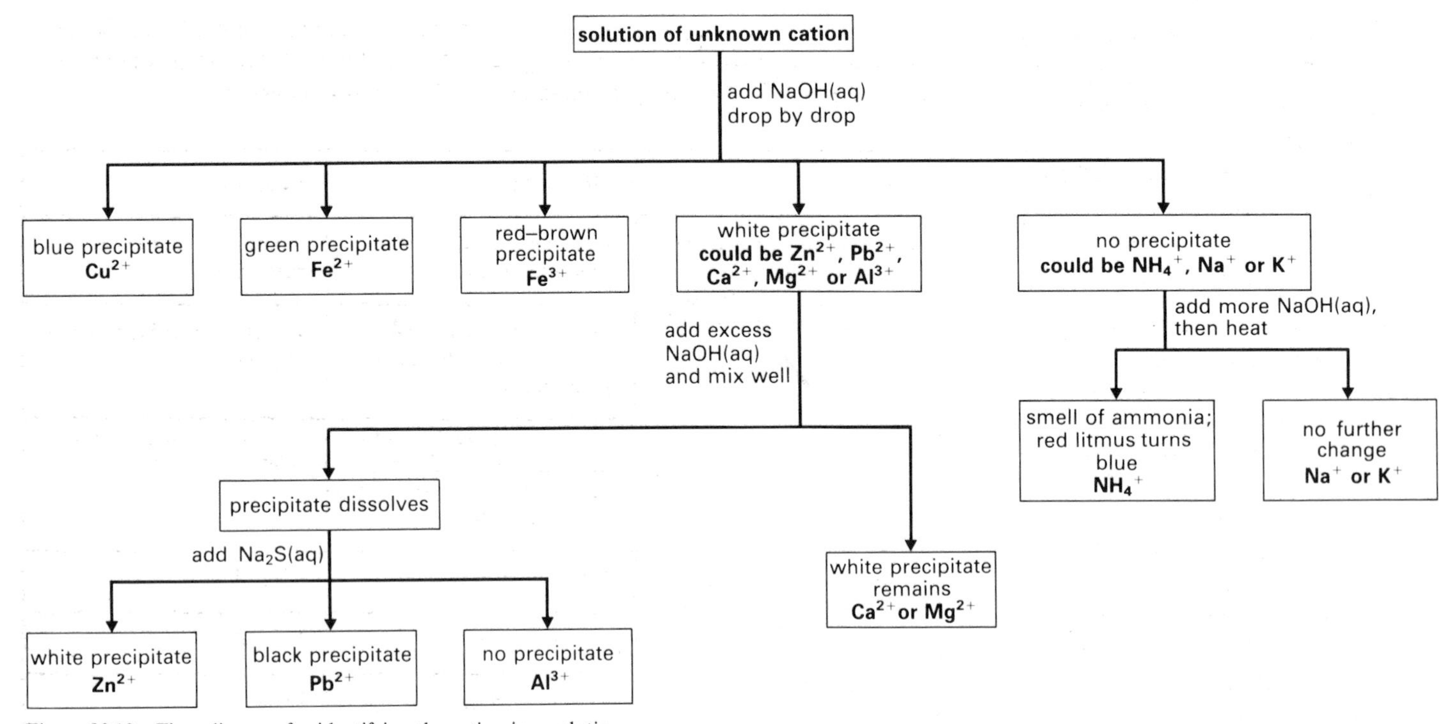

Figure 23.12 Flow diagram for identifying the cation in a solution

Some hydroxides will dissolve in an excess of sodium hydroxide solution. These are the **amphoteric** hydroxides.

The use of sodium sulphide is based on the fact that many sulphides are insoluble in water. Adding sodium sulphide to a solution of a metal salt will often produce a precipitate with a characteristic colour.

Flame tests can be used to distinguish metal ions which otherwise behave similarly. For example, both magnesium and calcium ions are precipitated by sodium hydroxide to give white precipitates which are insoluble in excess of the alkali. The procedure for a flame test is shown in figure 23.13. Common flame colours are shown in figure 23.14.

Figure 23.13 Flame test

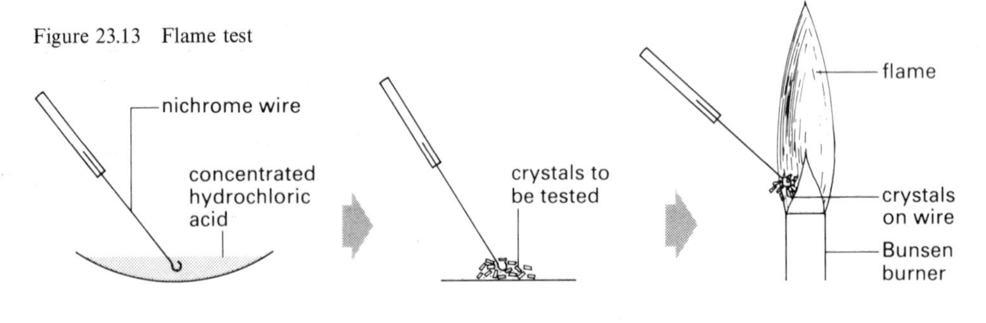

Metal ion	Flame colour
K^+	Pale mauve
Na^+	Bright yellow
Li^+	Bright red
Ca^{2+}	Orange–red
Cu^{2+}	Green with blue streaks

Figure 23.14

Questions

1 Identify these salts:
(a) A white solid which is insoluble in water. On heating, it gives off a gas which turns limewater milky. The residue after heating is yellow when hot, but white when cold.
(b) A white solid which, on heating, gives off a gas which turns dichromate paper green. No precipitate forms when sodium hydroxide solution is added to a solution of the salt. The salt gives a bright yellow flame.

2 Write ionic equations for
(a) the formation of a blue precipitate when sodium hydroxide is added to a solution containing copper(II) ions.
(b) the formation of a black precipitate when a solution of sulphide ions, S^{2-}, is added to a solution of lead(II) ions.

Questions

1 Write ionic equations for the precipitation of
(a) chloride ions with silver ions from silver nitrate.
(b) sulphate ions with barium ions from barium nitrate.
2 Identify these salts:
(a) This salt is soluble in water. A blue precipitate forms when aqueous sodium hydroxide is added to the solution. A white precipitate forms when nitric acid is added, followed by barium nitrate solution.
(b) This white solid sublimes on heating. An alkaline gas is evolved when it is warmed with sodium hydroxide solution. A white precipitate forms when nitric acid followed by aqueous silver nitrate is added to a solution of the salt.
3 Write word and symbol equations for the action of hydrochloric acid on
(a) magnesium carbonate,
(b) sodium sulphite and
(c) iron(II) sulphide.

A scheme for identifying negative ions (anions) is shown in figure 23.15. Some anions can be recognized by the gas given off when hydrochloric acid is added to the solid salt. The chemistry of these tests is described in chapter 21.

Test	Observation	Conclusion
1 Add dilute hydrochloric acid to the solid salt. Warm gently if there is no reaction in the cold.	Gas which turns limewater cloudy	Carbon dioxide from a **carbonate**
	Gas which smells foul and turns lead(II) ethanoate paper black	Hydrogen sulphide from a **sulphide**
	Gas which is acidic, has a pungent smell and turns acid dichromate paper from orange to green	Sulphur dioxide from a **sulphite**
2 **Test for halide ions** Make a solution of the salt. Acidify with nitric acid, then add silver nitrate solution.	White precipitate	**Chloride**
	Cream precipitate	**Bromide**
	Yellow precipitate	**Iodide**
3 **Test for sulphate ions** Take a fresh solution of the salt. Acidify with nitric acid then add a solution of barium nitrate or barium chloride.	White precipitate	**Sulphate**
4 **Test for nitrate ions** Take a fresh solution. Add an equal volume of iron(II) sulphate solution. Then pour concentrated sulphuric acid down the side of the tube.	Brown ring appears where the two layers meet	**Nitrate**

Figure 23.15 Tests for identifying the anion in a solution

Precipitation reactions are used to recognize chloride and sulphate ions. The identification of nitrates is based on the brown ring test. The procedure for this test is illustrated in figure 23.16.

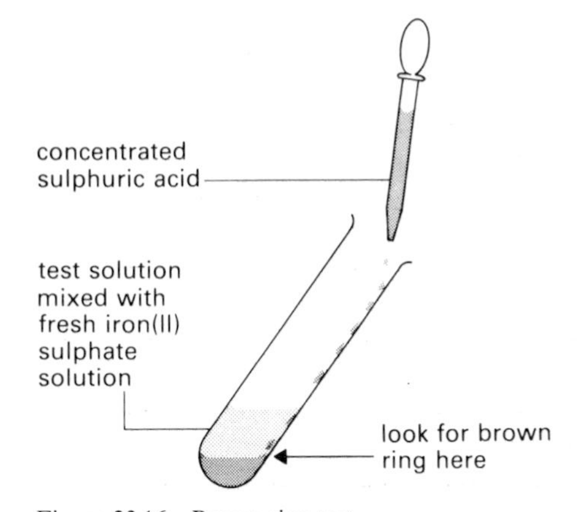

Figure 23.16 Brown ring test

Summary

1 What are salts?

How are they related to acids and bases? Name the parent acid and the parent base of each of these salts: potassium nitrate, copper(II) chloride, zinc sulphate.

2 How are salts made?

Which three reactions are commonly used to make soluble salts? What type of reaction is used to make insoluble salts?

3 Why are salts important?

Choose three salts and explain their everyday importance.

24 Acid-base theories

24.1 What makes an acid an acid?

Why do so many different compounds which we call acids have similar properties? Why do acids behave in the way that they do with metals, carbonates and bases?

Most of the reactions described in this theme have involved acids dissolved in water. What part, if any, does the water play? Does it matter which solvent is used?

The table in figure 24.1 compares the properties of a solution of hydrogen chloride in water with those of a solution of the same gas in methylbenzene. In the hydrocarbon solvent, methylbenzene, it shows none of the acid properties which it has in water. There are other examples. Citric acid dissolves in water and in propanone. In water it behaves as a typical acid. In propanone it does not. The conclusion is that acids only show their usual properties when dissolved in water.

Test	Observations with hydrogen chloride dissolved in water	Observations with hydrogen chloride dissolved in methylbenzene
Temperature change when the solution is made	Marked temperature rise	Very slight temperature change
Effect on dry litmus paper	Turns red	Little or no change
Reaction with magnesium ribbon	Hydrogen gas evolved rapidly	No reaction
Reaction with calcium carbonate	Carbon dioxide given off rapidly	No reaction
Electrical conductivity	Conducts electricity and hydrogen is evolved at the cathode	Non-conductor

Figure 24.1

Question

What will happen when
(a) magnesium ribbon is added to a solution of citric acid in propanone?
(b) sodium carbonate crystals are added to a solution of citric acid in water?

Acids do not simply mix with water when they dissolve in it. This is illustrated by the experiment shown in figure 24.2. There is little or no temperature change when hydrogen chloride dissolves in methylbenzene, but when it dissolves in water the temperature rises. This change suggests that there may be a chemical reaction.

A solution of hydrogen chloride in methylbenzene does not conduct electricity. The hydrogen chloride molecules have mixed with the methylbenzene molecules. There are no charged particles to carry an electric current.

A solution of hydrogen chloride in water does conduct electricity. Hydrogen is given off at the cathode. Hydrogen is always formed at the cathode when aqueous solutions of acids are electrolysed. This is an important clue. It shows that hydrogen ions are present in aqueous solutions of all acids.

All acids have hydrogen in their formulae. Examples are nitric acid, HNO_3, hydrochloric acid, HCl, and ethanoic acid, CH_3CO_2H. But not all substances

Questions

1 Write equations to show the ions formed when (a) nitric acid and (b) sulphuric acid react with water.

2 What are the spectator ions when
(a) magnesium reacts with sulphuric acid?
(b) sodium carbonate reacts with hydrochloric acid?
(c) copper(II) oxide reacts with nitric acid?
Name the salt formed in each of these three reactions.

which contain hydrogen are acids. Methylbenzene, $C_6H_5CH_3$, and ethanol, C_2H_5OH, are not acids.

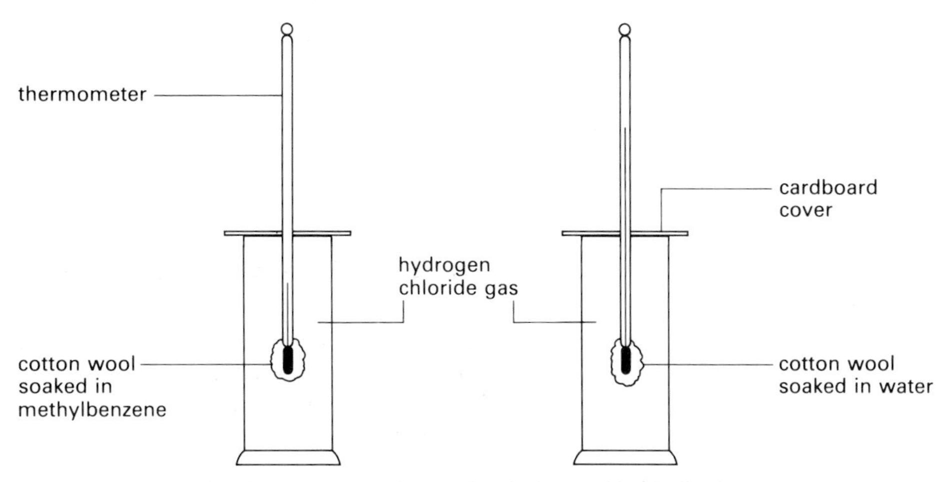

Figure 24.2 Measuring the temperature change when hydrogen chloride dissolves

Acids are substances containing hydrogen which react with water to produce hydrogen ions, e.g.

$$HCl(g) + Water \longrightarrow H^+(aq) + Cl^-(aq)$$

The typical reactions of acids are the reactions of the aqueous hydrogen ion, e.g. with metals:

$$Mg(s) + 2H^+(aq) \longrightarrow Mg^{2+}(aq) + H_2(g)$$

with carbonates:

$$CO_3{}^{2-}(s) + 2H^+(aq) \longrightarrow CO_2(g) + H_2O(l)$$

with bases:

$$O^{2-}(s) + 2H^+(aq) \longrightarrow H_2O(l)$$

But why is heat given out when acids react with water? Energy is needed to break chemical bonds. Energy is given out when new bonds are formed. This is discussed in more detail in chapter 37.

A hydrogen atom is the simplest of all atoms. It has one proton in its nucleus, and one electron in the first shell. When a hydrogen ion is formed, the atom loses its one electron. So a hydrogen ion is just a proton – a minute scrap of matter with an intense positive charge. Hydrogen ions do not exist free in water. They stick onto oxygen atoms in water molecules. It is the formation of these strong hydrogen–oxygen bonds that produces the heat energy. This is illustrated by the following equation:

$$H^+(\text{from the acid}) + H_2O(l) \longrightarrow H_3O^+(aq)$$

These hydrogen ions attached to water molecules are called *oxonium ions.*

When hydrogen chloride dissolves in water, the hydrogen chloride molecules give hydrogen ions to the water molecules. Hydrogen chloride is a strong acid. This means that it gives away hydrogen ions very readily. It reacts completely with water to form ions:

$$HCl(aq) + H_2O(l) \longrightarrow H_3O^+(aq) + Cl^-(aq)$$

An acid is a substance which will give away hydrogen ions. Ethanoic acid also gives away hydrogen ions, but not so readily. It is a weak acid. In a dilute solution of ethanoic acid only about one molecule in a hundred has reacted with water to form ions:

$$CH_3CO_2H(aq) + H_2O(l) \rightleftharpoons H_3O^+(aq) + CH_3CO_2{}^-(aq)$$

Questions

1 Which ion is formed when each of these acids gives away one hydrogen ion:
(a) hydrogen bromide, HBr,
(b) nitric acid, HNO_3,
(c) hydrogen sulphide, H_2S?
2 Which ion, or molecule, is formed when a hydrogen ion is added to each of these bases:
(a) the oxide ion, O^{2-},
(b) the carbonate ion, CO_3^{2-},
(c) the hydroxide ion, OH^-?

24.2 Alkalis and bases

A solution is alkaline if it contains more hydroxide ions, OH^-, than water does. The simplest way of making an alkaline solution is to dissolve a soluble hydroxide in water, e.g.

$$NaOH(aq) + Water \longrightarrow Na^+(aq) + OH^-(aq)$$

The hydroxide ions make the water alkaline.

Acids are neutralized by alkalis because hydrogen ions react with hydroxide ions to form water, which is neutral:

$$H^+(aq) + OH^-(aq) \longrightarrow H_2O(l)$$

Bases are the chemical opposites of acids. Acids give away hydrogen ions; bases take them. Ammonia is a base. A solution of ammonia is alkaline because ammonia molecules take hydrogen ions from water to leave hydroxide ions:

$$NH_3(aq) + H_2O(l) \rightleftharpoons NH_4^+(aq) + OH^-(aq)$$

The gain and loss of hydrogen ions is seen more clearly if this equation is split into two parts:

$$NH_3(aq) + H^+(\text{from water}) \longrightarrow NH_4^+(aq)$$
$$H_2O(l) \longrightarrow OH^-(aq) + H^+(\text{given to ammonia})$$

Summary

Copy the following sentences, filling in the missing words.

1 Acids only show their characteristic properties in the presence of ... .
2 A solution of hydrogen chloride in water is called ... .
3 A solution of hydrogen chloride in methylbenzene has ... effect on indicators and does not react with ... or ... .
4 Solutions of acids in water conduct electricity. When they conduct, ... gas is evolved at the cathode. All acid solutions contain ... ions.
5 Acids are substances which contain ... in their formulae and react with water to produce ... ions.
6 When a hydrogen atom loses an electron, all that is left is a ... .
7 An acid may also be defined as a ... donor.
8 A solution is alkaline if it contains ... ions.
9 During neutralization, the ... ions of an acid react with the ... ions of an alkali to form water.
10 A base is the chemical opposite of an acid. If an acid is a proton donor, a base is a proton ... .

Review questions

1

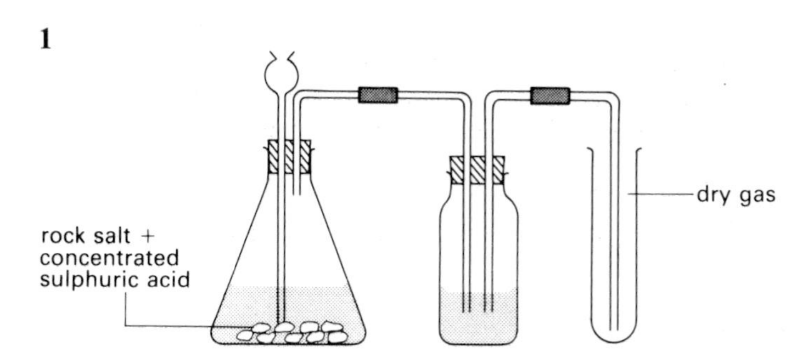

The apparatus shown above was used to prepare and collect test-tubes of dry hydrogen chloride gas. Concentrated sulphuric acid was added to lumps of rock salt (sodium chloride). The gas was then dried through concentrated sulphuric acid and collected by downward delivery.
(a) There is a mistake in the drying part of the diagram. Draw that part correctly to show you know what the mistake is.
(b) If the gas can be collected as shown, what does that tell you about the density of hydrogen chloride?
(c) A tube of the dry gas was tested with dry pH paper. Describe the result you would expect.
(d) A wet thermometer bulb was placed in another tube of dry hydrogen chloride. The temperature shown by the thermometer rose from 20 °C to 30 °C. What was the cause?
(e) The thermometer was then taken out and the drops of water left in the tube were tested with pH paper. What result would you expect this time?
(f) A tube of the dry gas was inverted with the mouth over water. What result would you see?
(g) Name the solution formed in (f). **(EMREB)**

2 This question is concerned with sodium hydroxide and with its manufacture from salt (sodium chloride) by an electrolytic process.
(a) (i) Sodium chloride is said to be an *electrolyte*. What do you understand by this term?
(ii) In what form is salt used in this process?
(iii) Name the materials used for the electrodes.
(iv) Give the changes which take place at the cathode and outline how the sodium hydroxide is obtained.
(b) For what reason is sodium hydroxide used in the manufacture of soaps from oils and fats?
(c) Briefly describe how you would use sodium hydroxide solution to distinguish (i) ammonium chloride from sodium chloride and (ii) an aqueous solution containing magnesium ions, Mg^{2+}, from one containing zinc ions, Zn^{2+}. **(AEB)**

3 Nitric acid and hydrogen chloride gas can be prepared in the laboratory by heating sodium nitrate and sodium chloride respectively with concentrated sulphuric acid.
(a) Write an equation for *each* of these preparations.
(b) Draw a labelled sketch of the apparatus you would use for *each* of these preparations.
(c) What property of concentrated sulphuric acid makes it particularly suitable for the preparation of nitric acid and hydrogen chloride?
(d) Name the gas formed in *each* case when concentrated sulphuric acid reacts with (i) methanoic acid (formic acid) and (ii) ethanol.
(e) What property of concentrated sulphuric acid makes it particularly suitable for the preparation of the gases named in (d)?
(f) Would it be possible to prepare hydrogen sulphide by the action of concentrated nitric acid on sodium sulphide? Give a reason for your answer. **(CLES)**

4 A student prepared a gas by using apparatus as follows:

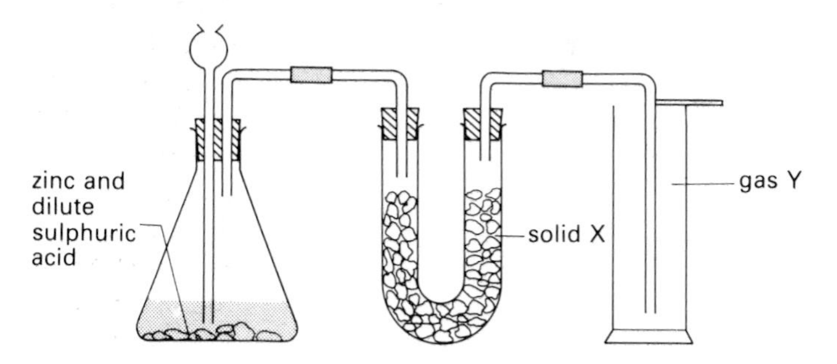

(a) Name gas Y.
(b) Name a suitable solid X for drying gas Y.
(c) There is an error in the student's method of collection. What is the error? How can dry gas Y be collected?
(d) Name the salt which is formed in the conical flask.
(e) Name *two* other chemicals which also react together to produce gas Y. What is the other product of this reaction?
(f) Give a *physical* property of gas Y not related to its method of collection.
(g) Give a *chemical* property of gas Y.
(h) Give a large-scale use of gas Y. **(WMEB)**

5 The following account describes how crystals of a salt were made.

Making sodium nitrate
30 cm^3 of a solution of sodium hydroxide were measured

out and placed in a beaker. A few drops of methyl orange solution were added, which gave the solution a yellow colour. Dilute nitric acid was then added, a little at a time, until the solution was coloured orange. A note was made of the volume of dilute nitric acid added, which was $28.0\,cm^3$.

The solution was then thrown away. A fresh $30\,cm^3$ of sodium hydroxide was measured into the beaker, but no methyl orange solution was added. Exactly $28.0\,cm^3$ of the dilute nitric acid was added.

The solution was then evaporated until it was saturated. When the saturated solution was cooled, crystals formed which were removed and dried.

(a) What apparatus would you use to measure out $30\,cm^3$ of sodium hydroxide solution?
(b) Why would it be incorrect to use a measuring cylinder for the dilute nitric acid?
(c) Suppose $29.0\,cm^3$ of dilute nitric acid had been added instead of $28.0\,cm^3$. What colour would you expect the methyl orange to be then?
(d) What kind of substance is methyl orange, when it is used in this way?
(e) Why do you think the first batch was thrown away and the experiment repeated?
(f) As sodium is in group I of the periodic table, what colour would you expect the crystals of sodium nitrate to be?
(g) Describe how to remove the crystals from the saturated solution without wasting any of them.
(h) Write the formula for sodium nitrate, if the nitrate ion is NO_3^-. **(EMREB)**

6 (a) A solution X of the nitrate of a metal M in water is colourless. On addition of a solution of sodium hydroxide to X, a white precipitate is formed which is soluble in excess of the alkali. On passing hydrogen sulphide through X, a black precipitate results. Identify the metal M, giving reasons for your answer and equations for the reactions involved.
(b) Suggest *one* confirmatory test for the metal M.
(c) A solution of sodium sulphite is believed to have undergone partial oxidation to sodium sulphate. Describe the tests you would carry out to show (i) that some sulphite ions are still present and (ii) that the solution also contains some sulphate ions. **(O & C)**

7 Temporary hard water is usually regarded as containing calcium hydrogencarbonate, $Ca(HCO_3)_2$, and permanent hard water is usually regarded as containing calcium sulphate, $CaSO_4$.
(a) Assuming that raindrops, at the moment they are formed, are absolutely pure water, explain the processes by which these two kinds of hard water are formed naturally.
(b) Describe experiments to show why these hard waters are called 'temporary' and 'permanent'.
(c) Describe *one* way by which you could soften either temporary or permanent hard water. **(L)**

8 A student was given aqueous solutions of five salts, A, B, C, D and E, and was told to carry out certain tests on each solution. These tests were to add to separate portions of each solution (i) aqueous sodium hydroxide, (ii) aqueous silver nitrate containing nitric acid and (iii) aqueous barium chloride containing hydrochloric acid. The student's results are given in the table below.

Solution	**Result of adding NaOH**	**Result of adding $AgNO_3$**	**Result of adding $BaCl_2$**
A	Blue precipitate	White precipitate	No change
B	White precipitate	No change	White precipitate
C	Red–brown precipitate	Pale yellow precipitate	No change
D	No change, but alkaline gas set free on warming	White precipitate	No change
E	Green precipitate	No change	White precipitate

Give the names of salts which could be present in the solutions, and explain why you choose that salt in each case. **(L)**

9 Carbon dioxide may be prepared by the action of hydrochloric acid on marble (calcium carbonate).
(a) Give the equation for the reaction.
(b) Why is hydrochloric acid used rather than sulphuric acid?
(c) (i) Apart from water vapour, what other impurity is present in the gas and how can it be removed?
(ii) How can the gas be dried?
(iii) Explain how the gas may be collected.
When carbon dioxide is passed over heated zinc some of the gas reacts as shown by the equation:

$$Zn + CO_2 \longrightarrow ZnO + CO.$$

(d) Sketch an apparatus by which, using this reaction, you could obtain a sample of carbon monoxide free from carbon dioxide.
(e) State and explain a chemical test for carbon monoxide.
(f) Name the products that would be formed if magnesium was used instead of zinc in the above reaction and suggest a reason for any difference in behaviour noted. **(AEB)**

THEME H
How much? How fast? How far?

The questions in the title of this theme are answered by the instruments in a control room

25 How much?

An important question to ask about reactions is, 'How much?' How much sulphur is needed to react with 1 g of iron to make iron sulphide? What volume of oxygen is needed to burn 5 g of magnesium, and how much magnesium oxide is formed? How many tonnes of iron could a manufacturer obtain from 100 tonnes of iron ore? The question 'How much?' is important to the industrial chemist who needs to work out whether an industrial process is economically worthwhile.

25.1 Counting atoms and molecules

When iron is heated with sulphur it gives iron sulphide. The equation for the reaction is:

$$Fe(s) + S(s) \longrightarrow FeS(s)$$

Figure 25.1 gives a clearer picture of what is happening.

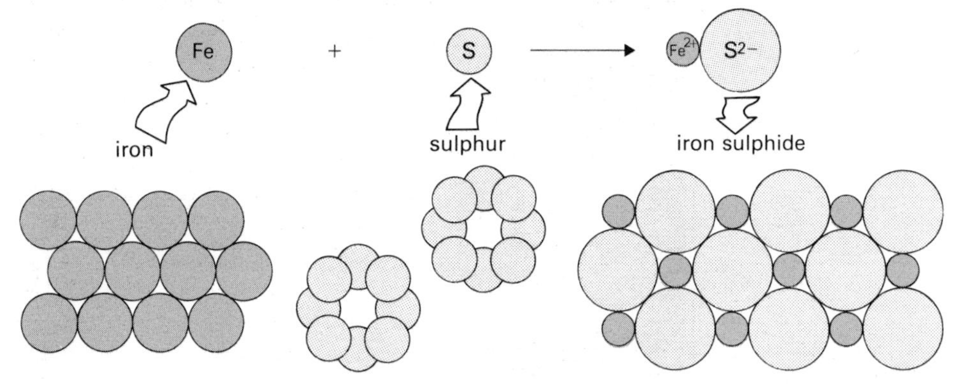

Figure 25.1 Model equation for the formation of iron sulphide

For every iron particle (ion) in iron sulphide, there is one sulphur particle (ion). This means that one iron atom reacts with one sulphur atom. To make iron sulphide, equal numbers of iron atoms and sulphur atoms must combine. It is tempting to think that '1 g of iron will combine with 1 g of sulphur', but that is wrong because different atoms have different masses. Dalton first suggested this idea, as described in chapter 3.

A hydrogen atom consists of just one proton and one electron (see chapter 17). Electrons have negligible mass, so the mass of a hydrogen atom is almost the same as that of a proton. Look at Dalton's list of atoms on p. 24. He gave hydrogen an 'atomic weight' of 1. Today we say that its **relative atomic mass** is 1. All other atoms are heavier than hydrogen atoms. Lithium has a relative atomic mass of 7, so each lithium atom is seven times heavier than a hydrogen atom. The usual abbreviation for relative atomic mass is A_r.

To make iron sulphide, equal numbers of iron and sulphur atoms are needed. But how can atoms be counted?

Bank clerks count large sums of money by weighing. It would be tedious and, perhaps, less accurate to count out large numbers of coins. Other people can

Question

From table 1 on p. 336, find the relative atomic masses of the following elements: hydrogen, carbon, nitrogen, oxygen, sulphur, bromine, calcium, magnesium, iron.

(a) Put these elements in order of increasing relative atomic mass. How does the order relate to their order in the periodic table?

(b) How many times heavier than hydrogen are atoms of (i) nitrogen and (ii) sulphur?

(c) How does the mass of a magnesium atom compare with the mass of a carbon atom?

(d) How does the mass of an iron atom compare with that of a nitrogen atom?

count by weighing. If a doctor prescribes 200 tablets for, say, backache, the pharmacist does not actually count out the tablets by hand. He uses an automatic counting machine or weighs them.

Atoms, too, can be counted by weighing, but atoms are so light that millions of them are needed to reach a mass which can be recorded on even the most sensitive of balances. A single hydrogen atom has a mass of about 1.7×10^{-24} g.

Figure 25.2 Each of these piles contains one mole of atoms

The counting number for atoms and molecules is about 602 000 000 000 000 000 000 000 (6.02×10^{23}). It is called the **Avogadro constant**. Avogadro was an Italian scientist who put forward theories as to how atoms and molecules combine. This number of hydrogen atoms has a mass of 1 g, and the amount is called one **mole**. This is convenient because the numerical value of the mass is the same as the relative atomic mass. Think of a mole as an amount which contains a fixed number, in the same way that a *dozen* always means '12 of …' and a *gross* always means '144 of …'. A mole of anything – atoms, molecules, ions or electrons – always contains 6.02×10^{23} of the particles. If the mass of one mole of hydrogen atoms is 1 g, the mass of one mole of carbon atoms is 12 g, because each carbon atom is twelve times heavier than a hydrogen atom.

Figure 25.2 shows one mole of various elements. Each contains the same number of atoms. Can you suggest why they have different volumes?

Every unit has a name and a symbol. A distance of ten metres is written as 10 m. A mass of fifty kilograms is written as 50 kg. In a similar way, an amount of two moles is written 2 mol.

A molecule of hydrogen, H_2, contains two atoms:

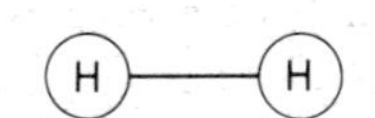

Each hydrogen molecule has a **relative molecular mass** of 2. The mass of 1 mol of hydrogen molecules is 2 g.

For compounds, the same ideas are used. In each water molecule, H_2O, there are two atoms of hydrogen and one atom of oxygen:

O
H H

The relative molecular mass of water is

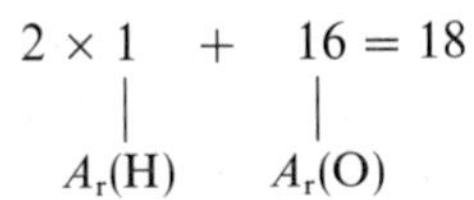

So the mass of 1 mol of water is 18 g.

Iron sulphide, FeS, is not molecular, but there is one iron ion for each sulphur ion. The mass of 1 mol of iron sulphide is

56 g	+	32 g	=	88 g
the mass of 1 mol of iron atoms		the mass of 1 mol of sulphur atoms		

Questions

1 What is the mass of
(a) 2 mol of lithium atoms?
(b) 0.5 mol of magnesium atoms?
(c) 0.02 mol of sulphur atoms?

2 What is the mass of
(a) 1 mol of oxygen molecules, O_2?
(b) 1 mol of tetrachloromethane, CCl_4?
(c) 0.1 mol of carbon dioxide, CO_2?

3 How many moles of atoms are there in
(a) 40 g of calcium?
(b) 560 g of iron?
(c) 4 g of bromine?

4 How many moles of molecules are there in
(a) 9 g of water?
(b) 64 g of oxygen?
(c) 8 g of methane, CH_4?

5 How many moles are there in
(a) 8 g of copper oxide, CuO?
(b) 117 g of sodium chloride, NaCl?
(c) 444 g of calcium chloride, $CaCl_2$?

6 About how many iron atoms are there in a typical two-inch nail, which has a mass of 5.6 g?

7 About how many molecules of water do you drink in a glassful? A small glass holds about 180 cm^3. 1 cm^3 of water has a mass of about 1 g.

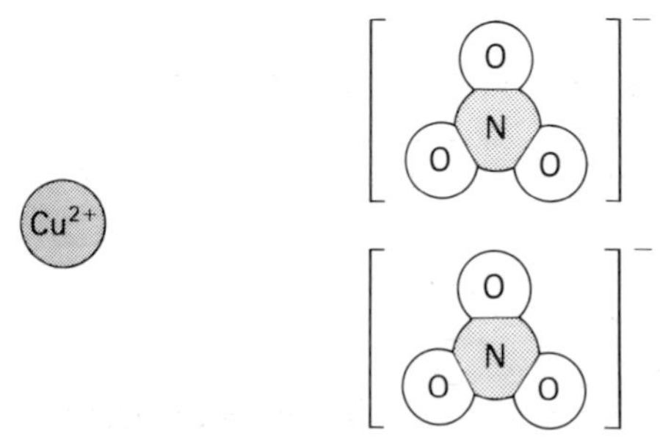

Figure 25.3 The ions in copper nitrate

In copper nitrate, $Cu(NO_3)_2$, for every copper ion, Cu^{2+}, there are two nitrate ions, NO_3^-. 1 mol of copper nitrate contains 1 mol of copper ions, Cu^{2+}, and 2 mol of nitrate ions, NO_3^-. The mass of 1 mol of copper nitrate is

$$64\,g + 2 \times 14\,g + 6 \times 16\,g = 188\,g$$

(the mass of 1 mol of copper atoms + the mass of 2 mol of nitrogen atoms + the mass of 6 mol of oxygen atoms)

The equations below are useful.

* Mass of a substance (g) = Number of moles × Mass of one mole (g)

* Number of moles of a substance $= \dfrac{\text{Mass of the substance (g)}}{\text{Mass of one mole (g)}}$

25.2 Finding formulae

Billy was a chemist, but Billy is no more,
For, what he thought was H_2O, was H_2SO_4.

H_2O and H_2SO_4 are the formulae for water and sulphuric acid. The formula tells us how many of each kind of atom there are in a molecule. In compounds which have **giant structures**, like calcium chloride, $CaCl_2$, it gives the simplest ratio between the numbers of each kind of atom or ion. In calcium chloride there are two chloride ions, Cl^-, for every calcium ion, Ca^{2+}. So one mole of calcium chloride contains one mole of calcium ions and two moles of chloride ions.

It is important to realize that all formulae have been confirmed by experiment. Scientists have found that 2 g of hydrogen (two moles of atoms) combine with 16 g of oxygen (one mole of atoms) to give 18 g of water (one mole of water molecules).

Question

Copy and complete the table in figure 25.4, which shows the formulae of some simple ionic compounds.

Compound	Formula	Number of moles of metal in one mole of compound	Number of moles of non-metal in one mole of compound
Calcium chloride	$CaCl_2$	1	2
Zinc oxide	ZnO		
Aluminium oxide	Al_2O_3		
Iron sulphide		1	1

Figure 25.4

To find a formula by experiment, the masses of elements which combine together must be found. In chapter 2, experiment 2e, the masses of oxygen and magnesium combining to form magnesium oxide are measured. The results can be used to find the formula of magnesium oxide. There are two ways of doing this. One is a graphical method and the other uses a calculation.

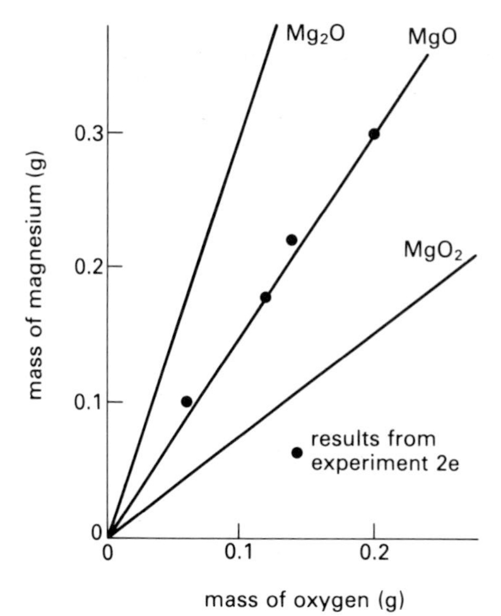

Figure 25.5 The masses of magnesium and oxygen that combine together: comparing the experimental results with various predictions

Using a graph

The formulae of many compounds involve simple numbers – usually 1, 2 or 3. Possible formulae for magnesium oxide are MgO, MgO_2 and Mg_2O.

If the formula is MgO, one mole of magnesium combines with one mole of oxygen atoms.

So 24 g of magnesium combines with 16 g of oxygen,
or 0.24 g of magnesium combines with 0.16 g of oxygen,
or 0.12 g of magnesium combines with 0.08 g of oxygen.

These points are plotted on the graph in figure 25.5. The line labelled 'MgO' is obtained when the points are joined to the origin. Any point on the line shows the masses of magnesium and oxygen which would combine if the formula were MgO.

If the formula is MgO_2, one mole of magnesium combines with two moles of oxygen atoms.

So 24 g of magnesium combines with 32 g of oxygen,
or 0.24 g of magnesium combines with 0.32 g of oxygen.

This information is also shown in figure 25.5. The line labelled 'MgO_2' gives the masses of magnesium and oxygen which would combine if the formula were MgO_2.

Work out combining masses for the formula Mg_2O. Do your values correspond with the line labelled 'Mg_2O' in figure 25.5?

Some of the results from experiment 2e have been plotted in figure 25.5. Copy the graph and add the remaining results. Do most points lie close to the 'MgO' line? The formula of magnesium oxide is MgO.

To find the formula of magnesium oxide, the elements were combined together. In experiment 25a, the formula of black copper oxide is found by removing oxygen from the oxide and weighing the copper left. Natural gas reduces copper oxide.

Question

Look again at the procedure given in experiment 2e.
(a) Why is it necessary to clean the magnesium before heating?
(b) Why is the crucible lid lifted from time to time during the heating?
(c) Why is it important not to let the smoke escape?

Experiment 25a
Finding the formula of copper oxide

copper(II) oxide

natural gas

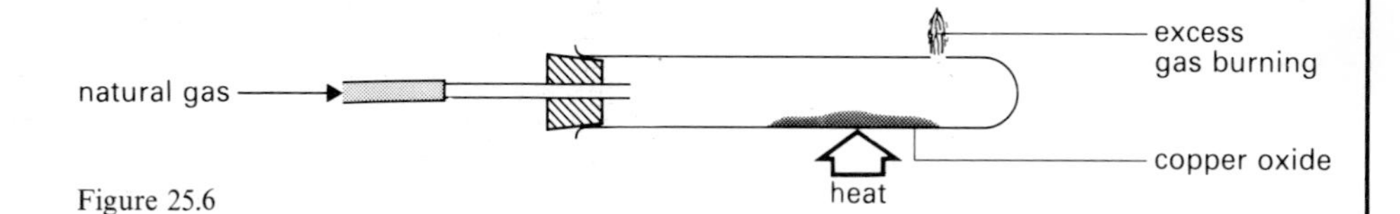

Figure 25.6

The apparatus is shown in figure 25.6. The dry test-tube is weighed accurately, first empty, and then with about 1 g of dry copper oxide. Natural gas is passed over the hot oxide. Unused gas burns at the hole in the test-tube. When all the oxide has been reduced to the metal, more gas is allowed to flow over the copper while it cools. Finally, the tube, with the copper, is weighed again.

Results

The black copper oxide becomes pink as copper is formed.

Typical results are:

Mass of test-tube empty	= 21.38 g
Mass of test-tube + copper oxide	= 22.51 g
Mass of test-tube + copper	= 22.29 g
Mass of copper oxide	= 1.13 g
Mass of copper	= 0.91 g
Mass of oxygen	= 0.22 g

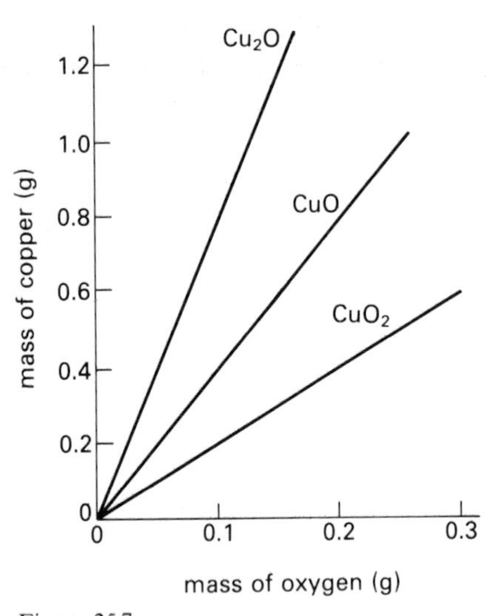

Figure 25.7

Discussion

Methane (natural gas) reduces copper oxide to copper. It is a compound of carbon and hydrogen. Carbon dioxide and water are formed during the reaction, but are swept out of the hot tube with the gas. Why is the copper cooled with gas flowing over it?

Figure 25.7 is a graph showing combining masses for the possible formulae CuO, CuO_2 and Cu_2O. Work through the calculations which lead to this graph. Figure 25.8 shows results from a series of experiments. Make a copy of figure 25.7, and include these experimental results. Use the measurements given above to add another point to your graph. Which results would you disregard? From the results of the experiment, what is the formula of copper oxide?

Mass of copper (g)	0.55	0.42	0.72	0.36	0.93	0.50	0.93
Mass of oxygen (g)	0.15	0.10	0.19	0.14	0.24	0.12	0.13

Figure 25.8

Question

Work out the formula of

(a) sodium chloride, if 2.3 g of sodium combines with 3.55 g of chlorine.

(b) ammonia, if 2.8 g of nitrogen combines with 0.6 g of hydrogen.

(c) nitrogen oxide, if 0.7 g of nitrogen combines with 1.6 g of oxygen.

(d) lead oxide, if 2.23 g of the oxide contains 2.07 g of lead.

(e) sodium nitrate, if 2.3 g of sodium, 1.4 g of nitrogen and 4.8 g of oxygen are combined.

Finding formulae using a graphical method is useful when a set of results are obtained by experiment. Points which are clearly in error can be disregarded.

Using a calculation

Formulae can be worked out by calculation. Masses of elements combining are converted to numbers of moles combining.

Example

In iron(III) oxide, it is found that 5.6 g of iron combines with 2.4 g of oxygen.

	Iron	Oxygen
Masses combining	5.6 g	2.4 g
Mass of one mole of the element	56 g	16 g
Number of moles combining	$\frac{5.6}{56} = 0.1$	$\frac{2.4}{16} = 0.15$
Simplest ratio of numbers of moles combining	1 or 2	1.5 3

The formula is Fe_2O_3.

The rules for finding formulae by calculation are shown in figure 25.9.

FINDING FORMULAE

1/ Write down the masses of elements combining.

2/ Write down the mass of one mole of each element.

3/ Work out amounts (numbers of moles) combining.

4/ Work out the simplest ratio between the amounts combining.

5/ Write down the simplest formula.

Figure 25.9 Finding chemical formulae by calculation

25.3 Investigating equations

An equation is a convenient shorthand way of describing what happens in a reaction. From it, we can work out how much chemical change can occur. An equation, like a formula, can be tested experimentally.

When the equation for the reaction of iron with sulphur is written:

$$Fe(s) + S(s) \longrightarrow FeS(s)$$

it tells us more than just, 'Iron reacts with sulphur to give iron sulphide' – it also tells us how much iron reacts with how much sulphur to give how much iron sulphide.

SOLVING PROBLEMS BASED ON EQUATIONS

1/ Write a balanced equation.

2/ In words, state what the equation tells you concerning the substances of interest.

3/ Change moles to masses.

4/ Scale the masses to those in the question.

Figure 25.10 Solving problems involving equations

The symbol 'Fe' means 'one iron atom', or 'one mole of iron atoms'. If it were changed to '2Fe' it would mean 'two iron atoms', or 'two moles of iron atoms'. So the equation tells us, '1 mol of iron reacts with 1 mol of sulphur to give 1 mol of iron sulphide.'

The mass of 1 mol of iron is 56 g, the mass of 1 mol of sulphur atoms is 32 g and the mass of 1 mol of iron sulphide is 88 g.

So 56 g of iron reacts with 32 g of sulphur to give 88 g of iron sulphide,
or 5.6 g of iron reacts with 3.2 g of sulphur to give 8.8 g of iron sulphide,
or any other masses such that the ratios stay the same.
How much iron sulphide could be made, starting with 14 g of iron?

Example

On heating limestone, $CaCO_3$, it decomposes to give calcium oxide, CaO, and carbon dioxide. What mass of calcium oxide can be formed from 25 g of limestone?

$$CaCO_3(s) \longrightarrow CaO(s) + CO_2(g)$$

From the equation, 1 mol of $CaCO_3$ reacts to give 1 mol of CaO.

The mass of 1 mol of $CaCO_3$ is

$$40\,g + 12\,g + 3 \times 16\,g = 100\,g$$

and the mass of 1 mol of CaO is

$$40\,g + 16\,g = 56\,g$$

So 100 g of $CaCO_3$ reacts to give 56 g of CaO
and 25 g of $CaCO_3$ reacts to give 14 g of CaO.
The steps in solving problems of this type are shown in figure 25.10.

Questions

1 In words, state what the following equations tell you:

(a) $C(s) + O_2(g) \longrightarrow CO_2(g)$

(b) $2Na(s) + Cl_2(g) \longrightarrow 2NaCl(s)$

(c) $Mg(s) + 2HCl(aq) \longrightarrow MgCl_2(aq) + H_2(g)$

2 What masses of (a) copper and (b) water are produced when 4 g of copper oxide is reduced by hydrogen?

3 What mass of sulphur is needed to just react with 13 g of zinc to give zinc sulphide?

4 In the blast furnace, haematite, Fe_2O_3, is converted to iron:

$$Fe_2O_3(s) + 3CO(g) \longrightarrow 2Fe(l) + 3CO_2(g)$$

(a) What mass of iron can be obtained from 16 tonnes of iron oxide?
(b) If 1 mol of carbon is needed to produce 1 mol of carbon monoxide, how much coke (carbon) is needed to reduce 16 tonnes of iron oxide?

In a process for extracting copper, the ore (copper pyrites) is converted to copper sulphate. Scrap iron is added to the copper sulphate solution and displaces the copper. An industrial process is of little value unless it is economically worthwhile. The success of the process depends on how much product is obtained. In experiment 25b, this reaction is investigated. By finding out how much copper is obtained using a known mass of iron, the equation for the reaction can be found.

Experiment 25b
Extracting copper from its salts using iron

h

copper(II) sulphate

The steps in the experiment are shown in figure 25.11.

Results
As the reaction occurs, the solution becomes hot and the blue copper ion solution becomes paler. Pink copper powder is formed.
Typical results are:

Mass of weighing bottle + iron	= 4.42 g
Mass of bottle (empty)	= 3.89 g
Mass of filter paper	= 0.84 g
Mass of filter paper + copper	= 1.46 g
Mass of iron used	= 0.53 g
Mass of copper formed	= 0.62 g

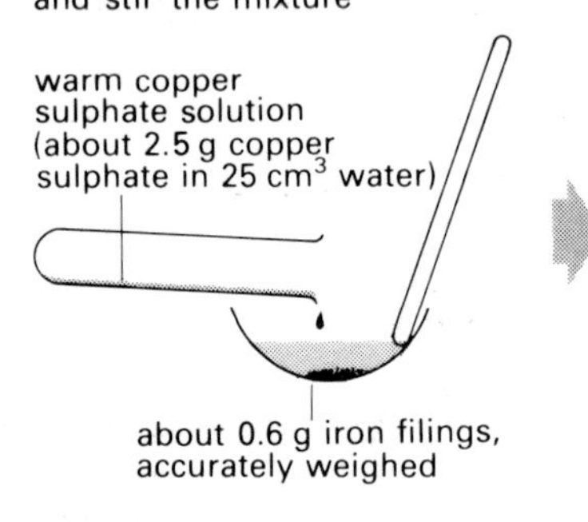
Slowly add warm copper sulphate solution to iron filings and stir the mixture
warm copper sulphate solution (about 2.5 g copper sulphate in 25 cm³ water)
about 0.6 g iron filings, accurately weighed

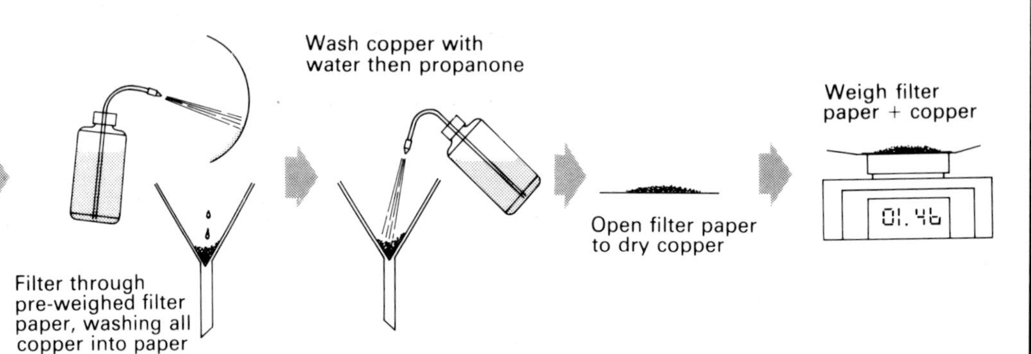
Wash copper with water then propanone
Weigh filter paper + copper
Open filter paper to dry copper
Filter through pre-weighed filter paper, washing all copper into paper
01.46

Figure 25.11

Discussion
Iron is more reactive than copper, so it displaces copper from its salts (see experiment 12d). What other metals could be used instead of iron? Why are these not used in the industrial process? Why does the blue colour become paler?
A word equation for the reaction is:

Iron + Copper(II) sulphate $\longrightarrow$ Copper + Iron sulphate

Iron can form two possible ions – Fe^{2+} and Fe^{3+} – in its salts (see chapter 3). The iron sulphate formed in the reaction could be iron(II) sulphate, $FeSO_4$, or iron(III) sulphate, $Fe_2(SO_4)_3$. There are two possible equations for the reaction:

$$Fe + CuSO_4 \longrightarrow Cu + FeSO_4 \quad [1]$$

$$2Fe + 3CuSO_4 \longrightarrow 3Cu + Fe_2(SO_4)_3 \quad [2]$$

From equation [1], how many moles of copper can be obtained from 1 mol of iron? What mass of copper can be obtained from 0.56 g of iron? How do the predictions change using equation [2]?
Show that the ratio

$$\frac{\text{Mass of copper obtained}}{\text{Mass of iron used}}$$

is 1.14 using equation [1], and 1.71 using equation [2].
What was the ratio obtained in the experiment? Is it closer to the predicted ratio from equation [1] or the one from equation [2]? Which equation is correct?
It is unlikely that the correct ratio will be obtained from one experiment, but an average of results from several experiments will give a clear indication of the correct equation. What are the main errors in the experiment? Can you see how you could express your results graphically, as in experiments 25a and b?

25.4 Gases in reactions

Questions

1 What is the volume of
(a) 1 mol of ethane?
(b) 0.1 mol of chlorine?
(c) 0.02 mol of ammonia?
2 How many moles of gas are present in
(a) 24 000 cm^3 of krypton?
(b) 120 cm^3 of carbon dioxide?
(c) 96 cm^3 of hydrogen?

Many chemical reactions involve gases. It is easier to measure the volume of a gas than its mass. The problem is to work out the number of moles from the volume.

The volume of a gas depends on the *number* of gas particles present, and not on their type or mass. Avogadro showed that equal numbers of moles of different gases have the same volume if the pressure and temperature are the same.

It has been found that one mole of any gas has a volume of about 24 000 cm^3 (24 dm^3) at room temperature (about 20 °C) and normal atmospheric pressure. Often, the conditions 0 °C and one atmosphere are chosen as the standard temperature and pressure (s.t.p.) when studying gases. Under these conditions, the volume of one mole of gas is 22 400 cm^3 (22.4 dm^3). By measuring the volume of a gas, the number of moles can be worked out easily. (The use of the unit 'dm^3' is explained on p. 344.) For example, at room temperature and pressure, the volume of 1 mol of carbon dioxide, $CO_2(g)$, is 24 000 cm^3, the volume of 0.5 mol of hydrogen, $H_2(g)$, is 12 000 cm^3, and the volume of 2 mol of chlorine, $Cl_2(g)$, is 48 000 cm^3.

The rules for relating volume to the number of moles are:

* Volume of gas (cm^3) = Number of moles × Volume of one mole (cm^3)

* $\text{Number of moles of gas} = \dfrac{\text{Volume of gas (cm}^3\text{)}}{\text{Volume of one mole of gas (cm}^3\text{)}}$

Experiment 25c
The reaction between magnesium and hydrochloric acid

dilute hydrochloric acid

Figure 25.12

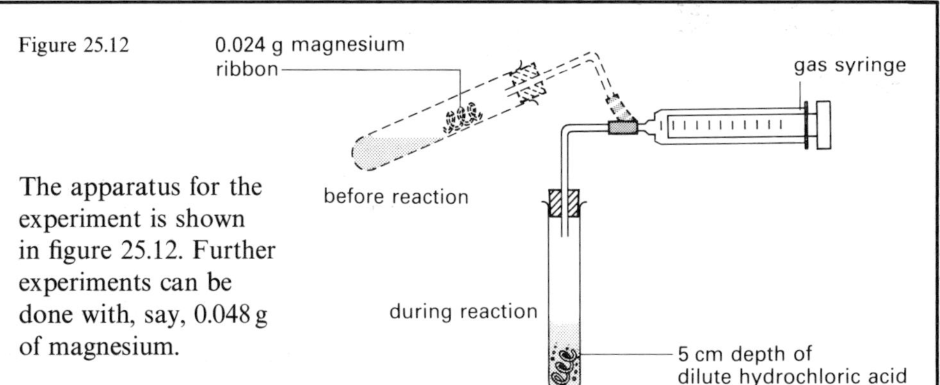

The apparatus for the experiment is shown in figure 25.12. Further experiments can be done with, say, 0.048 g of magnesium.

Results
A typical set of results is shown in figure 25.13. What is the average mass of magnesium used? What is the average volume of hydrogen obtained?

Mass of magnesium (g)	0.025	0.024	0.024	0.026	0.023	0.024	0.022
Volume of hydrogen (cm^3)	26.0	25.0	24.0	25.0	22.5	24.5	23.0

Figure 25.13

Discussion
The predicted equation for the reaction is

$$Mg(s) + 2HCl(aq) \longrightarrow MgCl_2(aq) + H_2(g)$$

From this, 1 mol of magnesium should give 1 mol of hydrogen. So, 24 g of magnesium (1 mol) should give 24 000 cm^3 of hydrogen, and 0.024 g of magnesium should give 24 cm^3 of hydrogen. Does the experiment confirm the equation? If the solution were evaporated away, what mass of magnesium chloride would you expect? Why is the volume of acid not measured accurately in the experiment?

Ammonia, NH_3, and hydrogen chloride, HCl, are both gases. When they meet, a white smoke of ammonium chloride is formed.

Experiment 25d
The reaction between ammonia and hydrogen chloride

ammonia gas
hydrogen chloride gas

hydrogen chloride gas

The apparatus is shown in figure 25.14. Syringe A contains 30 cm^3 of dry ammonia. Syringe B contains 50 cm^3 of dry hydrogen chloride. The gases are mixed via the 3-way tap.

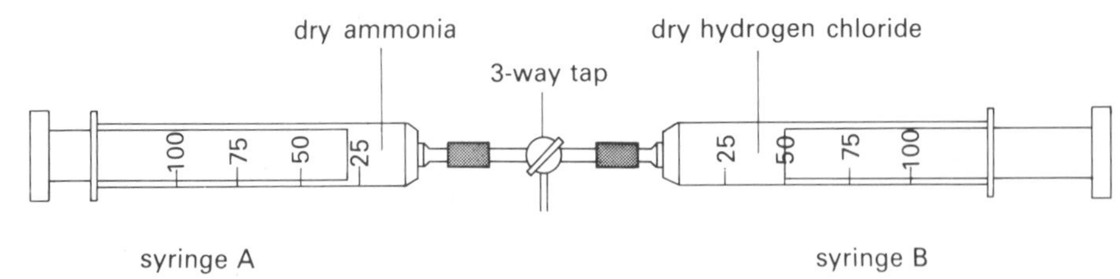

Figure 25.14

Results
Immediately the gases are mixed, there is a drop in the volume and a white solid forms. 20 cm^3 of gas is left. This gas is pushed out from the tap onto indicator paper, which turns red.

Discussion
The acid gas that remains at the end must be hydrogen chloride, because ammonia is alkaline. 30 cm^3 (50 cm^3 – 20 cm^3) of hydrogen chloride must have reacted. 30 cm^3 of ammonia reacts with 30 cm^3 of hydrogen chloride. Since equal volumes of gases contain equal numbers of moles, equal numbers of moles of the two gases must have reacted together. 1 mol of ammonia must react with 1 mol of hydrogen chloride. The left-hand side of the equation becomes:

$$NH_3(g) + HCl(g) \longrightarrow$$

The formula of ammonium chloride is NH_4Cl, so the equation can be completed.

Questions about equations involving gases are solved using the rules shown in figure 25.15. An example of their use is given on p. 185.

RULES
GASES IN EQUATIONS

1/ Write a balanced equation.
2/ In words, state what the equation tells you about the substances of interest.
3/ Change moles into masses or volumes (if gases.)
4/ Scale the masses or volumes to those in the question.

N.B. 1 mole of any gas has a volume of 24 000 cm^3 at room temperature and pressure (22 400 cm^3 at s.t.p.).
s.t.p. ≡ 0°C and 1 atm pressure.

Figure 25.15 Solving problems involving gases

Questions

1 What volume of carbon dioxide is produced when 25 g of calcium carbonate is heated?

2 Sulphur is burnt in air to give sulphur dioxide for making sulphuric acid:

$$S(s) + O_2(g) \longrightarrow SO_2(g)$$

What volume of oxygen is needed to burn 16 g of sulphur? What volume of air would be needed?

3 On heating, potassium nitrate gives oxygen:

$$2KNO_3(s) \longrightarrow 2KNO_2(s) + O_2(g)$$

What volume of oxygen is obtained from 5 g of potassium nitrate?

4 Methane (natural gas) burns in oxygen to give carbon dioxide and water vapour.
(a) Write an equation for the reaction.
(b) What volume of oxygen is needed to burn 10 m^3 of methane?
(c) What volume of carbon dioxide would be formed when 10 m^3 of methane burns?

Example
Calcium hydride, CaH_2, is a solid which reacts with water to give hydrogen. Calcium hydroxide is the other product. What volume of hydrogen is obtained from 2.1 g of calcium hydride, and what mass of water is required?

$$CaH_2(s) + 2H_2O(l) \longrightarrow 2H_2(g) + Ca(OH)_2(s)$$

From the equation, 1 mol of calcium hydride gives 2 mol of hydrogen. The mass of 1 mol of CaH_2 is

$$40\,g + 2 \times 1\,g = 42\,g$$

So 42 g of calcium hydride gives 48 dm^3 of hydrogen,
21 g of calcium hydride gives 24 dm^3 of hydrogen,
and 2.1 g of calcium hydride gives 2.4 dm^3, or 2400 cm^3, of hydrogen.
From the equation, 1 mol of calcium hydride reacts with 2 mol of water.
So 42 g of calcium hydride reacts with $2 \times 18\,g = 36\,g$ of water,
and 2.1 g of calcium hydride reacts with 1.8 g of water.

25.5 Reactions in solution

In many reactions, one or more of the reactants or products may be in solution. The concentration shows how much solute is dissolved in a known volume of solution. It may be expressed in terms of the mass dissolved in 1 dm^3 of solution, with units of g/dm^3. It is more useful to measure the concentration in moles of the substance in 1 dm^3.

A solution which contains one mole of substance in 1 dm^3 of solution is said to have a concentration of 1 mol/dm^3. This may also be written as 1 mol dm^{-3}. Chemists often use the term **molar solution** and the abbreviation 1.0 M.

A solution is called a **standard solution** if its concentration is accurately known. A standard solution is made up in a graduated flask. Figure 25.16 shows the steps in making up a standard solution. The example below shows how the mass needed for a solution can be calculated.

Figure 25.16 Preparing a standard solution

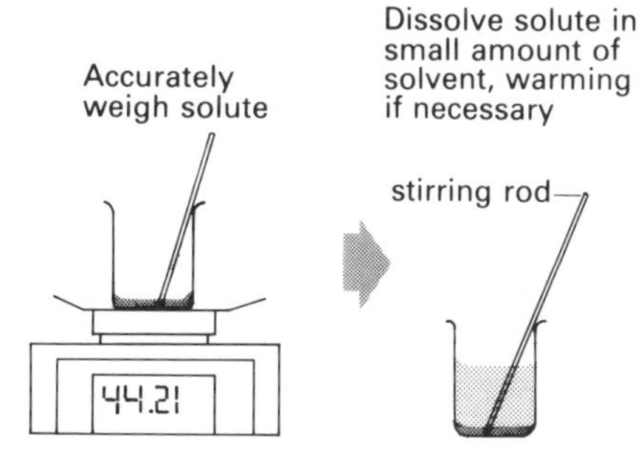
Accurately weigh solute
44.21
Dissolve solute in small amount of solvent, warming if necessary
stirring rod

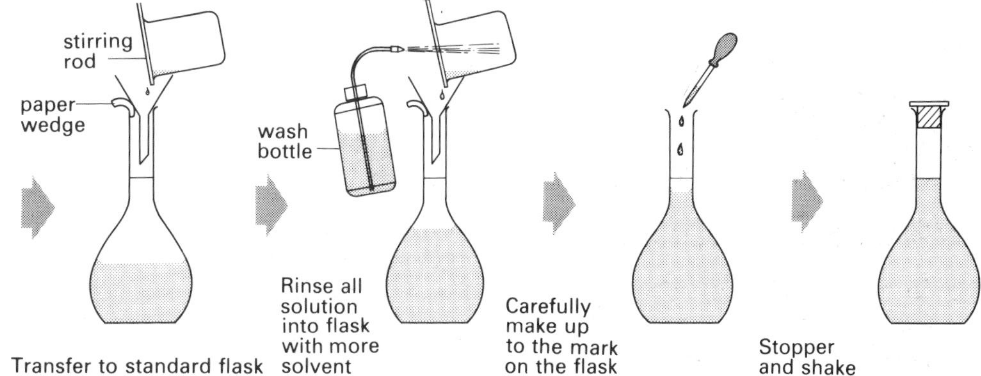
stirring rod
paper wedge
Transfer to standard flask
wash bottle
Rinse all solution into flask with more solvent
Carefully make up to the mark on the flask
Stopper and shake

Example
What mass of anhydrous sodium carbonate is needed to make 500 cm^3 of a 2 mol/dm^3 solution?

A solution of concentration 2 mol/dm^3 contains two moles of solute in 1 dm^3 (1000 cm^3) of solution. The mass of 1 mol of sodium carbonate, Na_2CO_3, is

$$2 \times 23\,g + 12\,g + 3 \times 16\,g = 106\,g$$

The mass of 2 mol of sodium carbonate is

$$2 \times 106\,g = 212\,g$$

So a 2 mol/dm^3 solution contains 212 g in 1 dm^3 (1000 cm^3) of solution, or 106 g in 0.5 dm^3 (500 cm^3).

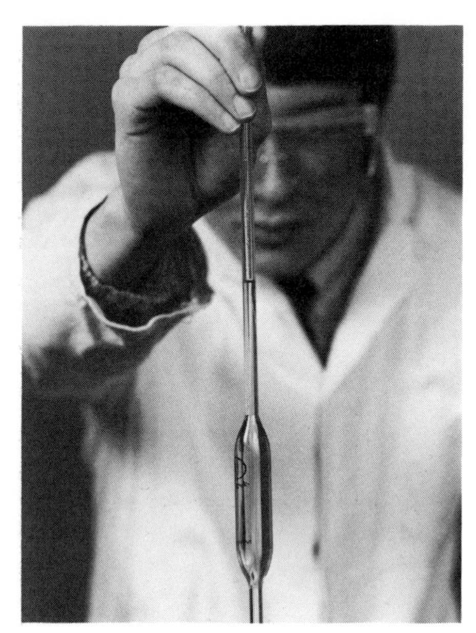

Figure 25.17 Using a pipette: the bottom of the meniscus is level with the graduation mark

Useful equations are:

✻Mass needed (g) = Volume of solution (dm^3) × Concentration (mol/dm^3) × Mass of 1 mol (g/mol)

✻Volume of solution (dm^3) $= \dfrac{\text{Volume of solution } (cm^3)}{1000}$

Knowing the concentration, the number of moles of solute in any volume of solution can be calculated:

✻Number of moles in a sample of solution (mol) = Volume of solution (dm^3) × Concentration (mol/dm^3)

Example
How many moles of hydrogen chloride are there in $40\,cm^3$ of a $0.2\,mol/dm^3$ solution?

In $1\,dm^3$ ($1000\,cm^3$) there is 0.2 mol of hydrogen chloride. The volume of solution is

$$\frac{40}{1000}\,dm^3$$

In this volume there is

$$\frac{40}{1000}\,dm^3 \times 0.2\,mol/dm^3 = 0.008\,mol$$

Questions

1 (a) What mass of sodium hydroxide, NaOH, is needed to make $1\,dm^3$ of a $1\,mol/dm^3$ solution?
(b) What mass of sodium carbonate, Na_2CO_3, is needed to make $500\,cm^3$ of a $0.5\,mol/dm^3$ solution?
2 What is the concentration of a solution which contains
(a) 0.1 mol of solute in $100\,cm^3$ of solution?
(b) 0.5 mol of solute in $250\,cm^3$ of solution?
3 How many moles of substance are present in
(a) $50\,cm^3$ of $0.01\,mol/dm^3$ sodium hydroxide solution?
(b) $5\,cm^3$ of $2.0\,mol/dm^3$ hydrochloric acid?

When equations for reactions in solution are being investigated, volumes must be measured accurately. Using a *pipette* (see figure 25.17), small fixed volumes, such as 5, 10 and $25\,cm^3$ are obtained. A *burette* allows a range of volumes to be delivered accurately. Correct readings are made on burettes and pipettes when the bottom of the liquid meniscus is in line with the graduation mark (see figure 25.17). Notice how these pieces of apparatus are used in experiments 25e and f.

Lead(II) nitrate and potassium iodide solutions react together to give a yellow precipitate of lead(II) iodide. In experiment 25e this reaction is investigated.

Experiment 25e
Investigating a precipitation reaction

lead(II) nitrate solution

Using a pipette or burette, $5\,cm^3$ samples of $1.0\,mol/dm^3$ potassium iodide are measured into each of eight test-tubes. From a burette, $0.5\,cm^3$ of $1.0\,mol/dm^3$ lead(II) nitrate is added to tube 1, $1.0\,cm^3$ to tube 2, $1.5\,cm^3$ to tube 3, etc. (see figure 25.18). After mixing, the tubes are centrifuged for 30 s.

Results
Some yellow precipitate is formed in each tube, but the height of it depends on the volume of lead(II) nitrate added. The results are shown in figure 25.19.

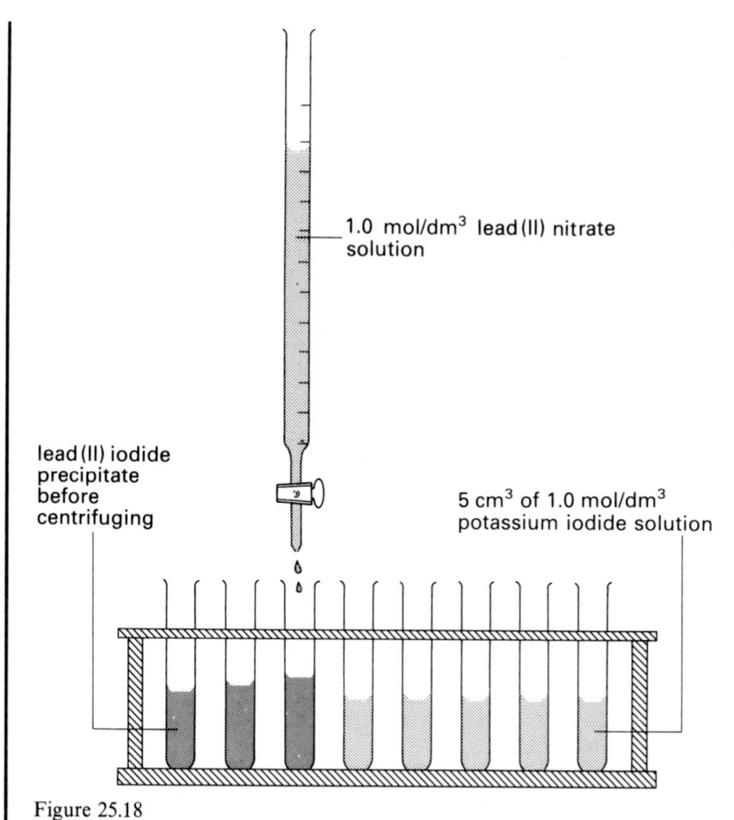

Figure 25.18

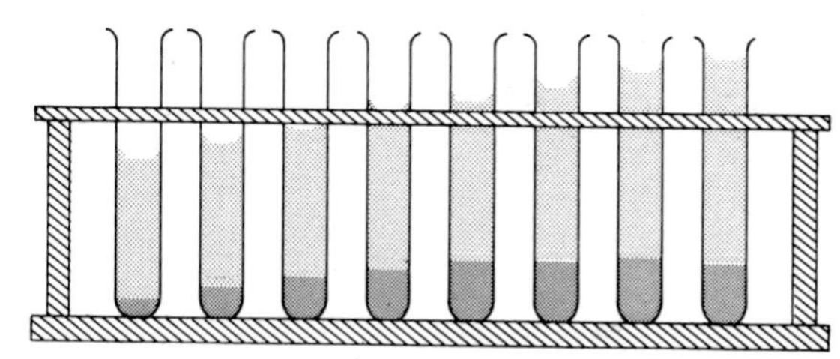

Figure 25.19

Discussion

There is a point when adding more lead(II) nitrate causes no more reaction. This is the 'end point' of the reaction. At the end point, $2.5\,cm^3$ of $1.0\,mol/dm^3$ lead(II) nitrate solution has been added. So $2.5\,cm^3$ of $1.0\,mol/dm^3$ lead(II) nitrate solution reacts with $5.0\,cm^3$ of $1.0\,mol/dm^3$ potassium iodide solution.

The number of moles of lead(II) nitrate used is

$$\frac{2.5}{1000}\,dm^3 \times 1.0\,mol/dm^3 = 0.0025\,mol$$

The number of moles of potassium iodide used is

$$\frac{5.0}{1000}\,dm^3 \times 1.0\,mol/dm^3 = 0.005\,mol$$

0.0025 mol of lead(II) nitrate just reacts with 0.005 mol of potassium iodide. The ratio is 1:2 so 1 mol of lead(II) nitrate reacts with 2 mol of potassium iodide.

The left-hand side of the equation is:

$$Pb(NO_3)_2(aq) + 2KI(aq) \longrightarrow$$

The products of the reaction are lead(II) iodide, $PbI_2(s)$, and potassium nitrate, $KNO_3(aq)$. Complete the equation.

What ions are present in solution when (a) $1\,cm^3$, (b) $2.5\,cm^3$ and (c) $4.0\,cm^3$ of lead(II) nitrate solution have been added?

In experiment 25e, the end point is indicated by no more precipitate being formed. In experiment 25f, there is no precipitate, but an added **indicator** shows the end point. Indicators like methyl orange and phenolphthalein have different colours in acidic and alkaline solutions. The end point, when the acid just neutralizes the alkali, is indicated by the indicator changing colour.

Experiments in which the volumes of solutions that react together are found, are called **titrations**.

Experiment 25f
Titrating sodium hydroxide with sulphuric acid

dilute sulphuric acid

dilute sodium hydroxide

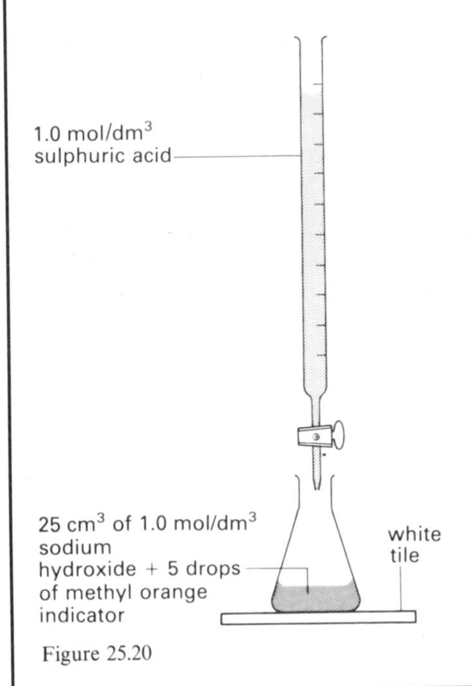

Figure 25.20

The flask in figure 25.20 contains 25 cm^3 of 1.0 mol/dm^3 sodium hydroxide solution. This is measured using a pipette. Methyl orange is the indicator. The burette contains 1.0 mol/dm^3 sulphuric acid. A trial titration gives an approximate end point before more accurate titrations are carried out.

Results
At the end point the indicator is a pink–orange colour. Methyl orange is red in acidic solution and yellow in alkaline solution. Typical burette readings are shown in figure 25.21.

	Trial	I	II
1st burette reading	0.00	13.00	25.60
2nd burette reading	13.00	25.60	38.10
Volume used (cm^3)	13.00	12.60	12.50

Figure 25.21

The average volume used is 12.55 cm^3. This volume of 1.0 mol/dm^3 sulphuric acid reacts with 25.0 cm^3 of 1.0 mol/dm^3 sodium hydroxide solution.

Discussion
The number of moles of sodium hydroxide in the flask is

$$\frac{25}{1000}\,\text{dm}^3 \times 1.0\,\text{mol/dm}^3 = 0.025\,\text{mol}$$

Work out the number of moles of sulphuric acid that just react with the sodium hydroxide. How many moles of sulphuric acid would be required to react with 1 mol of sodium hydroxide? What is the left-hand side of the equation for the reaction? What are the products? Complete the equation.

RULES

SOLUTIONS IN EQUATIONS

1/ Write a balanced equation if it is known.

2/ In words, state what the equation tells you about the substances of interest.

3/ Work out the amount of substance (number of moles) which *can* be found.

4/ Using the equation find the amount of the other substance in the titration.

5/ Find the mass, volume or concentration as needed.

6/ To find an equation, start at rule 3 and work backwards.

Figure 25.22 Solving problems involving solutions

In experiments 25e and f, the results are used to confirm equations for reactions. The following example shows one type of problem that can be solved if the equation is known.

Example
What volume of 0.5 mol/dm^3 hydrochloric acid is needed to neutralize 25 cm^3 of 2.0 mol/dm^3 sodium hydroxide solution?

The equation for the reaction is:

$$HCl(aq) + NaOH(aq) \longrightarrow NaCl(aq) + H_2O(l)$$

From the equation, 1 mol of hydrochloric acid reacts with 1 mol of sodium hydroxide. The number of moles of sodium hydroxide is

$$\frac{25}{1000}\,\text{dm}^3 \times 2.0\,\text{mol/dm}^3 = 0.05\,\text{mol}$$

so the number of moles of hydrochloric acid needed is 0.05 mol. The hydrochloric acid contains 0.5 mol in 1 dm^3, so 0.05 mol is contained in 0.1 dm^3 or 100 cm^3.

The rules shown in figure 25.22 may be helpful in solving most problems involving solutions.

Questions

1 What volume of 0.1 mol/dm^3 hydrochloric acid is needed to dissolve 0.24 g (0.01 mol) of magnesium? The equation is:

$$Mg(s) + 2HCl(aq) \longrightarrow MgCl_2(aq) + H_2(g)$$

2 In a titration involving nitric acid and sodium carbonate, 25 cm^3 of 0.8 mol/dm^3 nitric acid is found to react exactly with 10 cm^3 of 1.0 mol/dm^3 sodium carbonate. Work out the left-hand side of the equation.

3 What volume of 0.5 mol/dm^3 hydrochloric acid would be needed to remove 10 g of scale (calcium carbonate) from a kettle? The equation for the reaction is:

$$CaCO_3(s) + 2HCl(aq) \longrightarrow CaCl_2(aq) + CO_2(g) + H_2O(l)$$

Summary

Write a short rule book to show how to do the main types of chemical calculation. For each set of rules give a worked example. You should show (a) how to work out formulae from experimental results and (b) how to solve problems based on equations, including problems involving gases and solutions.

26 How fast?

Chemical reactions take place at a variety of speeds. Precipitation reactions are very fast and are complete in a fraction of a second. The rusting of iron and other corrosion processes are very slow and continue for years. Most reactions that you will study occur in a few minutes. It is important to know how quickly reactions occur, and how to change their speeds. If the reaction in an industrial process is too slow, it becomes uneconomic.

Chemical equations say nothing about how quickly the changes occur. The rate of a reaction is found by experiment.

26.1 Rate of reaction

The speed of a car is measured in kilometres (or miles) per hour. Your pulse rate is the number of times your heart beats in every minute. The production rate in a factory is a measure of how many articles are made in a particular time.

In chemical reactions, the amounts or concentrations of substances change. As products are formed, reactants are used up. For the reaction

$$Mg(s) + 2HCl(aq) \longrightarrow MgCl_2(aq) + H_2(g)$$

we could measure the rate of disappearance of magnesium, or the rate of disappearance of hydrochloric acid, or the rate of appearance of magnesium chloride, or the rate of appearance of hydrogen. Each of these would be a measure of the rate of reaction.

How the changes are followed depends on the nature of the reaction. Any property which changes as the amounts of reactants or products change can be monitored during the reaction.

Question

In the reaction

$$Mg(s) + H_2SO_4(aq) \longrightarrow MgSO_4(aq) + H_2(g)$$

if 36 cm^3 of hydrogen is formed in 20 s,

(a) what is the rate of formation of hydrogen, in cm^3/s?

(b) what is the rate of formation of hydrogen, in mol/s?

(c) what are the rates of appearance or disappearance of the other product and the reactants?

In most chemical reactions, the rate changes with time. Figure 26.1 shows the formation of a product with time. The graph is steepest at the start of the experiment, showing that the rate of reaction is greatest then. As the reaction continues, the rate decreases until, eventually, the reaction stops.

In drag racing, each car is timed over a fixed distance. The fastest car covers the distance in the shortest time. In experiments 26b and c, a similar idea is used to compare the rates of chemical reactions. The times for the formation of a certain

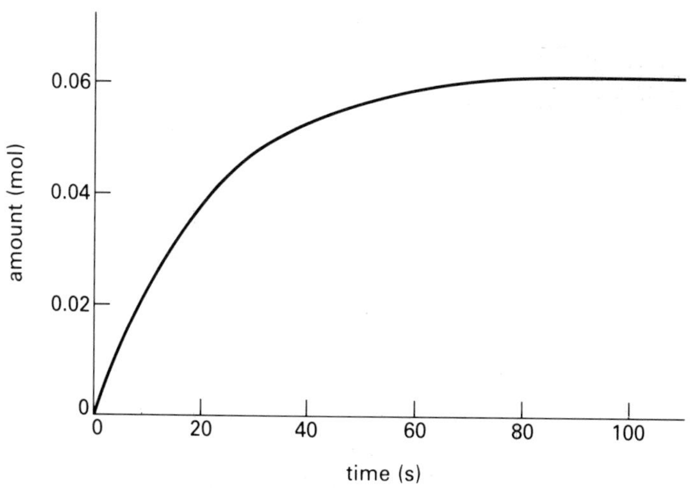

Figure 26.1

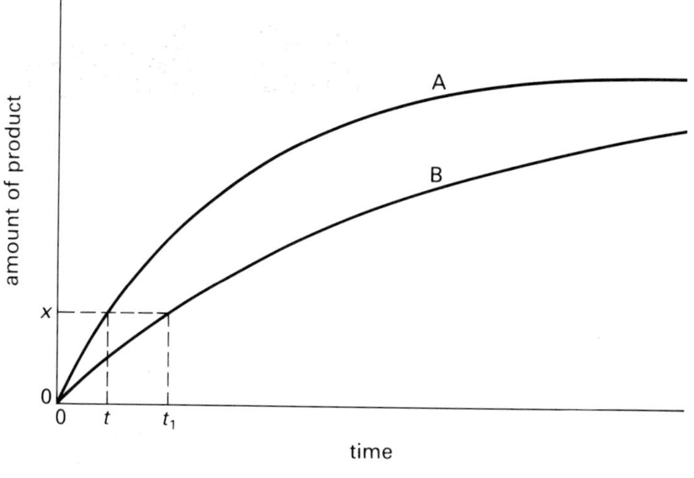

Figure 26.2

amount of product are found. In figure 26.2, line A shows the formation of a product under a particular set of conditions. The time, t, is for the formation of a certain amount of product, x. Line B shows a slower reaction. The time, t_1, is for the formation of the same amount of product, x. t_1 is greater than t.

The average rate of formation of product on line A is

$$\frac{x}{t}\ \text{mol/s}$$

and the average rate of formation of product on line B is

$$\frac{x}{t_1}\ \text{mol/s}$$

If the amount, x, is small, then the average rate is very nearly the same as the initial rate. So the initial rate of formation of product on line A

$$\propto \frac{1}{t}\ \text{mol/s}$$

and the initial rate of formation of product on line B

$$\propto \frac{1}{t_1}\ \text{mol/s}$$

A loaf of sliced bread goes stale faster than an unsliced loaf when exposed to air. Milk stored in a warm kitchen goes sour more quickly than milk kept in a refrigerator. The rates of these chemical reactions are altered by changing the conditions. Can you think of other similar examples?

Chemical reaction rates can be affected by

* changes in the surface area of solid reactants
* changes in the concentration of reactants in solution
* changes in pressure (if the reactants are gases)
* changes in temperature
* added substances called catalysts
* light (although this applies to a limited number of reactions).

26.2 Surface area

Marble is a form of calcium carbonate. It reacts with dilute hydrochloric acid, giving off carbon dioxide:

$$CaCO_3(s) + 2HCl(aq) \longrightarrow CaCl_2(aq) + CO_2(g) + H_2O(l)$$

In experiment 26a, the effect of the size of marble chips on the rate of formation of carbon dioxide is investigated.

Experiment 26a
The effect of particle size on a reaction rate

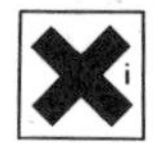

dilute hydrochloric acid

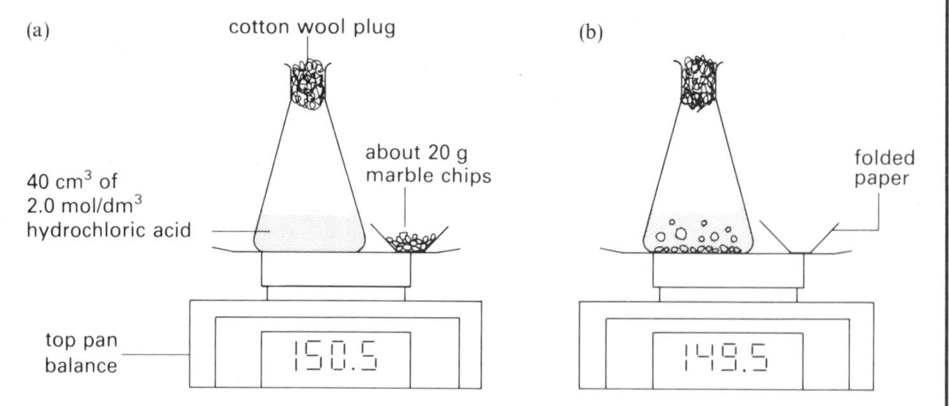

Figure 26.3

Figure 26.3(a) shows the apparatus before the reaction is started. In figure 26.3(b), the reaction is in progress. The mass of the apparatus is recorded at regular intervals for 6–7 minutes after adding the marble chips to the acid. Different grades (sizes) of marble chips are studied.

Results
Figure 26.4 shows typical results using two grades of marble chips. In both cases, the reaction started briskly.

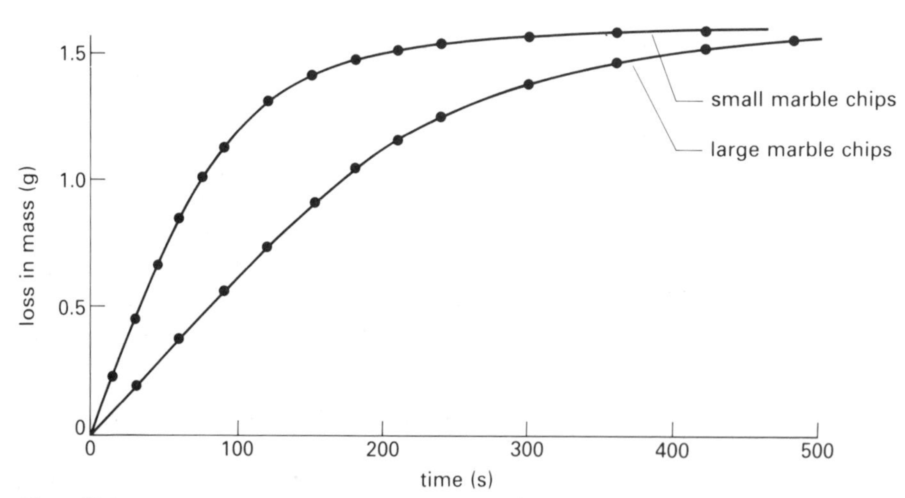

Figure 26.4

Discussion
The loss in mass is equal to the mass of carbon dioxide formed because the gas escapes through the cotton wool. So the graph in figure 26.4 shows the formation of carbon dioxide with time.

Which reaction has the greater initial rate? What happens to the rate during each reaction? Why does the reaction stop?

Particle size and surface area are related. For a given mass, the smaller the particles, the greater the surface area. The particle size, and therefore the surface area, of the chips does not change much during the reaction because excess marble is used. Very little is used up in the reaction. Altering the surface area of the marble by changing the size of the chips affects the rate, but not the extent, of reaction. This means that the total amount of carbon dioxide formed is the same in each experiment. Why?

Questions

1 Design another way of following the rate of reaction in experiment 26a.
2 Explain the following:
(a) Iron filings spark when thrown into a flame, whereas a lump of iron appears to do nothing.
(b) The labels on bottles of aluminium powder sometimes say, 'Danger! Dust explosion possible'.
(c) Sliced bread stales more quickly than unsliced bread.

From experiment 26a and others like it we see that

* increasing the surface area of a solid reactant increases the reaction rate.

26.3 Concentration

In experiment 26a, the reactions became slower with time. The reduction in their rates corresponded to a lowering of the hydrochloric acid concentration as it was used up.

In experiment 26b, the effect of concentration on the rate of a different reaction is studied. Dilute hydrochloric acid reacts with sodium thiosulphate solution to give a precipitate of sulphur:

$$Na_2S_2O_3(aq) + 2HCl(aq) \longrightarrow 2NaCl(aq) + H_2O(l) + SO_2(aq) + S(s)$$

The concentration of one reactant, sodium thiosulphate, is changed, and the rates are compared by recording the times for a certain amount of reaction to occur.

Experiment 26b
The effect of concentration on a reaction rate

dilute hydrochloric acid

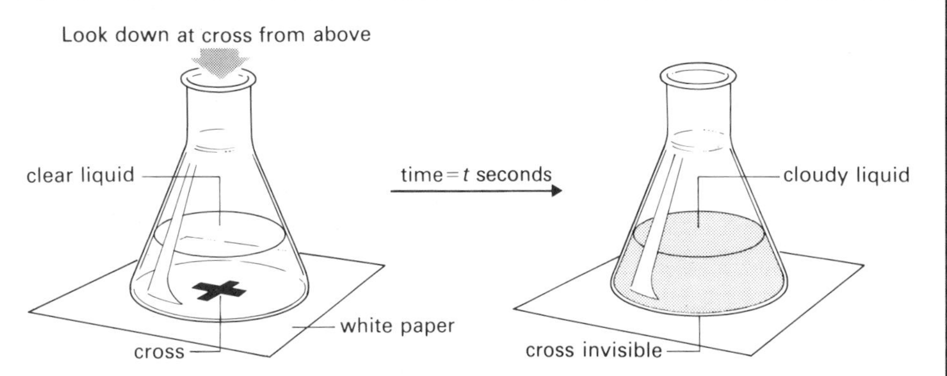

Figure 26.5

50 cm^3 of 0.15 mol/dm^3 sodium thiosulphate solution and 5 cm^3 of 2.0 mol/dm^3 hydrochloric acid are mixed in a flask over a paper marked with a cross (see figure 26.5). The cross is viewed from above, and the time for it to vanish is noted. The experiment is repeated with other concentrations of sodium thiosulphate, which are obtained by dilution (see figure 26.6).

Volume of 0.15 mol/dm^3 sodium thiosulphate solution (cm^3)	Volume of water (cm^3)	Concentration sodium thiosulphate solution (mol/dm^3)	Time, t, for cross to be obscured (s)	Rate of reaction, $1/t$(/s)
50	0	0.15	43	0.023
40	10	0.12	55	0.018
30	20	0.09	66	0.015
20	30	0.06	105	0.0095
10	40	0.03	243	0.0041

Figure 26.6

Results
In each experiment, the amount of creamy yellow precipitate of sulphur steadily

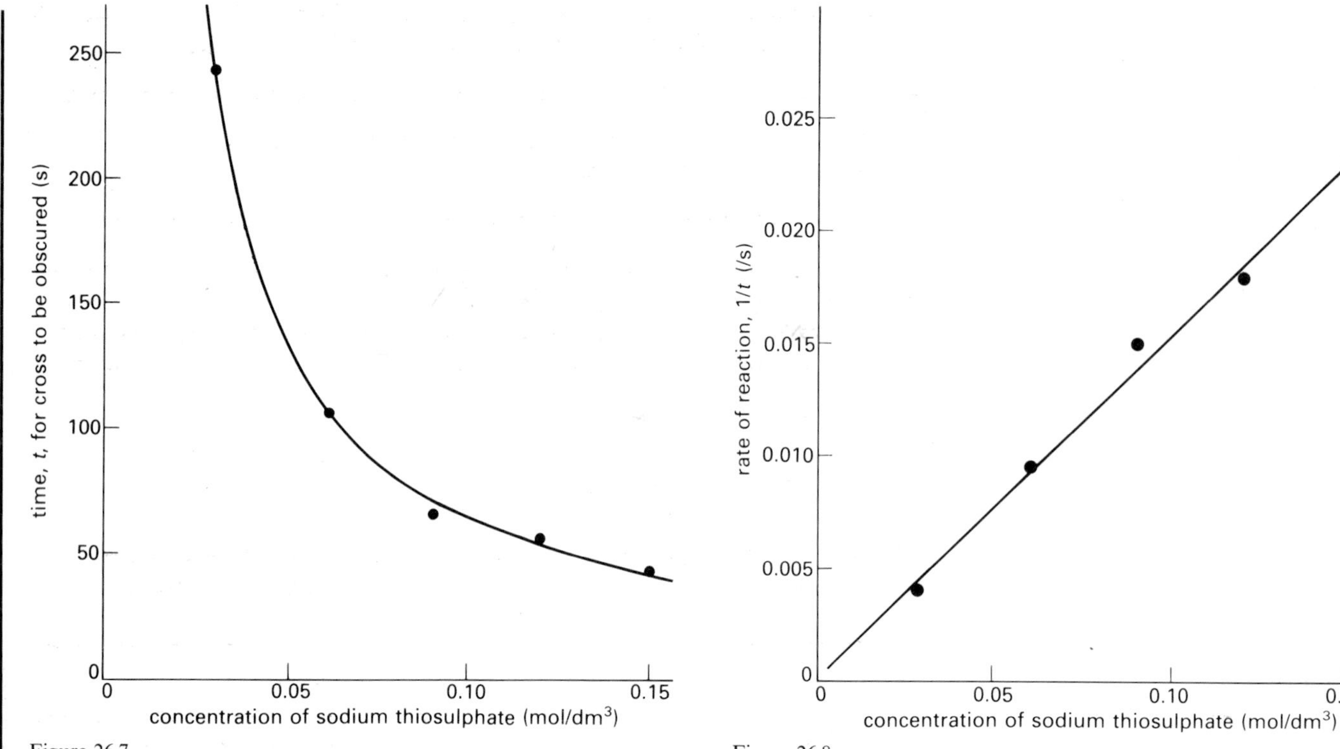

Figure 26.7

Figure 26.8

increases until the cross cannot be seen any more. Typical results are shown in figure 26.7 as a graph of the time for the cross to disappear against concentration.

Discussion

As the concentration increases, the time for the cross to disappear decreases, so the reaction becomes faster. The values of $1/t$ are proportional to the initial rates of reaction. These are plotted against concentration in figure 26.8. What does this graph show about the effect of the sodium thiosulphate concentration on the rate of this reaction? Does the reaction rate double if the concentration is doubled?

Question

Time (s)	0	15	30	45	60	75	90	120	150	180	210	240	300	360	420
Loss in mass (g)	0	0.21	0.45	0.67	0.85	1.01	1.13	1.31	1.41	1.48	1.51	1.54	1.56	1.58	1.59

Figure 26.9

The table in figure 26.9 shows the data used to plot the graph in figure 26.4 which shows the reaction of 20 g of small marble chips with 40 cm^3 of 2.0 mol/dm^3 hydrochloric acid.

(a) Plot the data, putting time along the x-axis. Label the line clearly.

(b) On your graph, sketch the lines you would expect if the experiment were repeated (i) with powdered calcium carbonate, (ii) with 40 cm^3 of 1.0 mol/dm^3 hydrochloric acid, (iii) with 80 cm^3 of 0.5 mol/dm^3 hydrochloric acid and (iv) with an equal volume of water added to the original mixture.

(c) What mass of carbon dioxide would you expect to be formed in the original experiment?

Experiment 26b shows that, as the concentration doubles, the reaction rate also doubles. This precise relationship is not always true, but, generally

* increasing the concentration of the reactants increases the rate of reaction.

In reactions between gases, increasing the pressure has the same effect as increasing the concentration, so

* in gas reactions, increasing the pressure increases the rate of reaction.

Some important industrial reactions use high pressures. The Haber process for making ammonia from nitrogen and hydrogen is operated at about 250 times atmospheric pressure. This process is described in chapter 34. Ethene combines with steam to give ethanol at 60 atmospheres pressure and a high temperature.

26.4 Temperature

Food can be kept longer in a refrigerator before it goes bad. Lowering the temperature slows the reactions which spoil milk, meat and other foods.

Experiment 26c
The effect of temperature on a reaction rate

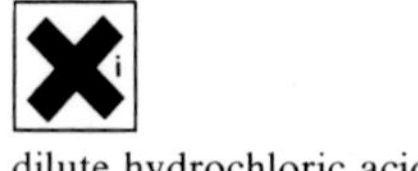

dilute hydrochloric acid

Experiment 26b is repeated, using just one concentration of sodium thiosulphate solution, but at different temperatures. The solutions are warmed to near the required temperatures before the acid is added. The temperature is recorded when the solutions are mixed.

Results
Typical results are shown in the table in figure 26.10. Plot a graph of time for the cross to disappear (on the *y*-axis) against temperature.

Temperature (°C)	Concentration of sodium thiosulphate solution (mol/dm^3)	Time, t, for the cross to be obscured (s)	Rate of reaction $1/t$ (/s)
20	0.03	280	0.0036
30	0.03	132	0.0076
40	0.03	59	0.017
50	0.03	31	0.032
60	0.03	17	0.059

Figure 26.10

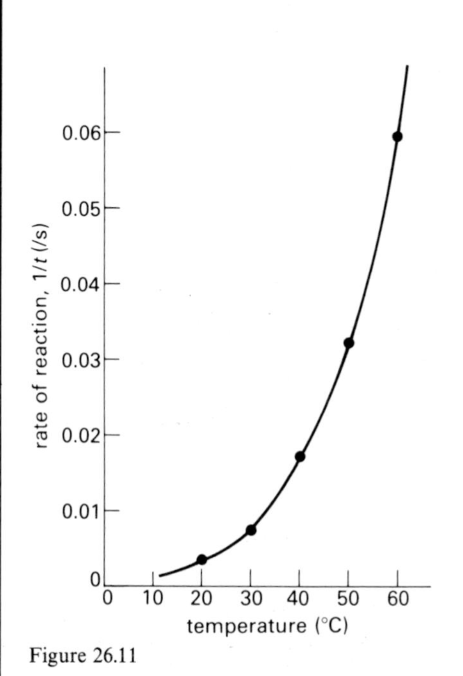

Figure 26.11

Discussion
Figure 26.11 shows how $1/t$ (i.e. rate of reaction) varies with temperature. Is the rate proportional to the temperature? What is the value for the rate of 20 °C? What temperature rise causes the rate to become doubled? What further temperature rise causes another doubling of the rate? The temperature rise that causes a doubling of the rate can also be calculated from your graph of time against temperature. How?

Temperature changes have a big effect on the rate of a reaction:

* a rise in temperature causes the rate to increase.

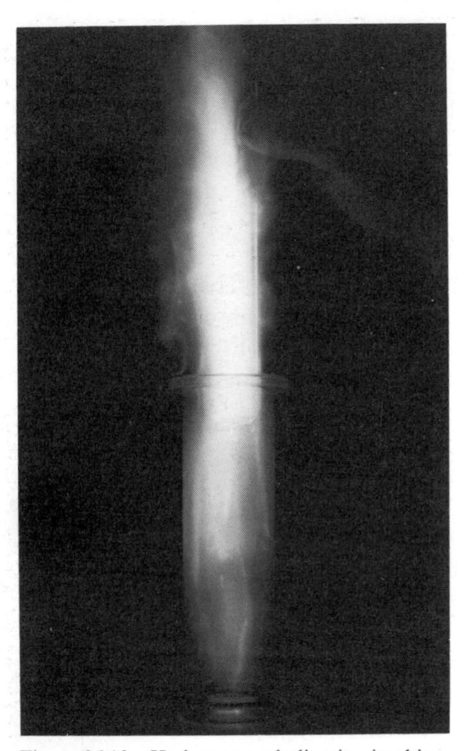

Figure 26.13 Hydrogen exploding in air: this reaction was started by holding a small coil of platinum wire in the mouth of the gas jar of hydrogen

For many reactions, the reaction rate is doubled for each 10 °C rise in temperature.

Because temperature changes have such a big effect on the rates of reaction, it is not surprising that many industrial processes are carried out at high temperatures. The Haber process for making ammonia and the contact process, in which sulphur dioxide is converted to sulphur trioxide for making sulphuric acid, are operated at about 500 °C. In the manufacture of nitric acid, ammonia is converted into nitrogen oxide at 900 °C. The contact process is described in chapter 28, the Haber process in chapter 34 and the manufacture of nitric acid in chapter 35. The reaction of ethene with steam to give ethanol occurs at over 300 °C (see chapter 31).

Questions

Time (s)	0	10	20	30	40	50	60	70	80	90	100	110	120	130	140	150
Volume of hydrogen (cm^3)	0	6	14	26	41	58	73	89	102	117	131	137	140	141	141	141

Figure 26.12

1 Figure 26.12 shows a table of results for the formation of hydrogen in the reaction of a calcium turning with excess water at 20 °C.
(a) Plot the results on a graph and sketch on it the lines you would expect if the water were kept at (i) 30 °C and (ii) 10 °C.
(b) Other than by raising the temperature, how could the rate be increased?
(c) Why is the reaction slower (i) at the beginning and (ii) near the end of the reaction?
(d) Draw the apparatus you would use to carry out the experiment.
2 How much faster will a reaction be at 50 °C compared with 20 °C, assuming that the rate doubles with every 10 °C rise in temperature?

26.5 Catalysts

When a gas jar of hydrogen is opened in air, nothing appears to happen. If mineral wool coated with platinum powder, or a coil of platinum wire, is held in the mouth of the jar, there is a squeaky explosion as hydrogen combines rapidly with oxygen to give water (see figure 26.13):

$$2H_2(g) + O_2(g) \longrightarrow 2H_2O(l)$$

Potassium chlorate decomposes on heating to give oxygen:

$$2KClO_3(s) \longrightarrow 2KCl(s) + 3O_2(g)$$

In the experiment illustrated in figure 26.14, one tube contains potassium chlorate only, and the other contains potassium chlorate with a little added copper(II) oxide. Both tubes are heated equally. The mixture with copper(II) oxide gives off oxygen much more quickly. A glowing splint relights at the mouth of this tube first. At the end, any unreacted potassium chlorate, together with the potassium chloride formed, can be dissolved in water, leaving the copper(II) oxide unchanged. Copper(II) oxide alone gives no oxygen when heated.

Platinum and copper(II) oxide are **catalysts**. They speed up these chemical reactions but appear to be unchanged themselves. Many chemical reactions are affected by catalysts.

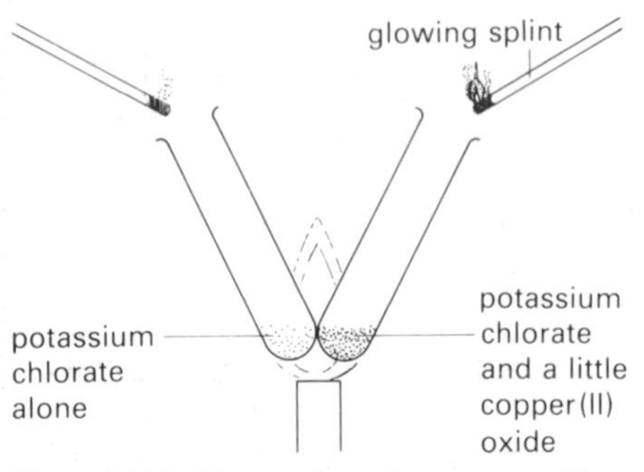

Figure 26.14 How to show that copper oxide speeds up the thermal decomposition of potassium chlorate

Hydrogen peroxide, H_2O_2, is a colourless liquid. At room temperature it decomposes very slowly to give water and oxygen:

$$2H_2O_2(aq) \longrightarrow 2H_2O(l) + O_2(g)$$

Experiment 26d
The effects of some metal oxides on the decomposition of hydrogen peroxide

metal oxides hydrogen peroxide

Either of the sets of apparatus in figure 26.15 can be used. A small measured amount of each metal oxide is added to the hydrogen peroxide solution in turn. A 'control' experiment without any catalyst is also carried out.

Results
The graphs in figure 26.16 show the volumes of oxygen formed. In the 'control' experiment, no oxygen is collected during the experiment. It may be several days before a tiny measurable volume is obtained.

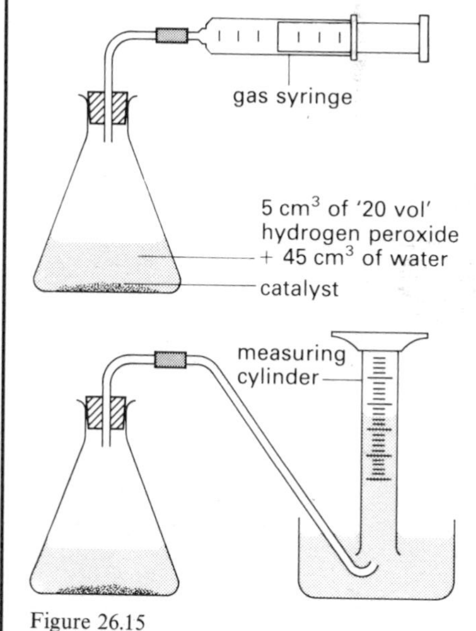

Figure 26.15

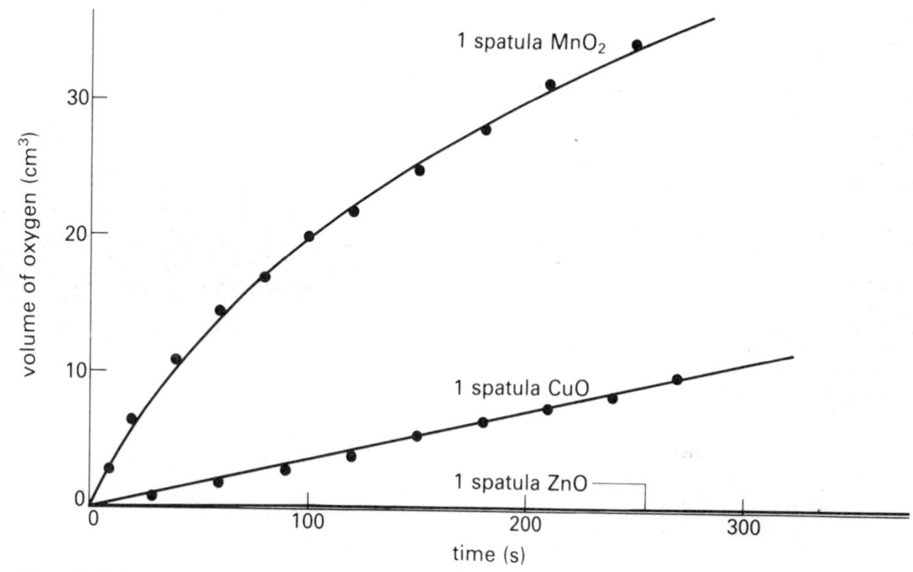

Figure 26.16

Discussion
Of the three metal oxides, one is clearly more effective as a catalyst than the others. Which one? What are the times for the formation of 10 cm³ of oxygen in each case? How would it be possible to work out how much faster the decomposition occurs with manganese(IV) oxide than with the 'control'? What other potential catalysts would you try?

Experiment 26e
Is manganese(IV) oxide used up when it catalyses the decomposition of hydrogen peroxide?

manganese(IV) oxide hydrogen peroxide

A weighed amount of manganese(IV) oxide is added to hydrogen peroxide solution. After the reaction, the mixture is passed through a weighed filter paper. When it is dry, the paper is reweighed.

Results
Typical results are:

Mass of weighing bottle + manganese(IV) oxide	= 6.64 g
Mass of weighing bottle empty	= 5.42 g
Mass of dry filter paper	= 0.53 g
Mass of filter paper + recovered manganese(IV) oxide	= 1.75 g

Discussion
How do you know when the reaction is over? How can you ensure that all the manganese(IV) oxide is recovered from the flask? Use the results to calculate the mass of manganese(IV) oxide before and after the reaction. Has it been used up during the reaction?

Question

	Catalyst		
	1 spatula of manganese(IV) oxide powder	2 spatulas of manganese(IV) oxide powder	1 spatula of manganese(IV) oxide granules
Time to produce $10\,cm^3$ of oxygen(s)	39	21	80

Figure 26.17

Extending the investigation of the effect of manganese(IV) oxide on the decomposition of hydrogen peroxide, a pupil recorded the time to obtain $10\,cm^3$ of oxygen under different conditions. The results are shown in figure 26.17. Using the results, answer the following questions:

(a) Does the rate depend on the amount of catalyst used?

(b) Does the rate depend on the surface area of the catalyst?

Figure 26.18 A catalytic converter for a car's exhaust system

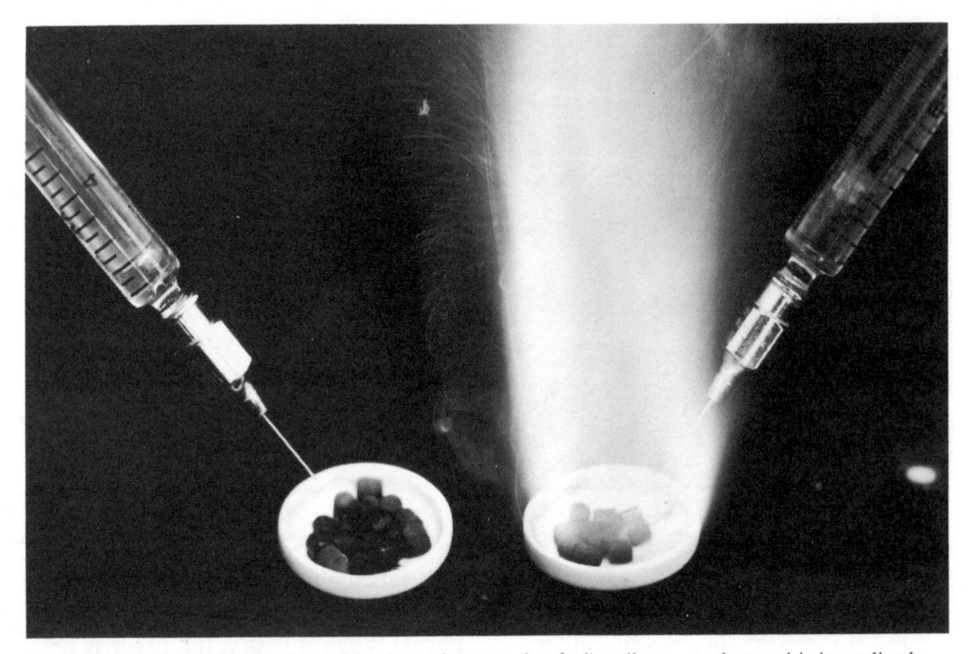

Figure 26.19 A jet of hydrazine (which is used as a rocket fuel) strikes a catalyst and is immediately oxidized by the air

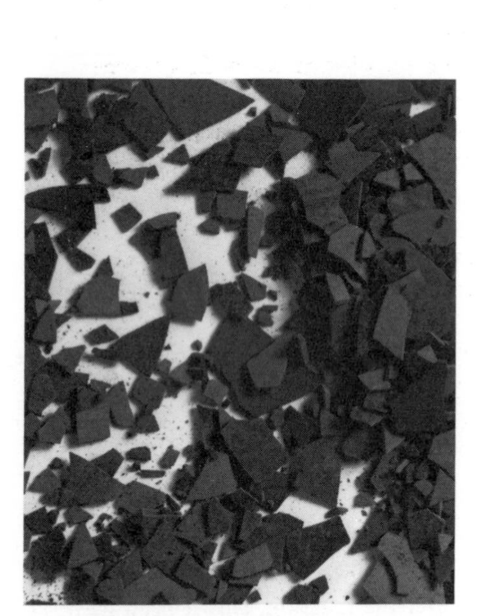

Figure 26.20 Flakes of the nickel catalyst that is used to convert edible oils into margarine

* Catalysts dramatically speed up reactions
* They are chemically unchanged at the end of the reaction
* They change the rate but not the extent of a reaction
* Higher temperatures and surface areas make them more effective.

Catalysts are widely used in industry. The Haber process uses an iron catalyst. The contact process uses vanadium(V) oxide pellets. One stage in the manufacture of nitric acid involves a platinum–rhodium alloy. A catalyst is used in margarine making (see chapter 30). Figure 26.18 shows an extra part that can be included in a car exhaust system. It contains a catalyst to convert harmful polluting gases into safer ones like carbon dioxide.

Figure 26.20 shows one example of an industrial catalyst. Many other reactions are also catalysed by **transition metals** or their compounds.

26.6 Enzymes

Figure 26.21 shows what happens when a drop of blood is added to hydrogen peroxide. A catalyst in the blood, called *catalase*, increases the rate of decomposition of hydrogen peroxide. It is much more effective than manganese(IV) oxide. A piece of liver, or liquidized celery, has the same effect. Catalase is an example of an **enzyme**. Enzymes are proteins. They are biological catalysts.

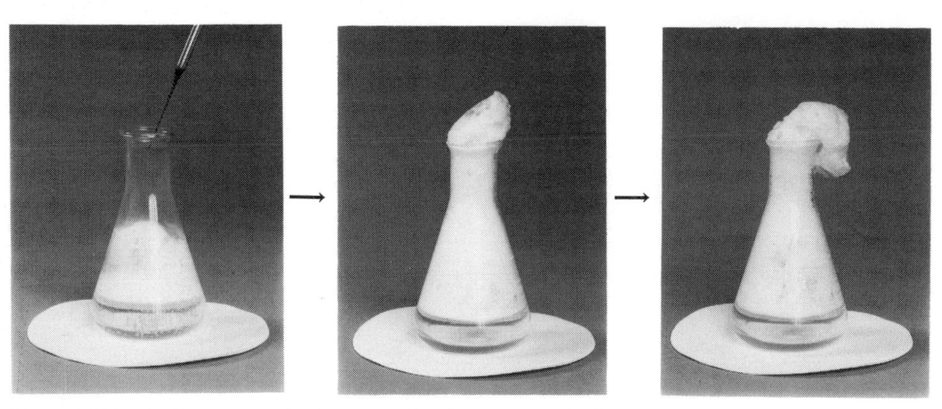

Figure 26.21 Hydrogen peroxide is a clear liquid that gradually decomposes at room temperature, giving water and oxygen: adding a drop of blood dramatically speeds up the reaction

Enzymes have been used for thousands of years, starting long before anyone knew they existed. **Fermentation** is catalysed by enzymes present in yeast. Making cheese, proving bread dough, brewing beer and tenderizing meat by hanging it, are all enzyme-catalysed processes which have been used for many years.

Most of the reactions taking place in the body are catalysed by enzymes. Each enzyme catalyses only one reaction. A living cell may contain as many as a thousand different enzymes.

Catalase is present in the blood to protect it from a buildup of dangerous peroxides.

One theory to explain how enzymes work pictures the enzyme and reactant molecules fitting together in a special way. Think of enzymes like jigsaw pieces. In figure 26.22, enzyme A will react with reactant A, but not with reactant B. When the enzyme and the reactant that fits it come together, the reaction can occur. Figure 26.23 shows a model for enzyme action.

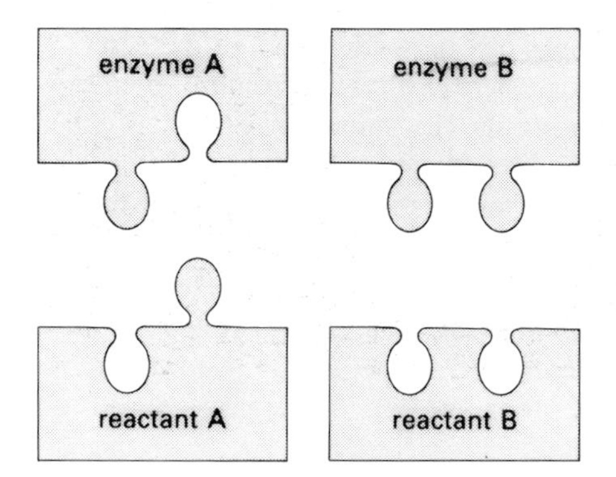

Figure 26.22 An enzyme and its reactants fit together rather like jigsaw pieces

Lysozyme is an enzyme which breaks down sugars in the cell walls of some bacteria. This destroys the bacteria. Figure 18.5 on p. 120 shows a model of the enzyme. The sugar molecule fits into the cavity.

Enzymes are powerful catalysts – it is reported that one molecule of catalase will decompose 40 000 molecules of hydrogen peroxide every second. However, they are very sensitive to conditions. Look at the graphs in figure 26.24. What do you conclude? How do enzymes differ from ordinary catalysts?

Despite their sensitivity and high cost, they are now widely used in industry. As well as obvious uses in fermentation, baking and cheese making, enzymes are extracted and used for predigesting baby food, for tenderizing meat, in toothpaste, for leather treatment and even for making soft-centred chocolates.

Perhaps their most familiar use is in biological washing powders. These contain

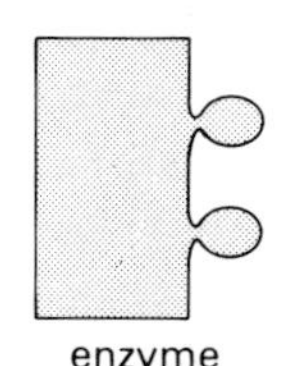
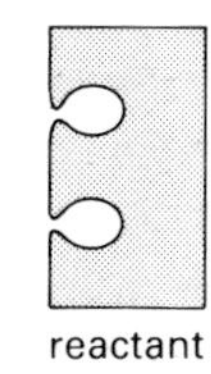

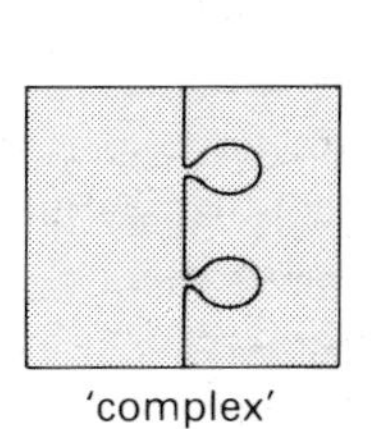

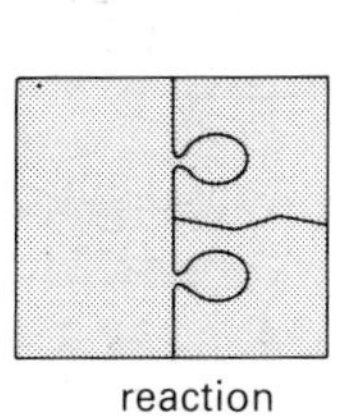

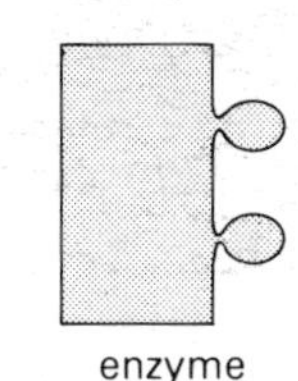
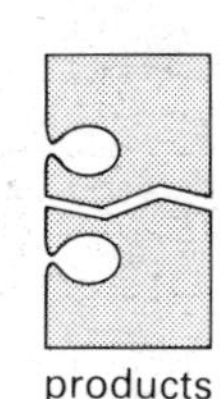

Figure 26.23 When the enzyme and the reaction are fitted together, the reaction can take place

Questions

1 Read the instructions on a packet of biological washing powder. Suggest reasons why you should
(a) never use with silk, wool or leather.
(b) always make sure that the powder is dissolved before adding clothes.
(c) use warm water for the best results.

2 Design an experiment you could carry out in the laboratory to compare the abilities of biological and ordinary washing powders to remove stains from cloth.

Figure 26.24 The effect of different conditions on an enzyme's effectiveness

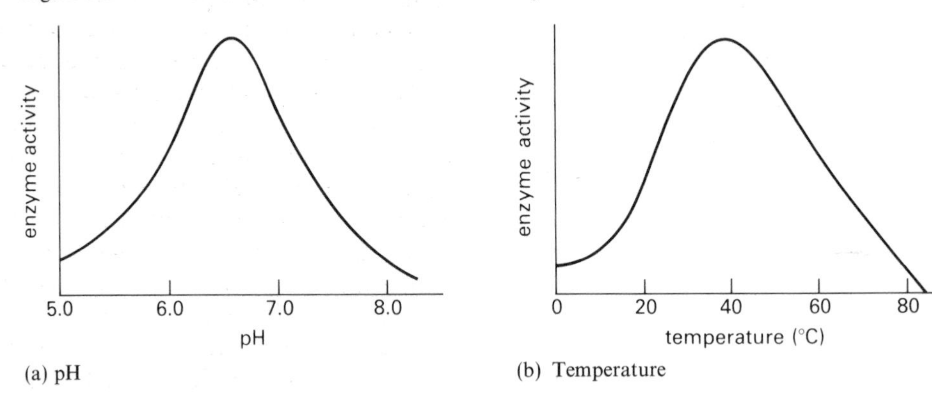

(a) pH (b) Temperature

enzymes called *proteases*. They break down protein materials and can be used to remove blood, gravy and other such stains from clothes. These washing powders have to be used with care because they can cause skin rashes.

The breakdown of starch to sugars, and the conversion of glucose to ethanol, are catalysed by enzymes. These two reactions are investigated in chapter 33.

26.7 Theories about reaction rates

In chemical reactions, atoms are rearranged. Figure 3.17 on p. 27 shows the reaction of hydrogen with oxygen to give water. Before any reaction between hydrogen and oxygen can occur, the molecules must come together and collide (see figure 26.25).

In gases and liquids, particles are in constant motion. Millions upon millions of collisions occur every second. If there were a reaction every time molecules collided, all chemical reactions would be over in a fraction of a millionth of a second. Iron would rust immediately on being brought into the air! From the reactions we have studied, it is clear that not all collisions are effective. Only a small fraction of them seem to be 'successful' collisions.

It is not enough for hydrogen and oxygen molecules to collide. Bonds between the atoms must be broken before new molecules can be made (see figure 26.26). This needs energy. For every chemical reaction, there is a certain minimum energy needed in the collisions before a reaction can occur. This minimum energy is called the *activation energy* of the reaction.

Particles in liquids and gases are moving at different speeds and colliding in different ways. Some collisions are 'head-on', while others are 'glancing' collisions (see figure 26.27). Think about dodgem cars at the fair – glancing collisions are quite painless, head-on collisions between slow-moving cars are not too bad, but head-on collisions between fast-moving cars *hurt*!

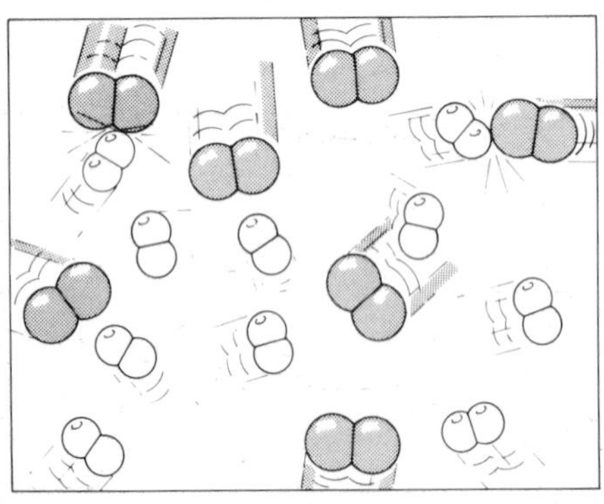

Figure 26.25 A mixture of hydrogen and oxygen gases: the molecules that are colliding may react to form water molecules

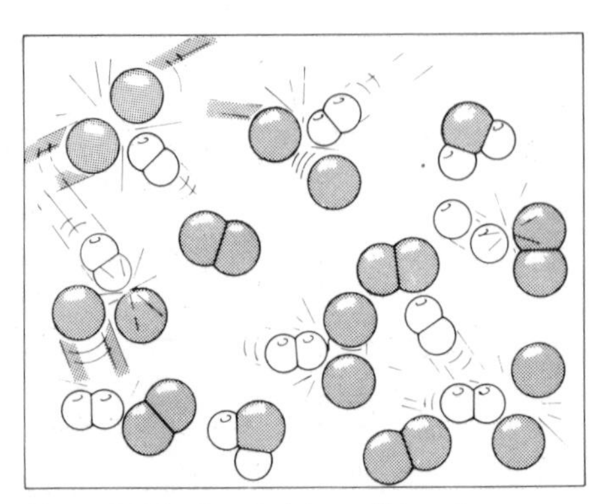

Figure 26.26 Some of the collisions have enough energy to break the bonds in the hydrogen and oxygen molecules

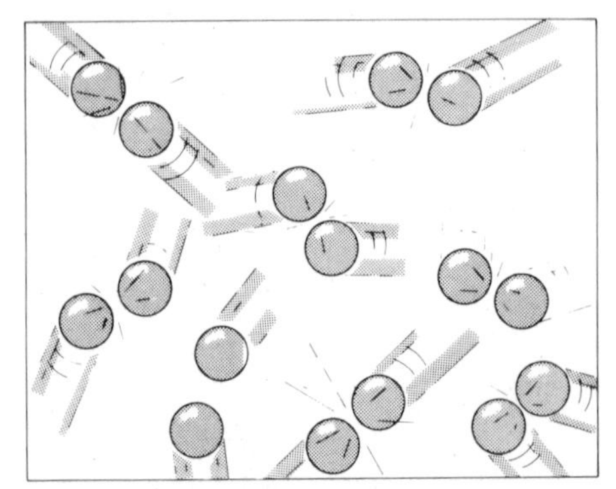

Figure 26.27 Particles colliding in different ways. Which of these collisions are likely to be 'successful'?

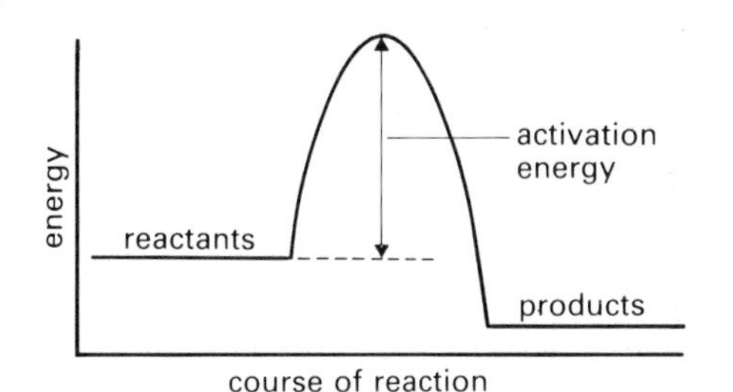

Figure 26.28 Energy level diagram for a reaction

The collisions between particles have a range of energies. Head-on collisions between fast-moving particles are the most energetic. If the colliding molecules have enough energy, the collision is 'successful', and a reaction occurs.

The course of a chemical reaction is like a high jump competition. The bar is set at a height such that only a few competitors with enough energy (and style!) can jump it and land safely on the other side. The chemical reaction equivalent is shown in figure 26.28. The height of the bar is represented by the activation energy and the landing area by the products of the reaction. If the high jump bar is low, many competitors are successful. If it is high the success rate is much less. In chemical reactions, if the activation energy is low, a high proportion of collisions will have enough energy and so the reaction is fast. Reactions in which the activation energy is high are very slow at room temperature, because only a very small fraction of collisions have enough energy to overcome the activation energy. The 'success rate' of collisions is low.

These ideas can be used to explain how changing conditions affect the rates of reactions.

Summary

Figure 26.29 summarizes the effects of surface area, concentration, temperature and catalysts on reaction rates. These effects are explained in terms of collisions between moving molecules.

Reaction rates depend on . . .

surface area

With a greater surface area of solid, collisions are far more frequent. Because there are more collisions, the reaction rate is greater

concentration

In solutions of higher concentration and in gases at higher pressure, particles are closer together. They have a greater chance of colliding. Because there are more collisions, the reaction rate is greater

temperature

At higher temperatures, particles are moving faster, so there are more collisions. Also (and more importantly), the collisions are more energetic. More collisions have an energy greater than the activation energy, so the reaction is faster

catalyst

The catalyst seems to lower the activation energy for the reaction. Many more collisions are 'successful', and the reaction is faster. Exactly how the catalyst achieves this remains a mystery for many reactions

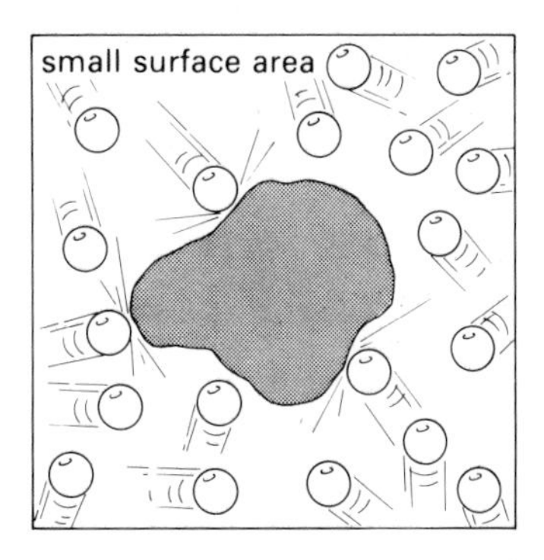

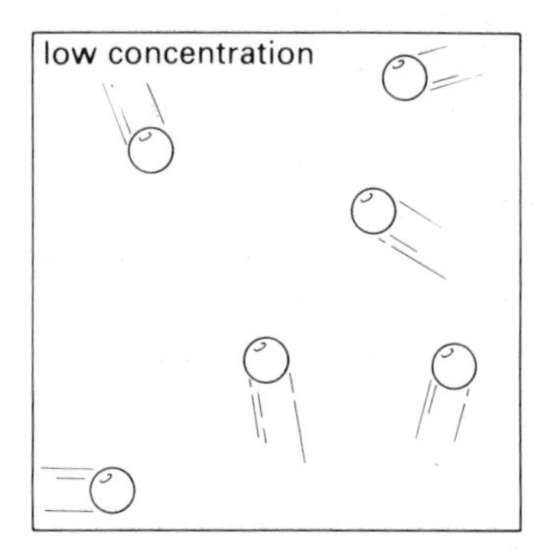

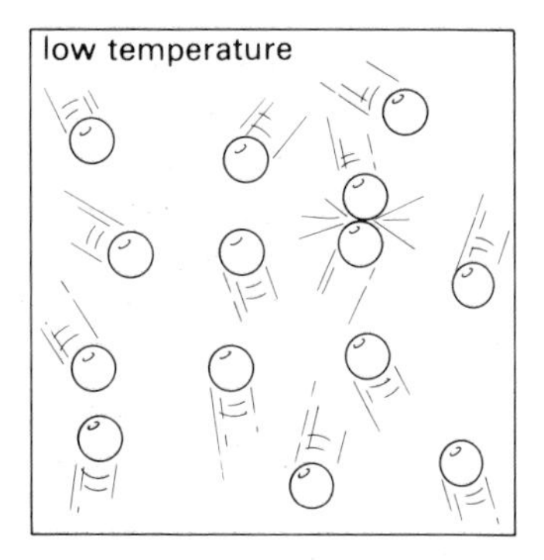

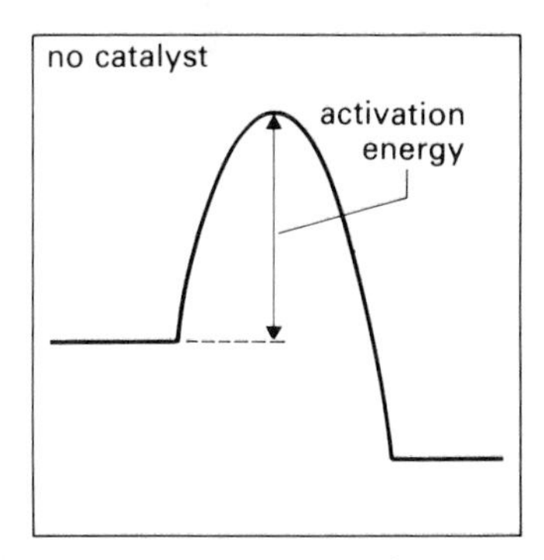

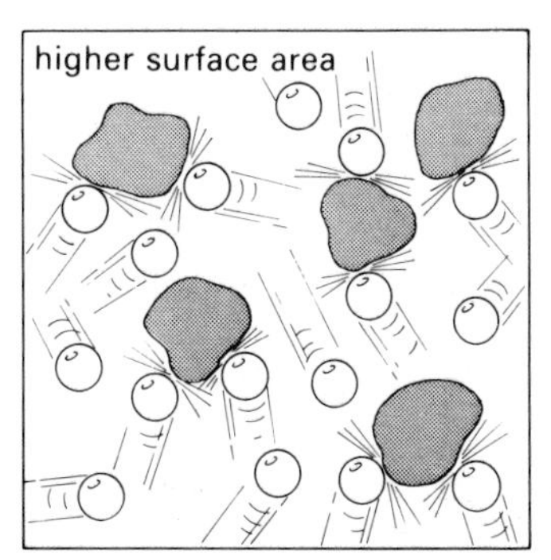

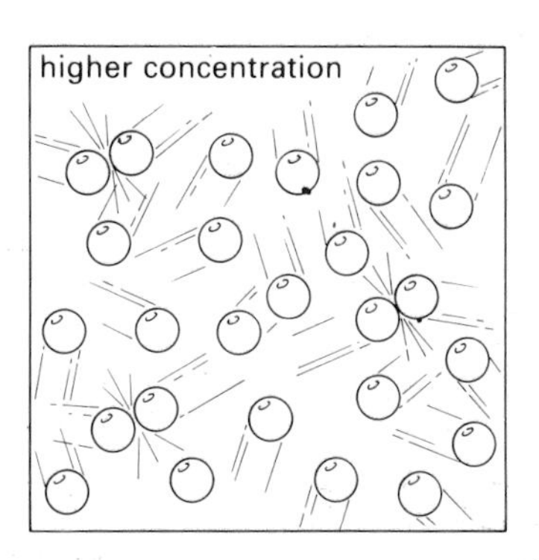

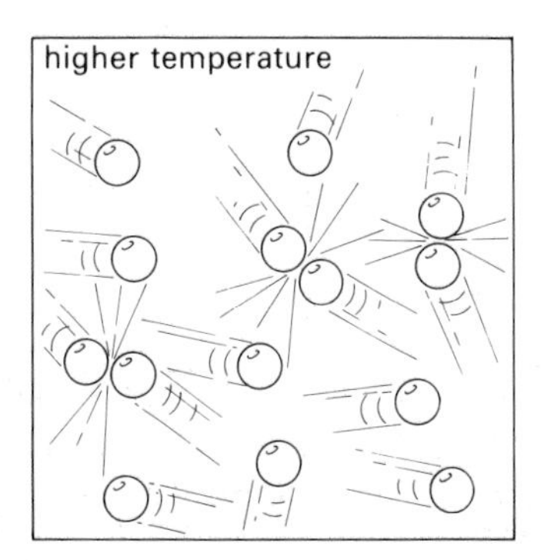

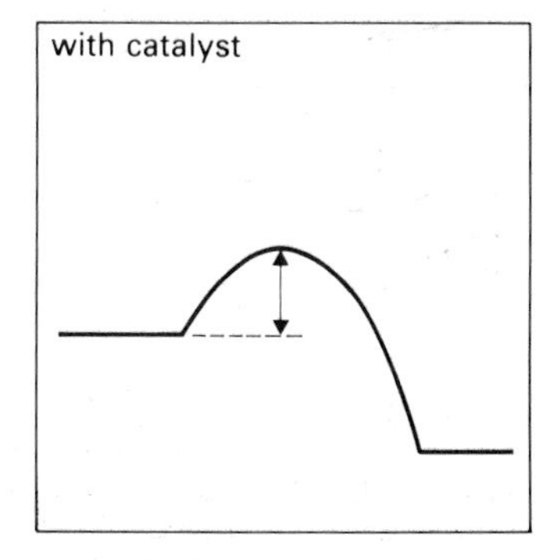

Figure 26.29

27 How far?

Questions

1 Write a symbol equation to show water turning into steam. Write another equation to show steam condensing to water. Combine them as a single reversible change.
2 Write an equation to show the reversible change. when iodine sublimes (see p. 10).

27.1 Reversible reactions

Heating lead nitrate is like frying an egg! They are both 'one-way' reactions. When an egg is being cooked, chemical changes occur which cannot be reversed. It is impossible to change the white part of a cooked egg back into a clear jelly. The effect of heat on lead nitrate is described in chapter 12. It decomposes to give lead(II) oxide, nitrogen dioxide and oxygen:

$$2Pb(NO_3)_2(s) \longrightarrow 2PbO(s) + 4NO_2(g) + O_2(g)$$

The products cannot recombine to give lead nitrate. The reaction is irreversible. Many other chemical reactions are irreversible.

Figure 27.1 Two familiar reversible processes.

Equally, many processes *can* be reversed. Jelly, set by cooling it, turns back to a liquid on warming. A familiar **reversible** change is shown in figure 27.1. Ice is changed into water by heating it:

$$H_2O(s) \xrightarrow{\text{heat}} H_2O(l)$$

Ice re-forms when water is cooled:

$$H_2O(l) \xrightarrow{\text{cool}} H_2O(s)$$

The two equations can be combined:

$$H_2O(s) \underset{\text{cool}}{\overset{\text{heat}}{\rightleftharpoons}} H_2O(l)$$

Chemical reactions can be reversible. The effect of heat on blue copper(II) sulphate is described in chapter 2. It loses its water of crystallization and turns white. **Anhydrous** copper(II) sulphate is formed. When water is added to anhydrous copper(II) sulphate at room temperature, the blue solid is re-formed:

$$CuSO_4.5H_2O(s) \rightleftharpoons CuSO_4(s) + 5H_2O(l)$$

A similar change occurs with cobalt(II) chloride, $CoCl_2.6H_2O$. Write an equation for this change.

Ammonia and hydrogen chloride combine at room temperature to give a white smoke of ammonium chloride (see figure 27.2):

$$NH_3(g) + HCl(g) \longrightarrow NH_4Cl(s)$$

Heating makes this reaction go the other way – it is reversible.

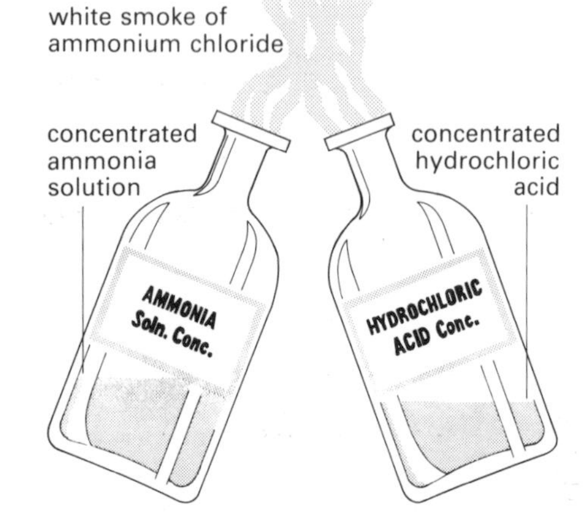

Figure 27.2 Ammonium chloride formation

Experiment 27a
The thermal dissociation of ammonium chloride

Ammonium chloride is heated in the apparatus shown in figure 27.3.

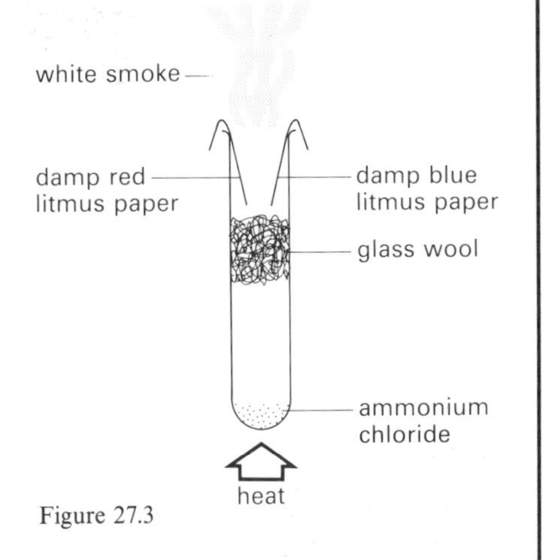

Figure 27.3

Results
After a short time, the red litmus paper turns blue. Later on, the blue litmus turns red. A white smoke forms above the tube.

Discussion
Ammonium chloride is decomposed. The gases produced are separated as they pass through the glass wool. Ammonia gas is alkaline and hydrogen chloride is acidic. Which gas is detected first? Why does this gas pass through the glass wool faster than the other? Why was the indicator paper damp? What is the white smoke above the tube, and how is it formed?

Ammonium chloride decomposes at high temperatures to give hydrogen chloride and ammonia while, at low temperatures, these gases recombine. This is an example of *thermal dissociation.*

$$NH_4Cl(s) \rightleftharpoons NH_3(g) + HCl(g)$$

This reaction is reversed by changing the temperature. Reactions can be reversed in other ways.

Hot iron reacts with steam to give an iron oxide and hydrogen (see chapter 12). This reaction is reversed by passing hydrogen over the hot iron oxide:

$$3Fe(s) + 4H_2O(g) \rightleftharpoons Fe_3O_4(s) + 4H_2(g)$$

When litmus is added to an acid solution it turns red. Adding excess alkali gives a blue colour. More acid gives the red colour again.

$$\text{Red litmus solution} \underset{\text{acid}}{\overset{\text{alkali}}{\rightleftharpoons}} \text{Blue litmus solution}$$

In an equation, the substances on the left-hand side are called the *reactants* and those on the right-hand side the *products.* The 'left-to-right' reaction is the 'forward' reaction and the 'right-to-left' reaction is the 'backward' reaction.

Figure 27.4 A reversible reaction

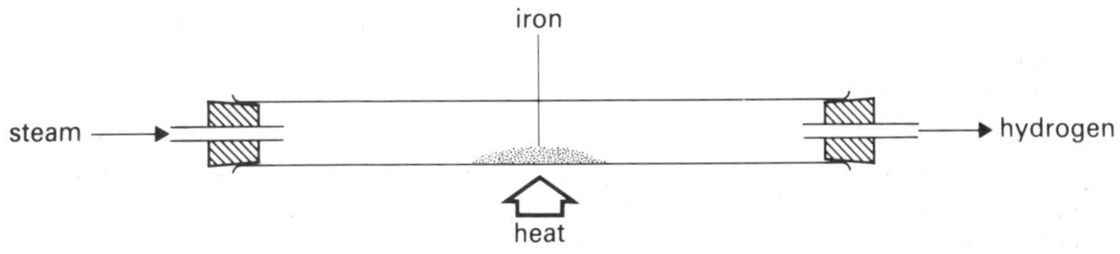

iron

steam → hydrogen

heat

(a) The 'forward' reaction $3Fe(s) + 4H_2O(g) \rightarrow Fe_3O_4(s) + 4H_2(g)$

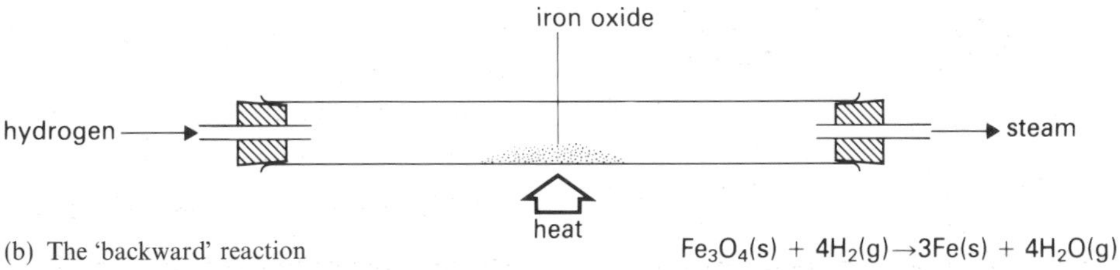

iron oxide

hydrogen → steam

heat

(b) The 'backward' reaction $Fe_3O_4(s) + 4H_2(g) \rightarrow 3Fe(s) + 4H_2O(g)$

Questions

1 Lavoisier heated mercury in air and obtained some mercury oxide. Priestley and Lavoisier heated mercury oxide and obtained mercury and oxygen. How can these two, apparently opposite, statements both be true?

2 In the reversible reaction

$$CO(g) + H_2O(g) \rightleftharpoons CO_2(g) + H_2(g)$$

which are (a) the reactants and (b) the products? Which is (c) the forward reaction and (d) the backward reaction?

27.2 Equilibrium

What happens to litmus in neutral solution – in pure water, for example? It has a purple colour, somewhere between red and blue. What happens when water is kept at exactly 0 °C for some time? There is a mixture of ice and water. In both these situations there is a balance position.

Balance points exist in most reversible reactions when neither the forward nor the backward reaction is complete. Reactants and products are present together and the reaction appears to have stopped. Reactions like this are in **equilibrium**.

To show when a reaction is in equilibrium, the sign '$\rightleftarrows$' in the equation is replaced by '$\rightleftharpoons$'. So, at 0 °C,

$$H_2O(l) \rightleftharpoons H_2O(s)$$

The question 'How far?' asks where the equilibrium point is in a reaction. At equilibrium, the reaction may be well to the right, well to the left, or at any point between these extremes.

Experiment 27b
Distributing iodine between organic and aqueous solvents

iodine
1,1,1-trichloroethane

Figure 27.5 shows the steps in the experiment.

Results
Iodine is slightly soluble in water but much more soluble in potassium iodide solution. The solution is yellow–brown. It is also soluble in the organic solvent, 1,1,1-trichloroethane. This solution is red. Water and the organic solvent do not mix. Figure 27.5 shows what happens when iodine is shaken in the two solvents.

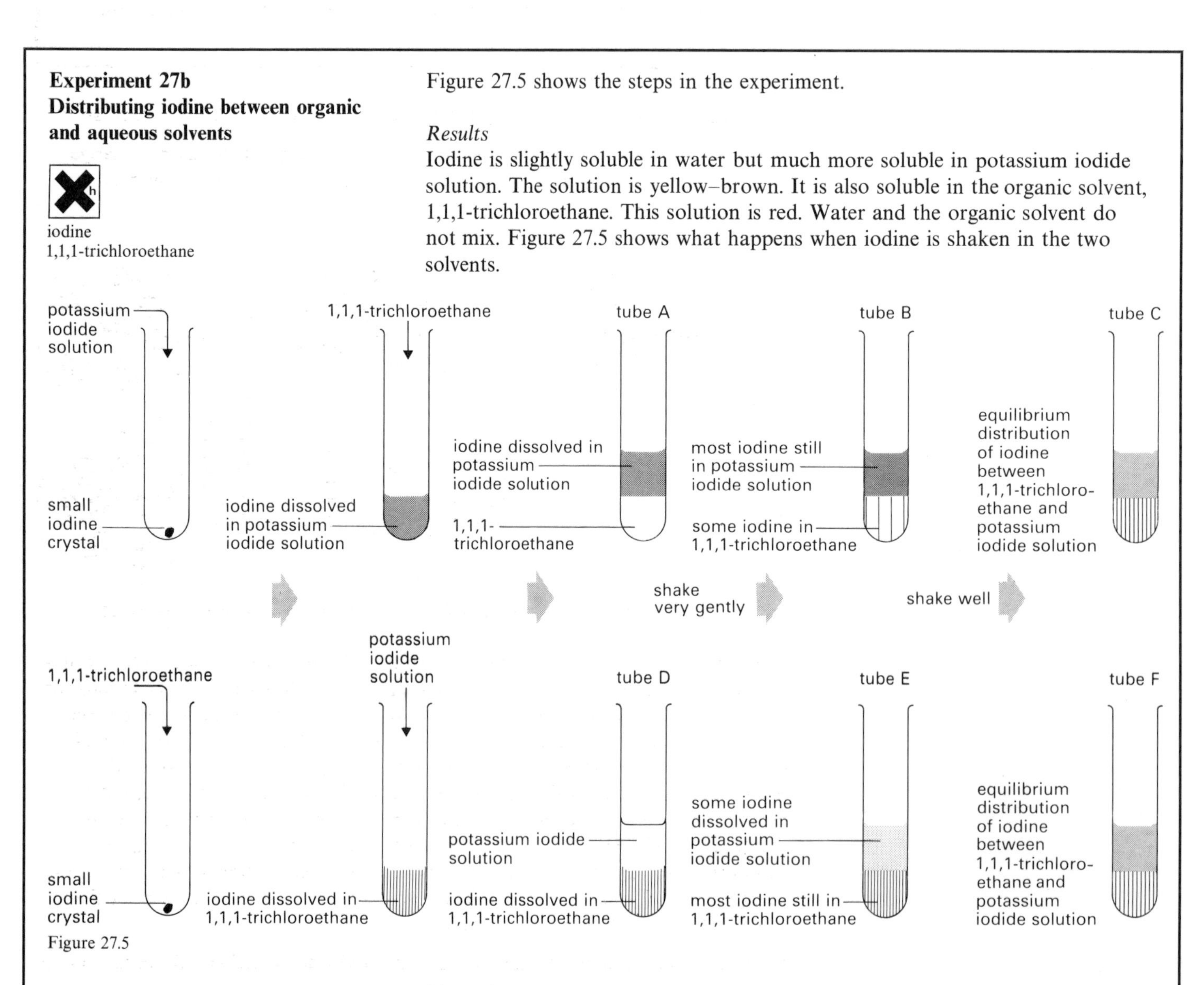

Figure 27.5

Discussion
The graphs in figure 27.6 show how the iodine concentrations in the two layers change with shaking. Is the iodine ever completely transferred from one layer to the other? In which solvent is iodine more soluble? In tube C, the iodine is

concentration of iodine
in CH_3CCl_3
in KI(aq)
time
tube A tube B tube C

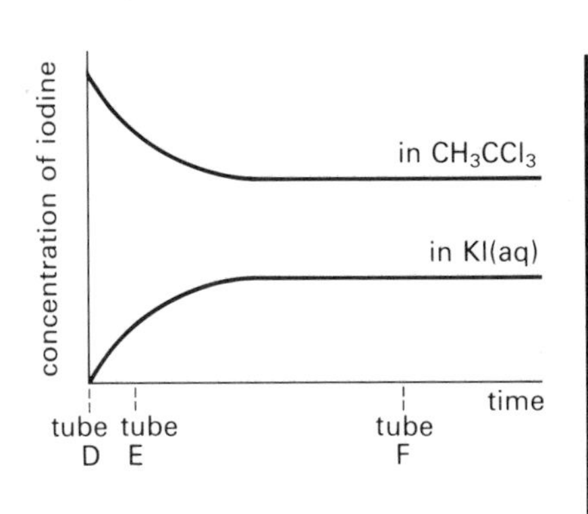

Figure 27.6

distributed between the organic and aqueous layers and there is no change with more shaking. In this tube there is an equilibrium:

$$I_2(aq) \rightleftharpoons I_2(organic)$$

Tube F looks like tube C. The equilibrium mixtures in the two tubes are the same.

Experiment 27b shows two important characteristics of equilibrium processes:

* at equilibrium, the concentrations of reactants and products do not change
* an equilibrium can be approached from either the 'reactant end' or the 'product end' of a reaction.

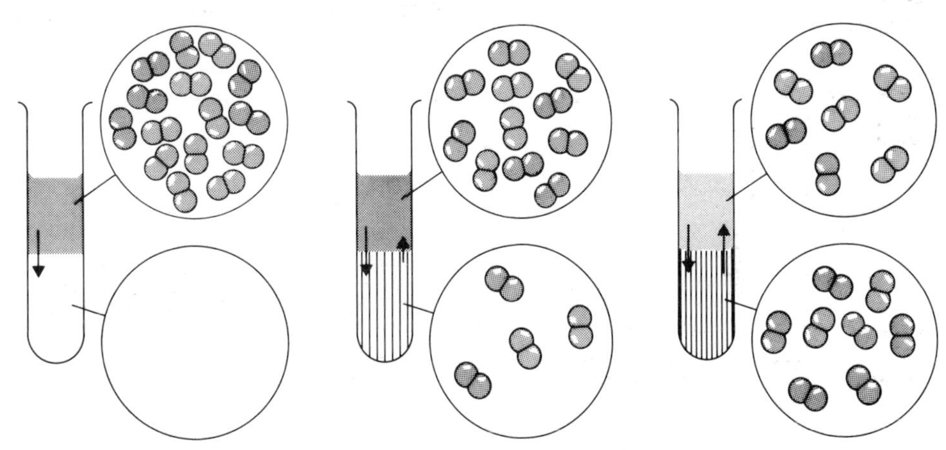

Figure 27.7 The distribution of the iodine molecules between the two solvents in experiment 27b

Figure 27.7 shows the movement of iodine molecules in experiment 27b. At first, all the iodine molecules are in the aqueous layer. On shaking, some move into the organic layer. Some of these molecules move back to the aqueous layer, but the rate is slow because the concentration in the organic layer is small. The rate of movement of molecules from the aqueous to the organic layer (the forward reaction) is greater than the rate of the backward reaction, so more molecules move into the organic layer. As the concentration in the organic layer rises, the backward rate increases until, at equilibrium, the two rates become equal. At equilibrium, the movement from one layer to the other continues, but there is no overall change because each layer is gaining and losing iodine molecules at the same rate. This is *dynamic equilibrium.*

* In dynamic equilibrium, the forward and backward reactions still occur, but at equal rates.

Figure 27.8 shows a 'down' escalator. On it someone is trying to walk up. The escalator keeps moving and the person keeps walking but makes no progress. This is like dynamic equilibrium.

Figure 27.8 Dynamic equilibrium!

Ideas about dynamic equilibrium can be tested if particles are 'labelled' and followed in a reaction. Compounds containing radioactive atoms can be used as tracers. They are detected using a Geiger–Müller tube and counter. Radioactivity and its detection are described in chapter 39.

In a **saturated solution**, undissolved solid is in equilibrium with dissolved solid. No more solid will dissolve if the temperature is constant. In saturated lead(II) chloride solution, the equilibrium is:

$$PbCl_2(s) \rightleftharpoons Pb^{2+}(aq) + 2Cl^-(aq)$$

In figure 27.9, labelled lead(II) chloride is added to non-radioactive saturated lead(II) chloride solution. After shaking, the solution contains some radioactive lead(II) ions even though no more lead(II) chloride could dissolve. The radioactive lead particles must have been exchanged for some inactive particles already in solution. What has happened to the inactive particles? These observations are evidence for the dynamic nature of the equilibrium.

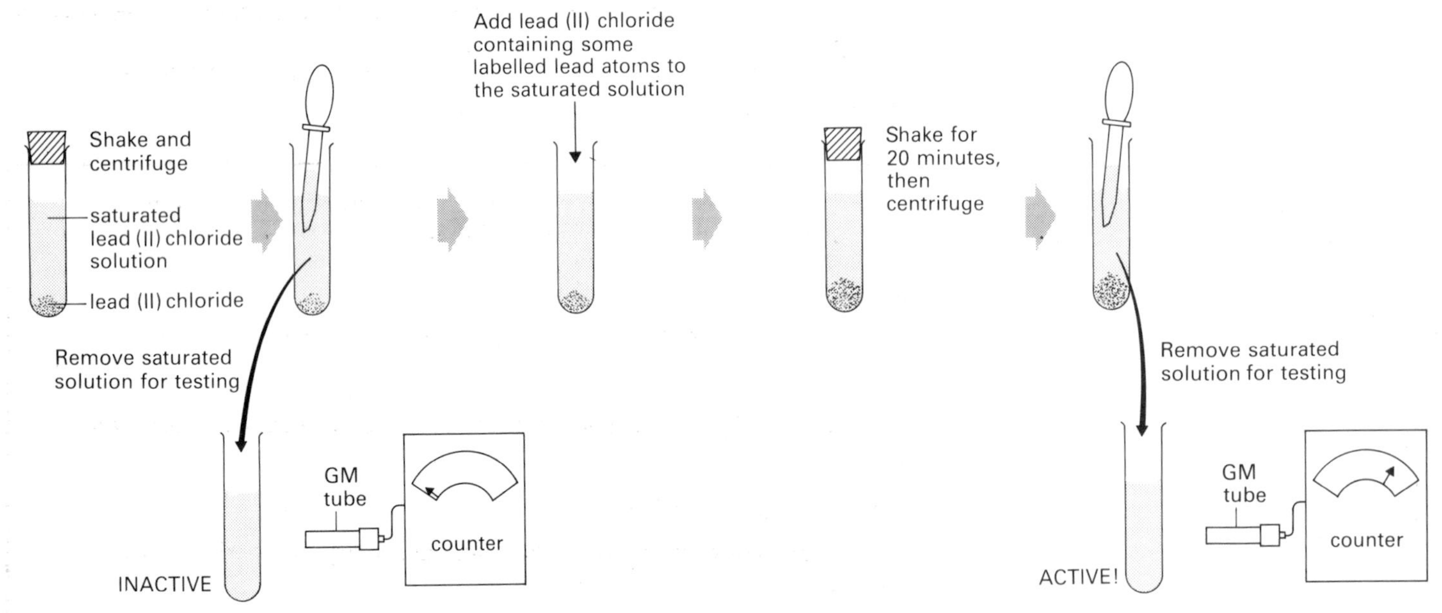

Figure 27.9 How to show that there is a dynamic equilibrium between a saturated solution and the undissolved solid

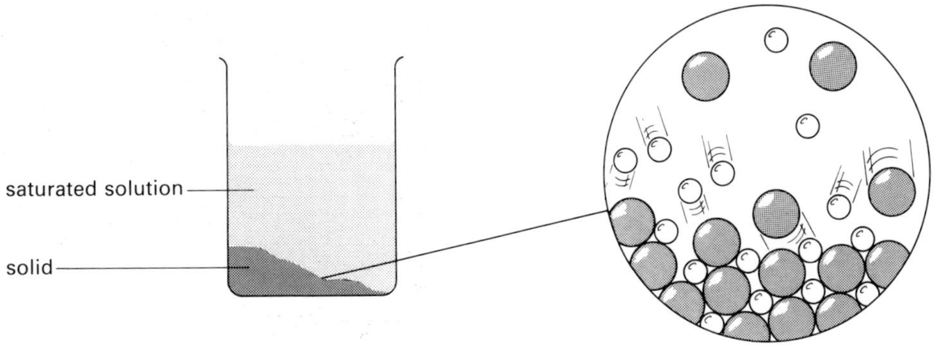

Figure 27.10 The behaviour of solute particles in a saturated solution

In a saturated solution, particles from the solid are going into solution and particles in solution are crystallizing out at the same rate (see figure 27.10).
How would you explain the dynamic equilibrium of ice and water at 0 °C?

27.3 Factors affecting equilibria

An equilibrium can be disturbed so that more products or reactants are formed. Figure 27.11 shows how more iodine can be transferred to the aqueous layer. Explain why this is happening, by considering the movement of the iodine molecules.

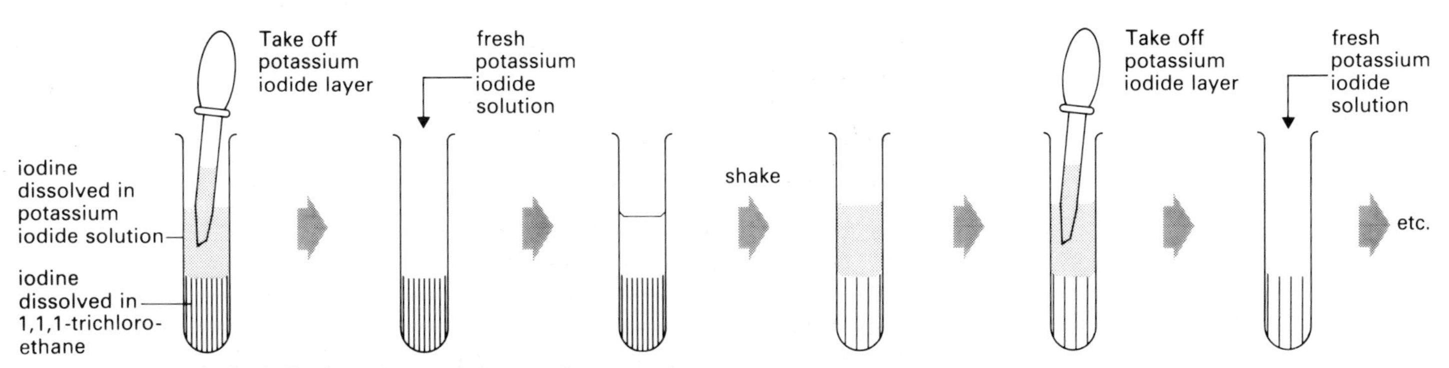

Figure 27.11 Transferring iodine from the organic layer to the aqueous layer

At equilibrium, the rates of the forward and backward reactions are the same. A change in either rate alters the position of equilibrium.
Experiment 27c shows how the position of an equilibrium can shift if the concentration of one of the reactants is changed.

Experiment 27c
An equilibrium involving iodine and chlorine

iodine
iodine monochloride

chlorine

Figure 27.12 overleaf shows the steps in the experiment. It is carried out in a fume cupboard because chlorine is poisonous.

Results
Figure 27.12 also shows the observations made during the experiment.

Discussion
The brown liquid is iodine monochloride, formed by combining chlorine and iodine:

$$I_2(s) + Cl_2(g) \longrightarrow 2ICl(l)$$

Iodine monochloride reacts with more chlorine, giving iodine trichloride, a yellow solid:

$$ICl(l) + Cl_2(g) \longrightarrow ICl_3(s)$$

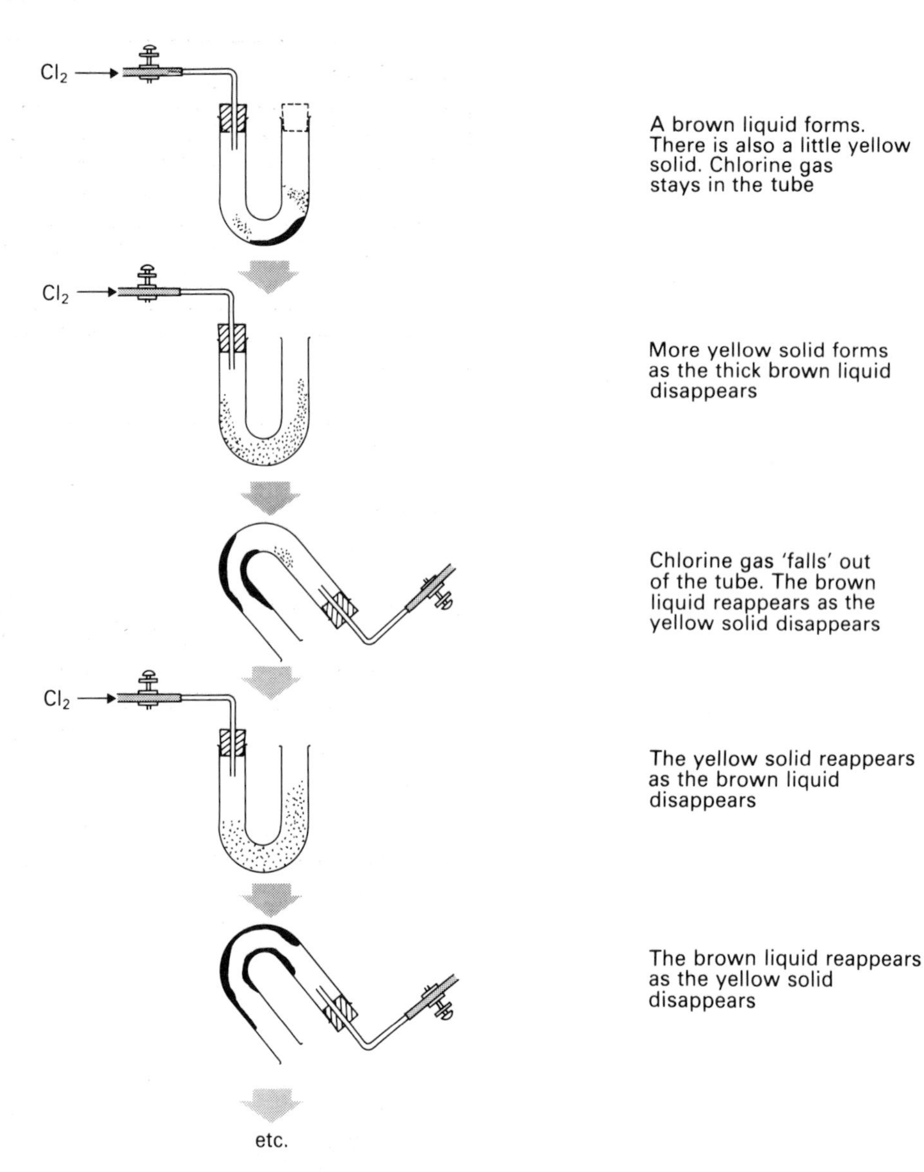

Chlorine is passed over iodine crystals. Then the chlorine supply is stopped and the U-tube stoppered
Cl2
A brown liquid forms. There is also a little yellow solid. Chlorine gas stays in the tube
More chlorine is passed through the U-tube
Cl2
More yellow solid forms as the thick brown liquid disappears
The chlorine supply is disconnected and the tube is inverted
Chlorine gas 'falls' out of the tube. The brown liquid reappears as the yellow solid disappears
More chlorine is passed through the U-tube
Cl2
The yellow solid reappears as the brown liquid disappears
The tube is inverted again
The brown liquid reappears as the yellow solid disappears
etc.

Figure 27.12

When the chlorine supply is stopped and the tube closed, iodine monochloride, iodine trichloride and chlorine are all present in equilibrium:

$$ICl(l) + Cl_2(g) \rightleftharpoons ICl_3(s)$$

Increasing the concentration of chlorine disturbs the equilibrium. The forward reaction rate increases and the reaction moves to the right. Some of the extra chlorine is used up, together with some iodine monochloride, giving more iodine trichloride. Removing chlorine, by pouring the heavy gas out of the tube, moves the reaction to the left. Iodine trichloride decomposes, replacing some of the removed chlorine and forming more iodine monochloride. How does removing chlorine affect the forward and backward rates? Does it explain what is observed?

The effects of changing conditions on an equilibrium can be predicted using Le Chatelier's principle. It states that

* if an equilibrium is disturbed by changing the conditions, the reaction moves to counteract the change.

Apply the principle to the equilibrium in experiment 27c. Do the observations fit the predictions?

Disturbance	How does the reaction respond?	The result
Concentration of A is increased	It moves to the right. Some A is used by reaction with B	More C and D are formed
Concentration of D is increased	It moves to the left. Some of the excess D is used up by reaction with C	More A and B are formed
Concentration of D is reduced	It moves to the right to make up for the lost D	There is more C and less A and B in the new mixture

Figure 27.13

The effects of changing concentrations in the equilibrium

$$A(s) + B(aq) \rightleftharpoons C(s) + D(aq)$$

are shown in figure 27.13. If the increase in the concentration of, say, B is small, a new equilibrium mixture is formed, containing more C and D. But if a vast excess of B is added, the reaction keeps moving to the right until almost all the other reactant, A, is used up. The reaction has moved almost completely to the right.

Bromine water is an equilibrium mixture:

$$Br_2(aq) + H_2O(l) \rightleftharpoons OBr^-(aq) + Br^-(aq) + 2H^+(aq)$$

The solution is yellow–orange because unreacted bromine molecules are present, shown on the left-hand side of the equation. All the ions on the right-hand side are colourless.

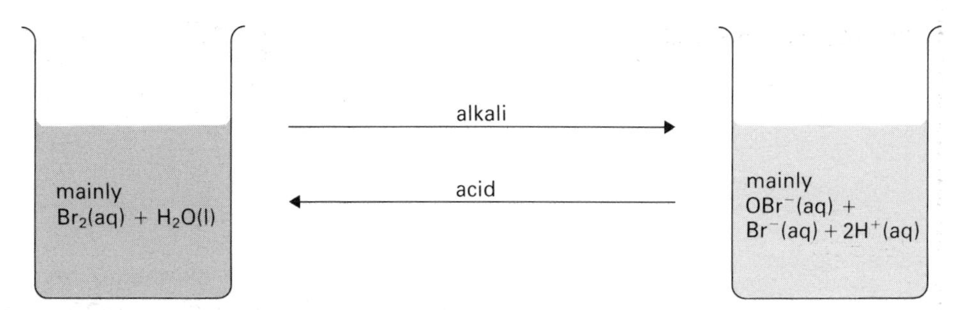

Figure 27.14 Shifting the equilibrium position in bromine water by adding acid or alkali

When an alkali is added to bromine water, it becomes colourless; when an acid is added, the yellow colour returns (see figure 27.14). Hydroxide ions in the alkali react with hydrogen ions in the equilibrium mixture:

$$H^+(aq) + OH^-(aq) \longrightarrow H_2O(l)$$

As the hydrogen ion concentration decreases, the reaction moves to the right to counteract the reduction. The forward reaction forms more hydrogen ions. If excess alkali is added, the reaction moves completely to the right. When extra acid, containing hydrogen ions, is added, the reaction moves to the left to reduce the hydrogen ion concentration again. Excess acid moves the reaction completely to the left. What would be the effect of adding excess sodium bromide to the equilibrium mixture?

An equilibrium can only be established in a 'closed' system – one from which neither reactants nor products can escape. Solids cannot escape from solutions, but reactions involving gases can only reach equilibrium in sealed containers.

In a closed container, the thermal decomposition of calcium carbonate (limestone) is an equilibrium:

$$CaCO_3(s) \rightleftharpoons CaO(s) + CO_2(g)$$

In the manufacture of quicklime from limestone, the carbon dioxide is swept away and the reaction goes to completion (see p. 15). Use Le Chatelier's principle to explain why.

Questions

1 When iron(II) ions, $Fe^{2+}(aq)$, are added to silver ions, $Ag^{+}(aq)$, a precipitate of silver is formed. An equilibrium mixture is formed:

$$Fe^{2+}(aq) + Ag^{+}(aq) \rightleftharpoons Ag(s) + Fe^{3+}(aq)$$

(a) What ions are present in solution in the equilibrium mixture?
(b) What would you expect to see if more iron(II) ions were added?
(c) What would happen to the silver precipitate if excess iron(III) ions were added?

2 If water is added to a solution of bismuth chloride in concentrated hydrochloric acid, the mixture becomes slightly milky.

$$BiCl_3(aq) + H_2O(l) \rightleftharpoons \underset{\text{bismuth(III) oxychloride}}{BiOCl(s)} + 2H^{+}(aq) + 2Cl^{-}(aq)$$

(a) What causes the milkiness?
(b) What would be the effect of adding more concentrated hydrochloric acid?
(c) What would be the effect of adding more water?

3 When iron and steam are heated in a closed container, a dynamic equilibrium is reached:

$$3Fe(s) + 4H_2O(g) \rightleftharpoons Fe_3O_4(s) + 4H_2(g)$$

(a) What does this statement mean?
(b) Explain why it is possible to make the reaction move completely to the left or to the right by using the apparatus in figure 27.4.

Summary

The following statements summarize the main ideas in this chapter:

* some chemical reactions are reversible
* reversible reactions may be made to go one way or the other by changing the conditions of temperature or pressure
* reversible reactions can come to a state of equilibrium
* chemical equilibrium is dynamic
* if a reaction at equilibrium is disturbed by changing the conditions, the position of equilibrium shifts to counteract the change.

Write out the statements. After each statement, give an example or a brief explanation.

28 The contact process

28.1 The importance of sulphuric acid

Sulphuric acid is manufactured on a huge scale. In 1982, world production was about 128 million tonnes. The acid is used at some point in the manufacture of almost all products. Its importance in industry and agriculture is shown by figure 28.1.

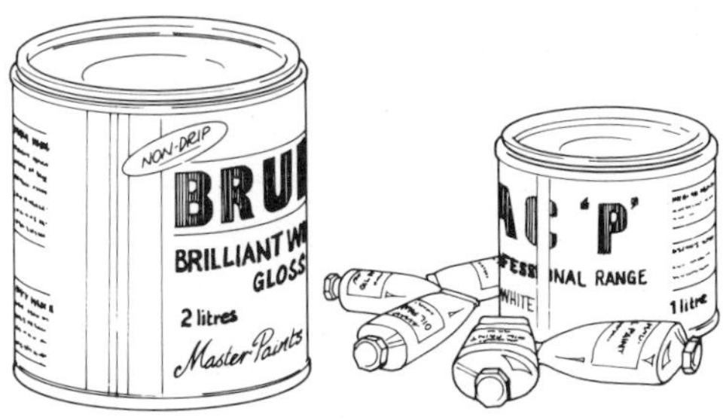

Paints and pigments Sulphuric acid is used to make titanium dioxide. This white powder is used in the manufacture of pigments

Agricultural chemicals. Ammonium sulphate and superphosphate fertilizers are made using sulphuric acid. About one-third of the sulphuric acid produced in the UK is used to make superphosphates

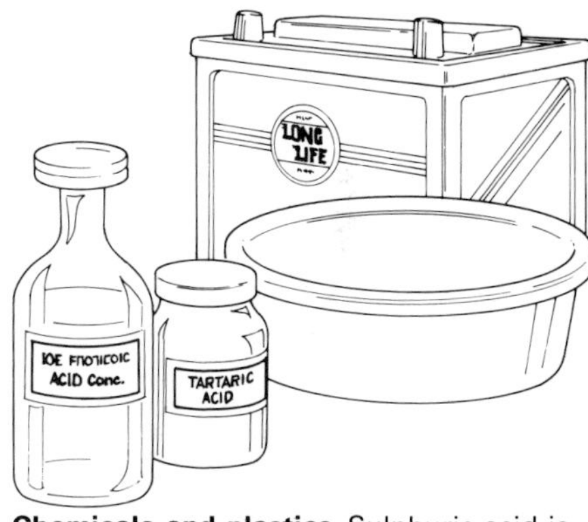

Chemicals and plastics Sulphuric acid is used to make plastics and many chemicals, including other acids such as tartaric acid. It is also used in car batteries

Detergents and soaps. By-products of oil refining are treated with sulphuric acid in the manufacture of washing powders, washing-up liquids, shampoos, etc.

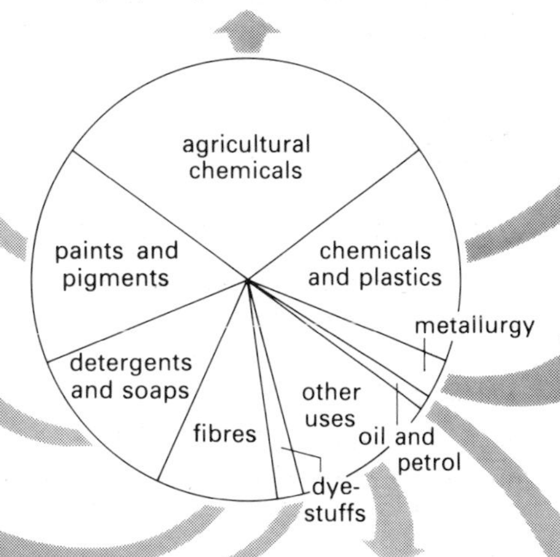

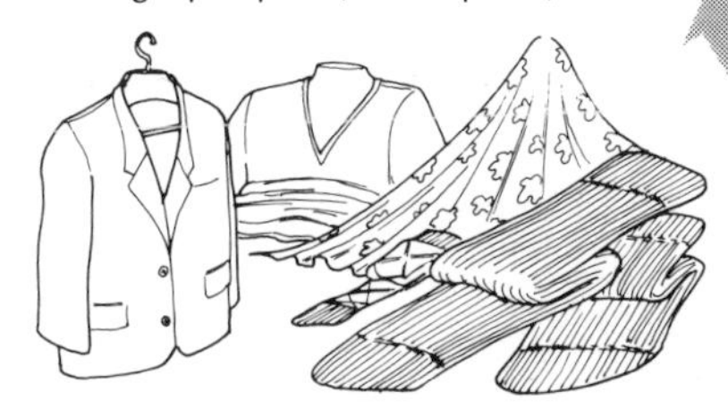

Fibres. Sulphuric acid is used in the manufacture of rayon and other synthetic fibres

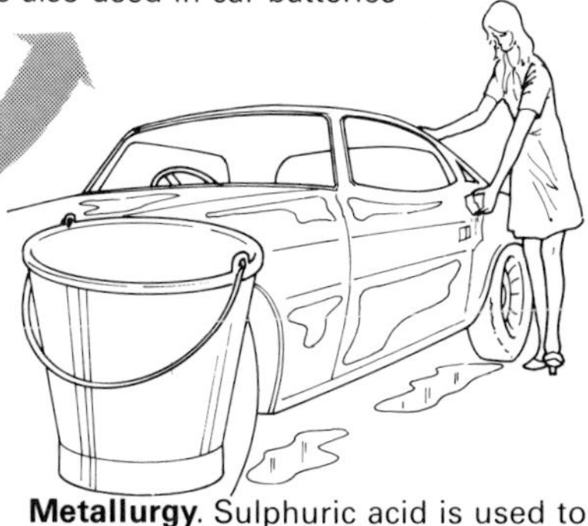

Metallurgy. Sulphuric acid is used to remove the oxide film from iron and steel products before they are given a protective coating to prevent rust

Dyestuffs. Sulphuric acid has been used to make synthetic dyestuffs since the middle of the nineteenth century

Other uses. Sulphuric acid is also important in many other industries, including leather tanning, pharmaceuticals and insecticides

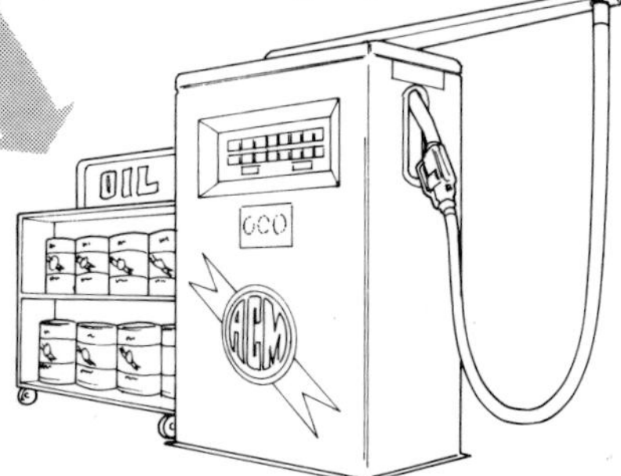

Oil and petrol Sulphuric acid is used to remove sulphur compounds and other impurities from crude oil

Figure 28.1 Uses of sulphuric acid

Figure 28.2 This photo shows just *part* of the sulphuric acid plant at Runcorn, Cheshire

28.2 Sources of sulphur dioxide

Figure 28.3 shows a highly simplified flow diagram of the contact process for making sulphuric acid. Sulphur dioxide is needed to make sulphuric acid. In the UK, the sulphur dioxide is made by burning sulphur. Liquid sulphur is sprayed into a furnace where it burns in dry air. The hot sulphur dioxide is cooled by being passed through a heat exchanger. The heat turns water into steam, which is then used to generate electricity. Saving energy in this way makes an important contribution to the economics of the process.

Sulphur is imported from Poland, Mexico and the USA, where it is extracted by the Frasch process, as described on p. 4. Another source of sulphur is impure natural gas. Natural gas in France and Canada may be contaminated with up to 25 per cent of hydrogen sulphide. The hydrogen sulphide must be removed before the gas is used as a fuel.

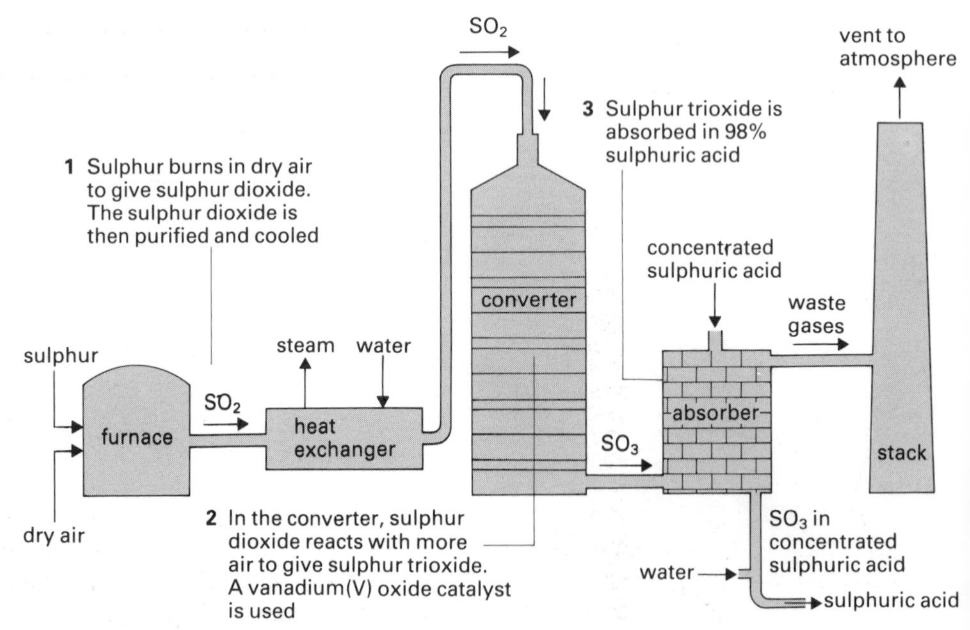

Figure 28.3 Flow diagram of the contact process for making sulphuric acid

Questions

1 Write an equation for the reaction of sulphur with oxygen.
2 Why must sulphur compounds be removed from natural gas before the gas is burned as a fuel?
3 Sulphur from Poland is imported in the liquid state in special tankers. Suggest why this is the most economical and convenient way of transporting this raw material.

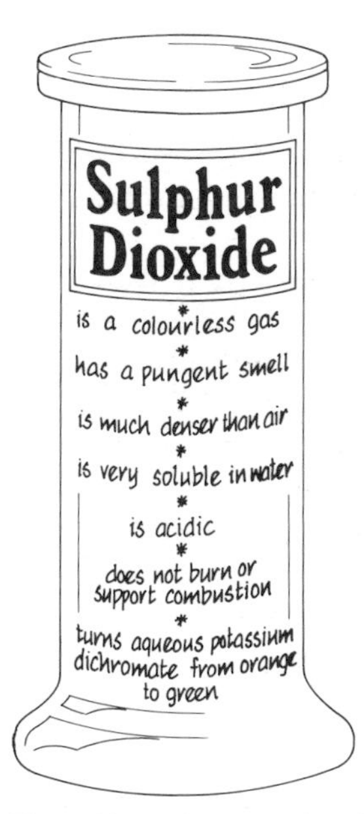

Figure 28.4 The properties of sulphur dioxide

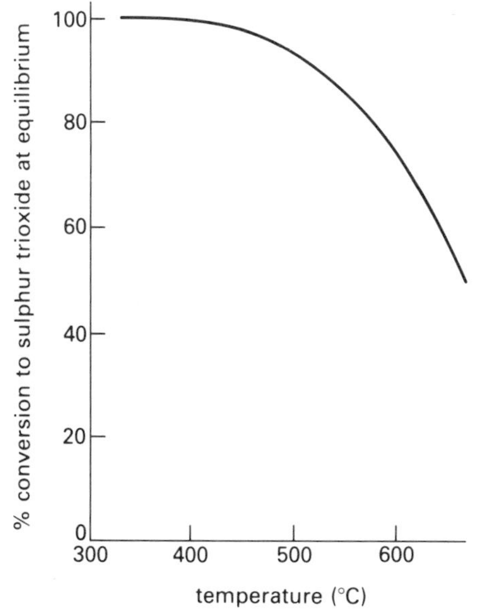

Figure 28.5 The yield of sulphur trioxide at different temperatures

Questions

1 How much sulphur dioxide can be made from 32 tonnes of sulphur?
2 How much sulphur trioxide can be made from 32 tonnes of sulphur, assuming that the yield is 100 per cent?
3 What would be the yield of sulphur trioxide (a) at 600 °C and (b) at 500 °C, assuming that the reaction mixture reaches equilibrium (see figure 28.5)?

Metals such as zinc, copper and lead are extracted from sulphide ores. These ores are converted to oxides by roasting them in air, e.g.

$$2ZnS(s) + 3O_2(g) \longrightarrow 2ZnO(s) + 2SO_2(g)$$

Worldwide, about 40 per cent of the sulphur dioxide used in the contact process comes from the roasting of sulphide ores.

The properties of sulphur dioxide are shown in figure 28.4.

28.3 How fast? How far?

In the contact process, a mixture of sulphur dioxide and air is passed into a converter in which there are four layers of the **catalyst**. The catalyst, which is vanadium(V) oxide, is in the form of pellets to increase its surface area. The process gets its name from the *contact* between the gases and the catalyst.

The sulphur dioxide combines with oxygen in the air to form sulphur trioxide:

$$2SO_2(g) + O_2(g) \rightleftharpoons 2SO_3(g)$$

This reaction is **reversible** and approaches a state of **equilibrium**.

Figure 28.5 shows how the percentage conversion to sulphur trioxide varies with temperature. The lowest temperature that can be used is 400 °C, because the catalyst does not work below this temperature. At 400 °C there is a good yield but the reaction gives out heat, so the temperature rises to about 600 °C as the gases pass through the first layer of catalyst. At this temperature the maximum yield is limited to about 70 per cent. To get round this problem, the gases leaving the first layer of catalyst are cooled back to about 450 °C before being passed through the second layer of catalyst. This process is repeated twice more to get an overall yield of about 98 per cent.

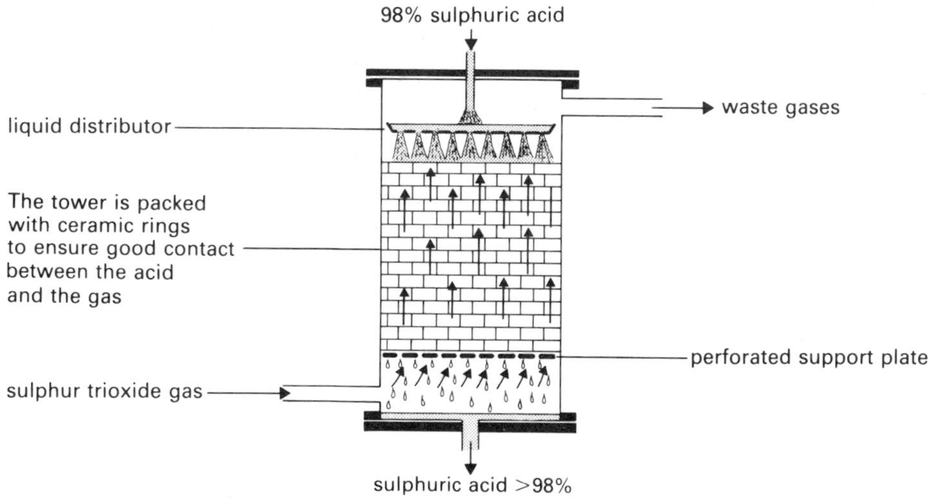

Figure 28.6 Absorption tower

Sulphur trioxide reacts with water to form sulphuric acid. This reaction is violent and gives out a lot of heat. In practice, sulphur trioxide cannot be combined directly with water because it reacts to form a mist of sulphuric acid droplets which cannot be collected. They would escape from the plant and pollute the atmosphere. To get round this problem, the sulphur trioxide is absorbed in 98 % sulphuric acid:

$$SO_3(g) + H_2O \text{ (in 98 \% sulphuric acid)} \longrightarrow H_2SO_4(l)$$

This happens in towers packed with ceramic rings. The gases pass up the tower and come into contact with a large surface area of 98 % sulphuric acid trickling down the tower. The acid concentration increases to about 99.5 %. Water is then added to dilute the acid back to 98.5 %.

28.4 Economic and social factors

If a sulphuric acid plant is run badly it can cause serious pollution. Expensive filters must be installed to remove any traces of acid spray or mist from the exhaust gases.

Sulphuric acid is made on a very large scale and it is relatively cheap. It is important that transport costs should add as little as possible to the price. Manufacturing plants are therefore sited close to major users of the acid.

The whole process is also much more economical if the heat given out during the chemical reactions is recovered and used. Heat is evolved when the sulphur burns and when the sulphur dioxide combines with oxygen. This heat can be used to make steam. The steam can then be sold for use in nearby factories or used to generate electricity. The income from these sources can be sufficient to cover the running costs of the sulphuric acid plant, including salaries and maintenance costs but not the cost of the sulphur.

Summary

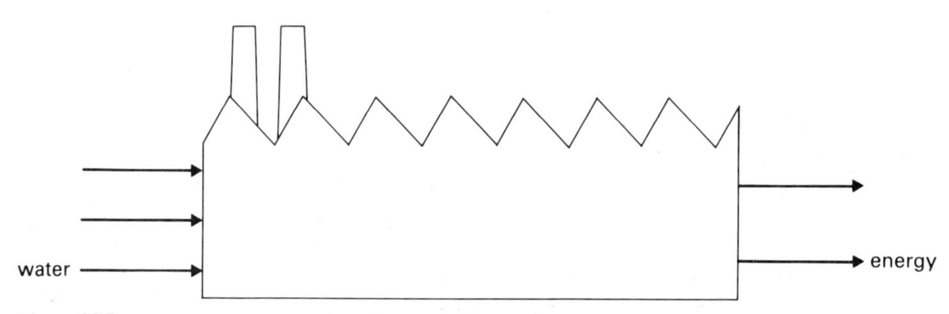

Figure 28.7

Copy and complete figure 28.7. Inside the outline of the factory, write symbol equations for the three main reactions used to make sulphuric acid.

Review questions

1 Copper(II) oxide can be reduced to copper by heating in a stream of hydrogen in the apparatus below.

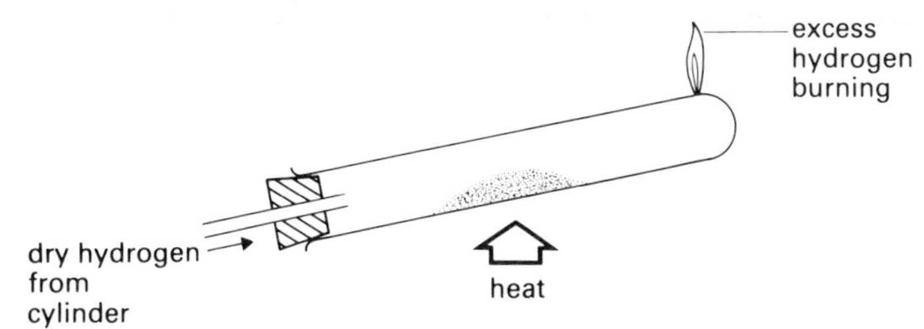

Dry copper(II) oxide was placed inside the test-tube. The hydrogen supply was turned on and the hydrogen was allowed to pass through the test-tube for some time before lighting. The test-tube was heated until all the copper(II) oxide was converted to copper. The apparatus was allowed to cool with hydrogen passing before dismantling.
(a) Mark on the diagram the most suitable place to clamp the test tube.
(b) Complete the word equation for the reaction:
Copper(II) oxide + Hydrogen $\longrightarrow$ Copper + ...
(c) Why was the hydrogen passed through the apparatus for some time before lighting?
(d) Why was the apparatus allowed to cool before dismantling?
(e) Which one of the following metal oxides is not reduced by hydrogen: iron(III) oxide, lead(II) oxide, silver oxide, zinc oxide? Give a reason for your answer.

A class carried out a series of reduction experiments with a dry oxide of copper to find its formula. Each group in the class used a different mass of oxide and the mass of copper produced was found in each case. The results are shown below.

Group	Mass of copper oxide (g)	Mass of copper produced (g)	Mass of oxygen lost (g)
1	0.62	0.55	0.07
2	0.90	0.80	0.10
3	1.12	1.00	0.12
4	1.69	1.50	. . .
5	1.80	1.60	. . .

(f) Complete the table of results above.
(g) Plot the mass of copper against the mass of oxygen on a graph. Draw the best straight line graph through the points and through the origin.
(h) From the graph, find the mass of oxygen which would combine with 1.28 g of copper.
(i) Calculate the mass of oxygen which would combine with 128 g of copper. (Remember $1.28 \times 100 = 128$.)
(j) How many moles of copper atoms does 128 g of copper represent?
(k) How many moles of oxygen atoms combine with 128 g of copper?
(l) What is the simplest formula of this oxide of copper? **(SREB)**

2 1 cm^3 of 3.0 M sodium hydroxide solution was added to 5 cm^3 of 1.0 M iron(III) chloride solution in a test-tube. After shaking to mix and allowing to stand for ten minutes, the height of the precipitate was measured. The experiment was repeated using different volumes of the 3.0 M sodium hydroxide solution. The results are shown below:

Volume of 3.0 M sodium hydroxide added (cm^3)	1	2	3	4	5	6	7
Height of the precipitate (mm)	4	8	12	16	20	20	20

(a) Draw a graph of these results.
(b) Explain why the height of the precipitate did not change from the 5th to the 7th cm^3.
(c) Calculate the number of moles of sodium hydroxide in 5 cm^3 of 3.0 M sodium hydroxide solution.
(d) The equation for the reaction is

$$3NaOH(aq) + FeCl_3(aq) \longrightarrow Fe(OH)_3(s) + 3NaCl(aq)$$

State the number of moles of iron(III) chloride required to react exactly with 5 cm^3 of 3.0 M sodium hydroxide solution.
(e) Show how the answer to (d) compares with the results of the experiment.
(f) Describe how you would attempt to obtain a pure dry sample of the precipitate formed in the experiment. **(L)**

3 (a) In the process for the industrial extraction of iron, iron(III) oxide (Fe_2O_3) is reduced by carbon monoxide. If 40 g of iron(III) oxide were completely reduced, calculate the mass of iron obtained, using the following stages:
(i) Write an equation for the reduction of iron(III) oxide with carbon monoxide.

(ii) How many moles of iron(III) oxide does 40 g represent?
(iii) From your answers to (i) and (ii) deduce the number of moles of iron formed.
(iv) Calculate the mass of iron formed.
(b) When iron is reacted with a dilute acid, iron(II) ions (Fe^{2+}) and hydrogen are the products. State and explain *one* test in *each* case that could be used to identify *each* of these products. **(WJEC)**

4 (a) State the volume (measured at s.t.p.) occupied by 16 g of each of the following gases: (i) oxygen, (ii) sulphur dioxide, (iii) methane.
(b) Write down a balanced equation to show how sulphur dioxide reacts with oxygen.
(c) By reference to the equation which you have written as your answer to (b) above, calculate the volume of oxygen which would be required to react completely with 100 dm^3 of sulphur dioxide, assuming that all measurements are made at the same temperature and pressure. **(O)**

5 A stick of lithium was removed from a bottle of oil in which it was stored and a piece was cut from it which was found to have a mass of 0.1 g. The piece was transferred to an apparatus containing an excess of water. After the reaction had started, the volume of hydrogen produced was measured at intervals. Some of the results are shown in the table below.

Time (minutes)	Volume of hydrogen (cm^3)
1	8
2	24
3	72
4	138
5	172
6	172

(a) Draw a diagram of a suitable apparatus for the experiment in which you could ensure that the lithium did not react until you were ready to start the reaction.
(b) Plot the results as a graph with time on the horizontal axis.
(c) What is the rate of the reaction in volume of hydrogen per minute between 3 and 4 minutes?
(d) Explain precisely the reasons for the change in rate between 4 and 5 minutes.
(e) It was thought that the results in the first 4 minutes were affected by oil on the lithium. So the experiment was repeated with 0.1 g of lithium from which all oil had been removed. Draw a *dotted* line on the graph to show the results you would expect from this second experiment.
(f) (i) When 1 mole (g-atom) of lithium has completely reacted, what volume of hydrogen would be produced at room temperature and pressure?
(Li = 7; 1 mole of molecules (g-molecule) of a gas at room temperature and pressure occupies 24 litres.)
(ii) Lithium hydroxide, LiOH, is also produced. Use the results above to complete and balance the following equation:

$$Li(s) + H_2O(l) \longrightarrow LiOH(aq)$$

(g) Give *one* property of the resulting solution. **(L)**

6 In an experiment, 0.5 g of manganese(IV) oxide was added to 50 cm^3 of 0.2 M hydrogen peroxide and the volume of oxygen evolved was measured at regular time intervals. The results of the experiment are shown in the graph below.

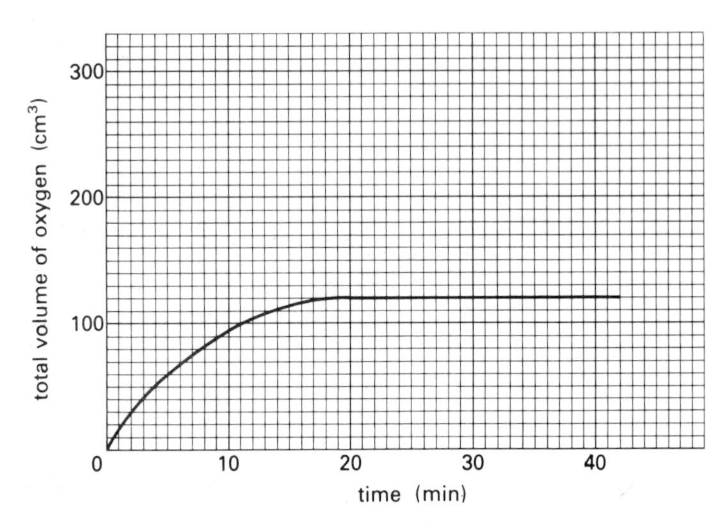

(a) Draw a sketch of the apparatus you would use to carry out the experiment.
(b) Use the graph to find the volumes of oxygen evolved (i) during the first 5 minutes (from 0 to 5 min), (ii) during the second 5 minutes (from 5 to 10 min) and (iii) during the fifth 5 minutes (from 20 to 25 min). Explain why the volumes produced in 5 minutes vary in this way.
(c) Using the same scales and axes, make a copy of the graph on squared paper: show clearly on your copy the approximate results you would have expected if the experiment had been carried out under the same laboratory conditions using (i) 100 cm^3 of 0.1 M hydrogen peroxide and 0.5 g of manganese(IV) oxide and (ii) 50 cm^3 of 0.4 M hydrogen peroxide and 0.5 g of manganese(IV) oxide. **(CLES)**

7 Some reactions are reversible, reaching a position of equilibrium after some time, while others appear to go completely in one direction.
(a) Select *two* reversible reactions which you have investigated in the laboratory. For each reaction, explain what happens to the chemicals involved as the reaction takes place and describe how changing the conditions can drive the equilibrium in one direction or the other.
(b) Select *one* reaction you have investigated that takes place completely in one direction. Write an equation for the reaction. Suggest *one* experiment that could be used to obtain information about the equation for the reaction. **(L)**

THEME I

Chemicals from oil and gas

Chemicals from oil are used to manufacture the nylon fibres from which this rope is being made

29 Introducing organic chemistry

Questions

1 To which group in the periodic table does carbon belong?
2 How many electrons are there in a carbon atom, and how are they arranged in shells?
3 Which groups in the periodic table contain elements which form ions? Is carbon likely to form ions?
4 What does the word 'organic' mean outside chemistry? Can you suggest why carbon chemistry is called 'organic chemistry'?

29.1 What makes carbon so special?

Carbon is an amazing element. It forms over a million compounds, and new carbon compounds are being discovered every day. There are more compounds of carbon than there are of all the other elements put together. The chemistry of carbon compounds is so important that it forms a separate branch of chemistry, called 'organic chemistry'. Organic chemistry includes the study of all carbon compounds except the very simplest ones, such as carbon dioxide, carbon monoxide and carbonates.

If a piece of toast is left under the grill too long, it chars and turns black. The black is carbon. Most substances from plants and animals also char and blacken if they are heated strongly. This simple test shows that they consist of carbon compounds. Wood turns to charcoal when heated away from air; so do bones and sugar.

For a long time it was thought that only living things could make the more complicated carbon compounds. The word 'organic' means 'living'. At first, organic chemistry was the study of carbon chemicals from plants and animals. Now it is known that complex carbon compounds can be made artificially. Organic chemistry includes the study of synthetic plastics, dyes and drugs.

Experiment 29a
Detecting carbon

copper(II) oxide

Many organic compounds char when heated, but this is not a very reliable way of testing to see whether a substance contains carbon. A better method uses copper(II) oxide as a source of oxygen to change the carbon to carbon dioxide. The carbon dioxide is then identified using limewater.

The substance to be tested is mixed with copper(II) oxide and heated in a test-tube. The fumes are bubbled through water in a second tube. This removes smoke and tarry fumes. Any carbon dioxide formed is detected with limewater.

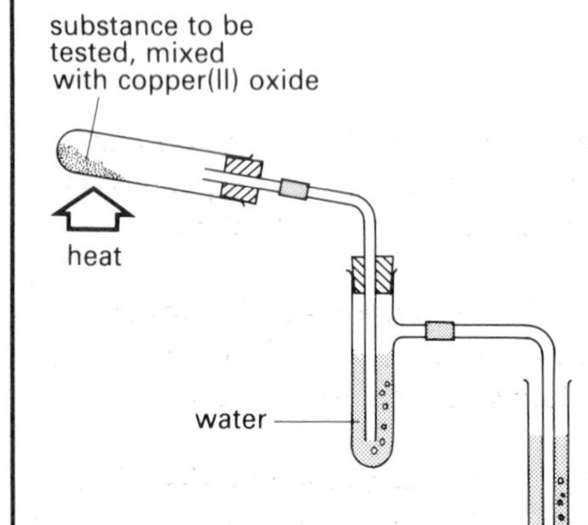

Figure 29.1

Results
This experiment shows that a wide range of substances, including wood, rice, sugar, fat, starch, coal, cotton and wool, all include carbon. It can also be shown that carbon is present in plastics such as polythene and polystyrene.

Discussion
What happens to copper(II) oxide when it is heated with a substance that contains carbon? What would you see if you scraped out the residue from the tube which was heated? Which substance is oxidized and which is reduced in this experiment?

Many carbon compounds also contain hydrogen. What will happen to the hydrogen when such a compound is mixed with copper(II) oxide and heated? What compound will form? Where will it appear in the apparatus shown in figure 29.1?

Questions

1 Using figure 3.15 to help you, draw structures for the following compounds:
(a) propene, C_3H_6,
(b) cyclobutane, C_4H_8,
(c) methanol, CH_4O,
(d) methanal, CH_2O,
(e) trichloromethane, $CHCl_3$,
(f) methylamine, CH_5N.
(You will find this question easier if you have a set of molecular models to work with.)

2 Work out structures for
(a) the two isomers with the formula C_2H_6O.
(b) the three isomers with the formula C_5H_{12}.
(c) the four isomers with the formula C_4H_8.

29.2 The structure and bonding of carbon compounds

Carbon can form so many compounds because carbon atoms can join together in different ways, forming chains, branched chains and rings. A great variety of molecules is possible. The carbon atoms form a skeleton to which atoms of other elements can be attached. In organic compounds, carbon often forms bonds with hydrogen, oxygen, nitrogen and halogen atoms.

The bonding in organic compounds is covalent. The structures are molecular. The structures can be worked out from knowing how many covalent bonds each type of atom can form. These 'combining powers' are shown in figure 3.15 on p. 25.

The table in figure 19.7 on p. 130 includes the formulae and structures of some common carbon compounds. As you look at the structures, check that each atom has formed the same number of covalent bonds as listed in figure 3.15. Note that the structures written flat on paper do not give the right idea of the shapes of the molecules. Molecular models give a better impression of the shapes of molecules in three dimensions. Ball-and-spring models show the bonding between atoms clearly, and give some idea of molecular shape. Space-filling models can be more realistic. The two types of model are compared in figure 19.7.

Even more compounds are possible than you might expect, because it is possible to join the same set of atoms in a variety of ways. Figure 29.2 shows two different compounds with the same molecular formula, C_4H_{10}. Each of the two compounds has its own set of values for properties such as boiling point and density. They are two distinct compounds. Compounds with the same molecular formula, but different structures, are called **isomers**.

Figure 29.2 Two compounds with the formula C_4H_{10}

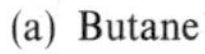

(a) Butane

(b) 2-methylpropane

29.3 The alkanes

There are so many carbon compounds that it could be hopelessly confusing to study them all separately. Fortunately, it is possible to group the compounds into families, and to learn about the families rather than about the individual compounds.

The simplest family is the group of **alkanes**. The alkanes are well-known because they are present in natural gas, petrol and paraffin wax. All the alkanes are **hydrocarbons**. This means that they are compounds of only two elements: hydrogen and carbon.

A family of carbon compounds is sometimes called a **homologous series**. This means that the members of the family are related. The compounds in a series have the same general formula. The formula of one member in the series differs from the formula of the next member by CH_2. All compounds in the series have similar chemical properties. There is a gradual change in the physical properties of the compounds in the series as the number of carbon atoms in the molecules increases.

Physical properties of some alkanes are given in table 3 on p. 340.

Questions

1 What do the names of all the alkanes have in common?
2 What is the general formula of the alkanes?
3 Draw a graph of boiling point (along the *y*-axis) against the number of carbon atoms in the molecule (along the *x*-axis) for the alkanes. Beside each point, show whether the alkane is a gas, a liquid or a solid at room temperature, 20 °C, using state symbols.
4 Use the graph to predict the boiling point of octane, C_8H_{18}. Will octane be a gas, a liquid or a solid at room temperature?

29.4 The chemical reactions of the alkanes

The alkanes do not react with many of the common laboratory reagents. They do not dissolve in water. They do not react with acids or alkalis. But they do have three important reactions:

* they burn
* they can be 'cracked'
* they react with chlorine and bromine.

Burning

Many common fuels contain alkanes. They burn in air to release much energy, together with carbon dioxide and water. Burning is discussed in chapter 6, and the energy changes involved are described in chapter 37.

Cracking

Cracking is one of the important processes used in oil refineries to get the best value from crude oil. Cracking involves breaking up molecules by using heat and catalysts. This is described in chapter 30.

Halogens

The reactions with chlorine and bromine are important because they are the first steps in making other valuable chemicals from the alkanes. Compounds made from the alkanes by replacing hydrogen atoms in them with halogen atoms have many uses:

* they are used as solvents
* they are used to control and put out fires
* they are used as anaesthetics and refrigerants.

Alkanes react with chlorine when the mixture is heated or exposed to ultraviolet light. Hydrogen atoms are replaced by chlorine atoms. This is called a **substitution reaction**. Any of the hydrogen atoms in the alkane may be replaced. The reaction can continue until all of the hydrogen atoms have been replaced. Thus, a mixture of products is obtained. This is illustrated in figure 29.3.

$$CH_4(g) + Cl_2(g) \longrightarrow CH_3Cl(g) + HCl(g)$$

Figure 29.3 Model equation for the reaction of methane with chlorine. What other products could be formed?

Questions

1 Can you think of other reactions which involve light energy? Photosynthesis and photography are two examples. Are there others?
2 Write an equation for the reaction of trichloromethane with chlorine. What is the product?
3 How many possible products are there when ethane reacts with chlorine? Draw the structures of the products of this substitution reaction.

Alkanes can react similarly with bromine. If a few drops of bromine are added to a tube containing hexane, the orange colour gradually fades when a bright light is shone onto the mixture. There is no reaction in dim light.

29.5 Natural gas

Natural gas consists almost entirely of methane. The Frigg gas field, which lies in the North Sea about 360 km north-east of the Scottish mainland, produces gas which is 95 per cent methane and 4 per cent ethane, with only traces of other hydrocarbons. Unlike the gas made from coal (see p. 280), natural gas does not contain carbon monoxide, so it is not poisonous. It has no smell, so an artificial odour is added before the gas is distributed to consumers, so that leaks may be detected before there is a risk of explosion.

Before the discovery of natural gas, the gas for lighting and heating was made from coal. This gas was expensive because its price included the costs of mining and transporting coal, manufacturing the gas and then distributing it. For a short period between 1955 and 1965, there was so much cheap oil available that it became worthwhile building plants to make 'town gas' from oil. With the discovery of natural gas, it was no longer necessary to make the gas. Today, British Gas is primarily concerned with distributing the natural gas from the North Sea.

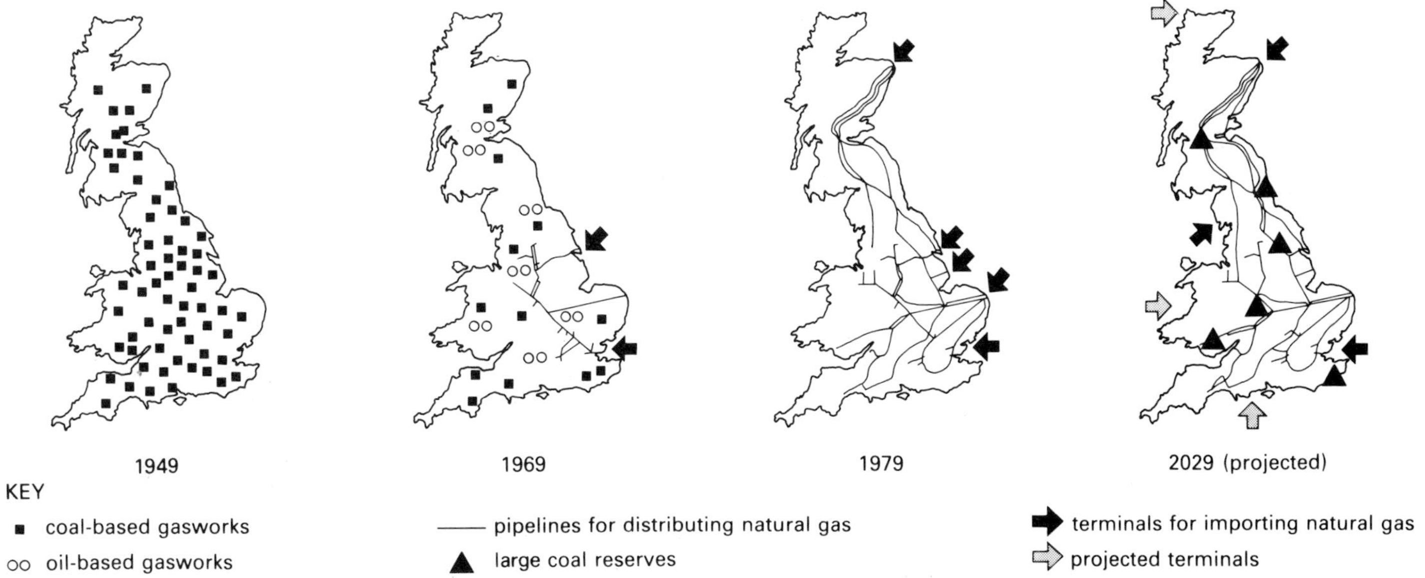

Figure 29.4 The changing role of British Gas: in 1948, there were 1050 gas-manufacturing plants in Great Britain but by 1968 the number had dropped to 192. Now the main function of British Gas is distribution of natural gas. It is hoped that the pipeline will be used to distribute a high-grade fuel made from coal when the supply of natural gas begins to decline

Natural gas supplies are expected to last into the twenty-first century, but the amounts will begin to decline noticeably after this. Fortunately, Britain has large reserves of coal. It is expected that new and more efficient processes for making gas from coal will be developed in the next 15–20 years.

Questions

1 In future, will it be better to use coal to generate electricity or to make gas? Which will be the more efficient method of distributing the energy stored in coal?

2 Write equations for burning methane (a) to form carbon monoxide and (b) to form carbon dioxide. Compare the amounts of oxygen in the two equations.

3 Gas board leaflets say that gas appliances must be able to 'breathe' (see figure 29.5 overleaf). In what ways is burning gas similar to breathing? Why is it dangerous to burn gas in too little air?

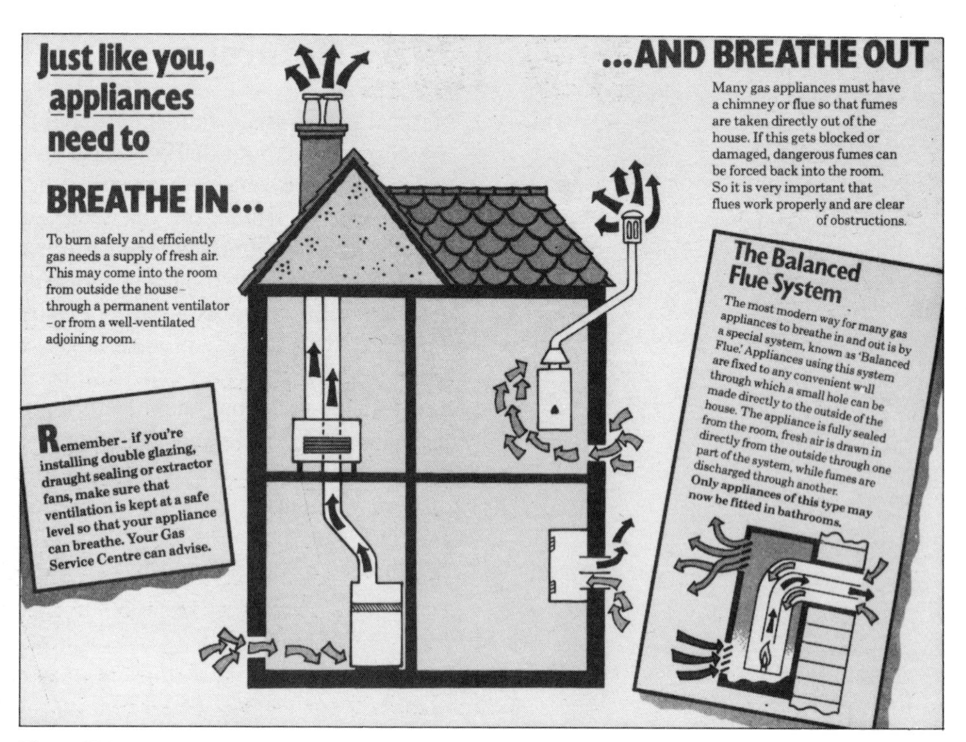

Figure 29.5

Summary

Write a paragraph summarizing the main ideas in this chapter. You should include the following terms: hydrocarbon, isomer, alkane, homologous series, burning, substitution reaction.

30 Oil

30.1 Refining

Crude oil is a complicated mixture of **hydrocarbons**. It includes molecules of many shapes and sizes. Some of the molecules may also contain sulphur. There are traces of other elements.

Crude oil must be refined to make useful fuels and chemicals. The first step in refining is the separation of the oil into fractions by **distillation**. This process is described in chapter 1. The table in figure 30.1 shows the nature of typical oil fractions. The table also shows the proportions of the fractions in two different crude oils. A crude oil with a higher proportion of the lighter fractions is more valuable than one with more of the heavier fractions.

Fraction	Number of carbon atoms in the molecules	Percentage of each fraction from the initial distillation		Approximate demand for each fraction (%)
		Oil from North Sea	Oil from Persian Gulf	
Gas	1–5	2	2	4
Petrol	5–10	8	5	22
Naphtha	8–12	10	9	5
Kerosine	9–16	14	12	8
Gas oil	15–25	21	17	23
Fuel oil	20–30	45	55	38

Figure 30.1

The problem in refining is to produce the various oil products in the amounts required by industrial and domestic users. The final column in the table shows the approximate demands for the different fractions. These are very different from the proportions produced by distillation.

After separating the fractions, the refinery must carry out a variety of conversion processes to produce saleable products in the required quantities. There are three main conversion processes:

* cracking – this breaks up larger molecules into smaller ones
* polymerization – this joins small molecules together to make larger ones
* reforming – this rearranges the atoms in molecules into different structures.

Both **cracking** and **polymerization** are used to increase the yield of petrol from crude oil. Catalytic cracking is used to break up the heavy fractions, such as gas oil, to produce molecules in the C_9–C_{10} range. Polymerization produces similar molecules by linking up small molecules from hydrocarbon gases.

Questions

1 Which of the crude oils will sell for the higher price, crude from the North Sea or crude from the Persian Gulf?
2 Which fractions are overproduced by fractional distillation?
3 Which fractions are underproduced by distillation?

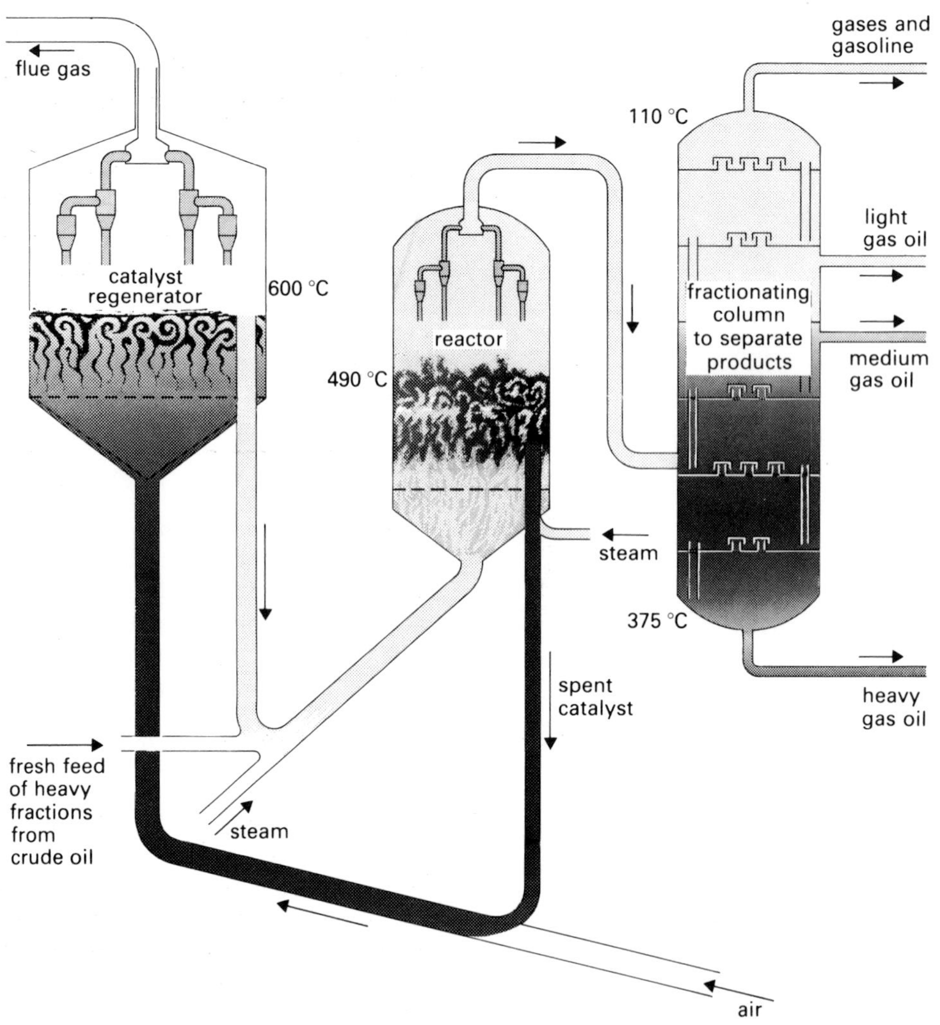

Figure 30.2 Catalytic cracking: the catalyst powder passes to the reactor, where the cracking takes place. The cracked vapours then pass on to the fractionating column and the used catalyst returns to the regenerator, where it is cleaned for reuse

Reforming is used to improve the quality of petrol. Modern high compression engines require high octane fuels. Changing the shape of the molecules can improve the octane rating.

Cracking is also used to make the basic building blocks for the petrochemical industry. This is illustrated on a small scale by experiment 30a.

Experiment 30a
Cracking

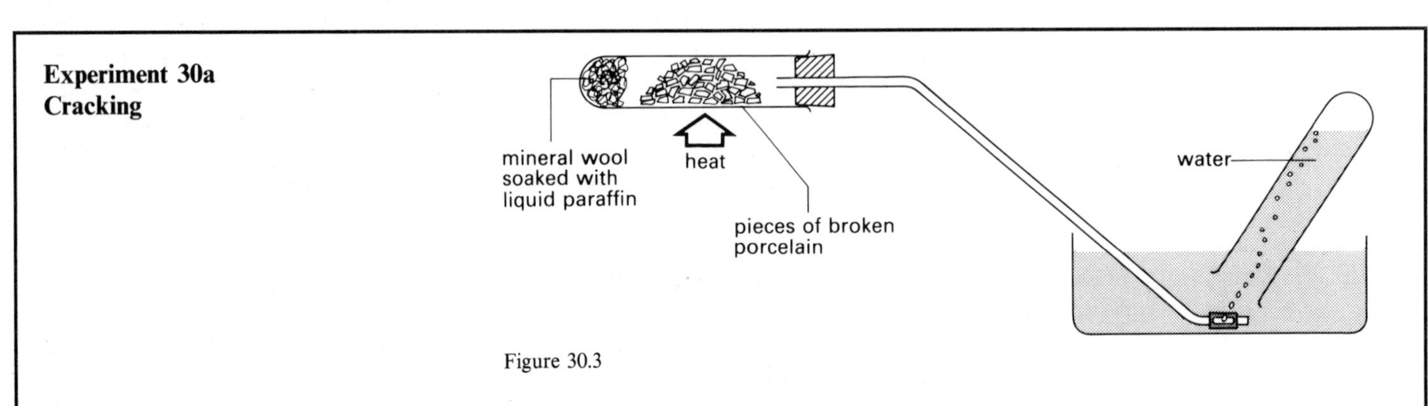

Figure 30.3

In this experiment, a long-chain hydrocarbon is cracked by passing its vapour over a hot catalyst. The burner is controlled to make sure that a steady stream of hydrocarbon vapour passes over the catalyst. Why is the porcelain heated strongly before the mineral wool is warmed? Why are the first bubbles of gas from the delivery tube not collected?

Results
The table in figure 30.4 compares the properties of the liquid hydrocarbon with those of the product obtained by cracking.

Property	Liquid hydrocarbon (paraffin oil)	Gas formed by cracking
Colour	Colourless	Colourless
Smell	No smell	Sweetish smell
Does it burn?	Burns on strong heating	Burns easily with a yellow flame
Reaction with aqueous bromine, which is orange	No reaction	The bromine solution becomes colourless on shaking
Reaction with an acid solution of potassium manganate(VII), which is purple	No reaction	The purple solution becomes colourless on shaking

Figure 30.4

Discussion
The only way to get a gaseous hydrocarbon from a liquid is to make the molecules smaller. The hot catalyst must have split (cracked) the molecules. What is the catalyst in this experiment?

The results show that the gas formed is much more reactive than the alkanes in the original hydrocarbon mixture. It reacts with bromine and acid potassium manganate(VII), which alkanes do not do. This reactive gas consists mainly of *ethene.*

Ethene is normally produced industrially by thermal cracking in the presence of steam. In Britain and Europe the naphtha fraction from oil has been the main source of hydrocarbons to be cracked to make ethene. However, ethene made by cracking ethane is much cheaper. Ethane is now being piped ashore from North Sea oilfields and is being used to make ethene.

The final stage of refining is purification – the removal of hazardous chemicals from oil products. Sulphur must be removed from fuels, because fuels containing sulphur will produce sulphur dioxide when burned. Sulphur dioxide pollutes the air, as explained in chapter 6. Some hydrocarbons (the arenes) burn with a very smoky flame. Benzene is an arene. These must also be removed from fuels. The sulphur and the arenes are not wasted – the sulphur is used to make sulphuric acid, and the arenes are used to produce solvents, plastics and dyes. The sale of these, and other, by-products is an important part of the economy of a refinery.

30.2 The alkenes

The ethene made by cracking is invaluable because of its ability to combine with other chemicals:

* it will react with water to form ethanol, which is an important solvent
* it combines with benzene to make the chemical needed for the manufacture of the plastic, polystyrene
* it reacts with chlorine to form the intermediate used to make the plastic, pvc
* its molecules can be polymerized to make polythene.

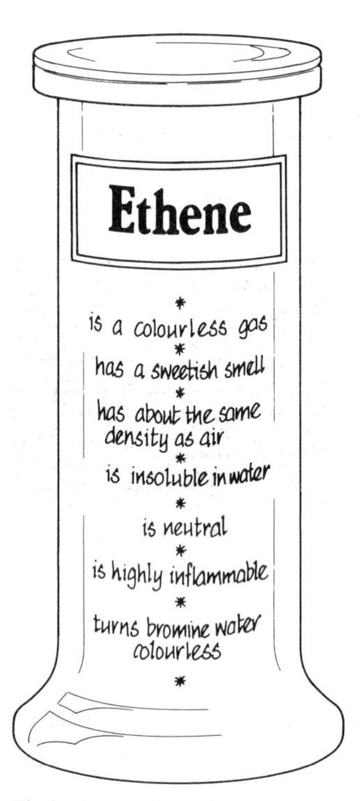

Figure 30.5 The properties of ethene

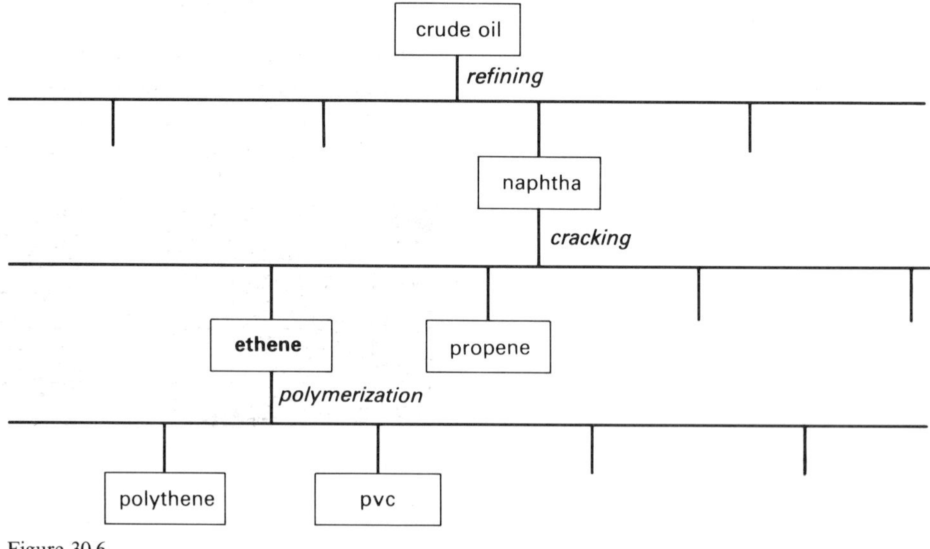

Figure 30.6

The properties of ethene are shown in figure 30.5.

The importance of ethene, the building brick of modern chemistry, is illustrated by figure 30.6.

Another important product of cracking is *propene*. Propene is the second member of the **alkene** family. Figure 30.7 shows the structures of ethene and propene. The names of alkenes are obtained by taking the name of the corresponding alkane and changing the ending to 'ene'.

Figure 30.7 Two alkenes

(a) Ethene

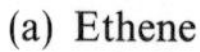

(b) Propene

Questions

The general formula of the alkenes is C_nH_{2n}.

1 Why is there no alkene for which $n = 1$?

2 Write down the structures of the three alkenes for which $n = 4$.

Question

The alkenes burn in a similar way to the alkanes. Write an equation for the combustion of ethene.

All alkenes have similar chemical properties because they all have a double bond in the molecule. The double bond is what makes them so much more reactive than the alkanes.

Compounds like the alkanes in which all the bonds are single bonds are called **saturated compounds**. There are no 'spare bonds'. The alkenes are called **unsaturated compounds** because of the double bond. The double bond gives the molecules 'spare bonding', which can be used to add new bits to the molecules. In their important reactions, the alkenes use the spare bonding of the double bond to add on more atoms to their molecules.

Reaction with bromine

When ethene gas is shaken with a solution of bromine, the colour of the bromine disappears. The ethene and bromine molecules add together. There is only one product. This is an example of an **addition reaction**. The reaction goes rapidly at room temperature, and is used as a test to detect unsaturated compounds.

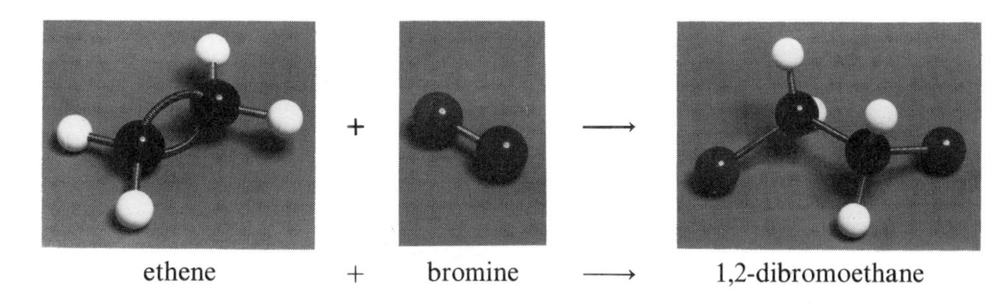

Figure 30.8 Model equation for the addition of bromine to ethene

Reaction with hydrogen

Hydrogen adds to alkenes on heating, in the presence of a nickel catalyst. There are important unsaturated compounds for which this reaction is useful. Margarine is made from vegetable oils. Many of these oils are unsaturated, with C=C bonds in the molecules. With the help of a nickel catalyst, it is possible to add hydrogen across the double bonds, thus making saturated compounds. The change in structure can convert a liquid oil into a solid fat which is suitable for making margarine.

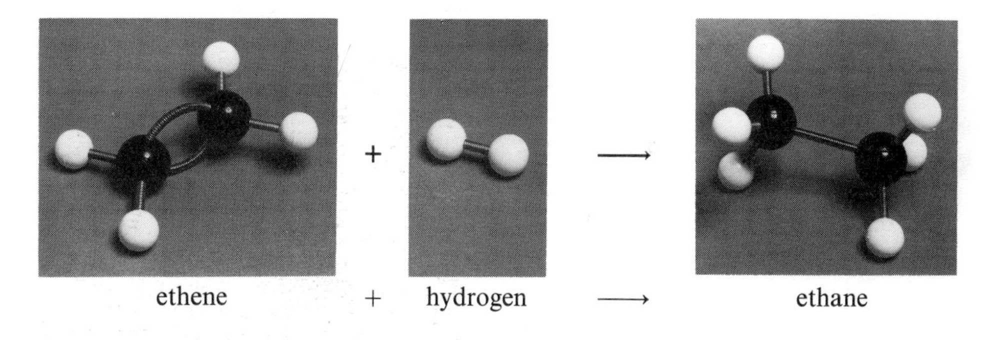

Figure 30.9 Model equation for the addition of hydrogen to an unsaturated compound – ethene

Figure 30.10 Hardening tanks for making vegetable oil into margarine. The margarine can be made hard or soft by adjusting the amount of hydrogen added

Questions

1 Write equations for the reactions of (a) bromine with propene and (b) hydrogen with propene.

2 Chlorine reacts with alkenes in a similar way to bromine. Draw the structures of the products of the reactions of chlorine with ethene and with propene.

There are other important reactions of alkenes. The reaction of ethene with steam to make ethanol is described in chapter 31. The polymerization of ethene to make polythene is explained in chapter 32.

Summary

1 An oil refinery has to separate, convert and then purify. Describe briefly what is involved in each of these three stages.
2 Copy and complete figure 30.11 to summarize the main reactions of ethene.

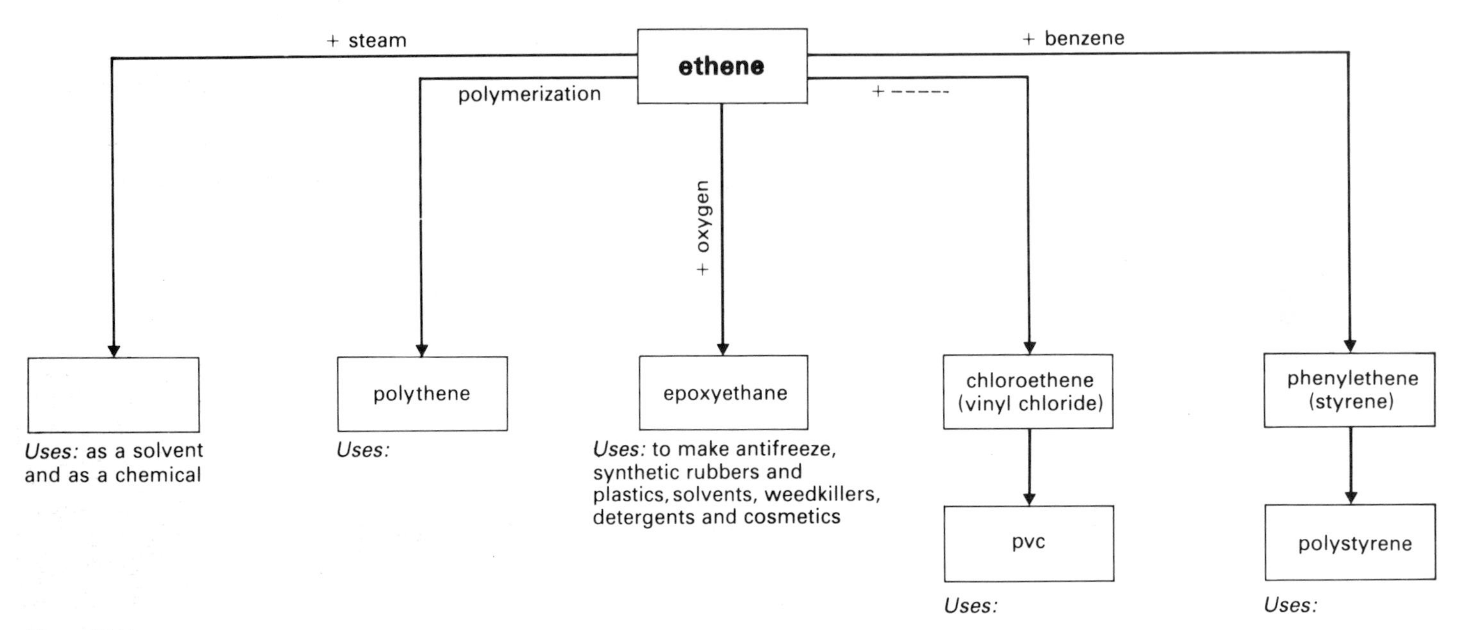

Figure 30.11

31 Alcohols, acids and esters

31.1 Ethanol

Ethanol is the best-known member of the family of **alcohols**. It is the alcohol in beer, wine and spirits. Ethanol is also a very useful solvent. It is used in the manufacture of varnishes, polishes, inks, glues and paints. It is a solvent which evaporates quickly, and for this reason it is used in deodorants, colognes and after-shave lotions.

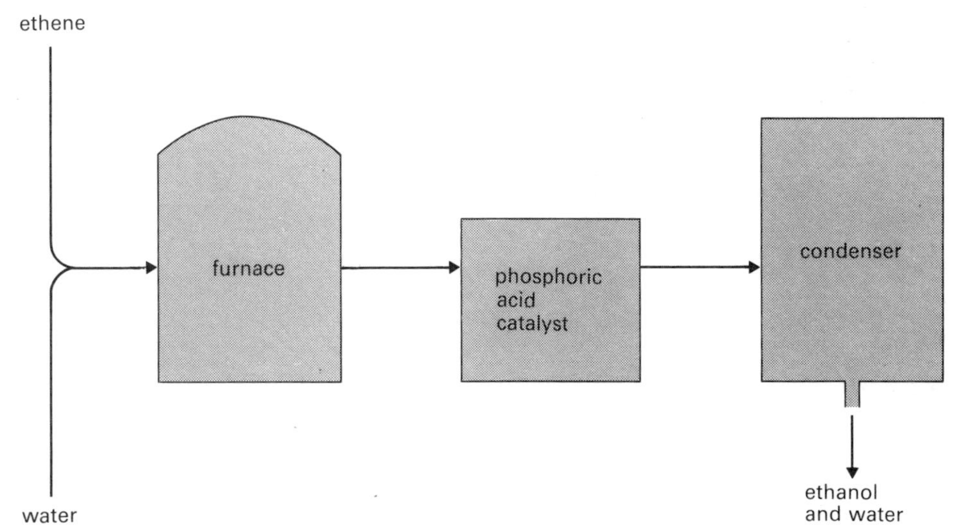

Figure 31.1 Flow diagram of the industrial process for making ethanol

The production of ethanol by **fermentation** is described in chapter 1. The large amounts of ethanol needed for industry are made by a faster method, starting from ethene (see figure 31.1). The ethene is mixed with steam and passed over a phosphoric acid catalyst at 300 °C. This reaction is illustrated in figure 31.2, which also shows the 'dog-like' shape of the ethanol molecule.

Figure 31.2 Model equation for the formation of ethanol from ethene

The industrial preparation of ethanol is not easy to demonstrate, but the reverse reaction can be done simply. It illustrates the connection between ethanol and ethene.

Experiment 31a
Cracking ethanol

ethanol

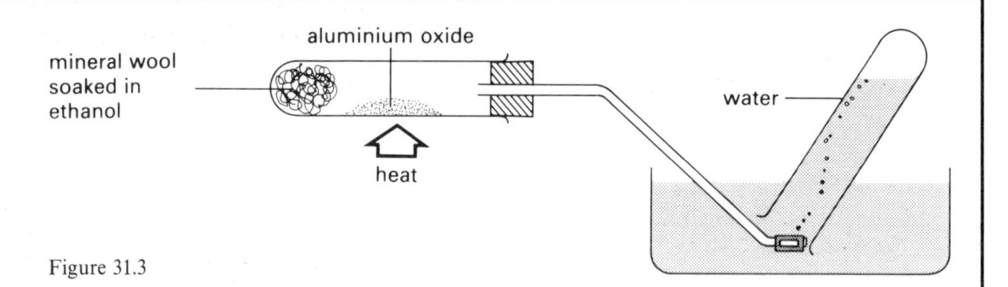

Figure 31.3

The apparatus is the same as that used to crack hydrocarbons in experiment 30a. Aluminium oxide powder is a good catalyst for the reaction.

Results
A gas collects in the tube over water. This gas burns. It reacts with bromine solution, turning it colourless. It also decolorizes a purple solution of acidified potassium manganate(VII).

Discussion
What is the gas formed in this experiment? Consider the formula of ethanol and the formula of the gas, then try to predict what other product is likely to be formed. What happens to the second product in the experiment? Work out the equation for the reaction, showing the structures of the carbon compounds.

Alcoholic drinks are highly taxed. Synthetic alcohol cannot be sold in the form of pure ethanol, because it might be used in drinks to evade these taxes. For most industrial uses, the ethanol is mixed with 5 per cent of methanol, which is highly poisonous. This makes it undrinkable. The mixture is called 'industrial methylated spirits' (IMS). The 'meths' bought in hardware shops is the same mixture with a violet dye added and traces of other chemicals to give it a foul taste. The mixture chosen is such that the ethanol cannot easily be recovered in a pure state by distillation.

31.2 The alcohols

The first three members of the alcohol series are methanol, ethanol and propanol. They are named by changing the end of the name of the corresponding alkane to 'ol'.

Figure 31.4 Three simple alcohols

(a) Methanol, CH_3OH

(b) Ethanol, C_2H_5OH

(c) Propanol (propan-1-ol), C_3H_7OH

The fact that the millions of carbon compounds can be grouped in series helps to make sense of their chemistry. It helps because the same group of atoms usually behaves in the same way in all compounds. The alkanes are unreactive because the C—C and C—H bonds in the molecules are unreactive. The alcohols are more

Questions

1 What is the general formula for the family of alcohols?
2 Write the structures of the isomers with the formula C_3H_8O. How many of the isomers are alcohols?
3 Write an equation for the reaction of methanol with sodium.
4 Give the names and formulae of the products formed when
(a) propanol is passed over hot aluminium oxide.
(b) ethanol is heated with concentrated phosphoric acid.
(c) propanol reacts with sodium.

reactive than the alkanes because C—O and O—H bonds are more reactive than C—C and C—H bonds. The alcohols share similar properties because they all have the $-\overset{|}{\underset{|}{C}}-O-H$ group of atoms in their molecules (see figure 31.5).

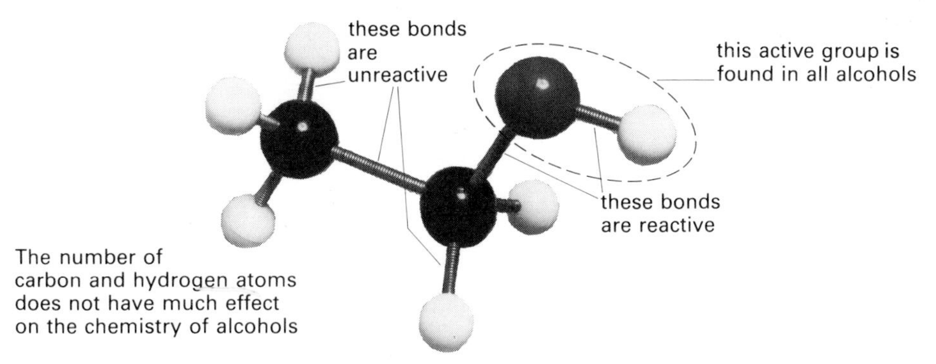

Figure 31.5 The reactivities of the bonds in ethanol

Reaction with sodium

Alcohols react with sodium in a similar way to water. This is because both water molecules and alcohol molecules include the —O—H group of atoms. With water, the products are sodium hydroxide and hydrogen; with ethanol the products are sodium ethoxide and hydrogen:

$$2\,H-\overset{H}{\underset{H}{\overset{|}{\underset{|}{C}}}}-\overset{H}{\underset{H}{\overset{|}{\underset{|}{C}}}}-O-H + 2\,Na \longrightarrow 2\,H-\overset{H}{\underset{H}{\overset{|}{\underset{|}{C}}}}-\overset{H}{\underset{H}{\overset{|}{\underset{|}{C}}}}-O^-Na^+ + H_2$$

The reaction is dangerously rapid with water, but much slower with the alcohols. The safe way of getting rid of waste sodium in the laboratory is to cut it into small pieces and then add it to ethanol, a little at a time.

Note that in the reaction with sodium, only the hydrogen atom joined to oxygen is involved. The hydrogen atoms linked to carbon are inert (unreactive), as they were in the alkanes.

Dehydration

The **dehydration** of ethanol to ethene has been described in experiment 31a. Other alcohols can be dehydrated in a similar way. An alternative method is to heat the alcohols with concentrated phosphoric acid, which is a dehydrating agent.

Oxidation

Alcohols are **oxidized** to organic acids. The oxidation of ethanol to ethanoic acid is described in the next section.

31.3 The organic acids

Organic acids form another series of carbon compounds. The reactive group of atoms in the molecules of the acids is

$$-C\begin{matrix}\nearrow\!\!\!\!\!\nearrow O \\ \searrow O-H\end{matrix}$$

These compounds are sometimes called **carboxylic acids**. As usual, the names of

Figure 31.6 Two simple carboxylic acids

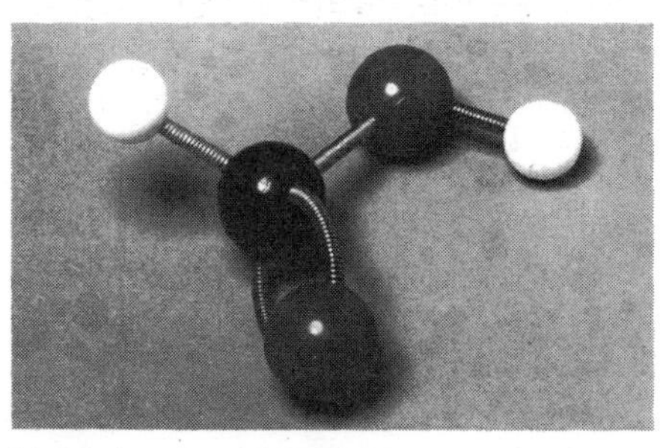

(a) Methanoic acid, HCO_2H

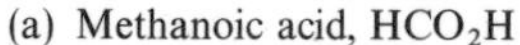

(b) Ethanoic acid, CH_3CO_2H

these acids are based on the names of the alkanes with the same number of carbon atoms. The ending 'ane' becomes 'anoic acid'.

The older names for these acids are still used quite commonly. They were based on the original sources of the acids. Methanoic acid used to be called formic acid because it was first obtained by distilling red ants, and *formica* is the Latin word for ant. On food labels, ethanoic acid is usually called acetic acid (from the Latin word for vinegar – *acetum*). Octadecanoic acid (with eighteen carbon atoms in the molecule) is better known as stearic acid (from the Greek word for tallow – *stear*).

Vinegar is made by exposing alcohol to the air in the presence of bacteria. The ethanol is oxidized to ethanoic acid:

$$\begin{array}{ccccccc} & H & H & & & H & \quad O \\ & | & | & & & | & \quad // \\ H- & C- & C-O-H & + \; O_2 \longrightarrow & H- & C- & C \qquad + \; H_2O \\ & | & | & & & | & \quad \backslash \\ & H & H & & & H & \quad O-H \end{array}$$

Wine becomes wine vinegar. Beer (made by fermenting sugars from malt) becomes malt vinegar.

In the laboratory, ethanoic acid is made by oxidizing ethanol using a reagent such as an acidified solution of potassium dichromate.

Experiment 31b
The oxidation of ethanol to ethanoic acid

ethanol

potassium dichromate

concentrated sulphuric acid

Ethanol, sulphuric acid and a solution of potassium dichromate, which is orange, are mixed in a flask and heated. Ethanol and ethanoic acid evaporate readily, so the apparatus used is designed to stop them escaping (see figure 31.7). Any vapour which escapes from the flask turns back to liquid in the condenser and runs back into the flask. A condenser used in this way is called a *reflux condenser*. 'Reflux' means 'flow back'.

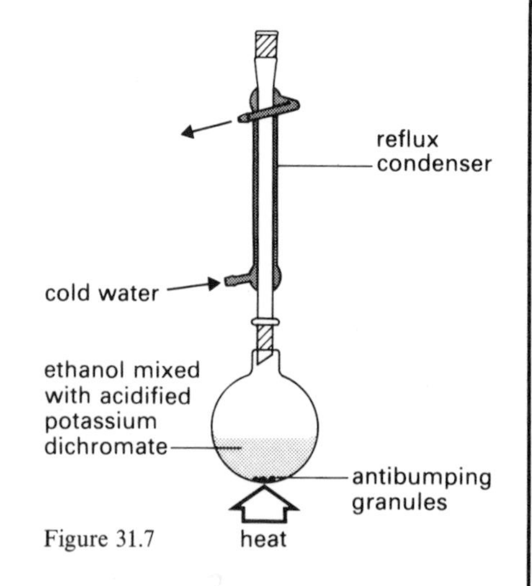

Figure 31.7

Results

On heating, the solution changes from orange to green. When the reaction is complete, the apparatus is rearranged for distillation, and a mixture of ethanoic acid and water is collected as the distillate. The distillate is colourless. It smells of vinegar and is acidic.

Discussion

How would you test the distillate to show that it contains an acid? Draw a diagram to show the apparatus reassembled for simple distillation.

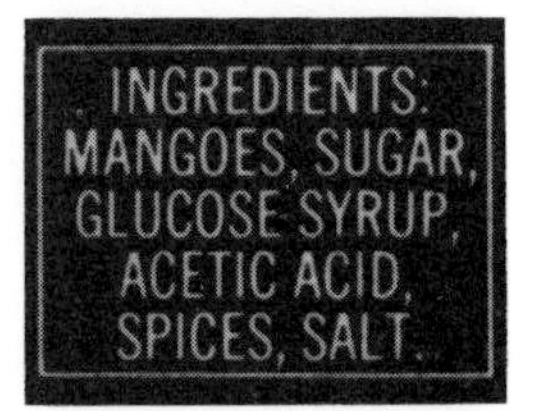

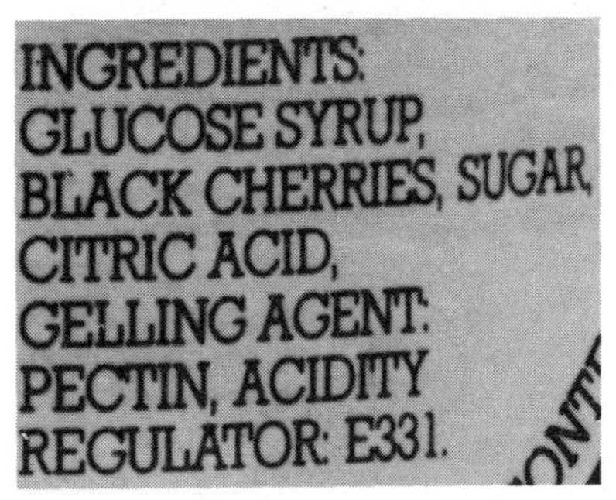

Figure 31.8 Can you find out the structures of the organic acids listed on these food labels?

You may see the names of some acids on food labels. These might include acetic acid, citric acid, fumaric acid and tartaric acid. These are all organic acids. The traditional names will probably continue to be used by everybody, except laboratory chemists, for many years. They are safe to use in foods because they are weak acids.

An acid is a compound which reacts with water to produce hydrogen ions. This is explained in chapter 24. In a molecule of ethanoic acid, there are four hydrogen atoms. Three are attached to carbon and one to oxygen. As with the alcohols, only the hydrogen atom attached to oxygen is reactive. In water:

$$CH_3COOH + H_2O \rightleftharpoons CH_3COO^- + H_3O^+$$

only this H atom is acidic (the H of the O–H group)

ethanoic acid ⇌ ethanoate ion

This is a reversible reaction and the equilibrium is over to the left-hand side of the equation. Ethanoic acid is a weak acid (see chapter 24).

Questions

The organic acids react like other acids with metals, bases and carbonates (see chapter 21).

1 What products would you expect when ethanoic acid reacts with (a) magnesium, (b) copper(II) oxide and (c) sodium carbonate?

2 (Difficult) Attempt to write equations for these reactions.

31.4 The esters

When you eat a banana, or suck a pear drop, or remove nail varnish with a solvent, you smell the strong and fruity smell of an **ester**. A ripe pineapple contains about 120 mg/kg of the ester, ethyl ethanoate, together with smaller amounts of other esters and 60 mg of ethanol. Together, these make up the flavour of the fruit.

Questions

Look at the structure of ethyl ethanoate.

1 Are the hydrogen atoms bonded to carbon or to oxygen?

2 Would you expect ethyl ethanoate to react with (a) sodium and (b) water?

3 Would you expect ethyl ethanoate to be an acid?

4 Write down the structures of ethyl propanoate and propyl ethanoate.

Esters are very common. Fats and vegetable oils, including butter, margarine and corn oil, are esters. Terylene is an ester and so are many of the laminated plastics and surface finishes on kitchen equipment.

Esters are made by combining an organic acid with an alcohol. The process is called *esterification*. The reaction is carried out by warming the acid and the alcohol with a little concentrated sulphuric acid.

$$CH_3CH_2O\text{–}H + HO\text{–}C(=O)CH_3 \longrightarrow CH_3CH_2\text{–}O\text{–}C(=O)\text{–}CH_3 + H_2O$$

ester link

ethanol | ethanoic acid | ethyl ethanoate

Many people find it difficult to remember the structures of esters. At first it is difficult to work out how to relate the names and structures of these compounds. Use a set of molecular models to make models of molecules of ethanoic acid and ethanol. Then carry out the **condensation reaction** with the models, to make ethyl ethanoate. The process is called a condensation reaction because a small molecule is split off as the molecules join together. Go on to make models of methyl propanoate and propyl methanoate so that you are clear about the way in which esters are named.

An important property of esters is that they can be split back into the acid and the alcohol. This is an example of **hydrolysis**. Water is used to split the molecule apart. Dilute acid or alkali is used to speed up the reaction, and the mixture is heated.

Ester + Water $\longrightarrow$ Acid + Alcohol

This type of reaction is important in soap making, as described in chapter 10.

Summary

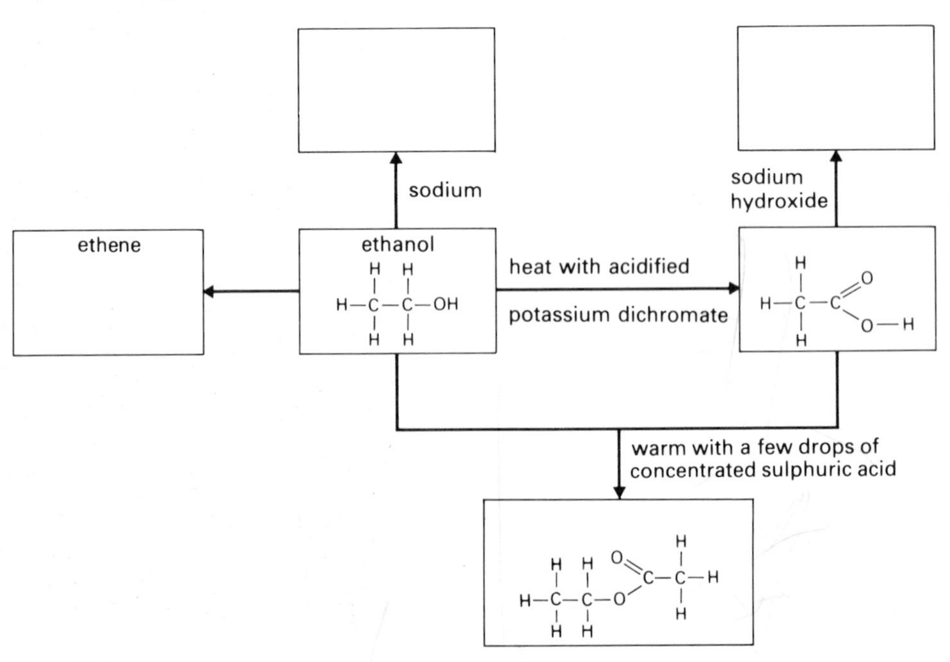

Figure 31.9

Copy and complete the diagram in figure 31.9 to summarize the reactions of alcohols, acids and esters.

32 Plastics

32.1 Polythene

Figure 32.1 How many different uses of plastics can you see in this picture?

There were many explosions when polythene was discovered. This was because ethene becomes dangerous and unpredictable under pressure. In 1933 a group of research chemists was investigating chemical reactions at high pressures. Working at pressures about 2000 times atmospheric pressure, they found that sometimes ethene turned into a white, waxy solid. The danger was that it was quite likely to decompose violently into carbon and hydrogen. It took a lot of research to find the right conditions to get the white solid safely.

The white solid was found to be a **hydrocarbon**. After analysis, it seemed that the high pressure and temperature, and the catalysts, had made the ethene molecules join together in a long chain. As more of the solid was made, it was found to have some very useful properties. It was an excellent electrical insulator and it could be made into thin, transparent films of considerable strength. It was soon being used in the development of radar and for the manufacture of submarine cables.

Now polythene is one of the best-known and cheapest plastics. It is used to make plastic bags, beakers, buckets, bowls and bottles for washing-up liquid.

Figure 32.2 Articles made from low density polythene

The type of polythene made from ethene under high pressure is called *low density polythene*. The ethene molecules join together in long chains. The chains may be made from as many as 50 000 ethene units. In the high pressure process

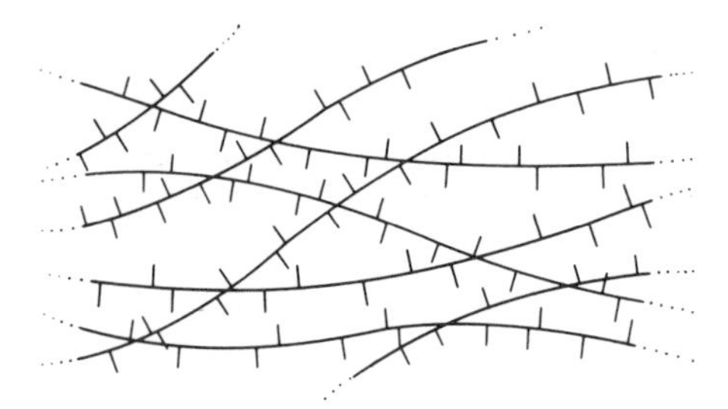

Figure 32.3 Hydrocarbon chains in low density polythene

Figure 32.4 Model equation for the formation of polythene

the chains are not straight, but have side branches (see figure 32.3). This means that the chains cannot pack closely together and this lowers the density of the solid. One of the disadvantages of low density polythene is that it begins to soften at the temperature of boiling water.

In 1953 a new method of polymerizing ethene was discovered. Professor Ziegler, in Germany, discovered that ethene could be polymerized at atmospheric pressure by using special catalysts. The new polythene was more rigid, had a higher density and did not soften at 100 °C. This makes it especially useful for hospital equipment which needs to be sterilized. Plastic bags can be made from very thin films of the plastic. They make a slight crackling noise when crushed. The molecules are unbranched, so they can pack together more closely (see figure 32.5). This is *high density polythene.*

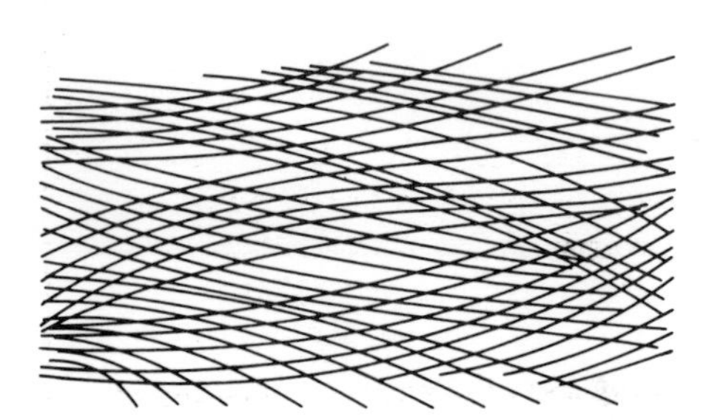

Figure 32.5 Hydrocarbon chains in high density polythene

Figure 32.6 Articles made from high density polythene

32.2 Addition polymerization

When ethene is made into polythene, the process is called **polymerization**. The single ethene molecule is the **monomer** (one-part). The polythene formed is the **polymer** (many-parts).

A catalyst starts the chains growing. Then the molecules of ethene add on one by one in rapid succession. There are no other products. The reaction is an example of **addition polymerization**.

The length of the carbon chains has an important effect on the properties of a polymer. This is illustrated by a comparison of paraffin wax with polythene. Wax consists of hydrocarbon molecules with 20–30 carbon atoms in the chain. Polythene molecules can be pictured as very, very long alkane molecules with about 20 000 atoms in the chain (see figure 19.25 on p. 137). The main difference between wax and polythene is the difference in chain length. Wax snaps when it is bent, whereas polythene is flexible. Pieces of wax are easily scraped off a candle with a knife. A knife will scratch a polythene bowl but the bowl is hard to cut. Wax melts and burns at a much lower temperature than polythene.

There is a family of compounds in which one or more of the hydrogen atoms in ethene is replaced by another atom or group of atoms. Each of these compounds is **unsaturated** and can be made into a polymer by addition polymerization.

Questions

1 What is the relative molecular mass of ethene, C_2H_4?
2 The average relative molecular mass of polythene is about 140 000. How many ethene molecules, on average, join together to make a molecule of polythene?
3 Draw short lengths of the polymer chains of (a) pvc, (b) ptfe and (c) polystyrene.

Figure 32.7 Some monomers for making plastics

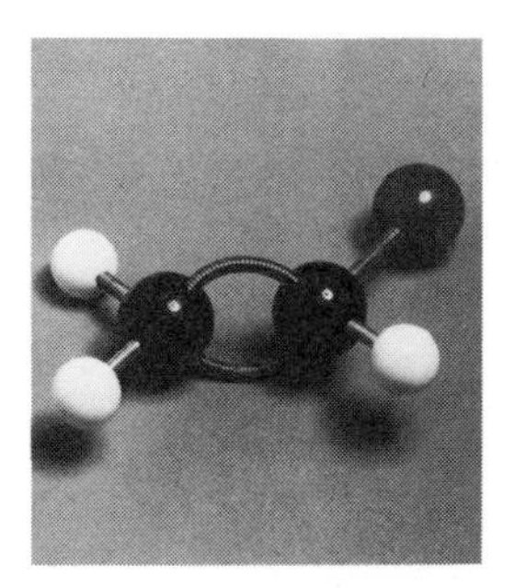

(a) Chloroethene (vinyl chloride), the monomer for pvc

(b) Tetrafluoroethene, the monomer for ptfe

(c) Propene, the monomer for polypropene

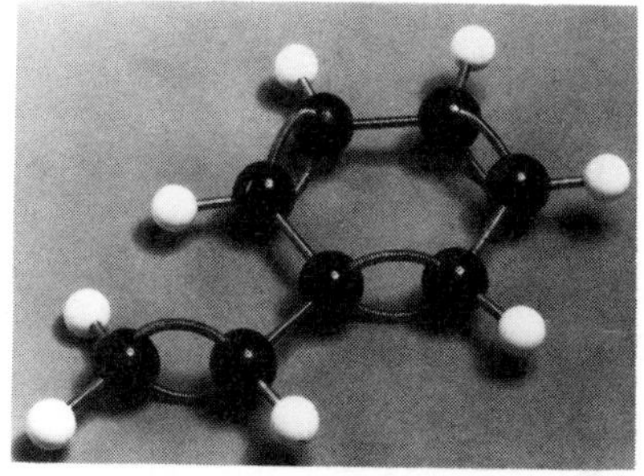

(d) Phenylethene (styrene), the monomer for polystyrene

Experiment 32a
Making an acrylic plastic

peroxide initiator

methyl 2-methylpropenoate

The monomer molecules for making acrylics are larger than ethene molecules. The monomer for this reaction is a liquid at room temperature. Polymerization will take place in a test-tube at about 60 °C in the presence of a peroxide catalyst. The monomer fumes are dangerous, so the experiment should be carried out in a fume cupboard.

One of the commercial names for an acrylic polymer is Perspex.

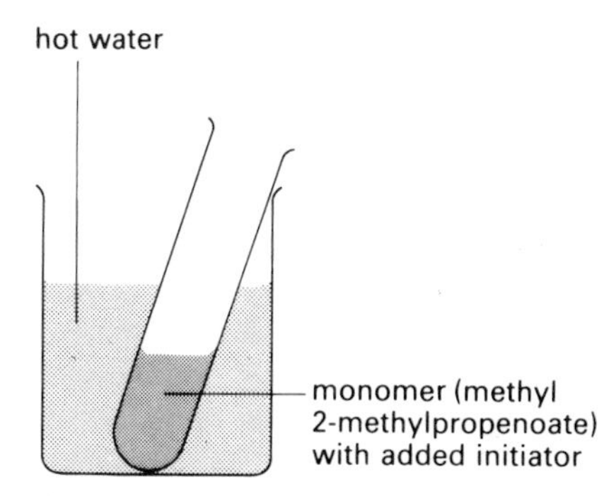

Figure 32.8

Results
The colourless liquid in the tube gradually becomes thicker and thicker. After about an hour, it can no longer be poured from the tube. Later still, the tube can be broken to obtain a clear lump of solid plastic.

Discussion
Why does the liquid in the tube get thicker and thicker? What is happening to the molecules?

32.3 Condensation polymerization

Nylon, Terylene and Bakelite are familiar examples of condensation polymers. The procedure for making nylon shown in experiment 32b is quite unlike the method used to make the polymer in industry. However, it is fun to do and it illustrates the principles of this type of polymerization.

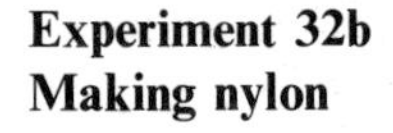

Experiment 32b
Making nylon

1,6-diaminohexane
hexane-1,6-dioyl chloride

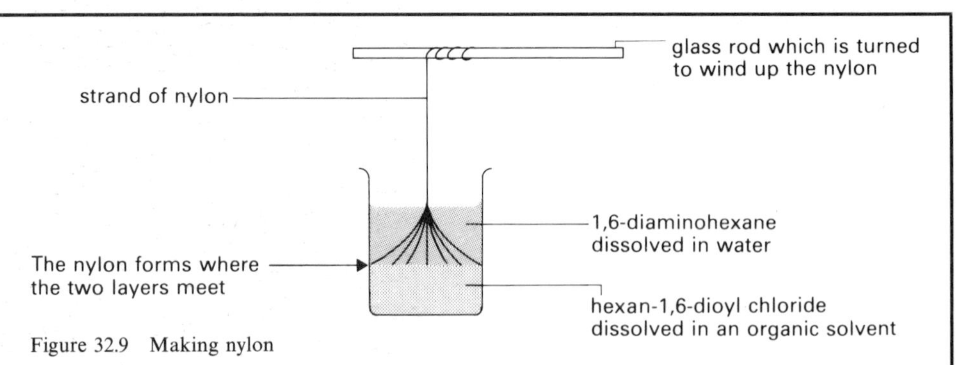

Figure 32.9 Making nylon

Making nylon molecules is rather like making a necklace with two different colours of beads popped together alternately. Two different chemicals are used. One is dissolved in a dense organic solvent, and this forms the lower liquid layer in the beaker in figure 32.9. The other chemical is dissolved in water, which forms the upper layer because the two liquids do not mix. The two chemicals join together and react when they meet. They can only meet at the boundary where the two solutions are touching.

Results
A continuous strand of white nylon can be pulled from the boundary between the two liquids. The fibres are weak and are wet with the two solvents.

Discussion
How is it possible for a continuous fibre to form at the boundary between the two liquids? Why, in the end, does the fibre stop forming? Why would this method of making nylon be unsuitable for large-scale manufacturing?

```
                   H  H H H H H H  H      O   H H H H  O
                    \ | | | | | | /        \\ | | | | //
1,6-diaminohexane   N-C-C-C-C-C-C-N    +    C-C-C-C-C-C      hexan-1,6-dioyl
                    / | | | | | | \        /  | | | |  \     chloride
                   H  H H H H H H  H     Cl   H H H H  Cl
                                   |
                                   v

This molecule has   H  H H H H H H H O H H H H  O          It can join with
reactive groups at   \ | | | | | | | || | | | | //          other molecules to
each end              N-C-C-C-C-C-C-N-C-C-C-C-C-C          make a long chain
                     / | | | | | |    | | | |  \
                    H  H H H H H H    H H H H  Cl

                                   +

                                  HCl
```

Figure 32.10 One step in the formation of a nylon molecule

The chemicals used to make nylon have reactive groups at both ends of their molecules. The reactive groups join by splitting off a small molecule of water or hydrogen chloride. In this way a long-chain polymer molecule is formed. The process is called *condensation polymerization.*

Question

Figure 32.11 illustrates some of the uses of nylon. Which properties of nylon make it suitable for making (a) the bristles of brushes, (b) the fibres of ropes, (c) pocket combs and (d) fabrics for clothing?

Figure 32.11

The links in the chain of a Terylene molecule are *ester links*. This type of polymer is called a *polyester*. The reaction is similar to that used to make ethyl ethanoate from ethanol and ethanoic acid (see chapter 31). However, the alcohol has two —OH groups and the acid has two acid groups, so the molecule can continue to grow in both directions.

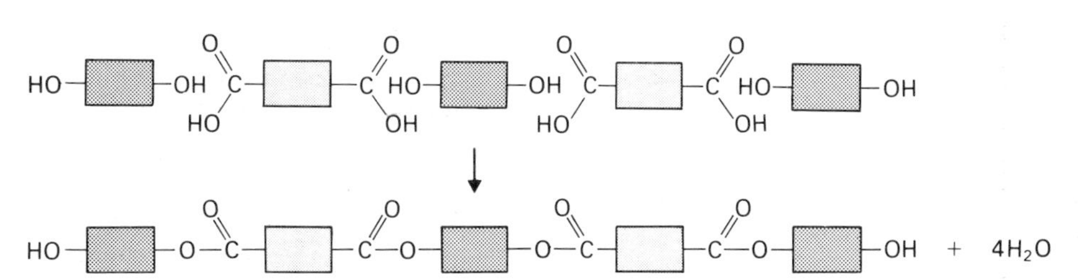

Figure 32.12 Formation of part of a polyester molecule

Questions

Compare plastics with other materials such as metals, glass and wood.

1 Which materials have good resistance to corrosion?

2 Which materials can be hit hard without shattering?

3 Which materials will support a heavy load?

4 Which materials are hard to scratch?

5 What is the range of densities for each class of materials? (Use table 5 on p. 342.)

6 Which materials will melt or burn?

7 Which materials are conductors and which are insulators?

32.4 Shaping plastics

All plastics can be moulded on heating. This is the property that gives them their name.

Some plastics soften on heating and harden again when cooled. Polythene, polystyrene, pvc and nylon do this. They are called **thermoplastics**. These plastics can be moulded by several methods, including extrusion, blow moulding, injection moulding and vacuum forming (see figures 32.13–14).

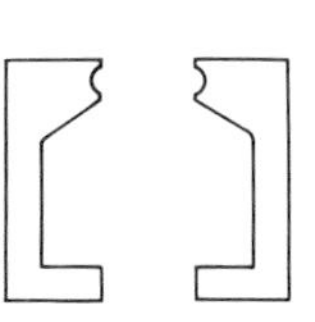

Blow moulding is used to make hollow objects, such as bottles

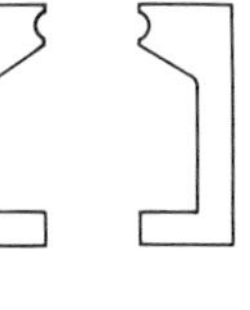

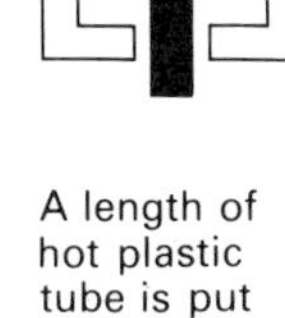

A length of hot plastic tube is put into the mould

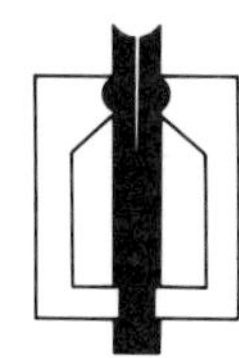

As the mould closes, it seals the bottom of the tube

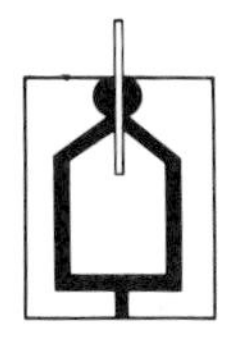

Air is blown into the tube until it takes the shape of the mould

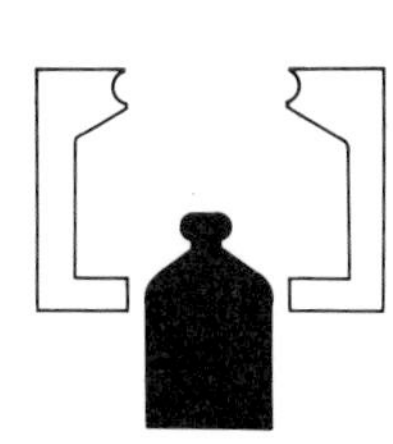

The mould opens and the finished bottle drops out

Figure 32.13 Blow moulding

Questions

1 Suppose that you have samples of polythene, Perspex (acrylic), nylon and Bakelite. Which plastic would you use to make (a) a saucepan handle, (b) a squeezy bottle, (c) a safety screen, (d) an ashtray and (e) a plastic bag?

2 Which method of shaping plastics would you use to make (a) guttering, (b) a polythene bag, (c) a door handle, (d) a bottle for shampoo, (e) a bath and (f) a comb?

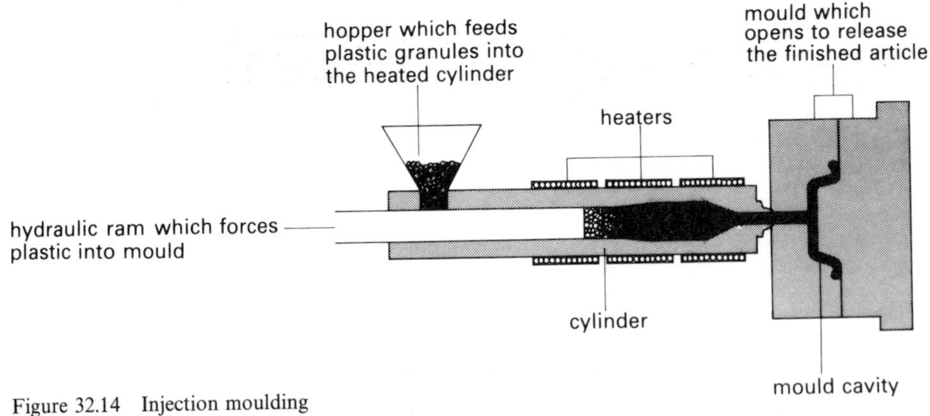

Figure 32.14 Injection moulding

Figure 32.15 The arrangement of the polymer chains after heating

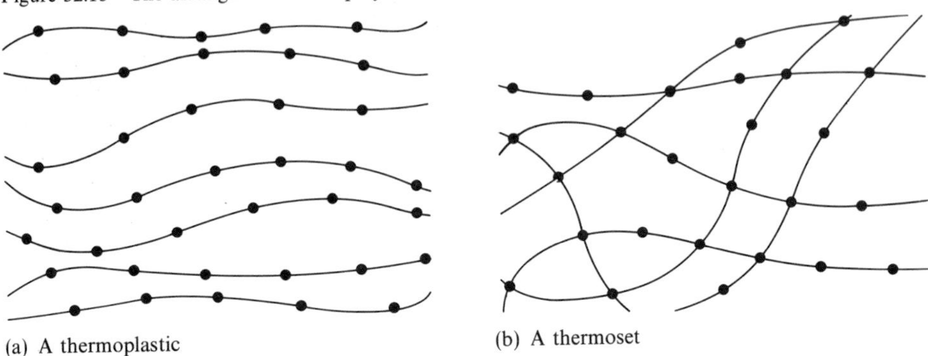

(a) A thermoplastic (b) A thermoset

Some plastics can be heated and moulded only once. These plastics set hard when hot and they will then not melt again. Strong chemical bonds cross-link the polymer chains into a rigid network during the moulding process. These plastics are called **thermosets**. Examples include Bakelite, melamine, urea–methanal and polyester resins.

The commonest method for shaping thermosets is compression moulding.

Summary

Write a paragraph summarizing the main ideas in this chapter. You should include the following terms: monomer, polymer, addition polymerization, condensation polymerization, thermoplastic, thermoset.

Review questions

1 Crude oil consists of a complex mixture of hydrocarbons. During refining it is separated into fractions with different physical properties and uses. An experiment was carried out to separate crude oil into fractions using the apparatus shown below. The different fractions are summarized in the table.

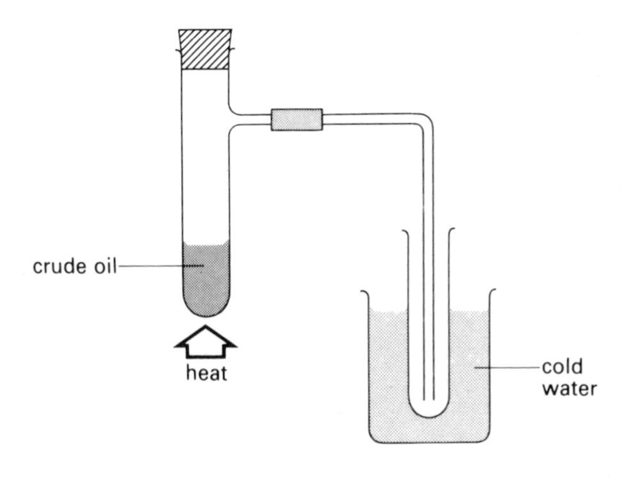

Fraction	Boiling point range (°C)
1	30–70
2	70–120
3	120–170
4	170–250
5	250–360

(a) Complete the diagram by drawing a thermometer in the correct place for taking the temperature measurements.
(b) What name is given to this process for refining crude oil?
(c) What was the purpose of the cold water?
(d) Give *two* differences in physical properties other than boiling point between fraction 1 and fraction 5.
(e) Fraction 2 contains the hydrocarbon octane. The structural formula for octane is

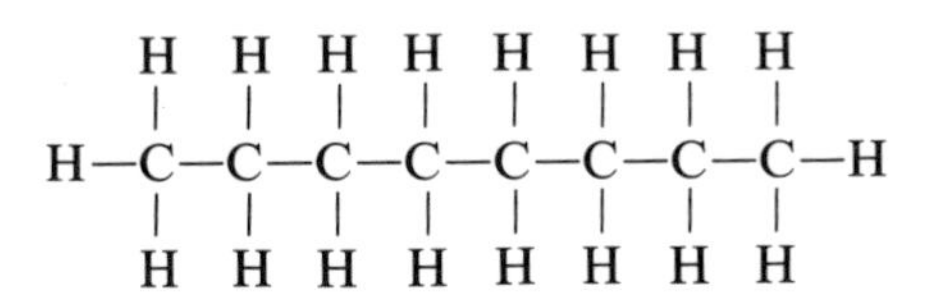

(i) What is the molecular formula for octane?
(ii) Give *one* use for the mixture of hydrocarbons in fraction 2.
(iii) Octane is a *saturated hydrocarbon.* What is the meaning of the words hydrocarbon and saturated?
(f) When the vapour from fraction 5 was passed over a heated broken pot, a colourless gas was produced. This gas decolorized both bromine water and acidified potassium manganate(VII) solution.
(i) What process was taking place when the vapour was passed over a heated pot?
(ii) What can be concluded about the gas from the reactions with bromine water and acidified potassium manganate(VII)?
(iii) Give *one* important industrial use for the gas produced. **(EAEB)**

2 Information about some hydrocarbons is given in the table below.

Compound	Formula	Boiling point (°C)
A	CH_4	−162
B	C_2H_4	−104
C	C_2H_6	−89
D	C_3H_6	−48
E	C_3H_8	−42
F	C_4H_8	−6
G	C_4H_{10}	0

(a) These hydrocarbons are members of one or other of two homologous series. Name these homologous series and give the letters of the compounds in each series.
(b) Predict the formula and boiling point of the next member of *one* of these series.
(c) Some of the hydrocarbons in the table can exist in isomeric forms. Choose *one* compound in the table for which this is so and draw the structural formulae of the *two* isomers of this formula.
(d) Some of the hydrocarbons in the table react with halogens by *addition.* Choose *one* hydrocarbon which reacts in this way. Show by means of structural diagrams or an equation what happens when the hydrocarbon reacts with the halogen bromine.
(e) Some of the hydrocarbons in the table can react with halogens by *substitution* but cannot react by addition. Choose *one* hydrocarbon which reacts in this way. Show by means of structural diagrams or an equation what happens when the hydrocarbon reacts with the halogen bromine.
(L)

3 This question is about the preparation and properties of ethyl ethanoate, $CH_3CO_2C_2H_5$.

Preparation of ethyl ethanoate

The apparatus used is shown in the diagram below. A mixture containing *equal* numbers of moles of ethanol, C_2H_5OH, and ethanoic acid, CH_3CO_2H, is placed in the flask. Five drops of concentrated sulphuric acid are also added. The contents of the flask are then boiled until no further reaction occurs.

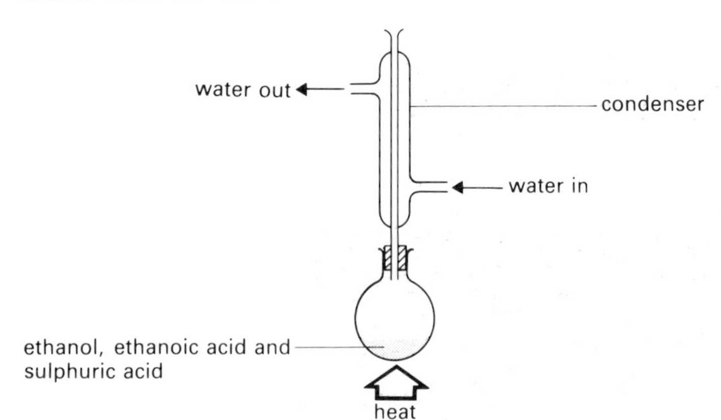

The reaction taking place is

$$C_2H_5OH(l) + CH_3CO_2H(l) \rightleftharpoons CH_3CO_2C_2H_5(l) + H_2O(l)$$

When the reaction has finished, the mixture is cooled. Addition of water to this mixture causes two liquid layers to separate. The layers are separated from each other and the layer containing ethyl ethanoate is treated to remove impurities. The table below gives some data for the substances referred to in this question.

Substance	Density (g/cm^3)	Relative molecular mass	Boiling point (°C)
Ethanol	0.79	46	78.5
Ethanoic acid	1.05	60	118.0
Ethyl ethanoate	0.90	88	77.0
Water	1.00	18	100.0

(a) What is the purpose of the condenser shown in the apparatus?

(b) Which of the layers (upper or lower) is more likely to contain the ethyl ethanoate? Explain your choice.

(c) (i) Explain why the impure ethyl ethanoate still contains some of the original reactants.
(ii) Name a chemical which could be used to neutralize the ethanoic acid impurity.
(iii) The amount of sulphuric acid remaining at the end of the reaction is the same as that added at the beginning. Suggest why the acid was added.

(d) If the mixture to start with contained 0.1 mol of ethanol, what *mass* of ethanoic acid would have been used?

Properties of ethyl ethanoate

Ethyl ethanoate is hydrolysed by water to form ethanol and ethanoic acid. A research project was carried out to discover how this reaction took place. The water used contained molecules in which 'heavy' oxygen ($^{18}_{8}O$) had replaced 'normal' oxygen atoms ($^{16}_{8}O$). Analysis of the two products showed that only one of them contained the atoms of 'heavy' oxygen from the water. The result is shown in the diagram below.

water containing $^{18}_{8}O$ atoms and showing where the bond breaks (-----)

ethyl ethanoate containing 'normal' oxygen atoms

(a)

(b)

ethanoic acid

ethanol

(e) (i) By studying the diagram above, state which of the two bonds (labelled (a) and (b)) is broken during the reaction of ethyl ethanoate molecules with water molecules.
(ii) Explain how the information shown in the diagram above has enabled you to make your choice.
(iii) In what way does the nucleus of an atom of oxygen-16 differ from that of an atom of oxygen-18? **(L)**

4 (a) State how the processes of 'fractional distillation of crude oil' and 'catalytic cracking' are used to provide a range of hydrocarbon fuels.
(b) For *each* of the following types of reaction in organic chemistry, give an example, state the reaction conditions and write an equation: (i) addition, (ii) substitution (replacement), (iii) polymerization.
(c) How do you explain the fact that the organic compound tetrachloromethane (CCl_4) is a volatile liquid, whereas sodium chloride is a high melting point solid? **(WJEC)**

5 (a) Name, and give one corresponding use of (i) a thermoplastic and (ii) a thermosetting plastic.
(b) Describe, giving all essential experimental details, how you would distinguish between two similar looking pieces of plastic, one of which is thermoplastic, and the other a thermosetting plastic. **(EMREB)**

6 (a) Explain what is meant by *addition polymerization* and describe, with practical details, how you would make a small sample of an addition polymer.
(b) Explain what is meant by *condensation polymerization* and describe, with practical details, how you would make a small sample of a condensation polymer. **(L)**

THEME J
Chemistry and food

Cheese and onion? Sodium chloride and ethanoic acid? Here crisps are being flavoured and checked

33 Carbohydrates

33.1 Thinking of food

* What is food? Food is something which contains nutrients.
* What are nutrients? The six main groups of nutrients are carbohydrates, proteins, fats, vitamins, minerals and water.
* What are nutrients for? Nutrients are used for the growth and repair of the body. They are used to produce energy for movement, warmth and growth. They are needed because they help to control the chemical processes in the body. This is illustrated in figure 33.1.

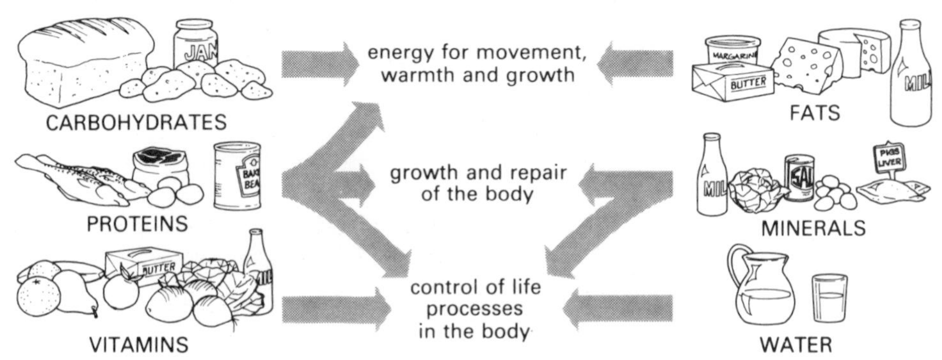

Figure 33.1 The functions of nutrients

33.2 Introducing carbohydrates

Starch, glucose, sucrose and cellulose all belong to the family of **carbohydrates** (see figure 33.2). These compounds consist of the elements carbon, hydrogen and

Figure 33.2 Some familiar carbohydrates.

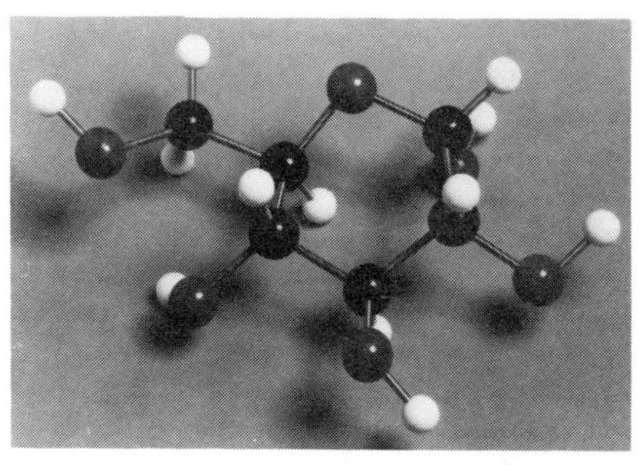

Figure 33.3 A model of a glucose molecule

oxygen. They are called carbo*hydrates* because in their formulae, the hydrogen and oxygen are always present in the same ratio as in water. The formula of glucose is $C_6H_{12}O_6$, which can be rewritten as $C_6(H_2O)_6$. Neither way of writing the formula gives any idea of how the atoms are arranged in a glucose molecule. The structure of glucose is shown in figure 33.3.

Sugars are carbohydrates which are soluble in water and taste sweet. The chemical name for ordinary sugar is *sucrose. Lactose* is the sugar found in milk, and *fructose* occurs, with other sugars, in fruit juices and honey.

Question

Consider the formulae of the following compounds and decide which of them might be carbohydrates: sucrose, $C_{12}H_{22}O_{11}$; ethanol, C_2H_6O; maltose, $C_{12}H_{22}O_{11}$; fructose, $C_6H_{12}O_6$; propanone, C_3H_6O.

Experiment 33a
Test-tube tests for starch and sugars

Benedict's solution
Iodine solution

The two test solutions are iodine solution and Benedict's solution. (Fehling's solution is similar to Benedict's solution and gives the same result. It is more strongly alkaline.)

Iodine solution is made by dissolving iodine in potassium iodide solution. When dilute, the reagent has a yellow–brown colour. A few drops of the reagent are added to a little of the carbohydrate solution to be tested.

Benedict's solution is a deep blue colour. It is added to the solution of the carbohydrate to be tested and the mixture is warmed.

Results
The table in figure 33.4 shows the results of carrying out these tests on four common carbohydrates.

Carbohydrate	Iodine test	Benedict's test
Glucose	No colour change	Turns green, and then an orange–red precipitate forms
Sucrose	No colour change	No colour change
Maltose	No colour change	Turns green, and then an orange–red precipitate forms
Starch	Deep blue–black colour forms	No colour change

Figure 33.4

Discussion
The results show that the iodine solution reacts with only one of the carbohydrates. Which one? This test is very sensitive.

Sugars which react with Benedict's solution are called *reducing sugars*. The copper(II) compound in the reagent is reduced to copper(I) oxide, which is insoluble and appears as an orange–red precipitate. Which of the sugars in the table are reducing sugars?

33.3 Carbohydrates in the carbon cycle

Leaves can be thought of as chemical factories. The green chlorophyll in the leaves absorbs light energy from the sun and uses it to make (synthesize) sugars from

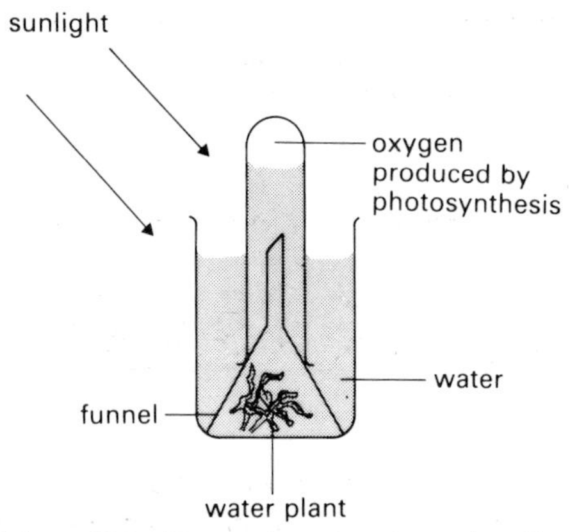

Figure 33.5 How to show that plants give off oxygen in sunlight

carbon dioxide and water. The carbon dioxide comes from the air and the water from the soil through the roots and stem of the plant (see figure 1.13 on p. 7).

$$6CO_2(g) + 6H_2O(l) \xrightarrow{\text{sunlight}} C_6H_{12}O_6(aq) + 6O_2(g)$$

This process is called *photosynthesis.* As well as sugar, it produces oxygen which is released into the air (see figure 33.5). The plant uses its sugar in various ways. It puts some into store by converting it into starch. Some is converted into cellulose which the plant needs in order to grow. Cellulose makes up the cell walls and the woody structure in the stems. Also, the sugar can be regarded as a reserve of chemical energy, either for the plant or for an animal which eats the plant. The energy is released by *respiration.* This is the reverse of photosynthesis and is an **exothermic** process. Energy is released.

$$C_6H_{12}O_6(aq) + 6O_2(g) \longrightarrow 6H_2O(l) + 6CO_2(g)$$

Plants which are grown as sources of carbohydrates include sugar beet, potatoes and corn. We eat the foodstuffs made from these crops to get the carbohydrates which our bodies cannot make for themselves.

The processes of photosynthesis and respiration are vital in the carbon cycle (see figure 33.6).

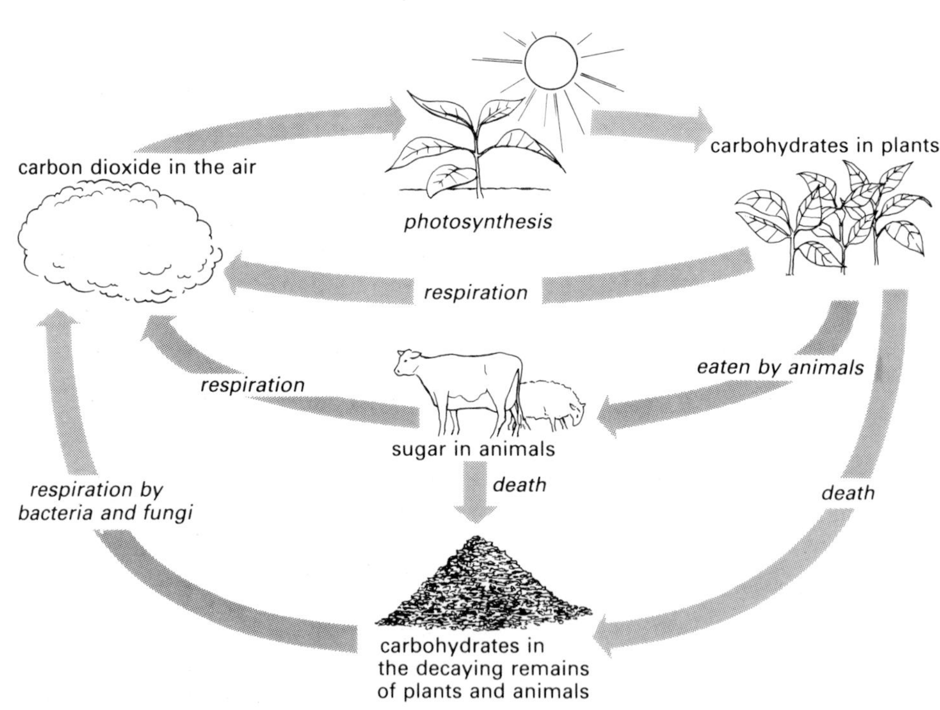

Figure 33.6 The carbon cycle

33.4 The breakdown of starch

In plant foods, much of the carbohydrate is present as starch. Starch molecules are much too large to pass through the walls of the intestines into the blood. So to get the benefit of these foods, our bodies have to break down the starch into smaller molecules. This process of digestion takes place in the presence of **enzymes**.

There are enzymes in saliva which speed up the reaction of starch with water. The starch is split up into smaller molecules of sugar. The reaction is an example of **hydrolysis** (from the Greek words *hydro-*, meaning water, and *-lysis*, meaning splitting). You can test the idea by chewing a piece of dry bread for some time. It begins to taste sweet.

It is also possible to break down starch by boiling it with dilute hydrochloric acid.

Questions

1 Draw a picture to show how a leaf acts as a 'factory'. Label your diagram to show what goes into the factory, the machinery in the factory and the outputs.

2 Why do the processes of photosynthesis and respiration help to keep the oxygen content of the atmosphere approximately constant? What other processes may affect the proportion of oxygen in the air? How might the widespread destruction of the tropical rain forests upset the balance of nature?

Experiment 33b
Using enzymes to break down starch

Benedict's solution
Iodine solution

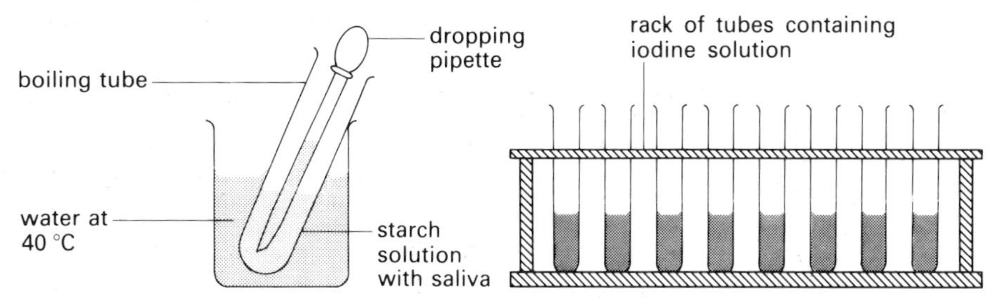

Figure 33.7

Figure 33.7 shows an experiment to investigate the effect of saliva on starch solution. Digestion takes place in the boiling tube. Why do you think that the tube is kept at 40 °C? Every minute or so, a few drops of the mixture in the boiling tube are added to one of the tubes containing iodine solution. Any colour changes are noted.

Results
The first few samples from the boiling tube give a blue–black colour with iodine. Later samples give a reddish colour. Finally, after about 10 minutes, the samples have no effect on the iodine solution.

At the start of the experiment, the starch solution does not react with Benedict's solution, but after about 10 minutes it does react, giving an orange–red precipitate.

Discussion
Which substance gives a blue–black colour with iodine? Why do the final samples from the boiling tube have no effect on the colour of iodine solution?

What do the results of the tests with Benedict's solution tell you about the products of the hydrolysis of starch?

Experiment 33c
Using acid to break down starch

dilute hydrochloric acid

Starch solution has to be boiled with dilute hydrochloric acid for about 30 minutes to make sure that it is fully broken down. Draw and label a diagram of suitable apparatus for the experiment. It should prevent the solution from evaporating to dryness. What test would you use to show that all the starch has been hydrolysed?

Results
In time, it is found that all the starch is hydrolysed and that a solution containing reducing sugars is formed. Which test is used to detect reducing sugars?

Discussion
Which is the better catalyst for breaking down starch – saliva or acid? Which catalyst is more likely to be used to get sugar from starch on an industrial scale?

33.5 What are the products when starch is hydrolysed?

Iodine solution and Benedict's solution can tell the difference between starch and reducing sugars. They cannot tell the difference between glucose and maltose. Both of these are reducing sugars. A technique which can be used to identify similar sugars is **chromatography**.

Experiment 33d
Using paper chromatography to identify the hydrolysis products of starch

The diagrams in figure 33.8 show the main stages in this experiment. Spots G and M on the paper are samples of pure glucose and pure maltose. Spot E is a sample of starch which has been hydrolysed using the enzyme in saliva. Spot A is a sample formed by hydrolysing starch with acid.

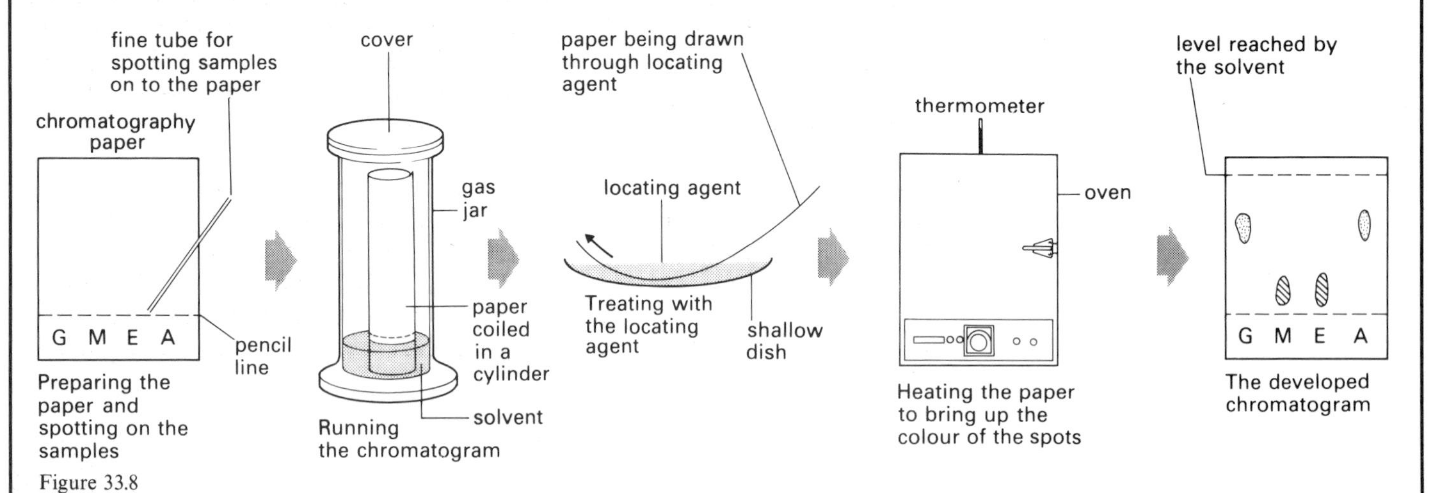

Figure 33.8

Results
Figure 33.8 also shows the developed chromatogram.

Discussion
What is the purpose of the locating agent? Why is no locating agent needed when inks or grass pigments are analysed by chromatography?

A pure sugar shows up as a single spot on the developed chromatogram. A mixture of sugars splits into two or more spots. Any one sugar, for example glucose, will move the same distance up the paper, whether it is pure or part of a mixture.

Which of the solutions spotted on the paper was a mixture of sugars? Which sugar moves further up the paper – glucose or maltose? What is the main product when starch is hydrolysed by saliva? What is the main product when starch is hydrolysed by acid? Figure 33.9 summarizes the changes, showing the structures of the molecules.

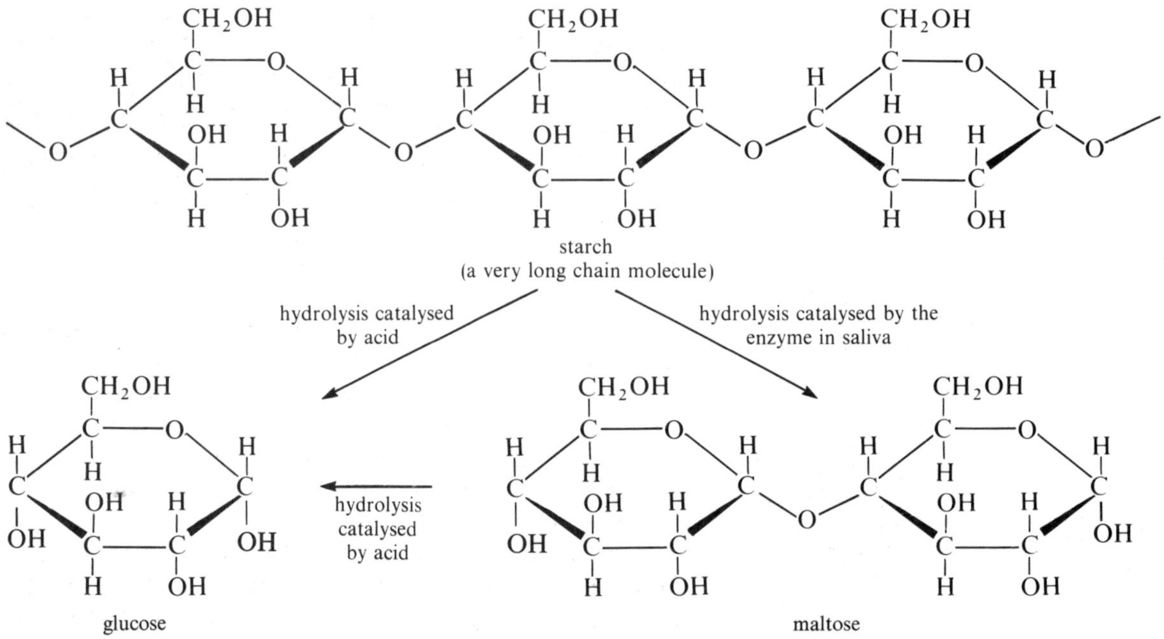

starch
(a very long chain molecule)

hydrolysis catalysed by acid

hydrolysis catalysed by the enzyme in saliva

hydrolysis catalysed by acid

glucose

maltose

Figure 33.9

33.6 Carbohydrates as polymers

Figure 33.9 shows that a starch molecule consists of a long chain of glucose molecules. Glucose (a **monomer**) can be joined up in a long chain to form starch (a **polymer**).

Cellulose is also a polymer of glucose, but the monomer molecules are joined in a different way. It is more difficult to digest cellulose than starch. Cows and other animals that feed on plants rely on bacteria in their intestines to help break down cellulose to glucose. Cows chew the cud to give the bacteria time to digest the cellulose in grass. We gain little food value from eating grass and leaves because we cannot digest cellulose.

Cellulose fibres from wood are used to make paper. The wood is pulped by being heated under pressure with a solution of sodium sulphite. The pulp is filtered, washed and bleached, and then beaten to break down the fibres. A continuous sheet of paper is made by pouring a thin slurry of wood pulp (mixed with various additives) onto a rapidly moving wire gauze. The paper forms as the water drains away. The damp sheet is then pressed between rollers and dried in ovens (see figure 33.10).

Figure 33.10 Paper making: pressing the damp paper to remove water and make it firm. The paper is travelling from left to right, supported by a sheet of wire gauze

Figure 33.11 The dangers of handling cellulose nitrate film: a frame from a 1948 Admiralty training film, called *This film is dangerous*

Chapter 32 describes how synthetic polymers are used as plastics and fibres. However, the first plastics were not made from chemicals produced from oil. They were prepared by modifying natural polymers, such as cellulose. One of the first, based on cellulose nitrate, was made by Alexander Parkes in 1865. Celluloid was later developed from Parkes's product and was used to make the early motion picture films until it was replaced by cellulose acetate, which is much safer. Cellulose nitrate is highly inflammable and can be explosive – it is used as guncotton. The chemical similarity between cellulose and guncotton is well known to those responsible for keeping the archives of World War II films. These were mostly made using cellulose nitrate. As nitrate film ages, it deteriorates and becomes increasingly hazardous (see figure 33.11).

Figure 33.12 Taking a sample from a fermentation vat in a brewery

33.7 Can sugars be broken down further?

Fermentation has been used for thousands of years to convert sugars to ethanol (alcohol) and carbon dioxide. The process is described in chapter 1. Brewers and winemakers use fermentation to make alcoholic drinks. Bakers use fermentation because the carbon dioxide formed makes the bread rise.

Question

The fermentation of glucose by yeast is described by this equation:

$$C_6H_{12}O_6(aq) \longrightarrow 2C_2H_5OH(aq) + 2CO_2(g)$$

Calculate (a) the mass of ethanol and (b) the volume of carbon dioxide which can be obtained from 36 g glucose at room temperature and pressure.

Summary

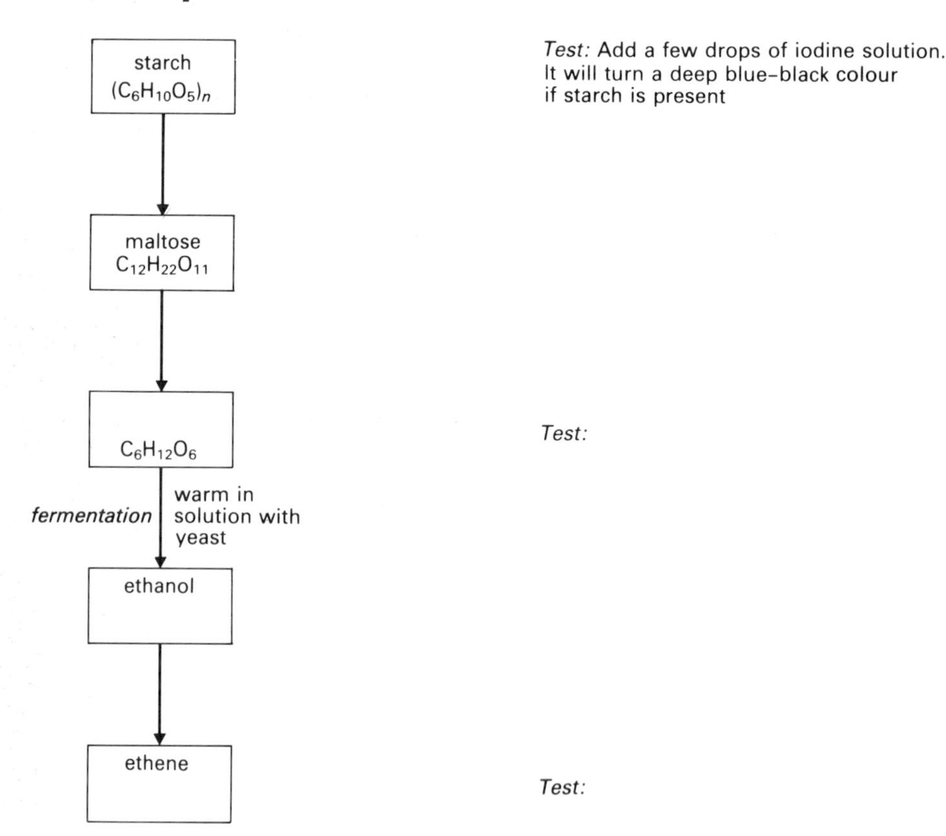

Figure 33.13

Copy and complete figure 33.13, which is a flow diagram showing the stages in the breakdown of starch to ethene. The final step, from ethanol to ethene, is described on p. 230. Each box contains the name and formula of a compound. Beside each arrow write the method used to make one compound into the next, naming any chemicals used and stating the conditions. Also give the type of change taking place.

Where the word 'test' is printed beside a box, give a test which can be used to detect the presence of the compound named in the box. State how the test is carried out, and the result of the test if the compound is present.

34 Proteins

34.1 What are proteins?

About 15 per cent of the human body is **protein**. There are many types of protein molecule, each doing a special job in the body. Skin, muscles and hair are made of fibrous proteins. Similar proteins hold the body together by tying muscle to bone, and bone to bone. The **enzymes**, which catalyse all the body's chemistry, are also proteins. Insulin, a hormone which controls the use of sugar in the body, is a protein. Haemoglobin is a protein: it colours the blood red and carries oxygen round the body.

If the body is to work properly, it must make the proteins it needs. The raw materials for making proteins come from protein foods.

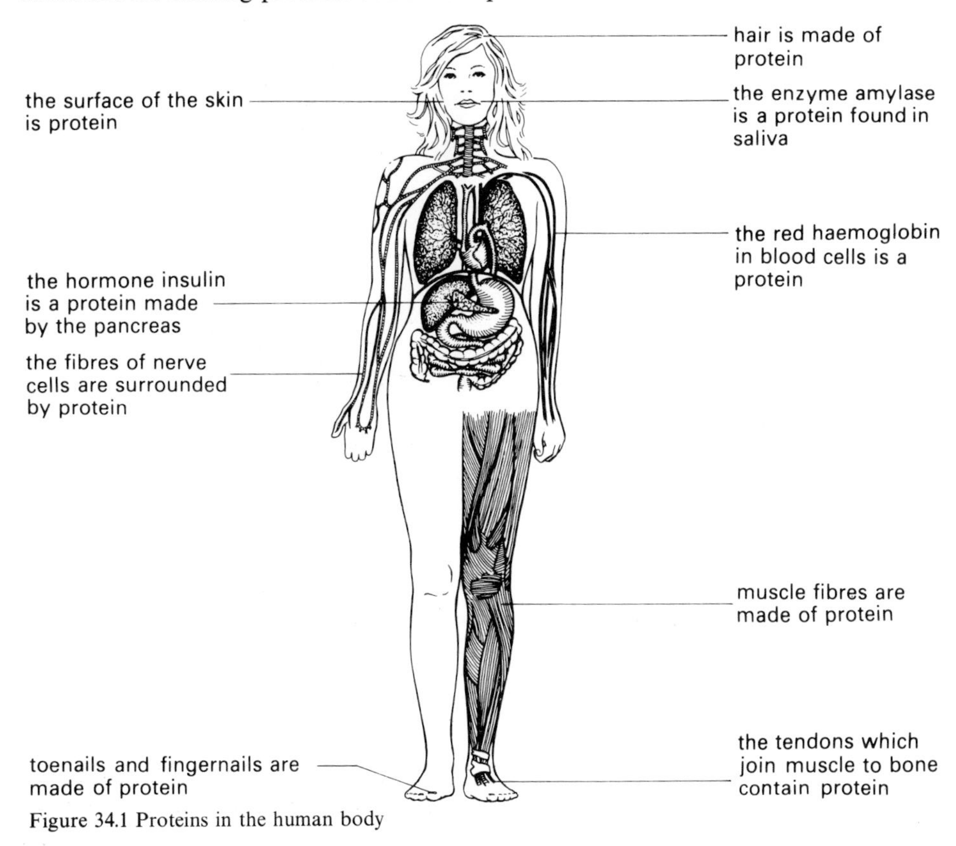

hair is made of protein
the surface of the skin is protein
the enzyme amylase is a protein found in saliva
the red haemoglobin in blood cells is a protein
the hormone insulin is a protein made by the pancreas
the fibres of nerve cells are surrounded by protein
muscle fibres are made of protein
the tendons which join muscle to bone contain protein
toenails and fingernails are made of protein

Figure 34.1 Proteins in the human body

Experiment 34a
Which foods contain proteins?

h

copper(II) sulphate solution

dilute sodium hydroxide

There are several tests which can be used to detect proteins in food. One of them is the biuret test. The food to be tested is crushed and shaken with water. A few drops of copper(II) sulphate are added, and then the mixture is made alkaline with sodium hydroxide solution.

Results

The table in figure 34.2 (overleaf) shows the result of doing the biuret test on a variety of foods.

Food	Biuret test
Milk powder	Solution becomes violet
Lean bacon	Particles become a dark colour
Sugar	No colour change
Egg albumen	Solution becomes violet
Salt	No colour change

Figure 34.2

Discussion
A violet colour shows that the food contains protein. Which of the foods tested contain protein?

Figure 34.3 Some foods that are rich in proteins

34.2 What are proteins made of?

We can learn something about the composition of foods simply by heating them. More information comes from heating protein foods with alkali.

Experiment 34b
Heating protein foods with alkali

soda lime

A small amount of the food is mixed with soda lime and heated in the apparatus shown in figure 34.4. The soda lime is an alkali. It is a mixture of sodium hydroxide and calcium hydroxide. The gases given off are tested with damp indicator paper.

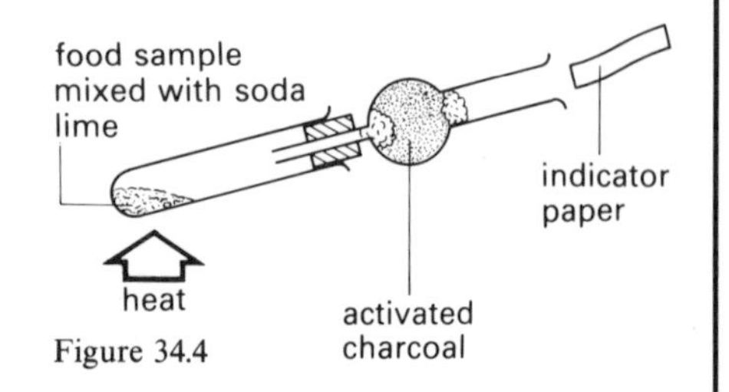

Figure 34.4

Results
Some results for this experiment are shown in the table in figure 34.5 opposite.

Food	Smell of gas given off	Effect of gas on universal indicator paper
Milk powder	Pungent	Turns blue
Lean bacon	Pungent	Turns blue
Sugar	Caramel	No effect
Egg albumen	Pungent	Turns blue
Salt	No gas detected	No gas detected

Figure 34.5

Discussion
Why is activated charcoal included in the apparatus? Are the gases given off in the experiment acidic or alkaline? The only common alkaline gas is ammonia. Which of the foods tested give off ammonia? Which of the foods tested contain protein (see figure 34.2)?

Ammonia is given off when proteins are heated with alkali. In the laboratory it is easier to produce ammonia by heating an ammonium salt with alkali.

Experiment 34c
What elements are present in ammonia?

copper(II) oxide

calcium oxide

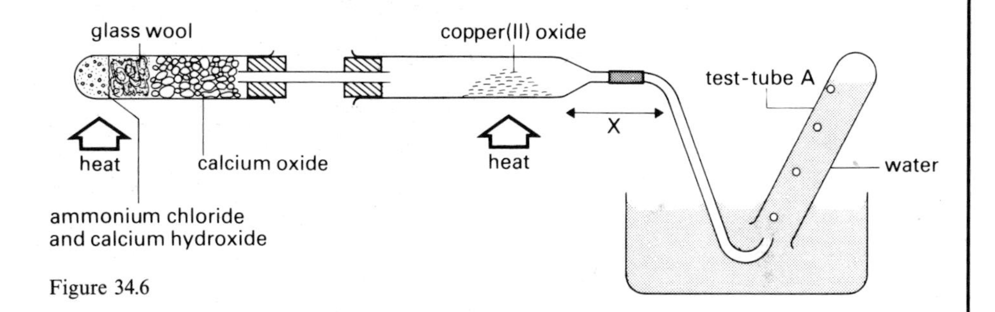

Figure 34.6

A stream of ammonia gas is passed over hot copper(II) oxide in the apparatus shown in figure 34.6. The ammonia is made by heating ammonium chloride with calcium hydroxide. The gas is dried with calcium oxide before it passes over the hot copper(II) oxide.

Results
The copper(II) oxide turns pink and droplets of clear liquid collect in the part of the apparatus labelled X. The clear liquid turns cobalt chloride paper from blue to pink.

A colourless gas collects in test-tube A. It extinguishes a burning splint and has no effect on indicator paper or limewater.

Discussion
The clear liquid which condenses at X is water. Why must the water have been formed by the reaction of ammonia with copper(II) oxide? Where did the hydrogen in the water come from? Where did the oxygen come from?

The gas collected does nothing with any of the usual gas tests. There is no simple way of proving it, but the gas is nitrogen.

The experiment shows that ammonia is a compound of nitrogen with hydrogen (and maybe oxygen). The results of experiment 34d show that it is a compound of nitrogen and hydrogen only.

Proteins give off ammonia when they are heated with alkali. This shows that they must contain nitrogen. Proteins are natural polymers. Protein molecules are very large. Insulin is one of the simplest proteins, and its formula is $C_{254}H_{377}N_{65}O_{75}S_6$! What is the mass of one mole of insulin molecules? A model of the protein lysozyme is shown in figure 18.5 on p. 120.

Proteins are built up from small molecules called **amino acids**. Amino acids are the monomers from which proteins are made. About twenty different amino acids are used to make natural proteins. Two of the simplest ones are shown in figure 34.7.

Figure 34.7 Two simple amino acids

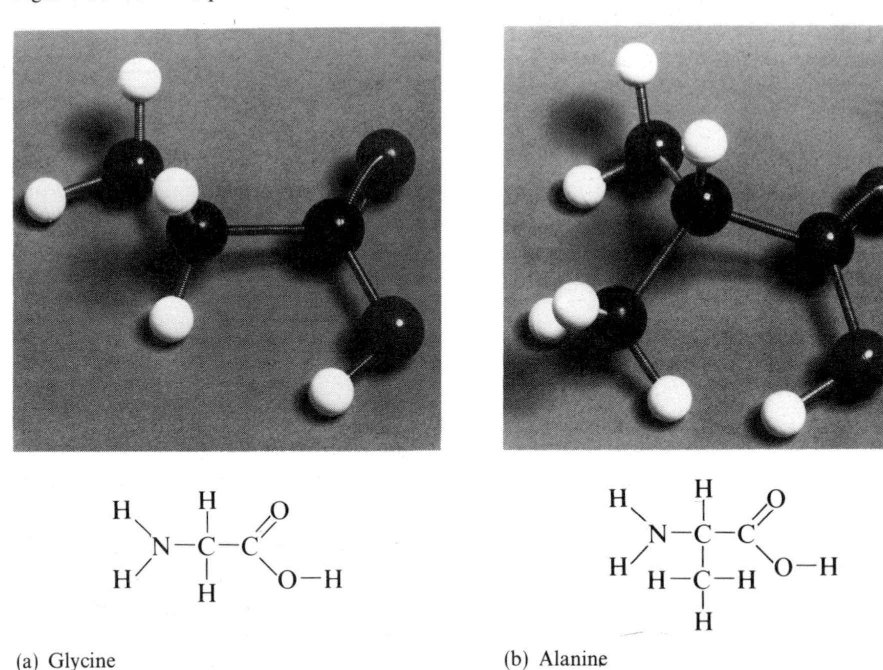

(a) Glycine

(b) Alanine

Every amino acid has a reactive amine group, $-NH_2$, and a reactive acid group, $-CO_2H$. When proteins are formed, an amine group from one amino acid reacts with the acid group from another. This process can continue repeatedly to produce a long chain. The bonding between the amino acids in a protein is similar to the bonding in nylon.

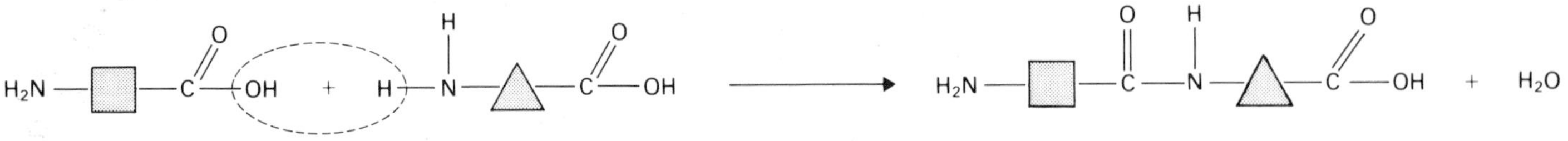

$$H_2N-\square-C(=O)-OH + H-N(H)-\triangle-C(=O)-OH \longrightarrow H_2N-\square-C(=O)-N(H)-\triangle-C(=O)-OH + H_2O$$

Figure 34.8 Formation of part of a protein chain

In polymers such as polythene and starch, one monomer repeats over and over again in the polymer. In proteins, any of the twenty amino acids can be combined together in any order. This means there is a vast number of possible combinations. There are thousands of different proteins.

Questions

1 Write an equation to show two glycine molecules joining with one alanine molecule, to give a new molecule in which the order of amino acids is glycine–alanine–glycine.

2 How is the structure of protein (a) similar to and (b) different from that of nylon (see p. 238)?

Figure 34.9 Nodules containing nitrogen-fixing bacteria on the roots of a pea plant

Proteins are broken down by **hydrolysis** into their amino acids. This process occurs when protein foods are digested. Some of the amino acids are reused in the body for building new proteins. Fairly concentrated acid, or **enzymes** called *proteases*, are needed as catalysts (see chapter 26). Most reactions which take place in plants and animals need catalysts. Biological catalysts (enzymes) are all proteins. Every reaction occurring in the body has its own 'tailor-made' catalyst.

34.3 Proteins in the nitrogen cycle

Like other animals, we depend on plants for our food. We eat plant foods and we eat animals which feed on plants. Plants can survive without animals, but animals cannot survive without plants. Plants produce all the amino acids and proteins they need. Animals are unable to do this and rely on plants for several amino acids.

Plants need nitrogen to produce the proteins they need for growth. However, plants cannot use nitrogen directly from the air. They need soluble nitrogen compounds, which can be taken in from the soil through their roots. This is 'fixed' nitrogen. Nitrogen compounds affect the yield of a crop, especially leafy crops such as cereals and cabbage.

Nitrogen compounds removed from the soil must be returned to it if more plants are to grow in the same soil. They are returned when plants die and rot into the ground. Some nitrogen is also returned to the soil through animal death and excretion. However, not all the nitrogen is returned: sewage is lost to the sea, rain causes drainage of nitrogen compounds from the soil, and some bacteria reconvert useful nitrogen compounds into nitrogen itself, which is lost to the air.

The losses are made up in two ways. During thunderstorms, some nitrogen and oxygen in the air combine to give nitrogen oxides. These oxides dissolve in rain-water and enter the soil as very dilute nitric acid. Secondly, some plants, such as clover, peas and beans, have roots which contain special bacteria (see figure 34.9). These bacteria convert nitrogen from the air directly into usable nitrogen compounds. Crops like clover, if ploughed into the soil, can be used to restore the nitrogen level.

These processes can be represented in a natural *nitrogen cycle* (see figure 34.10).

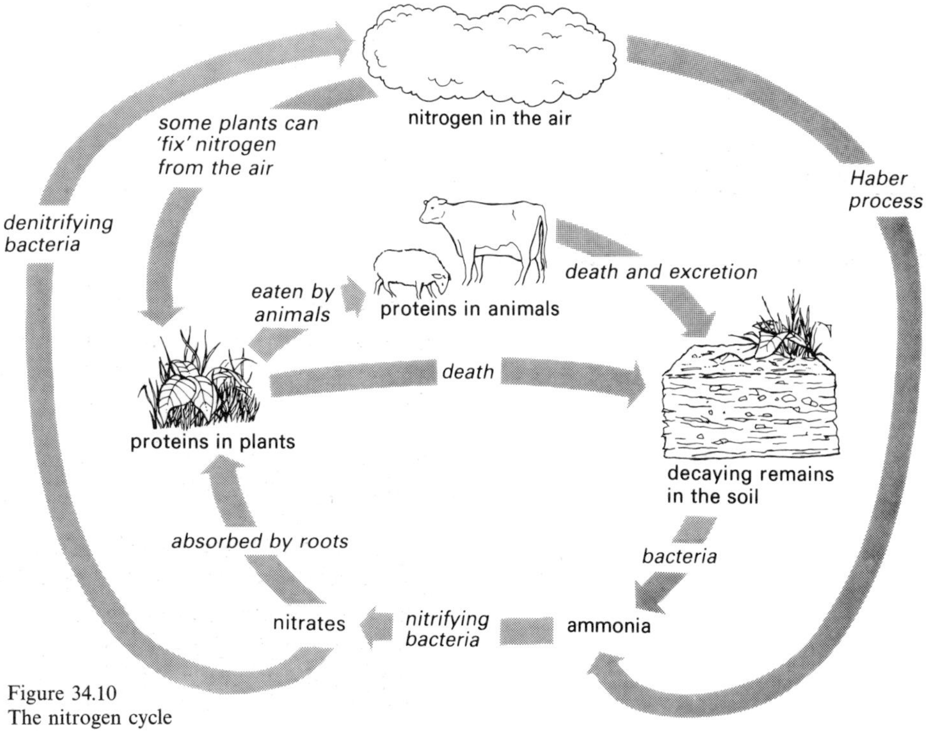

nitrogen in the air
some plants can 'fix' nitrogen from the air
Haber process
denitrifying bacteria
eaten by animals
proteins in animals
death and excretion
death
proteins in plants
decaying remains in the soil
absorbed by roots
bacteria
nitrates
nitrifying bacteria
ammonia

Figure 34.10
The nitrogen cycle

Questions

1 How do modern farming methods upset the natural nitrogen cycle?

2 Many gardeners follow a simple rotation of crops: peas and beans, followed by greens such as cabbage, followed by root crops. What are the advantages of crop rotation?

34.4 The nitrogen problem

For centuries, farmers relied on animal manure and crop rotation to maintain the nitrogen balance in the soil. As the population of cities increased rapidly, more food was required and crop rotation became uneconomical. There was a need to add to the natural reserves of nitrogen.

About a century ago, guano was imported from Peru. Guano is a natural fertilizer consisting of seabird droppings. These supplies were soon exhausted and new fertilizers had to be found.

Towards the end of the 1800s, sodium nitrate from Chile became the important new nitrogen fertilizer. Reserves were high, but the demand rose rapidly.

About this time there was also a growing need for nitrogen compounds for making dyestuffs and explosives. Industry and agriculture were competing for the limited supplies of nitrogen compounds. New, large-scale, sources of these compounds were needed. An obvious source of nitrogen was the atmosphere, containing 78 per cent of nitrogen. In 1904, a German chemist, Fritz Haber, started work on trying to combine nitrogen and hydrogen to form ammonia.

Experiment 34d
Ammonia from nitrogen and hydrogen

hydrogen

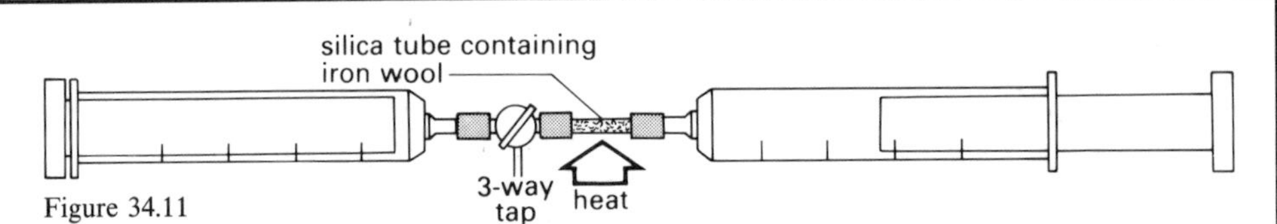

Figure 34.11

The apparatus is shown in figure 34.11. First, it is flushed out with nitrogen to remove all the air. Next, some hydrogen is put into the apparatus and passed over the heated iron. This removes any iron oxide from the surface of the iron.

After this preparation, a mixture of $20\,cm^3$ of nitrogen and $60\,cm^3$ of hydrogen is put into the syringes. The iron is heated and the gas mixture passed to and fro over the iron.

After cooling, the volume of gas remaining at the end is measured. The final gas mixture is tested with moist indicator paper.

Results

There is little visible change. The final gas volume is only slightly less than $80\,cm^3$. However, the indicator paper turns blue, showing that an alkaline gas is present at the end.

Discussion

Why is the air flushed from the apparatus at the start? Why is the surface of the iron freshly reduced with hydrogen? What is the purpose of the iron?

How do nitrogen and hydrogen affect damp indicator paper? What is the alkaline gas? The equation for the reaction is:

$$N_2(g) + 3H_2(g) \rightleftharpoons 2NH_3(g)$$

This shows that four moles of gas (three moles of hydrogen and one mole of nitrogen) will give two moles of ammonia if the reaction is complete. What would the final volume have been in the experiment, if all the nitrogen and hydrogen had reacted? What can you say about the position of this equilibrium under the conditions of the experiment?

34.5 The Haber process

After years of work, Haber found that a high temperature, high pressure and a catalyst are needed to make even modest amounts of nitrogen and hydrogen combine. There were many practical problems, but eventually the Haber process

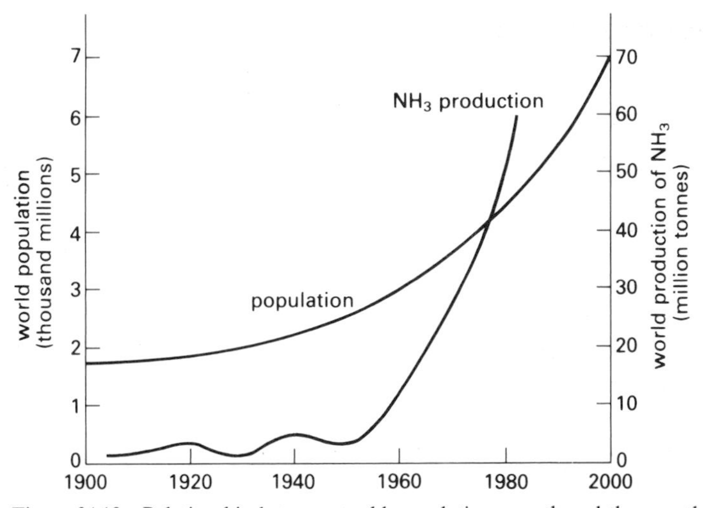

Figure 34.12 Relationship between world population growth and the growth of ammonia production

Figure 34.13 An ammonia plant at Botany, New South Wales

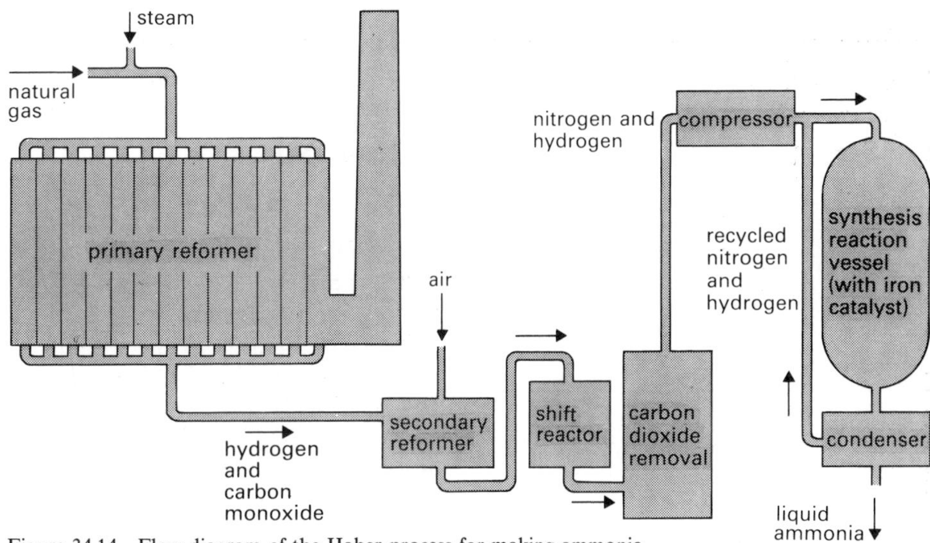

Figure 34.14 Flow diagram of the Haber process for making ammonia

for making ammonia became the most important method of fixing nitrogen.

Today, ammonia is produced on a vast scale by the Haber process. A simplified flow scheme for a modern ammonia plant is shown in figure 34.14.

Hydrogen is now obtained from methane (natural gas) or from naphtha obtained from oil distillation (see p. 11). Methane and steam react at high temperatures and pressures over a nickel catalyst in the primary reformer to give a mixture of carbon monoxide and hydrogen:

$$CH_4(g) + H_2O(g) \longrightarrow CO(g) + 3H_2(g)$$

Air is added at this stage. It reacts with some of the hydrogen formed in the primary reformer, and removes it. The amount of air is controlled so that the final gas composition is one mole of nitrogen to three moles of hydrogen.

Carbon monoxide would 'poison' the catalyst. It is removed by converting it to carbon dioxide by reaction with more steam at high temperatures over another catalyst in the shift reactor:

$$CO(g) + H_2O(g) \xrightarrow{\text{iron oxide}} CO_2(g) + H_2(g)$$

The carbon dioxide is absorbed in hot potassium carbonate solution. Nitrogen and hydrogen then pass to the reaction vessel, where they react at a pressure of about 150–300 atmospheres and a temperature of 350–450 °C over an iron catalyst. The conditions vary from plant to plant.

Ammonia is easily liquefied and removed for storage.

Questions

1 Figure 34.15 shows the yield of ammonia at different temperatures and pressures. What percentage of ammonia is expected in the gas mixture leaving the reactor in a plant operating at 450 °C and at 150 times atmospheric pressure?
2 What are the ideal conditions for getting the maximum yield of ammonia? Why are these conditions not used in practice?
3 Is the unreacted nitrogen and hydrogen wasted? What happens to it?

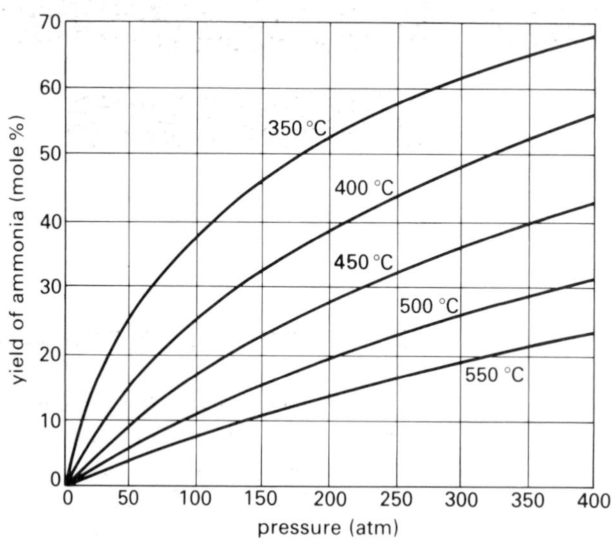

Figure 34.15 The yield of ammonia at different temperatures and pressures

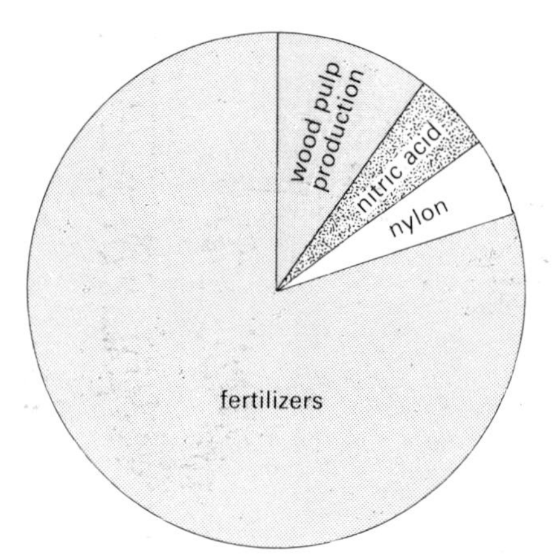

Figure 34.16 Uses of ammonia

Summary

By answering the following questions, you will make a summary of the main ideas covered in this chapter.
1 What evidence is there that all proteins contain nitrogen, in addition to carbon, hydrogen and oxygen?
2 Why are there thousands of different proteins?
3 Enzymes are biological catalysts and are proteins. In what ways are enzymes different from ordinary chemical catalysts?
4 'Plants can survive without animals, but animals cannot survive without plants.' Why is this true?
5 What do you understand by the 'fixing' of nitrogen?
6 How is nitrogen 'fixed' in nature?
7 Why is the series of changes undergone by nitrogen in an undisturbed system referred to as the nitrogen *cycle*?
8 In what ways have human beings disturbed the nitrogen cycle?
9 What do you understand by the 'nitrogen problem'?
10 What contribution did Haber make to solving the nitrogen problem?
11 What is ammonia?
12 Why is it difficult to make ammonia from nitrogen and hydrogen in the laboratory?
13 What are the main steps in the Haber process for producing ammonia?

35 Fertilizers from ammonia

Ammonia
*
is a colourless gas
*
has a pungent smell
*
is less dense than air
*
is very soluble in water
*
is alkaline
*
burns in oxygen but not in air
*
gives a white smoke with hydrogen chloride gas
*

Figure 35.1 The properties of ammonia

35.1 The properties of ammonia

Ammonia is used for making important industrial and farm chemicals. To understand how these chemicals are made we need to know the properties of ammonia itself. These are summarized in figure 35.1.

Figures 35.2 to 35.6 illustrate some of the experiments which can be carried out in the laboratory to investigate the properties of ammonia. Use the facts about ammonia given in figure 35.1 to explain the observations made during the experiments.

Dry ammonia is prepared by heating ammonium chloride with calcium hydroxide and dried by passing through calcium oxide. Explain why

* concentrated sulphuric acid (a good drying agent) is *not* used to dry ammonia
* the gas is *not* collected by displacement of water
* ammonia is collected by *downward* displacement of air as shown
* ammonia could be produced simply by boiling ammonia solution.

Test-tubes of ammonia are opened, one upside-down and the other the right way up. After 20 seconds a piece of damp indicator paper is held at the mouth of each tube. Explain why indicator B turns blue but A does not.

A test-tube of ammonia is opened close to a tube containing concentrated hydrochloric acid. Explain why a white smoke is produced.

Equal concentration solutions of sodium hydroxide and ammonia are each tested with universal indicator paper. Explain why the sodium hydroxide solution turns the indicator paper a darker purple than the ammonia solution does.

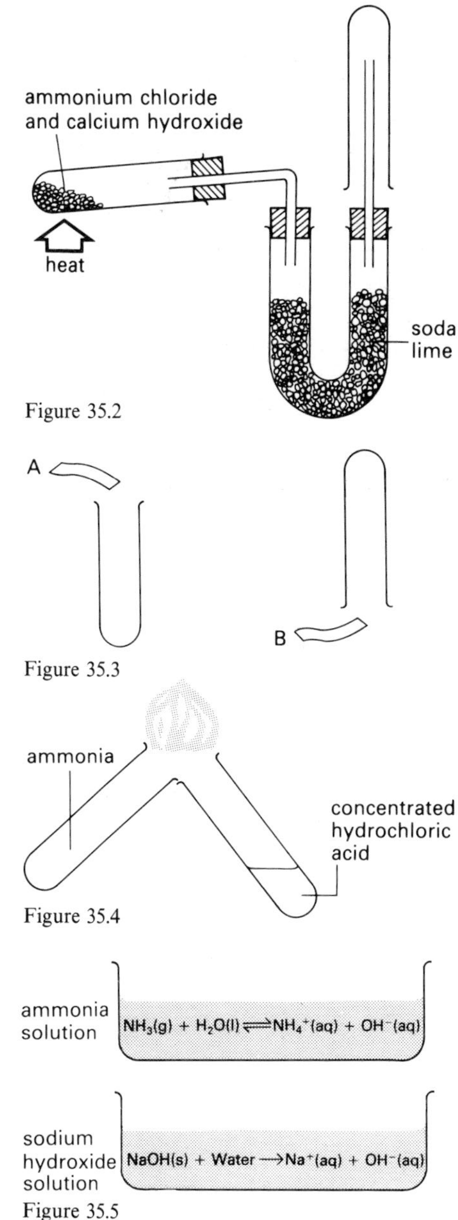

Figure 35.2

Figure 35.3

Figure 35.4

Figure 35.5

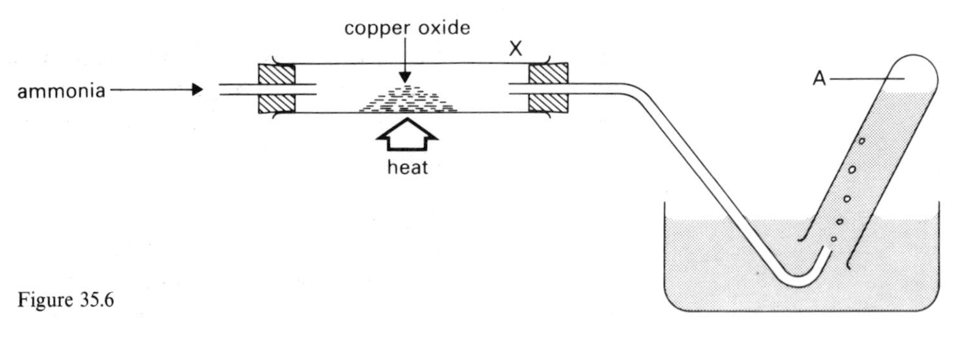

Figure 35.6

Dry ammonia is passed over heated black copper oxide. The gas collected at A is nitrogen. Explain why

* the copper oxide changes to a pink colour
* droplets of a clear colourless liquid are formed at X
* the same reaction would occur with lead oxide, but not with magnesium oxide.

35.2 From ammonia to nitric acid – base to acid!

When ammonia is oxidized in the presence of a platinum catalyst it is converted to oxides of nitrogen. The oxides can be dissolved in water to form nitric acid.

Experiment 35a
The catalytic oxidation of ammonia

concentrated ammonia solution

A coil of platinum wire is gently heated and then lowered into a flask containing ammonia solution until it is just above the surface, as shown in figure 35.7.

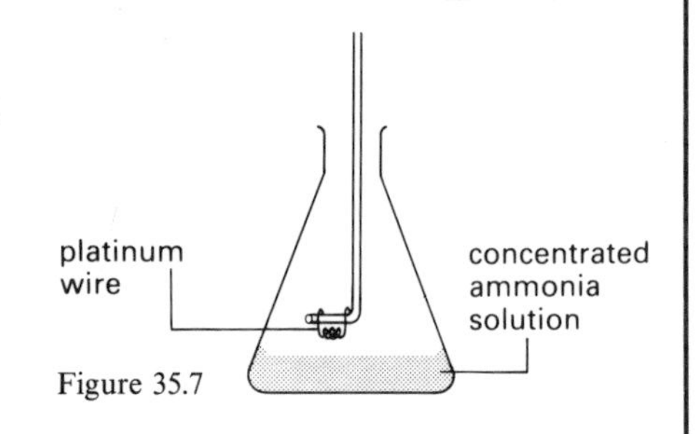

Figure 35.7

Results
The platinum wire glows red-hot and the flask fills with fumes.

Discussion
The wire glows because the ammonia reacts with oxygen on the metal surface. The reaction gives out heat. It is an **exothermic** reaction. The products of the reaction can be seen more clearly in experiment 35b.

Experiment 35b
Making oxides of nitrogen

concentrated ammonia solution

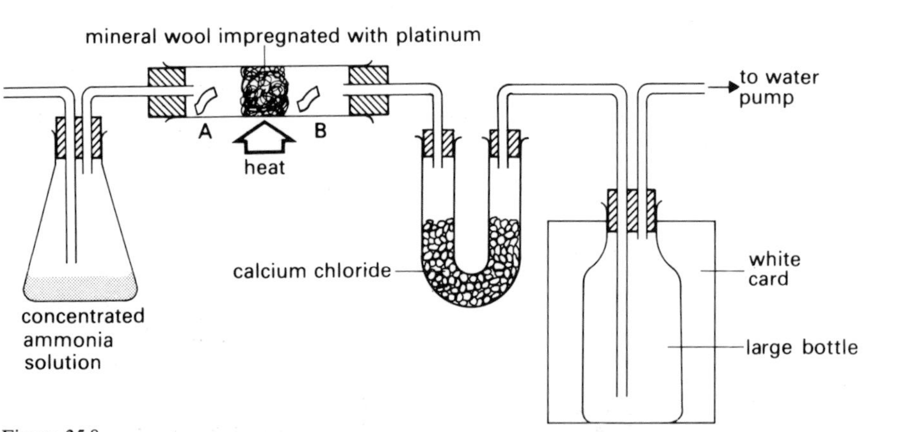

Figure 35.8

In this experiment, the platinum catalyst is a fine powder on the surface of mineral wool (see figure 35.8). The water pump draws a mixture of air and ammonia over the catalyst. The mineral wool is heated to start the reaction.

Results

Once the reaction starts, the platinum continues to glow without further heating. A brown gas appears in the large aspirator. The indicator paper at A stays blue. The indicator at B turns red. The brown gas dissolves in water to produce an acidic solution.

Discussion

The indicator papers show that an alkaline gas has been converted to an acidic gas. The brown gas is nitrogen dioxide. When nitrogen dioxide is dissolved in water in the presence of oxygen, it is turned into nitric acid.

Nitrogen dioxide is formed from ammonia in a two-step process. First the ammonia reacts with air in the presence of platinum to give a colourless gas – nitrogen oxide, NO:

$$4NH_3(g) + 5O_2(g) \longrightarrow 4NO(g) + 6H_2O(g)$$

This reaction is **exothermic**. Enough heat is produced to keep the platinum glowing. Then, as the mixture of gases cools, nitrogen oxide reacts with oxygen directly to give nitrogen dioxide:

$$2NO(g) + O_2(g) \longrightarrow 2NO_2(g)$$

On shaking with water, nitric acid, HNO_3, is formed:

$$4NO_2(g) + O_2(g) + 2H_2O(l) \longrightarrow 4HNO_3(aq)$$

The three reactions leading to nitric acid have one reagent in common. What is it?

This same sequence of reactions is used in the industrial manufacture of nitric acid, which is illustrated in figure 35.9. The uses of nitric acid are illustrated in figure 21.6 on p. 152.

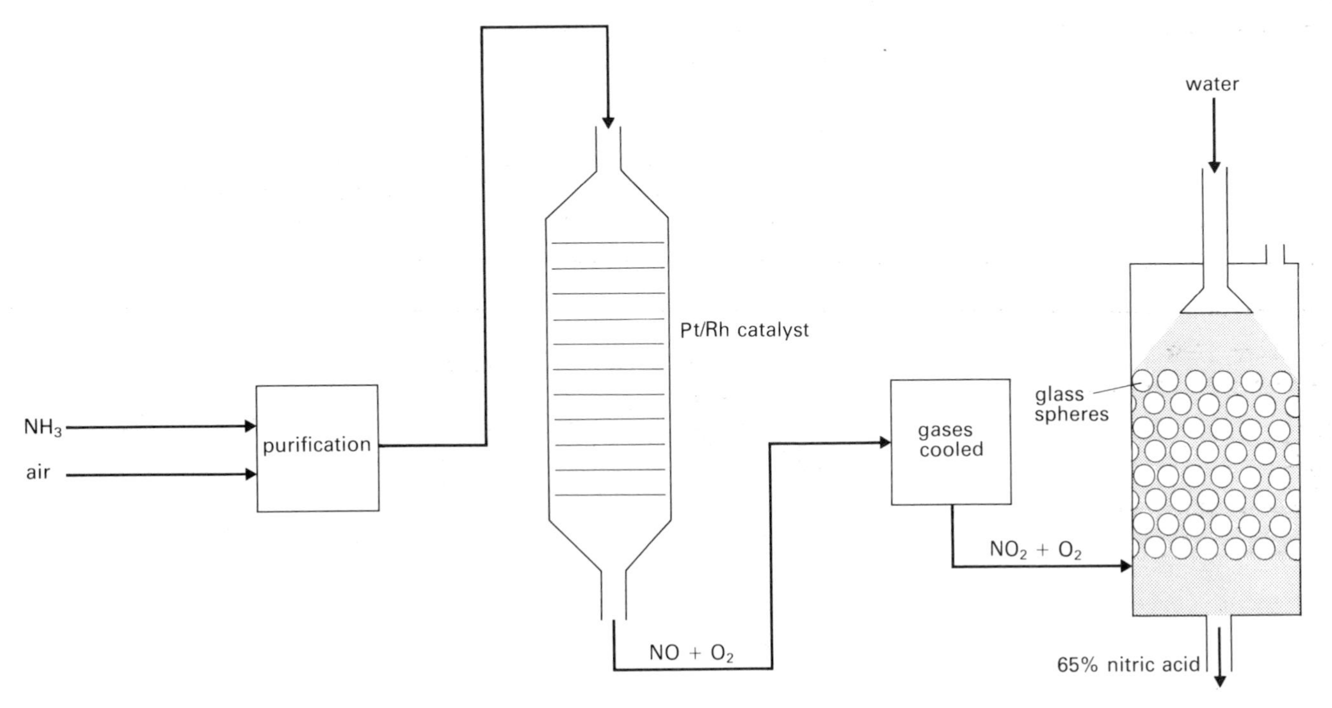

1 Raw materials
Air and ammonia from the Haber process

2 Purification
To remove dust and gases which would poison the catalyst

3 Conversion
The catalyst consists of several sheets of platinum/rhodium gauze. The conversion is started by heating the gauze electrically. Nitrogen oxide is formed. (Why is electrical heating *not* continuous?)

4 Cooling
Nitrogen oxide reacts with oxygen to give nitrogen dioxide

5 Absorption
Nitrogen dioxide and oxygen pass up the tower as water trickles down it. (Why is the tower packed with glass spheres?)

Figure 35.9 Flow diagram of the industrial process for making nitric acid

35.3 Fertilizers

Plants need nutrients from the soil if they are to grow well. These nutrients can be divided into two groups.

1 Trace elements

Very small amounts of elements such as boron, copper and manganese must be present in the soil. In Britain, the concentration of these elements in soils is usually high enough. Only minute amounts of them are needed for healthy plant growth, and it would be dangerous to add too much because they are poisonous in larger amounts.

2 The major plant nutrients

These include nitrogen, phosphorus, potassium and calcium. They are needed in large quantities. Phosphorus aids the growth of roots and speeds up the ripening of crops. Nitrogen is used for making the proteins in stalks and leaves. One nitrogen fertilizer is liquid ammonia, which can be injected directly into the soil. More usually, ammonia is converted into compounds such as ammonium nitrate, urea and ammonium sulphate. These are supplied to the farmer as solids.

Figure 35.10 Injecting liquid ammonia direct into the soil as a fertilizer

Figure 35.11 Bulk storage of a fertilizer

Questions

1 What problems are involved in handling, transporting and using fertilizers? What properties will make a fertilizer safe and convenient to use?
2 What difficulties are likely to be involved in using liquid ammonia as a fertilizer? What are the advantages?

Experiment 35c
Making ammonium sulphate

dilute sulphuric acid

Ammonium sulphate is a soluble salt. Ammonia is a soluble base. Refer to chapter 23 and choose the method you would use to make ammonium sulphate. Describe, in words or diagrams, how you would make a sample of ammonium sulphate crystals.

Discussion
Write a balanced equation for the reaction of ammonia, NH_3, with sulphuric acid, H_2SO_4, to make ammonium sulphate, $(NH_4)_2SO_4$. Use the equation to calculate the maximum amount of ammonium sulphate that could be made from 20 cm^3 of 1.0 M sulphuric acid. Why would you be unlikely to obtain the calculated mass of crystals if you did the experiment?

Question

Tomato plants are sensitive to the amounts of added fertilizer. Describe how you would carry out an experiment to investigate the effects of ammonium sulphate on plant growth, given a packet of tomato seeds.

In Britain the most common nitrogen fertilizer is ammonium nitrate. Figure 35.12 shows a modern fertilizer plant. Figure 35.13 shows the main stages in converting ammonia to ammonium nitrate. After neutralization, the water is

Figure 35.12 A modern fertilizer plant

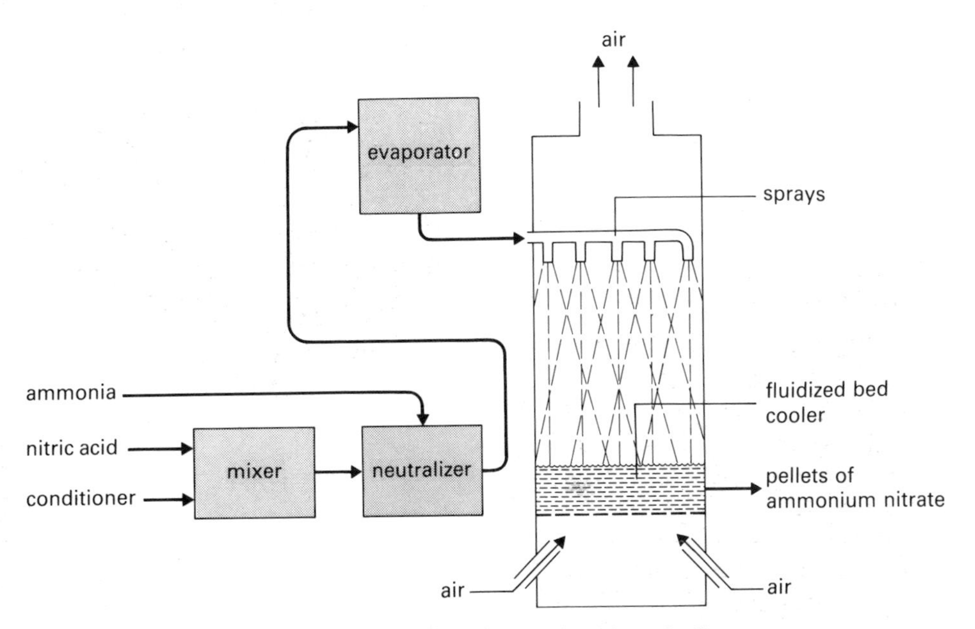

Figure 35.13 Flow diagram to show how pellets of ammonium nitrate fertilizer are made

Figure 35.14 A prilled fertilizer

evaporated. The solid ammonium nitrate is melted and then sprayed down a tall tower. The falling droplets meet an upward current of air. The fertilizer solidifies as tiny, hard pellets called *prills*. Prilled fertilizers are free of dust, and easy to handle and to spread on the fields.

The higher the nitrogen content of a fertilizer, the better. The higher the proportion of nitrogen, the smaller the amount of useless mass which has to be transported from factory to field. The percentage of nitrogen in a compound can be calculated from its formula, as shown in figure 35.15. It can also be measured by experiment.

Compound	Formula	Mass of one mole of compound	Mass of nitrogen in one mole of compound	Percentage of nitrogen
Ammonia	NH_3	17 g	14 g	$\frac{14}{17} \times 100 = 82\%$
Ammonium sulphate	$(NH_4)_2SO_4$	132 g	28 g	$\frac{28}{132} \times 100 = 21\%$
Ammonium nitrate	NH_4NO_3			
Urea	$CO(NH_2)_2$			

Figure 35.15

Question

Complete the table in figure 35.15. The relative atomic masses needed are given in table 1 on p. 336.

35.4 The world food problem

It is estimated that two-thirds of the 4000 million people in the world are not properly nourished, and that one-tenth of the world population faces starvation. There is a dramatic need to increase food production, but it is also necessary to make sure that those who are in greatest need are not too poor to buy the food if it is available. Not everyone who lives in an underdeveloped country is short of food – the rich may have plenty to eat. Some countries may be exporting foodstuffs to earn foreign currency while part of their population is affected by famine. Thus the solution to the food problem requires political as well as scientific action. It may be more effective to give poor farmers the resources they need to grow their own food on their own land than to increase food production in the richer parts of the world for export to the Third World. Economic changes can also be important: if the raw materials exported from underdeveloped countries were given higher value on world markets, then these countries would have more resources to solve their own problems.

There are no simple solutions to the world food problem. Some of the possible approaches to solving it, and the problems associated with them, are mentioned below.

Controlling the population

Some estimates suggest that the world population will rise to 6000 million by the year 2000. How are these estimates made, and are they reliable? Will the increase in population be spread evenly throughout the world, or will it be concentrated in some areas? What are the present trends of population growth in Europe and the United States? Some people argue that we are not facing a world food problem, but a population problem. Population control suggests birth control. What part have scientists to play in providing methods of control that are reliable and sufficiently cheap? How are people's attitude to controlling the size of their families affected by custom and belief?

Growing more food

At present, only about 10 per cent of the earth's land area is cultivated. How much more land could be used for the production of food? What plans are there for watering the deserts and bringing areas with poor soil into cultivation? There are several ways of growing more food on the land now being farmed. This can be done by preventing soil erosion, increasing irrigation, and using fertilizers and other chemicals such as herbicides, insecticides and fungicides. Another approach is to develop new varieties of crops which will produce more grain from a given area, and to breed better livestock. Two-thirds of the earth's surface is covered with water, and we can look to the sea as a source of more food. What schemes are there for farming the sea to produce more food? Industry is also developing methods of making protein foods by growing yeasts and bacteria on inedible materials such as natural gas, methanol, hydrocarbons and cellulose. What problems are there in getting such unconventional sources of food accepted for human consumption? Figure 35.16 shows how successfully industrial processes can obtain edible protein from grain.

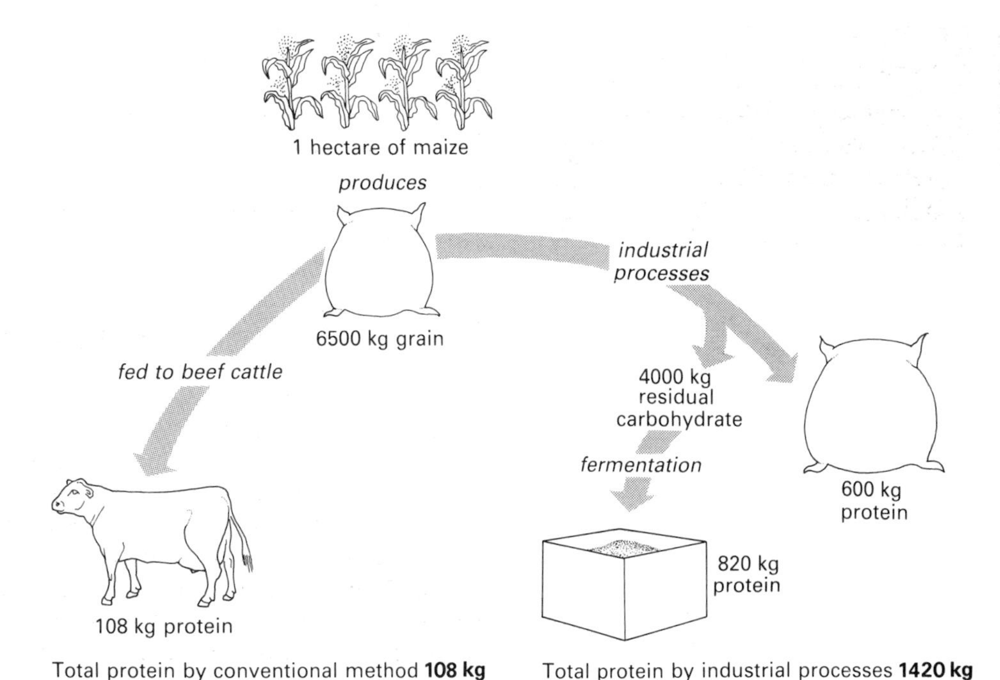

Figure 35.16 Two methods of producing edible protein

Storing and distributing food better

Stored foods can become inedible because of attack by pests and diseases. These can be controlled by using pesticides. Political and economic action can bring about a fairer distribution of food within countries and between countries.

Summary

Copy and complete the passage below, which summarizes this chapter. The following words fill the gaps (they are listed in alphabetical order):
ammonium chloride, ammonium nitrate, boron, blue, catalyst, hydrogen, less, manganese, nitric acid, nitrogen, phosphorus, potassium, pungent, reducing agent, salts, soluble.

Ammonia is a compound of ... and .... It is a gas which is ... dense than air and very ... in water. Ammonia is easily recognized because it has a ... smell and is the only common gas which turns full range indicator ....

Ammonia is a base, so it will react with acids to form .... One example is the reaction of ammonia with hydrogen chloride gas to form solid .... The equation for this reaction is:

...

Ammonia acts as a ... when it reacts with oxygen and hot copper(II) oxide. When it reacts with excess oxygen in the presence of a platinum ..., the product is nitrogen dioxide. Nitrogen dioxide is a brown gas which, in the presence of oxygen, dissolves in water to form ....

Ammonia reacts with nitric acid to form ..., which is a common fertilizer. Ammonium nitrate provides nitrogen for plant growth. Other major plant nutrients are ... and .... Plants also need traces of other elements such as ... and ....

Review questions

1 A white powder was made by crushing the inside parts of some seeds. The following tests were carried out on the powder:

Test 1

Some of the powder was heated in a test tube. It went black. A vapour was formed which turned anhydrous copper(II) sulphate blue. A gas was also evolved which, when bubbled through limewater, turned it milky.

(a) Considering that the powder came from seeds, what was the black solid likely to be?

(b) What colour was the anhydrous copper(II) sulphate to start with?

(c) What was the vapour which turned it blue?

(d) What was the gas which turned the limewater milky?

Test 2

Some more of the white powder was then warmed with some very dilute hydrochloric acid. The mixture was then made neutral and tested with Fehling's solution, which gave a positive result.

(e) Describe how you would do the test with Fehling's solution (or Benedict's solution).

(f) What would you see when the test gave a positive result?

(g) What new substance did this test detect?

Test 3

A fresh supply of this new substance was dissolved in water. The solution was placed in a flask with some yeast. The mixture was kept warm for a few days. The mixture frothed up and a new liquid was formed. This liquid, however, was mixed with water, so it was separated. When the new liquid was tested it was found to burn with a blue flame?

(h) What gas caused the mixture to froth up?

(i) Name the method which would have to be used to separate the new liquid from the water.

(j) Name the new liquid which had been formed.

After considering all three tests,

(k) name the white powder obtained from the seeds.

(l) name one kind of seed from which the white powder is obtained on a large scale. **(EMREB)**

2 (a) Wines containing low quantities of ethanol (alcohol), after long periods of storage taste of vinegar, or are said to have gone sour.

(i) Write down the chemical name of the acid found in vinegar.

(ii) Give *one* reason why the wines go sour.

(b) Ethanol can be produced by the fermentation of sugar solution. The sample obtained contains much water.

(i) Name the other essential ingredient needed in this process.

(ii) State *one* essential condition for the process to take place.

(iii) Name *one* other product of fermentation.

(iv) Draw a labelled diagram to show how you would separate the ethanol from the mixture with water.

(c) Ethanol reacts with organic acids to form compounds called esters.

(i) Write the name of one ester.

(ii) State one use of esters. **(EMREB)**

3 Two white powders A and B, each a pure compound, were tested as follows:

Test 1

When a dry sample of A or B was heated with dry copper(II) oxide, carbon dioxide and water were formed.

Test 2

On boiling with water a small quantity of each dissolved: a solution of A had no action on Fehling's solution but a solution of B gave a positive test with the Fehling's solution.

Test 3

On boiling solutions of A and B with dilute hydrochloric acid for two minutes, both gave a positive reaction with Fehling's solution.

Test 4

A chromatogram of these acid hydrolysates of A and B and also of three simple sugars was made with the result shown below.

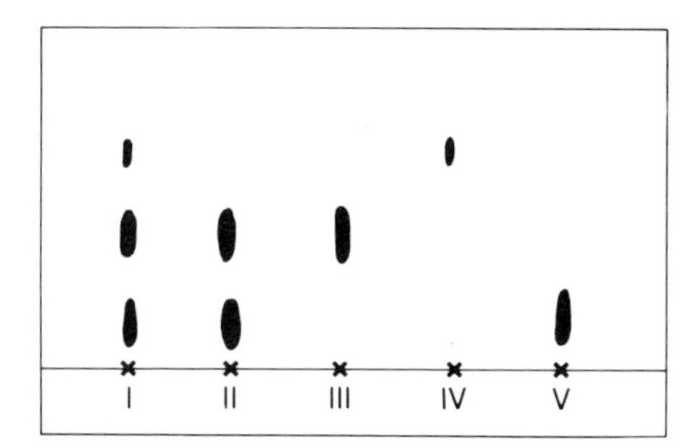

(a) What two elements must be present in powders A and B as shown by test 1?

(b) Devise an experiment to show that these powders are carbohydrates and not hydrocarbons.

(c) What is observed in a positive result to Fehling's test?

(d) Explain why the results of tests 2 and 3 are different.
(e) How do you know that A and B are not simple sugars?
(f) Identify A and B and state which of these two is the reducing sugar. It is known that:

Lactose hydrolyses to glucose and galactose.
Maltose hydrolyses to glucose.
Sucrose hydrolyses to glucose and fructose.
Raffinose hydrolyses to glucose, fructose and galactose.
(L)

4 (a) Give the equation for the decomposition of ammonia into nitrogen and hydrogen. Calculate the change in volume which occurs when 100 cm^3 of ammonia is completely decomposed. (Assume all the measurements are made at the same temperature and pressure.)
(b) A pupil attempted to decompose ammonia by passing the gas over heated iron wool. Although much of the gas decomposed, a little ammonia remained unchanged.

(i) What was the purpose of the iron wool?
(ii) By what simple chemical test could unchanged ammonia be detected?
(iii) Suggest why the decomposition was incomplete.
(iv) State a chemical reaction that may be used to remove hydrogen from a mixture of hydrogen and nitrogen.

(c) One stage in the manufacture of nitric acid involves the oxidation of ammonia by the reaction:

$$4NH_3 + 5O_2 \longrightarrow 4NO + 6H_2O.$$

State the conditions under which this reaction is carried out and outline the further stages in the production of nitric acid. (No diagram is required.)
(d) 0.1 mol of nitric acid is neutralized by ammonia. Calculate the mass in grams of the product contained in the resulting solution. **(AEB)**

5 This question is about chemical aspects of the nitrogen problem, the problem of producing the large quantities of nitrogen compounds needed for fertilizers.
(a) Explain why it is necessary to add fertilizers to cultivated ground where crops are regularly harvested.
(b) Why is there a nitrogen problem when there is so much nitrogen available in the atmosphere?

The best known solution to the nitrogen problem is the Haber process which combines nitrogen from the air with hydrogen to make ammonia. The equation for the reaction is:

$$N_2(g) + 3H_2(g) \rightleftharpoons 2NH_3(g); \quad \Delta H = -103\,kJ$$

The reaction is carried out at high pressure in the presence of a catalyst at a temperature between 350 °C and 550 °C. To maintain a constant temperature, the bed of catalyst particles must be cooled.
(c) Name *one* material which is available on a large scale and can be used as the source of hydrogen for the Haber process.
(d) Give *one* reason why the reaction of nitrogen with hydrogen cannot be used to produce ammonia at room temperature and pressure.
(e) Why would the temperature of the catalyst rise if it were not cooled?

The catalyst used is iron. It is fed into the reaction vessel in the form of iron oxide, Fe_3O_4, but this is converted to minute crystals of iron when it is exposed to the hot mixture of nitrogen and hydrogen gases. If the catalyst is allowed to overheat, the iron crystals rapidly join to form larger particles.
(f) Describe in words, or with an equation, the reaction which converts the iron oxide to iron.
(g) Suggest an explanation for the fact that overheating leads to a reduction in the activity of the catalyst.

The graph below shows how the proportion of ammonia in an equilibrium mixture of nitrogen, hydrogen and ammonia varies with the pressure at two different temperatures.

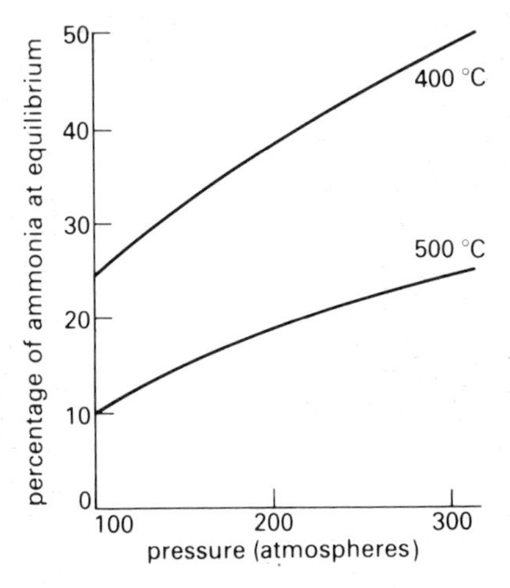

(h) Use the graph to determine the effect of pressure and temperature on the equilibrium mixture of gases. Explain how you arrive at your answers.

(i) What happens to the equilibrium mixture when the pressure is raised?
(ii) What happens to the equilibrium mixture when the temperature is raised?

(i) Ammonia is commonly converted to one of its salts before being used as a fertilizer. One of the most widely used salts is ammonium nitrate, which contains 35% of nitrogen by mass.

(i) Work out the percentage by mass of nitrogen in ammonia.
(ii) Suggest *one* disadvantage of using pure ammonia as a fertilizer. **(L)**

THEME K
Chemistry and energy

Extracting and refining metals involves vast amounts of energy

36 Energy and structure

Figure 36.1 The demolition of a block of flats. Where has all this energy come from?

Energy gets things done. It can warm, move or change matter. Every physical or chemical change is accompanied by an energy change. Sometimes energy, in the form of heat, electricity or light, is supplied to make a change take place. At other times, a physical or chemical change may occur, and heat, electricity or some other form of energy be produced. These changes can be dramatic (see figure 36.1).

Energy cannot be created or destroyed. It simply changes from one form to another.

Question

Some forms of energy are: chemical (contained in all substances), heat, electrical, light, gravitational potential, kinetic. Trace the energy changes that occur when

(a) a 1 kg mass falls from a shelf.
(b) a match is lit.
(c) a steam-engine is working.
(d) the process in figure 36.1 takes place

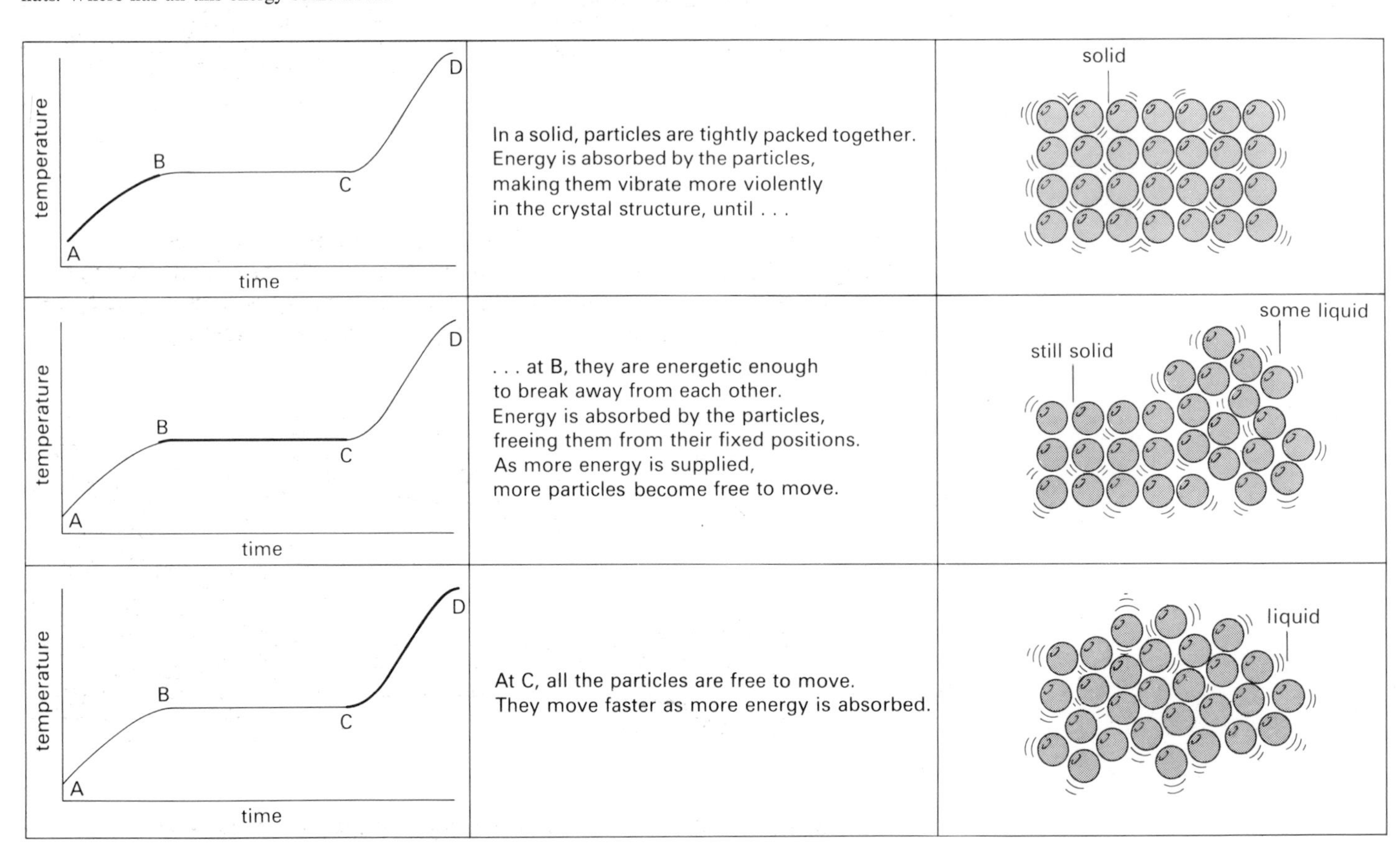

Figure 36.2 The behaviour of particles in a solid as it melts

Figure 1.9 on p. 5 shows the effect of supplying heat energy to stearic acid. The temperature of the solid rises until the melting point is reached. At the melting point, it stays constant and energy is used to melt the substance. When melting is complete, the temperature rises again until it matches the temperature of the surroundings. All solids which melt have heating curves like this. The behaviour of the particles as heat is supplied is described in figure 36.2.

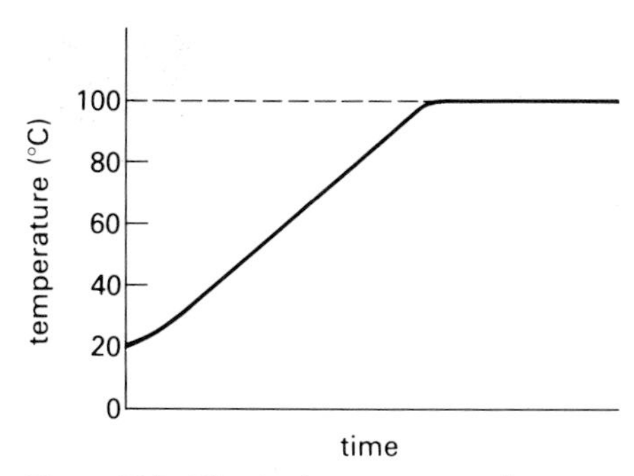

Figure 36.3 The rise in temperature of water as it is heated up to its boiling point

When water is heated, the temperature rises until the boiling point is reached. Then it remains constant and energy is used to **vaporize** the liquid (see figure 36.3). The behaviour of the particles is described in figure 36.4.

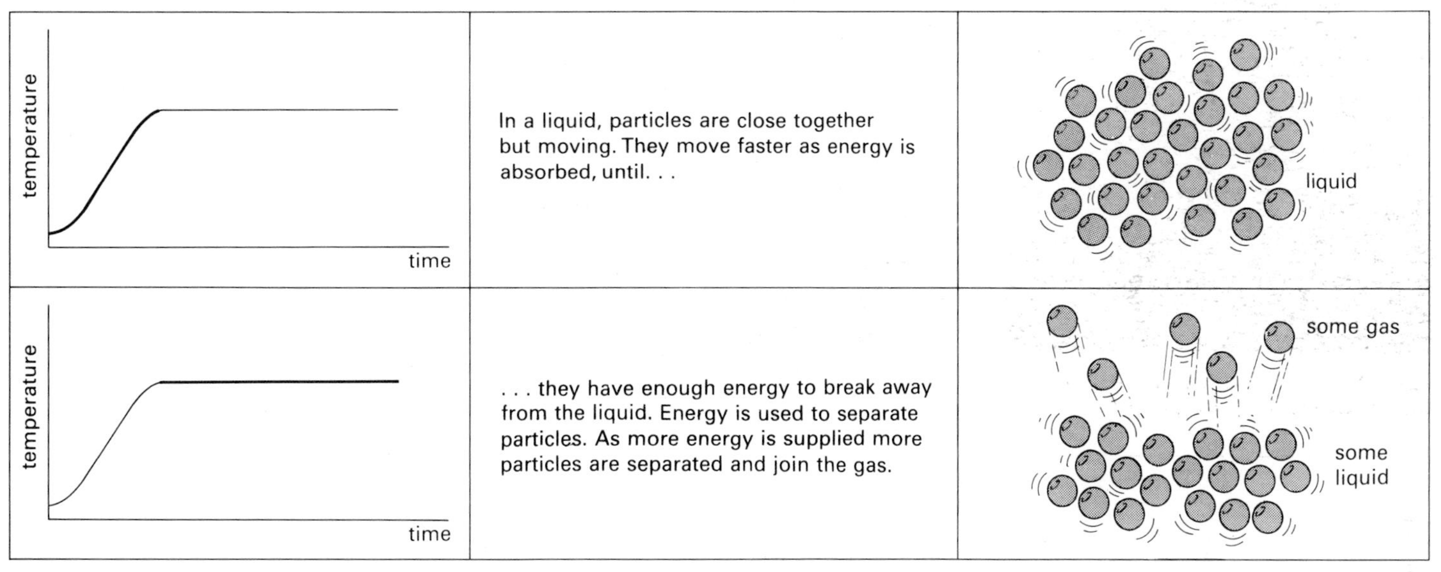

temperature
time
In a liquid, particles are close together but moving. They move faster as energy is absorbed, until. . .
liquid
temperature
time
. . . they have enough energy to break away from the liquid. Energy is used to separate particles. As more energy is supplied more particles are separated and join the gas.
some gas
some liquid

Figure 36.4 The behaviour of particles in a liquid as it boils

During these changes, what kind of energy is gained by the particles?

Energy is absorbed by a substance to warm it, melt it or vaporize it. During the reverse processes, of cooling, freezing and condensing, energy is released (see figure 36.5). In a cooling curve, the temperature stays constant while freezing and condensing take place.

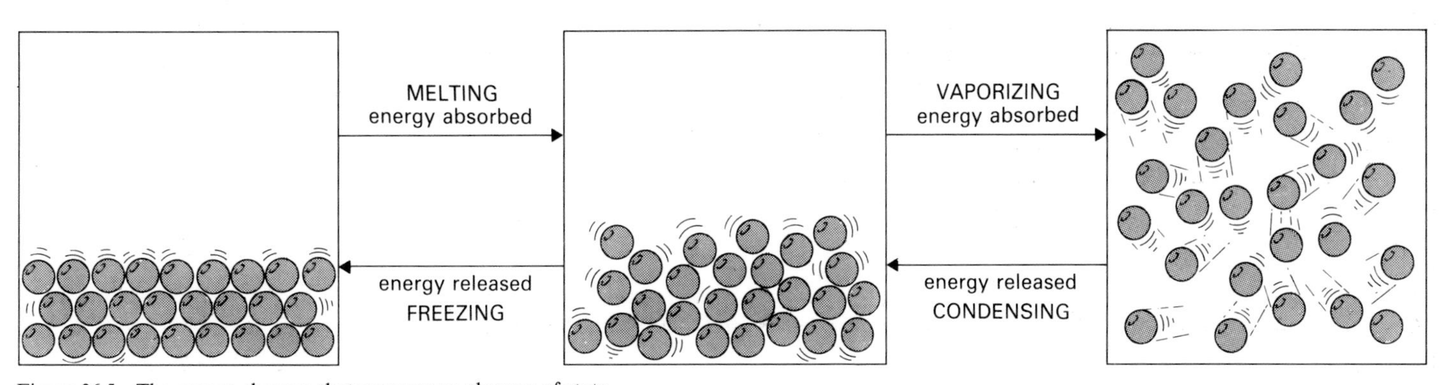

Figure 36.5 The energy changes that accompany changes of state

It is possible to take advantage of the energy changes that occur when substances change state. In a refrigerator, a liquid with a low boiling point is pumped around the coolant circuit. It vaporizes in the coils around the ice compartment. Energy is absorbed from the cabinet, making the air cool. Outside the refrigerator, the vapour condenses under pressure to give liquid again. Does this require energy, or is energy released? Where does the warm air at the back of a refrigerator come from?

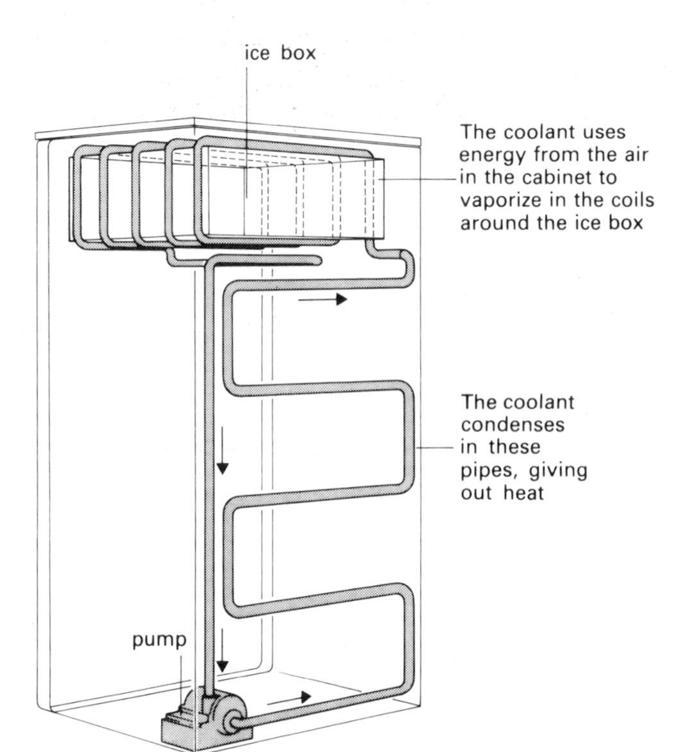
ice box
The coolant uses energy from the air in the cabinet to vaporize in the coils around the ice box
The coolant condenses in these pipes, giving out heat
pump

Figure 36.6 The coolant circuit of a refrigerator

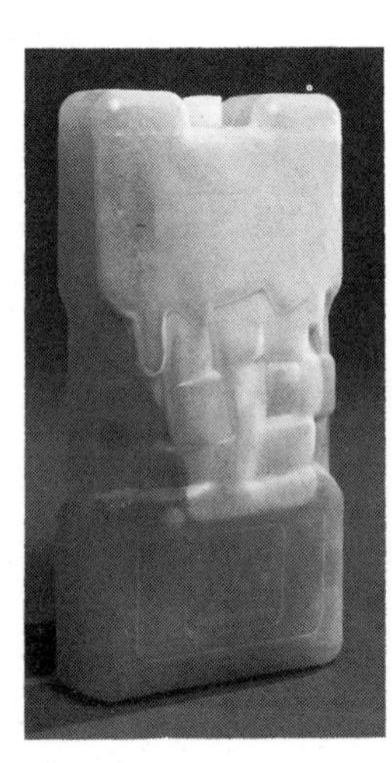

Figure 36.7 A cold pack

A cold pack, like the one in figure 36.7, can be used to keep the contents of an insulated 'cool box' cool. It contains a liquid. When not in use, the cold pack is stored in a freezer where the viscous liquid freezes. In the cool box it slowly melts. What energy changes take place as it melts? The energy comes from the air inside the box, keeping it cool.

Can you think of ways in which a chemical substance might be used as a 'heat store', so that heat energy absorbed during the day might be released at night? What kind of substance is needed? Where would a chemical heat store of this kind be most useful?

The amount of energy that is needed to warm, melt or vaporize a substance, can be measured by experiment. The unit of energy is the joule (J). It is rather a small amount of energy, so energy changes in most processes in chemistry are measured in kilojoules (kJ). These units are described on p. 344. Figure 36.9 gives you an idea of how much energy there is in 1 kJ.

The energy used in an electrical appliance can be found either by connecting it to its power supply through a joulemeter or by knowing its power. Power is the rate at which energy is transferred. It is measured in watts (W).

* $1\,\text{W} = 1\,\text{J/s}$

A 1 kW heater uses 1 kJ of electrical energy every second.

* $$\text{Power (J/s)} = \frac{\text{Energy transferred (J)}}{\text{Time taken (s)}}$$

* $\text{Power (J/s)} \times \text{Time (s)} = \text{Energy transferred (J)}$

In the experiment shown in figure 36.10, a 40 W immersion heater raises the temperature of 100 g of water by 10 °C in 120 s.

Questions

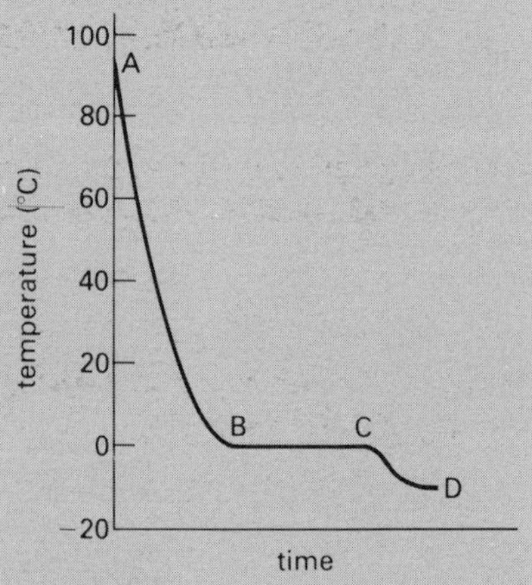

Figure 36.8

1 Figure 36.8 shows the cooling curve of a test-tube of water cooling from 90 °C.
(a) What is happening to the water particles as the water cools along line AB?
(b) Why is there no change in temperature between B and C?
(c) What is in the tube after point C?
(d) What is the temperature of the freezing mixture surrounding the tube?

2 The melting point of naphthalene is 80 °C. Sketch the heating curve you would expect when a tube of naphthalene at 20 °C is placed in boiling water. Mark the important temperatures on the scale of your graph.

Questions

In these questions, take the specific heat capacity of water as 4.2 J/(g °C).

1 How much energy is needed to raise the temperature of 200 g of water from 20 °C to 25 °C?

2 8.4 kJ of energy is transferred to 500 g of water, initially at 25 °C. What is the final temperature of the water?

3 An electric kettle has a power of 3 kW. How long will it take to heat 1 dm^3 (1 kg) of water from 20 °C to its boiling point?

4 1210 J of energy raises the temperature of 100 g of ethanol by 5 °C. What is the specific heat capacity of ethanol?

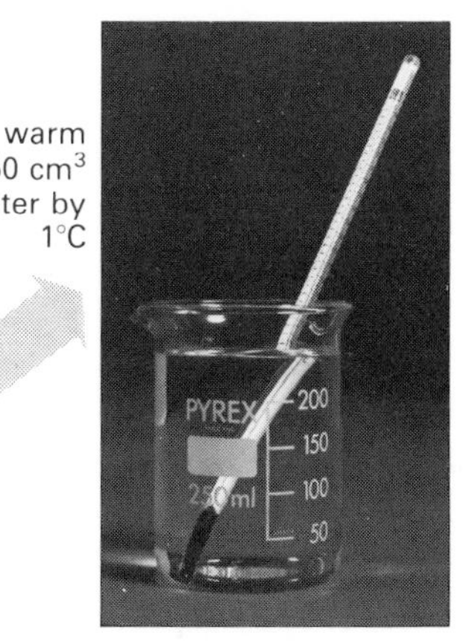

Figure 36.9 Some processes that use 1 kJ of energy

If all the electrical energy is converted into heat energy given to the water, then the amount of energy transferred to the water in 1 s is 40 J. In 120 s, the amount of energy transferred to the water is

$$120\,\text{s} \times 40\,\text{J/s} = 4800\,\text{J}$$

So the energy needed to raise the temperature of 100 g of water by 10 °C is 4800 J and the energy needed to raise the temperature of 1 g of water by 1 °C is

$$\frac{4800\,\text{J}}{100\,\text{g} \times 10\,^{\circ}\text{C}} = 4.8\,\text{J/(g}\,^{\circ}\text{C)}$$

This amount of energy is greater than it should be, because some heat energy is absorbed by the beaker and some is transferred to the surrounding air.

It is accepted that the energy needed to raise the temperature of 1 g of water by 1 °C is 4.17 J. This is the *specific heat capacity* of water. The unit of specific heat capacity is J/(g °C), or kJ/(kg °C).

$$* \quad \text{Specific heat capacity (J/(g °C))} = \frac{\text{Energy transferred (J)}}{\text{Mass (g)} \times \text{Temperature rise (°C)}}$$

$$* \quad \text{Energy transferred} = \text{Mass} \times \text{Specific heat capacity} \times \text{Temperature change}$$

The energy needed to vaporize water at its boiling point can also be found using the apparatus in figure 36.10. The time taken to heat the liquid to its boiling point is not needed in the calculation. In this experiment, a 200 W heater vaporizes 20 g of water at its boiling point in 240 s.

The amount of energy transferred to the water at its boiling point is

$$240\,\text{s} \times 200\,\text{J/s} = 48\,000\,\text{J}$$

So the amount of energy needed to vaporize 20 g of water at 100 °C is 48 000 J and the amount of energy needed to vaporize 1 g of water at 100 °C is

$$\frac{48\,000\,\text{J}}{20\,\text{g}} = 2400\,\text{J/g}$$

Again, this value is higher than it should be because of heat 'losses'. The accepted value is 2260 J/g.

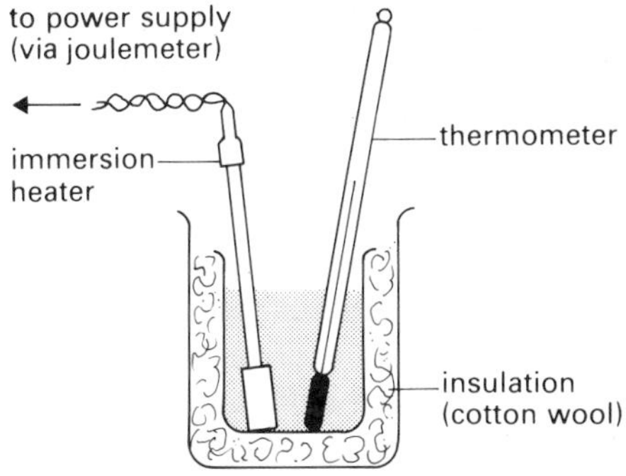

Figure 36.10 Measuring the energy supplied to a liquid as it is heated

When comparing one substance with another, chemists use the *molar heat (energy) of vaporization*. This is also known as simply the *heat of vaporization*, and is the energy needed to vaporize one mole of the substance at its boiling point.
The heat of vaporization of water is

$$2260\,J/g \times 18\,g/mol = 40\,680\,J/mol$$
$$= 40.68\,kJ/mol$$

The apparatus in figure 36.10 is unsuitable for liquids which are **volatile**, inflammable, expensive or have harmful vapours. A better apparatus is used in experiment 36a.

Experiment 36a
Finding the heat of vaporization of ethanol

ethanol

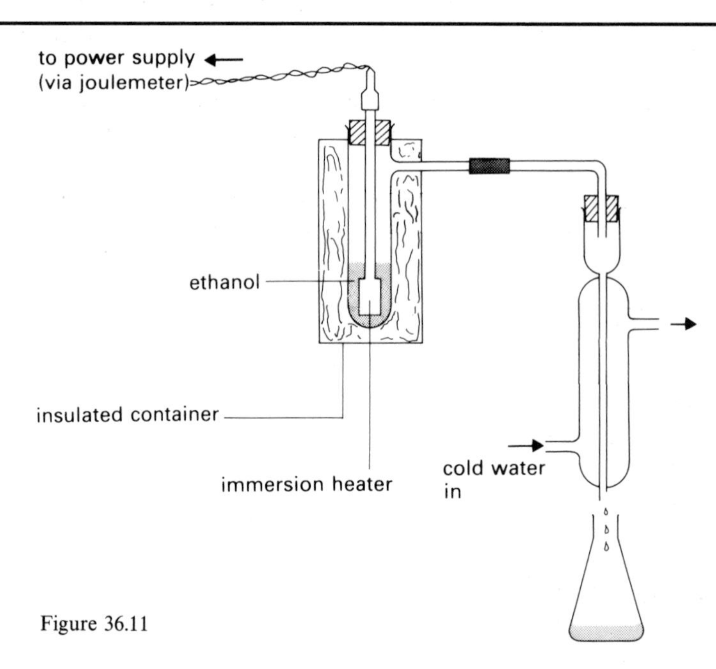

Figure 36.11

The apparatus is shown in figure 36.11. Condensed liquid is collected for a known time after the liquid has reached its boiling point and is condensing steadily. If a joulemeter is used, there is no need to record the boiling time. The meter is reset as the flask is placed into position to collect the liquid.

Results
Typical results are:

Mass of flask empty	=	52.4 g
Mass of flask with collected liquid	=	59.5 g
Mass of ethanol collected	=	7.1 g
Energy supplied (read direct from joulemeter)	=	6049 J
Mass of one mole of ethanol	=	46 g

Discussion
How much energy is used to vaporize 7.1 g of ethanol? How much energy would be needed to vaporize 1 g of ethanol? How much energy would be needed to vaporize one mole of ethanol? The accepted value for the molar heat of vaporization of ethanol is 39 kJ/mol. Is the value from the experiment close to this?

Heat 'losses' are less in this apparatus than in that shown in figure 36.10, because the heating tube is completely surrounded with polystyrene, which absorbs very little heat. However, there are still some sources of error: vaporized liquid can recondense in the heating tube, collected liquid may evaporate, and some energy is still absorbed by the tube. Less energy is needed to vaporize one mole of ethanol than one mole of water. Can you suggest why?

Particles – molecules, atoms or ions – are completely separated when a liquid vaporizes. The *heat of vaporization* is the energy needed to overcome the forces holding one mole of particles together in the liquid. The same amount of energy is released when one mole of vapour condenses to a liquid.

The *heat of fusion* is the energy needed to melt one mole of a substance at its melting point. It is used to overcome the forces holding particles together in fixed positions in the solid, allowing the particles to move. The same amount of energy is released when one mole of liquid freezes.

If particles are weakly attracted to each other, little energy is needed to separate them. Heats of fusion and vaporization are low. If the attractive forces are high, the heats of fusion and vaporization are high.

Look up the heats of fusion and vaporization of methane, water, magnesium, iron, sodium chloride and zinc chloride in table 4 on p. 341. Water and methane are molecular substances. Their heats of fusion and vaporization are low. The attractions between the molecules are small. Iron, magnesium, sodium chloride and zinc chloride have giant structures of atoms or ions. Their heats of fusion and vaporization are high. The attractions between the particles are high.

Why are heats of vaporization always higher than heats of fusion?

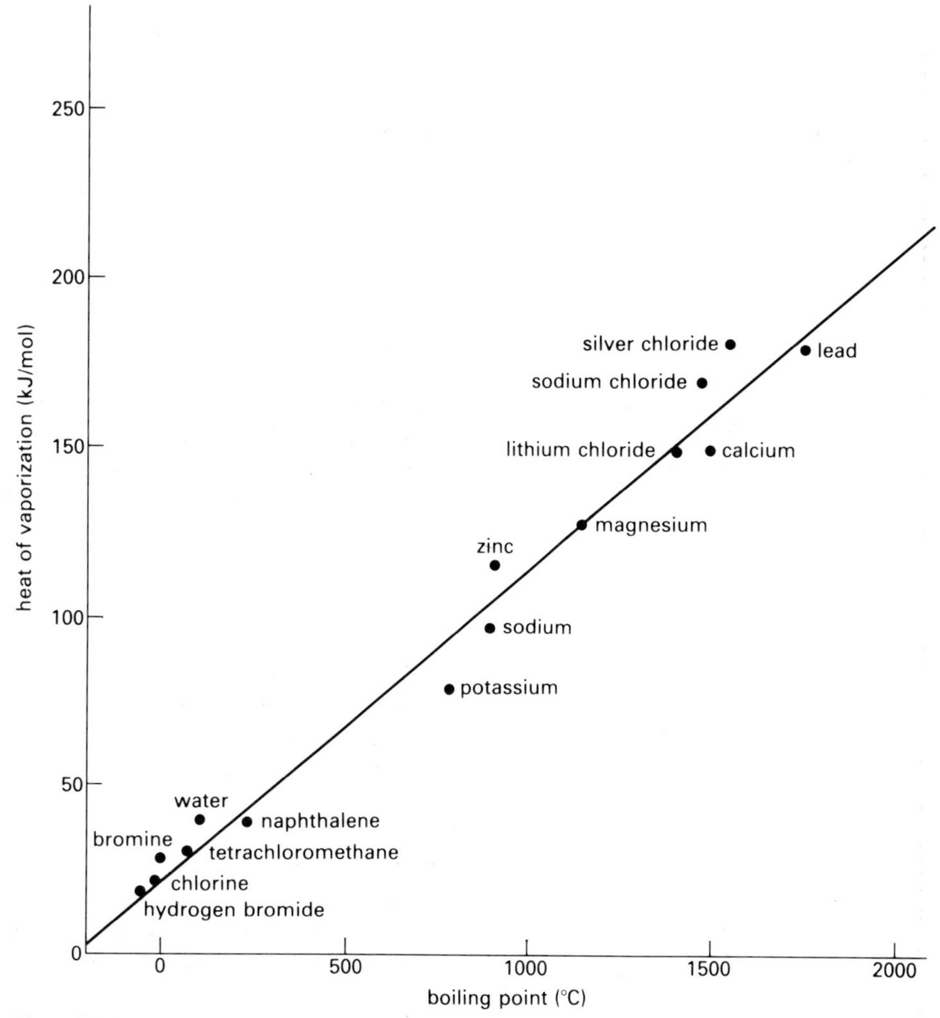

Figure 36.12

Figure 36.12 shows the boiling points and heats of vaporization of a variety of substances. Is there a link between the two values? Are there any substances which appear not to fit in the pattern?

Using table 4 on p. 341, plot a graph of heat of fusion against melting point. Is there a link between the two?

Questions

1 What is the heat of vaporization of
(a) hexane, C_6H_{14}, if 10 g is vaporized by 3320 J of energy?
(b) trichloromethane, $CHCl_3$, if 20 g is vaporized by a 60 W heater in 110 s?

2 Using the apparatus in figure 36.11, 3 g of ethanol is collected in 50 s, while 3 g of hexane is collected in 20 s.
(a) Which liquid needs the most energy to vaporize 1 g?
(b) If 2700 J are needed to vaporize 3 g of ethanol, how much energy is needed to vaporize 3 g of hexane?
(c) How much energy is needed to vaporize 1 g of ethanol?
(d) What is the heat of vaporization of ethanol?

3 50 g of ice in a beaker was melted over a small Bunsen flame in four minutes. In a separate experiment, it was found that the burner supplied energy at the rate of 4250 J/min.
(a) How much energy was transferred to the beaker in four minutes?
(b) How much energy is needed to melt 1 g of ice?
(c) What is the heat of fusion of ice?
(d) How could you measure the energy output of the burner? (Assume that you know the specific heat capacity of water.)

4 Tetrachloromethane and potassium chloride both contain chlorine.
(a) Use table 4 on p. 341 to find the heats of fusion and vaporization of the two compounds.
(b) Why are the values for the two compounds so different?

Summary

1 Copy and complete the table in figure 36.13.

Process	**Name of process**	**Is energy absorbed or released?**
Ice changes to water	Melting	
Water changes to steam		Absorbed
Water changes to ice		
Steam changes to water		

Figure 36.13

2 Copy and complete the following statements:

(a) Energy transferred = Mass × Specific heat capacity × ...

(b) Heat of vaporization = Energy needed to vaporize 1 g of a substance at its boiling point × ...

(c) ... = Energy needed to melt 1 g of a substance at its melting point × Mass of 1 mol of substance

3 Copy and complete the table in figure 36.14.

Substance	**Type of structure**	**Type of bonding**	**Is the heat of vaporization high or low?**
Ammonia	Molecular		
Diamond		Covalent	
	Giant structure of ions		High
		Metallic	

Figure 36.14

37 Energy from fuels

37.1 What is a fuel?

Most of the energy we use to keep warm, to cook food and to get us from place to place comes from fuels. Fuels release energy when they burn in air or oxygen. Because they usually contain carbon and hydrogen, the products are carbon dioxide and water – together with lots of energy, of course. For example,

$$\underset{\text{methane in natural gas}}{CH_4(g)} + 2O_2(g) \longrightarrow CO_2(g) + 2H_2O(g) + \text{Energy}$$

All combustion reactions are **exothermic**. They all release energy. Nuclear fuels release energy in a different way, which is described in chapter 39.

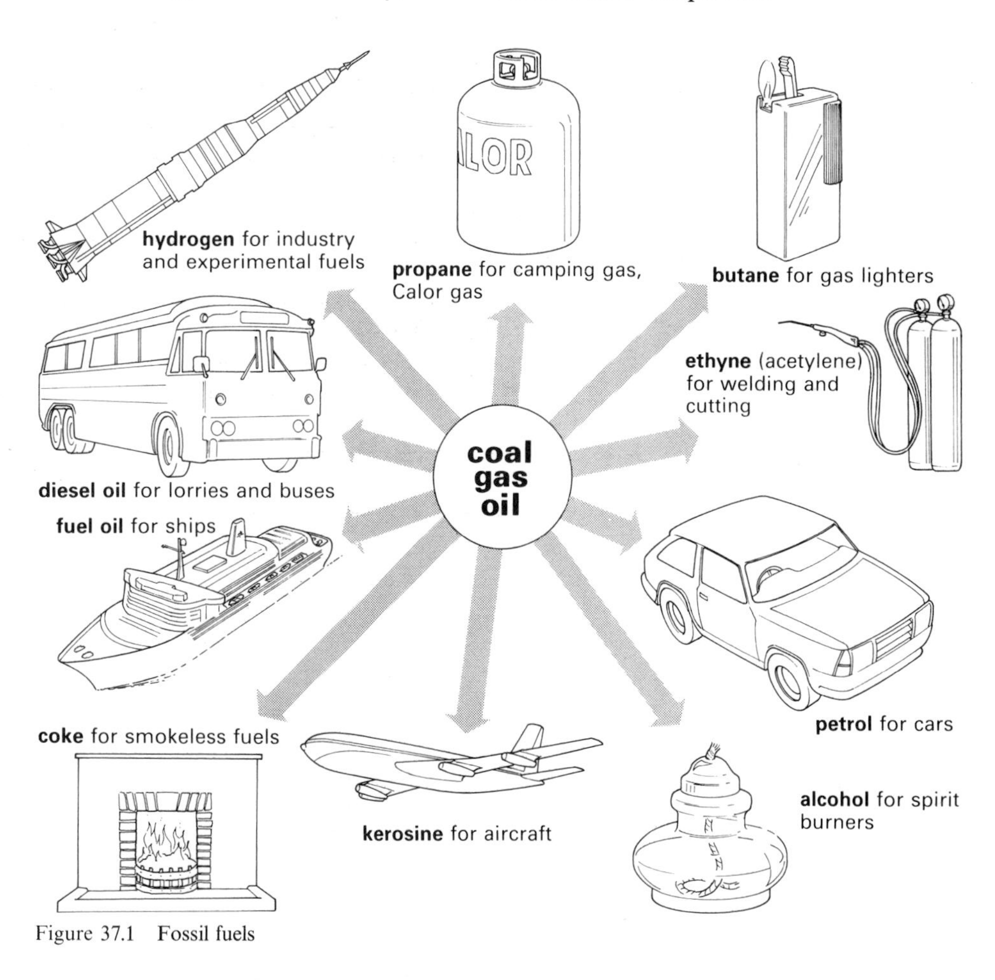

Figure 37.1 Fossil fuels

Fossil fuels were formed from the decaying remains of plants and tiny marine animals which lived millions of years ago. The remains became buried under deep layers of sediment. Intense heat and pressure resulted in the formation of coal, gas and oil. The energy that is released when we burn fossil fuels originally came from the sun. It became chemical energy through *photosynthesis* in the leaves of plants.

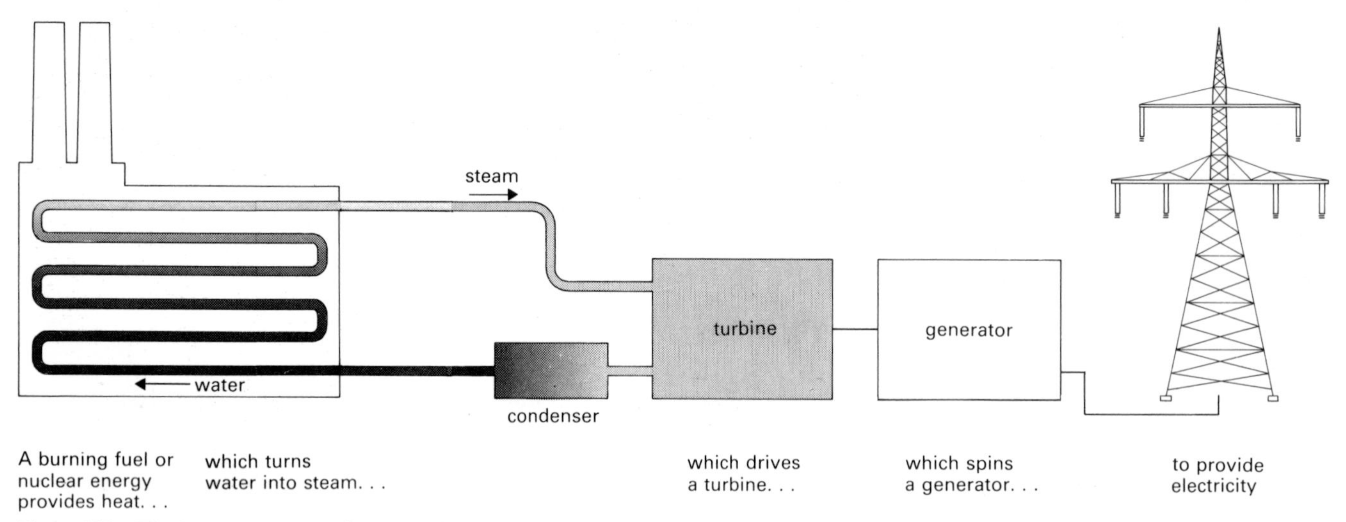
steam
turbine
generator
water
condenser
A burning fuel or nuclear energy provides heat. . .
which turns water into steam. . .
which drives a turbine. . .
which spins a generator. . .
to provide electricity

Figure 37.2 The four-stage process for generating electricity

Electricity is not a fuel. It is a secondary form of energy, obtained from primary fuels. It is, perhaps, the most convenient way of harnessing energy for domestic use, but it is difficult to store. Electricity is generated in a four-stage process (see figure 37.2). During the process some energy from the primary fuel is wasted.

From figure 37.3, calculate what percentage of the original energy is converted into useful electrical energy. What happens to the 'lost' energy?

At present, over 75 per cent of electricity in the UK is obtained from coal. Out of the 121 million tonnes of coal that were mined in the UK in 1982–3, 86 million tonnes were used for generating electricity. The proportion generated from nuclear

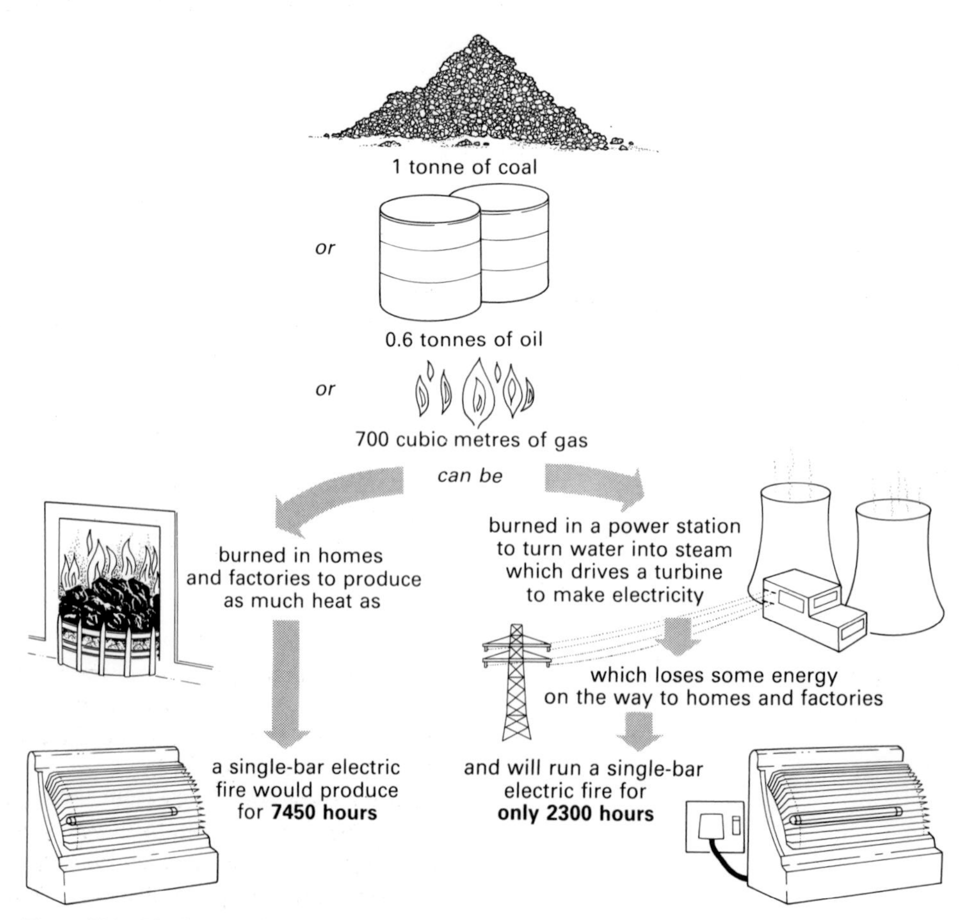
1 tonne of coal
or
0.6 tonnes of oil
or
700 cubic metres of gas
can be
burned in homes and factories to produce as much heat as
burned in a power station to turn water into steam which drives a turbine to make electricity
which loses some energy on the way to homes and factories
a single-bar electric fire would produce for 7450 hours
and will run a single-bar electric fire for only 2300 hours

Figure 37.3 Much energy is 'lost' when a primary fuel is burnt to make electricity

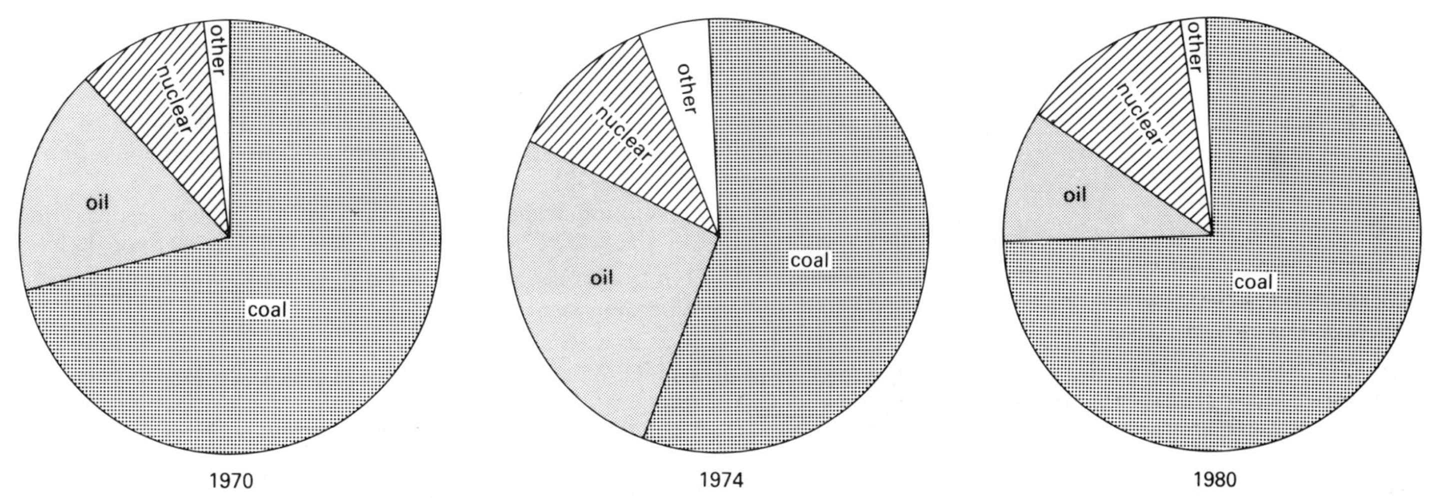

Figure 37.4 Fuels used for generating electricity in the UK, 1970–80. Can you explain some of the trends that have taken place? What do you think the 'other' sources might be?

energy has increased steadily since 1960, and was about 14 per cent in 1982. By 1990, it could be 20 per cent.

37.2 Coal – a fuel and a source of chemicals

Oil and gas are not just fuels. Theme I describes how oil can be refined to give useful chemicals which are used as raw materials for other products.

Coal is a complex mixture of substances, and the composition varies from one type of coal to another. It depends on the coal's age and the conditions under which it was formed. *Anthracite* is a very hard, black coal found in South Wales. It contains a high proportion of carbon. *Lignite* is a younger, soft, brown coal.

Coal burns in air with a yellow smoky flame. It is oxidized, and any useful chemicals in it are burnt away. When it is heated in the absence of air it is distilled destructively, giving different products.

Experiment 37a
Destructive distillation of coal

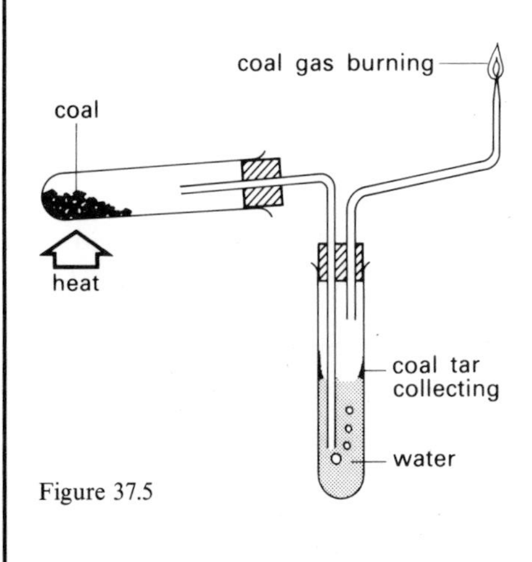

Figure 37.5

Coal is heated in the apparatus shown in figure 37.5. What precautions are necessary when the heating is stopped?

Results

Products of interest are (a) the gases given off, (b) the condensed liquids and (c) the solid residue left in the tube that contained the coal.

(a) The gas is often pale yellow and has a 'sulphury' smell. It burns with a yellow smoky flame and does not dissolve in water.

(b) In the vertical tube, a brown tarry gum condenses just above the water. The aqueous layer is often cloudy and alkaline.

(c) The coal changes from being black and shiny into a grey, light, porous solid. When this solid burns, it glows brightly. There are no flames.

Discussion

The *fractional* distillation of oil (see chapter 1) simply involves separating substances already present in the oil. In this *destructive* distillation, substances in the coal are broken down before they are distilled and collected.

After destructive distillation, the coal becomes *coke*, which is an impure form of carbon. When heated in air, it gives carbon dioxide. *Coal gas* is a mixture of hydrogen with smaller amounts of methane and carbon monoxide. These gases burn. The tarry gum, called *coal tar*, is a complicated mixture of organic chemicals. The aqueous alkaline solution contains dissolved ammonia. In the industrial destructive distillation, this was called the 'ammoniacal liquor'.

Until about 1960, coal gas was an important domestic fuel. Now, all the gas used in the home is natural gas, from the North Sea (see p. 221). However, some coal gas from coke ovens is still used in nearby industries.

The 'ammoniacal liquor', too, was once important for making fertilizers by reaction with sulphuric acid. Now, most fertilizers are made using ammonia from the Haber process (see chapter 35).

Coke and coal tar are still important. About 6 per cent of the coal output of the UK – 8 million tonnes in 1982–3 – is converted into coke. Each tonne of coal gives about two-thirds of a tonne of coke. Figure 13.4 on p. 90 shows a coke oven, in which coal is heated to obtain coke.

The main use of coke is as a reducing agent in the blast furnace for making iron (see chapter 13). Other metal ores can also be reduced with coke. Which ones? Some coke is used in the home as a smokeless fuel.

Coal tar is distilled again, giving a wide range of chemicals, such as benzene, xylene and phenol. These can be used for making disinfectants, insecticides and wood preservatives. Some of the useful products from the chemicals in coal tar are shown in figure 1.20 on p. 11.

The main use of coal is as a fuel. But is it a good fuel?

37.3 Good and bad fuels

Figure 37.6 shows Leeds during the industrial revolution, when coal was the most important source of energy. What is the obvious disadvantage of coal as a fuel? If you have a coal fire at home, or if you have watched coal burning, you can probably think of several more.

Many substances burn and release energy, but do they necessarily make good fuels? What are the qualities of a good fuel? Some questions to ask about potential fuels are:

- Is it easy to ignite?
- Does it keep burning once lit?
- Is much smoke formed?
- Does it give off dangerous fumes?
- Does it burn away rapidly or steadily?
- Does it release a lot of energy?
- How much ash is left after burning?
- Is it naturally-occurring or synthetic?
- Is it expensive?
- Is it safe and easy to store and transport?
- Is it plentiful?
- Could it be better used for other purposes than as a fuel?

Think about a fuel for the home or for a power station. What answers would you give to each question for an ideal fuel? Is there such a thing as a perfect fuel?

Questions

1 Some possible fuels for heating are: paper, wood, magnesium, coal, coke, candle wax, alcohol, petrol, natural gas, wood, sugar. Think about each of these substances in the light of the questions on the right. Make three lists: (a) good fuels, (b) fuels that are acceptable, but far from perfect and (c) poor fuels.
2 What special qualities must a rocket fuel have?
3 What are the advantages of a camping stove that uses bottled gas over one that runs on paraffin?
4 Could a motor car run on coal dust?
5 Why is it not a good idea to use waste-paper and cardboard as domestic fuels?

Figure 37.6 An old engraving showing Leeds in the nineteenth century

Figure 37.7 A car that uses hydrogen as a fuel

The qualities needed from a fuel depend on its use. Fuel for a car engine must ignite easily and burn rapidly. For other uses, the same fuel would be too dangerous.

Burning fuels can cause serious pollution. Some of the problems are described in chapter 6. The car shown in figure 37.7 runs on hydrogen. It causes no pollution because the only product of combustion is water. What disadvantages does this fuel have? What other attempts to reduce pollution from fuels have you heard of?

37.4 The energy value of fuels

The amount of energy released by a fuel can be found by allowing it to heat water. In experiment 37b, three fuels – a solid, a liquid and a gas – are compared.

Experiment 37b
Comparing the energy output of candle wax, ethanol and butane

ethanol

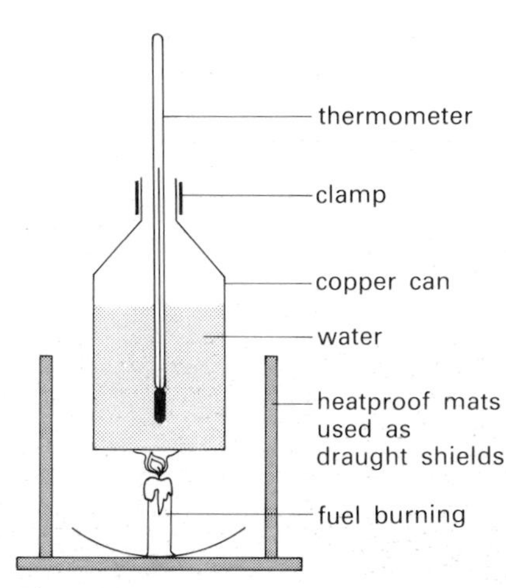

Figure 37.8

Figure 37.8 shows the apparatus used. Ethanol is contained in a crucible. A small wax candle is stuck on a watch glass. A gas cigarette lighter is used as a butane burner. Each burner is weighed before and after it is used to heat the water by about 10 °C.

Results
Figure 37.9 shows a table of typical results. The copper container becomes black when a candle is used. Incomplete combustion is occurring. How does this affect the results?

Discussion
Using the results for ethanol as an example, and assuming that 4.2 J of energy are needed to heat 1 g of water by 1 °C, the energy absorbed by the water is

$$150\,\text{g} \times 15\,°\text{C} \times 4.2\,\text{J/(g °C)} = 9450\,\text{J}$$
$$= 9.45\,\text{kJ}$$

Assuming that there were no heat losses, this energy came from 0.90 g of ethanol. So the energy released by 1 g of ethanol is

$$\frac{9.45\,\text{kJ}}{0.90\,\text{g}} = 10.5\,\text{kJ/g}$$

Repeat this calculation, with candle wax and butane as the fuels.

Would you expect the experiment to give reliable results? Why might the results of one experiment differ from another? It is accepted that 1 g of ethanol

Ethanol		**Candle wax**		**Butane**	
Mass of crucible + ethanol	= 21.84 g	Mass of candle before experiment	= 30.10 g	Mass of burner before use	= 41.43 g
Mass of crucible	= 20.94 g	Mass of candle after experiment	= 29.95 g	Mass of burner after use	= 41.25 g
Mass of ethanol used	= 0.90 g	Mass of candle used	= 0.15 g	Mass of gas used	= 0.18 g
Mass of water	= 150 g	Mass of water	= 150 g	Mass of water	= 150 g
Initial temperature of water	= 19 °C	Initial temperature of water	= 19 °C	Initial temperature of water	= 19 °C
Highest temperature of water	= 34 °C	Highest temperature of water	= 29 °C	Highest temperature of water	= 28 °C
Temperature rise	= 15 °C	Temperature rise	= 10 °C	Temperature rise	= 9 °C

Figure 37.9

can release 29.7 kJ of energy. Why is this value much higher than the one from the experiment?

Although the experiment does not give particularly reliable results, it shows the principle of how the energy output from fuels can be measured and compared.

The *molar heat of combustion*, or simply the *heat of combustion*, of a fuel is the amount of energy released when one mole of a substance burns in air.

1 g of ethanol releases 29.7 kJ of energy when it is burnt. The (molar) heat of combustion is

$$29.7\ \text{kJ/g} \times 46\ \text{g/mol} = 1366\ \text{kJ/mol}$$

The apparatus in figure 37.8 always gives low values for heats of combustion, because there are many ways in which heat energy can be 'lost'. Figure 37.11 shows an apparatus which can give more accurate values. It is used for solids, liquids or gases. Hot gases from the burning fuel pass through the copper spiral.

Questions

1 Calculate the heat of combustion of butane, using the results of experiment 37b.

2 Read the value of the heat of combustion of butane from figure 37.12. This is the accepted value. Compare it with your answer to **1** and comment on any difference.

3 Why is it nonsense to talk about the (molar) heats of combustion of candle wax and coal?

4 The results in figure 37.10 were obtained using apparatus like that in figure 37.8. The alcohols were contained in small wick burners.

(a) Calculate the heat of combustion of butan-1-ol and include it in a copy of the table.

(b) Why are the results different from the data book values?

(c) The structures of ethanol and propan-1-ol are:

```
   H  H                 H  H  H
   |  |                 |  |  |
H—C—C—O—H    and    H—C—C—C—O—H
   |  |                 |  |  |
   H  H                 H  H  H
```

What is the structure of butan-1-ol?

(d) Predict a value for the heat of combustion of pentan-1-ol, $C_5H_{11}OH$.

Question

The apparatus in figure 37.11 needs 3.5 kJ of energy to raise the temperature of the water by 1 °C. In a separate experiment, 2.5 g of propanone was burnt in the apparatus, and the temperature of the water rose by 22 °C.
(a) How much energy was released by the propanone?
(b) How much energy would be released by 1 g of propanone?
(c) What is the heat of combustion of propanone, C_3H_6O?
(d) Find the accepted value for the heat of combustion from table 3 on p. 340.

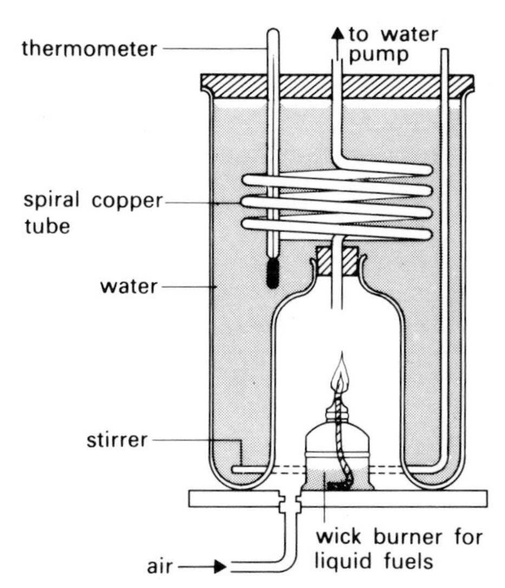

Figure 37.11 Measuring the heat of combustion of a fuel

	Ethanol	**Propan-1-ol**	**Butan-1-ol**
Formula of the alcohol	C_2H_5OH	C_3H_7OH	C_4H_9OH
Mass of water in can (g)	100	100	100
Rise in temperature (°C)	20	20	20
Mass of alcohol used (g)	1.00	0.90	0.74
Mass of one mole (g)	46	60	74
Heat of combustion, from experiment (kJ/mol)	380	560	
Heat of combustion, from data book (kJ/mol)	1330	2020	2680

Figure 37.10

This heats the water around the coil. The energy needed to raise the temperature of the apparatus and water by 1 °C (its *heat capacity*) must be known.

37.5 Where does the energy come from?

Figure 37.12 shows the heats of combustion of a series of **alkanes**. They increase from one alkane to the next. What is the difference in structure between one alkane and the next? The increased heat of combustion seems to be associated with an increased number of atoms in the fuel. Is there a similar pattern for the series of alcohols in figure 37.10?

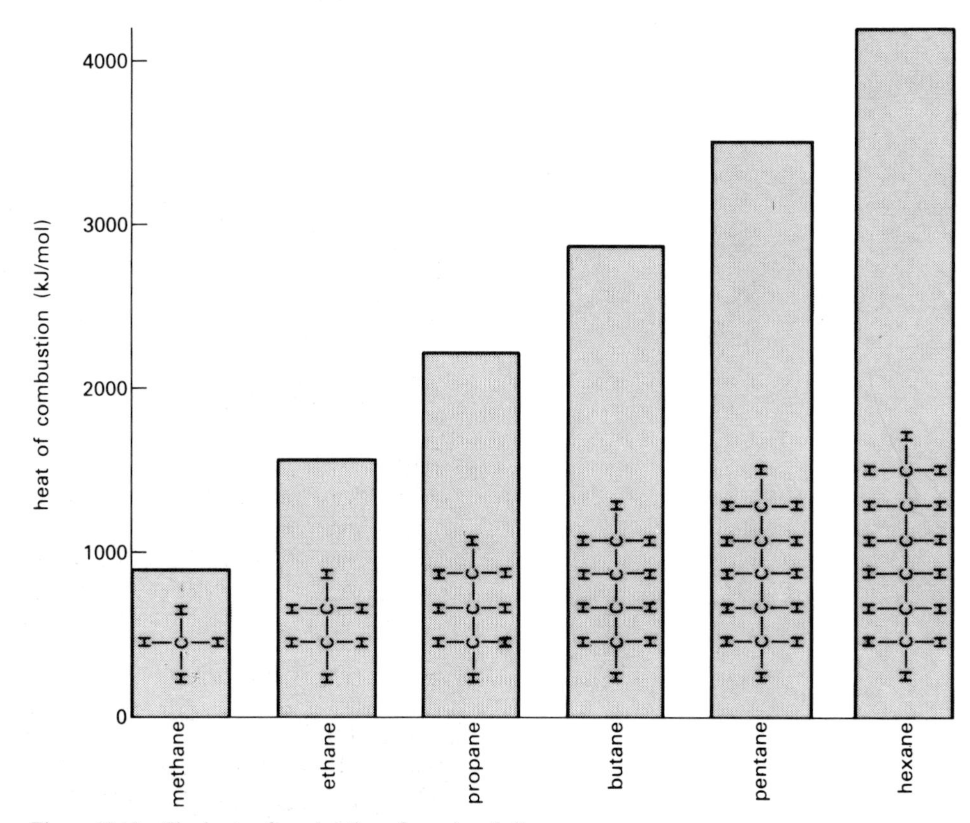

Figure 37.12 The heats of combustion of a series of alkanes

In a chemical reaction, bonds in the reactant molecules are broken and new ones in the products are made. Atoms are rearranged. Energy has to be put in to break bonds, and energy is given out when bonds form.

Each type of bond in a molecule has its own 'bond energy'. When hydrogen burns in oxygen to give steam

$$2H_2(g) + O_2(g) \longrightarrow 2H_2O(g)$$

bonds in hydrogen and oxygen molecules are broken. New bonds are formed as the atoms recombine to form water molecules.

The reaction is exothermic. It gives out heat because the total energy given out in bond making is greater than the energy needed for bond breaking. This is shown in figure 37.13.

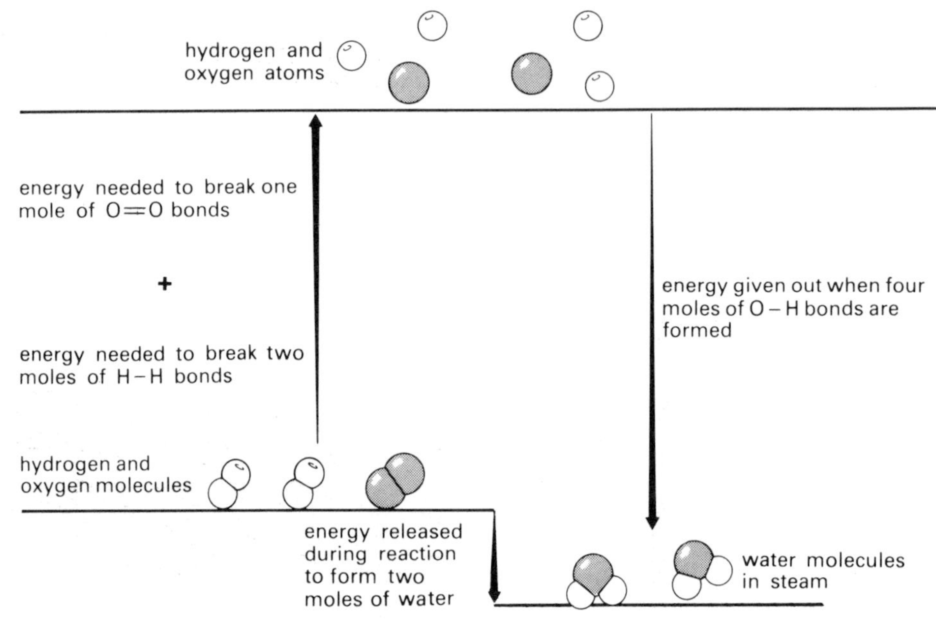

Figure 37.13 Energy level diagram for the reaction of hydrogen and oxygen molecules to give steam

Question

Hydrogen burns in chlorine:

$$H_2(g) + Cl_2(g) \longrightarrow 2HCl(g)$$

(a) What bonds are broken in the reaction?
(b) What bonds are made?
(c) Using the information in figure 37.14, find out whether heat is absorbed or released in the overall reaction.

Process	**Energy**
Breaking 1 mol of H—H bonds	436 kJ/mol needed
Breaking 1 mol of Cl—Cl bonds	242 kJ/mol needed
Making 1 mol of H—Cl bonds	431 kJ/mol released

Figure 37.14

(d) How much energy is absorbed or released in the reaction?
(e) Draw an energy level diagram for this reaction, like the one in figure 37.13.

Question

Put the energy resources shown in figure 37.15 into two lists: (a) continuous energy sources and (b) limited, irreplaceable energy resources.

Energy changes in chemical and physical processes result from bond making and bond breaking. The 'bonds' may be covalent bonds between atoms in molecules, attractions between ions and atoms in giant structures, or the smaller forces which hold molecules together.

37.6 Alternative energy sources

Figure 37.15 shows the energy resources available for us to use. Some, like the energy from the sun, are so vast that they provide a continuous supply of energy. Other resources, like coal, which is contained in the earth's crust and was formed over millions of years, are limited. Once they have been used, they cannot be replaced.

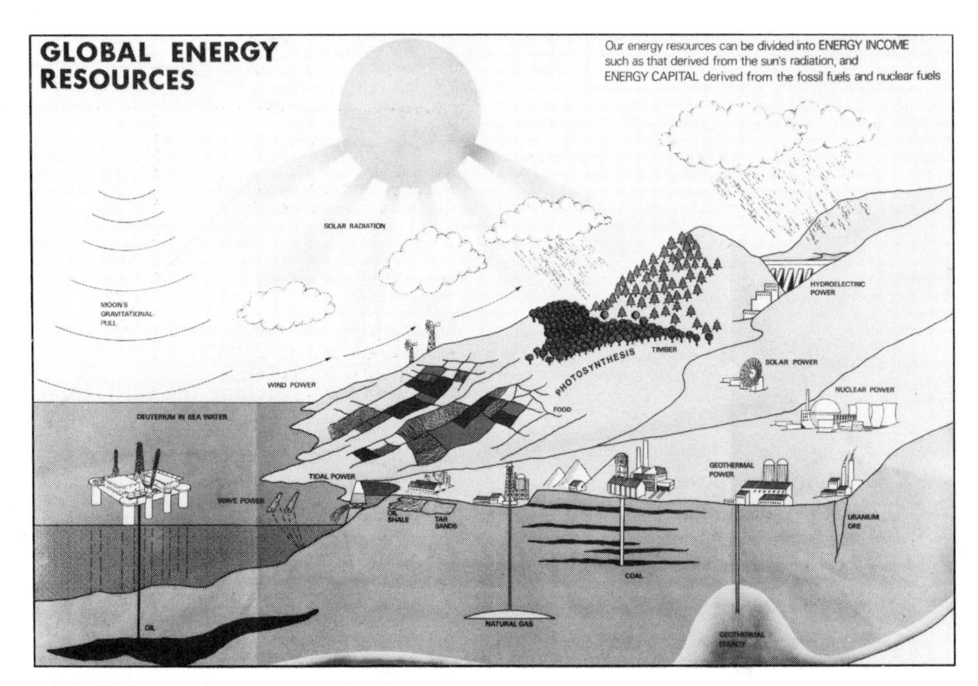

Figure 37.15

The world's demand for energy has been rising rapidly and only in the last few years has the rate of increase slowed (see figure 37.16). The demand has mainly been met by using irreplaceable resources. Of the energy used in 1981, over 91 per cent came from the fossil fuels, coal, gas and oil. We have become dependent on these fuels for energy and for new chemicals, and the reserves are running out.

It is difficult to know exactly how long fossil fuel reserves will last. New reserves may be discovered, and future energy demand is uncertain. At the present rate, coal could last for at least 150–200 years, but oil and gas wells will run dry much sooner. Oil reserves could be used up in as little as 30 years (see figure 1.10 on p. 6).

Where will our energy of the future come from? Coal will become more important as a fuel because its reserves will last longer. There are projects to make liquid fuels from coal, as well as from oil sands and shale. But even coal will not last for ever.

Already, nuclear energy is making a small contribution to the energy supply, and it could become an important source of electricity in the future. However, uranium is another non-renewable resource. What are the social problems involved in a massive expansion of the nuclear energy industry?

We could make more use of the continuous energy sources. Energy from winds, tides, the sun and moving water, and from heat stored in the ground, could all help our fossil fuels last longer.

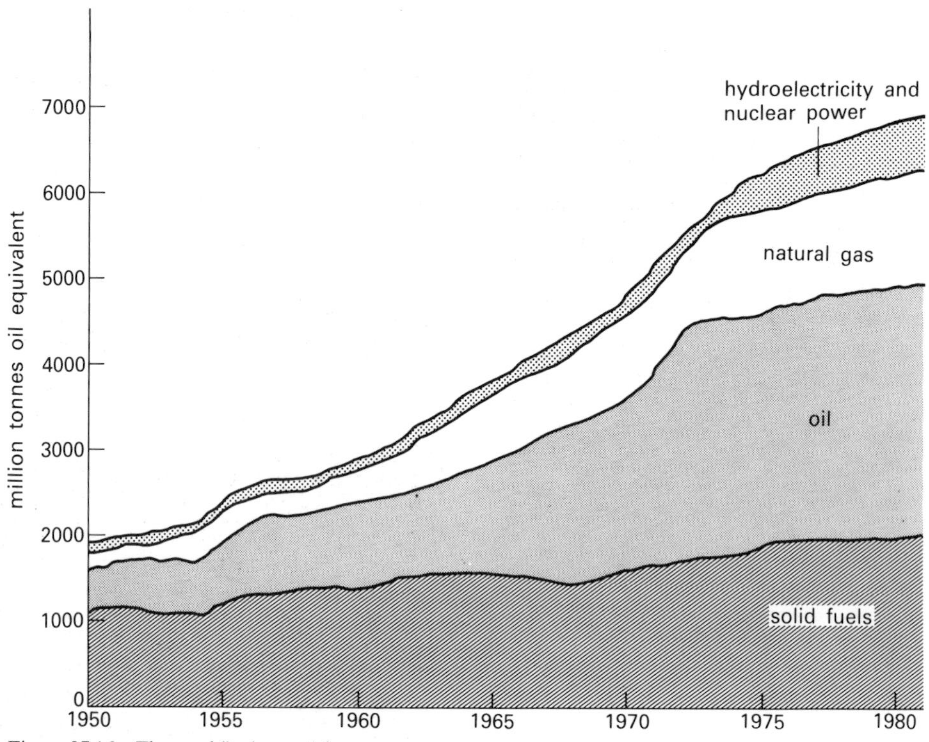

Figure 37.16 The world's demand for energy

The energy reaching the earth from the sun is thousands of times greater than the world's total energy demand. If only part of it could be trapped and used! The heat from the sun evaporates water. It falls as rain on high ground. Its potential energy can be changed into electricity in *hydroelectric* power stations. In the UK, less than 0.7 per cent of our energy comes from hydroelectric power. How does a hydroelectric power station work?

Plants collect and store energy from the sun. Rotting vegetable waste is a source of energy. It releases methane which can be used as a fuel. Some brick kilns in Bedfordshire are fired by methane piped from rotting waste in a disused clay pit. Animal waste, too, is an energy source. A farmer can milk his cows using milking machines which are powered by the animals' own waste! In many sewage treatment works, methane formed in the 'sludge digester' is used to provide electricity for the works.

Figure 37.17 A sludge digester at Rye Meads sewage purification works, Essex

Figure 37.18 Checking the solar panels on two satellites before they are carried into space by the space shuttle *Columbia*

Very little use is made of solar energy directly. Solar cells convert the sun's energy into electricity for satellites, spacecraft and electronic equipment in remote parts of the world. These are small uses. We need more efficient ways of converting solar radiation into usable energy.

Solar panels on houses, offices and swimming pools can reduce the need for fossil fuels. The solar energy is used for heating water. Why isn't solar heating more common in the UK? Why can the use of solar panels only supplement conventional energy sources, not replace them?

Figure 37.19 Over a whole year, the solar panels on this house in Switzerland provide 70% of the energy needed for heating and hot water

Yet another idea is to make a miniature 'sun' on earth. The sun gets its energy by the *fusion* (joining together) of light atoms. At very, very high temperatures, the nuclei of these atoms will join to form heavier ones. This

Figure 37.20 Diagram of the JET experiment, showing the toroidal vacuum chamber

Figure 37.21 The Rance tidal dam in France

process can release an enormous amount of energy. One possibility is to fuse two **isotopes** of hydrogen, $^{2}_{1}H$ and $^{3}_{1}H$ (see chapter 39) to form helium. The problem is to get the gas hot enough. The required temperatures are so high – over 100 million degrees – that any container would be vaporized. However, at these high temperatures the light atoms are ionized, and it should be possible to trap the ions in an enclosed magnetic field. A major European project to develop a fusion reactor is based in England. The project is called JET, which stands for Joint European Torus. As shown in figure 37.20, the magnetic field is shaped like a doughnut with a hole in it. This shape is called a *torus*. A successful fusion reactor could produce limitless amounts of energy. However, even if all the engineering and technological problems can be solved, it will be well into the next century before a commercial fusion power station can be built.

Figure 37.21 shows the tidal power station across the estuary of the Rance, in France. As the tide rises and falls, the water flows through turbines, generating electricity. It saves about 400 000 tonnes of coal each year. There were plans for a tidal energy scheme across the Severn estuary, but it may never be built because of the cost. There are many experimental designs for devices to get energy from the tides.

Using wind energy is not a new idea, but there is now a new generation of windmills to convert wind energy into electrical energy. Fifty or a hundred of these together might replace a small conventional power station and supply local needs. However, harnessing wind energy is expensive and unreliable.

Figure 37.22 Artist's impression of the giant windmill (60 m high) that is to be constructed on Orkney

Questions

1 Energy is wasted in our homes. It is lost through the windows, walls and roof, and through gaps around doors, unused fireplaces, etc. How can these losses be reduced?
2 Energy waste occurs when fossil fuels are used to make electricity.
(a) List ways in which electricity is used in your home.
(b) Alongside each use, give ways in which you could save electricity in this use.
3 What kind of heating do you use at home? How could energy economies be made?
4 Why do the following cooking hints save energy?
(a) Cut vegetables small for cooking.
(b) Do not overfill electric kettles.
(c) Descale kettles regularly.
(d) Use a steamer containing a food to be cooked over another pan.
Add more 'cooking energy economies' to this list.
5 Do you think that coal, gas and oil should be made more expensive to prevent us wasting them? Give some arguments for and against this idea.

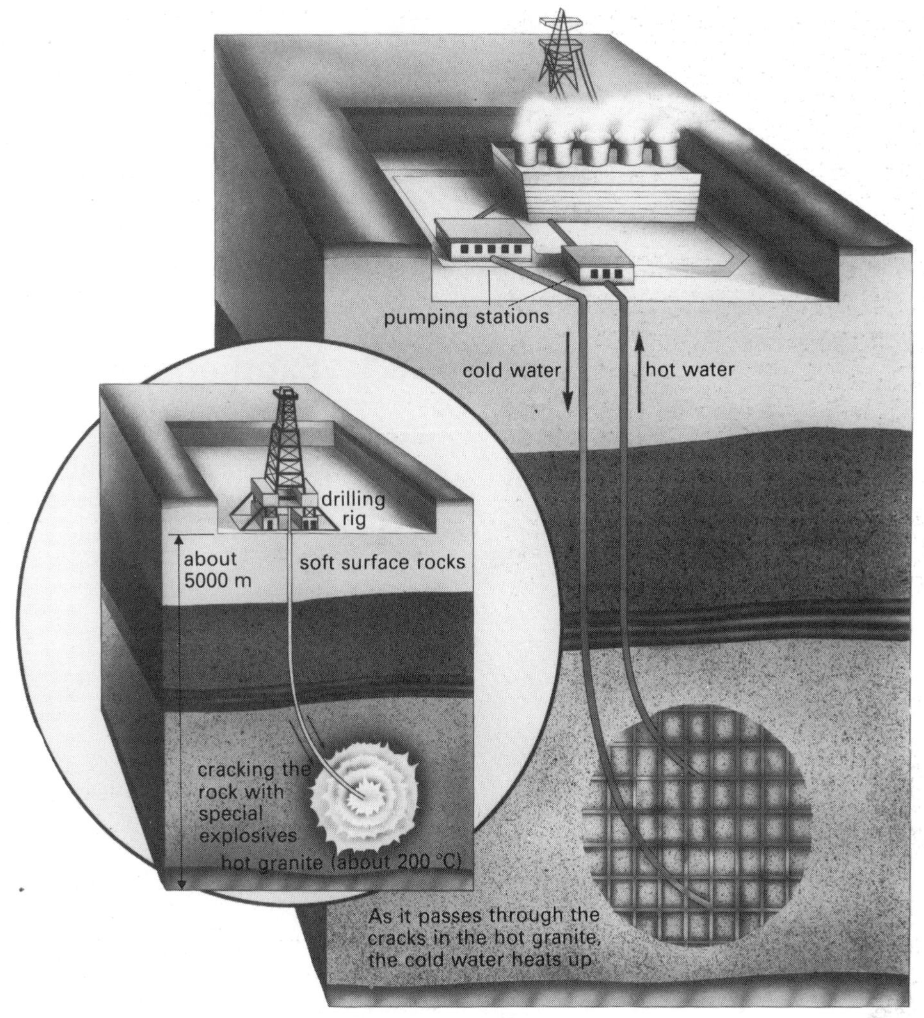

Figure 37.23 Geothermal energy: the hot granite deep below the surface is cracked to increase the surface area available for heating water

Under the surface of the earth, the rock is hot. The deeper we go, the hotter it becomes. This *geothermal* energy comes from the molten rock underneath the crust. In Britain there are experiments to use geothermal energy. At Rosemanowes in Cornwall, there is a plan to drill 5 km into the granite crust, where the temperature is 200 °C. Geothermal energy has almost no limit. It could be important in the future.

It seems unlikely that the 'new' energy sources can make a large contribution to our energy supply – certainly not during this century. Development is expensive, and no one source could solve our energy problem. However, they could help conserve our fossil fuel reserves.

Summary

Answer the following questions to compile a summary of this chapter.

1 What is a fuel?
2 What is a fossil fuel?
3 What are the stages in making electricity?
4 What are the main uses of coal?
5 What are the characteristics of a good fuel for domestic heating?
6 What is the *heat of combustion* of a fuel?
7 Why is energy released in all combustion reactions?
8 What other energy sources are there as alternatives to fossil fuels?

38 Energy changes in chemical reactions

38.1 Exothermic reactions

Exothermic reactions give out heat!

The reaction of magnesium with dilute hydrochloric acid shown in figure 38.1(a) is exothermic. The reaction tube feels warm because the reaction gives out heat energy to the water, the tube and your hand.

Figure 38.1 Exothermic reactions

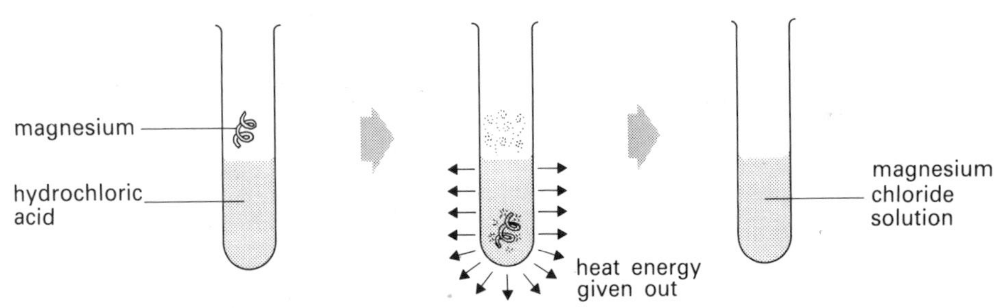

(a) An example of an exothermic reaction

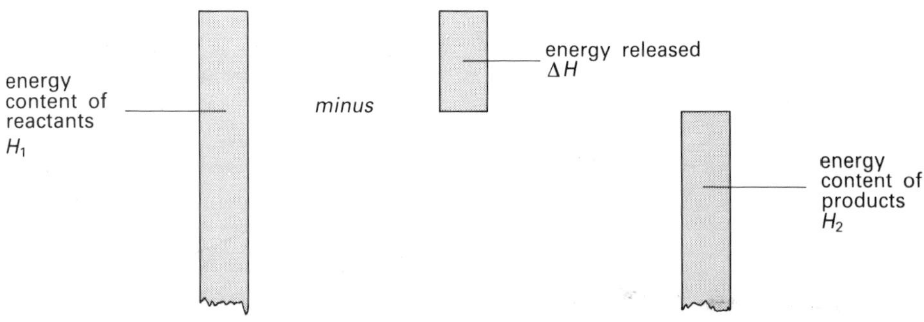

(b) The energy changes involved

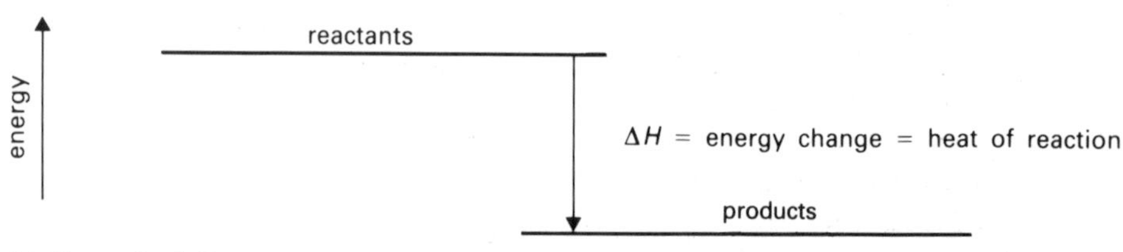

(c) Energy level diagram

The energy released by the amounts of reactants shown in an equation is called the **heat of reaction**.

Chemical energy can be released as heat. In figure 38.1(b), the reactants have an energy content, H_1, and the products have an energy content, H_2. During the reaction, some energy is given out by the reactants as they react. So the energy content of the products, H_2, is lower than that of the reactants, H_1. The heat of reaction is the difference between the energies of the reactants and products. It can be written as 'ΔH'. 'ΔH' means 'the change in energy content'. It must have a sign

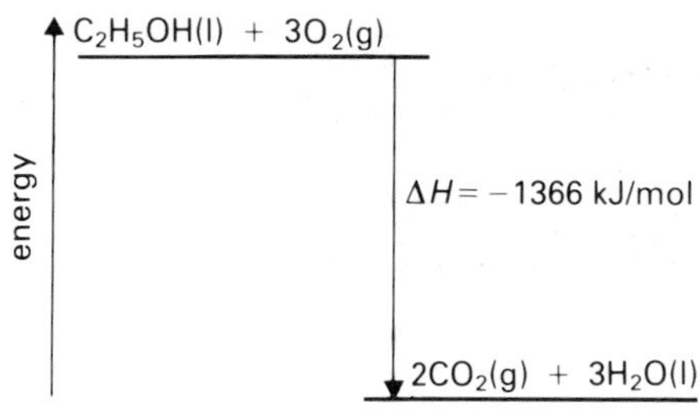

Figure 38.2

to show whether the energy content has increased or decreased. For an exothermic reaction, ΔH is negative. The reactants *lose* energy to give the products.

Energy level diagrams like the one in figure 38.1(c) can be used to show the energy changes.

All **combustion** reactions are exothermic. The *heat of combustion* of ethanol is 1366 kJ/mol. This information can be included in the equation:

$$C_2H_5OH(l) + 3O_2(g) \longrightarrow 2CO_2(g) + 3H_2O(l); \quad \Delta H = -1366\ \text{kJ/mol}$$

heat energy given out

The energy change is for the amounts shown in the equation. Figure 38.2 shows the energy level diagram for this reaction.

38.2 Endothermic reactions

Endothermic reactions take in heat!

Dissolving ammonium nitrate in water, as shown in figure 38.3(a), is endothermic. The tube feels cold because the process takes in energy from the water and your hand.

Figure 38.3 Endothermic reactions

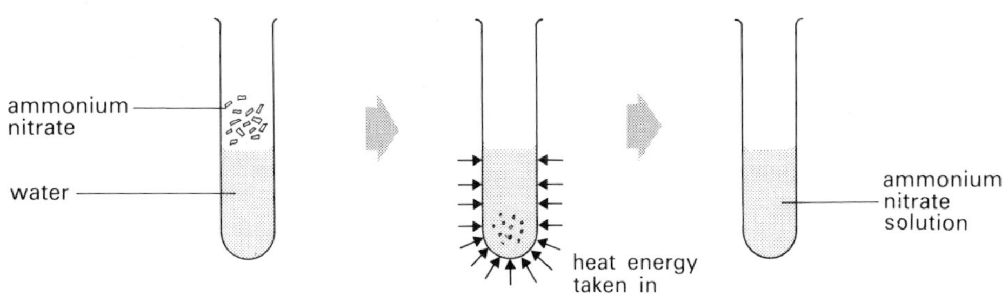

(a) An example of an endothermic reaction

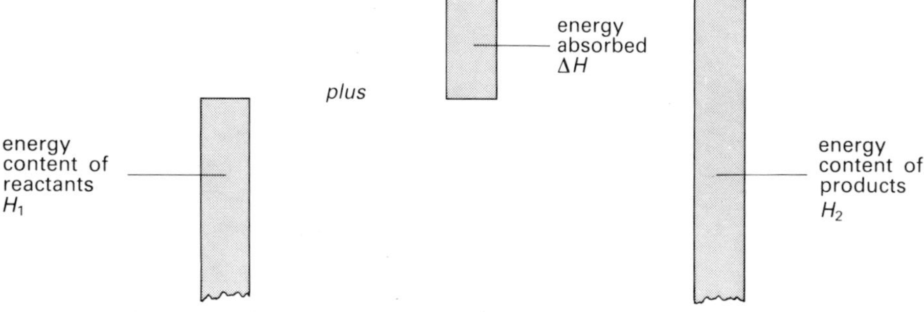

(b) The energy changes involved

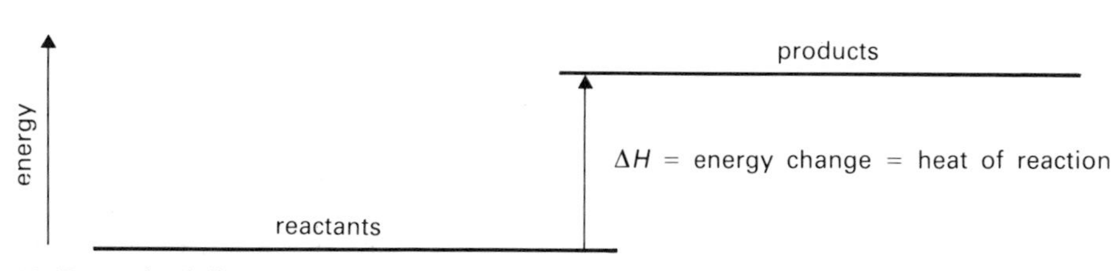

(c) Energy level diagram

The energy taken in when the amounts of reactants shown in an equation are used is the **heat of reaction**.

In figure 38.3(b), energy is taken in by the reactants, so the energy content of the products, H_2, is greater than that of the reactants, H_1. ΔH is positive. In figure 38.3(c), an energy level diagram shows the energy changes.

Questions

1 6.0 kJ of energy is needed to convert 1 mol of ice at 0 °C into water at 0 °C.
(a) Write a complete equation that gives this information.
(b) Draw an energy level diagram for the process.

2 When 1 mol of magnesium reacts with dilute hydrochloric acid, 463 kJ of energy is released.
(a) Write a complete equation that shows this information.
(b) Draw an energy level diagram for the reaction.

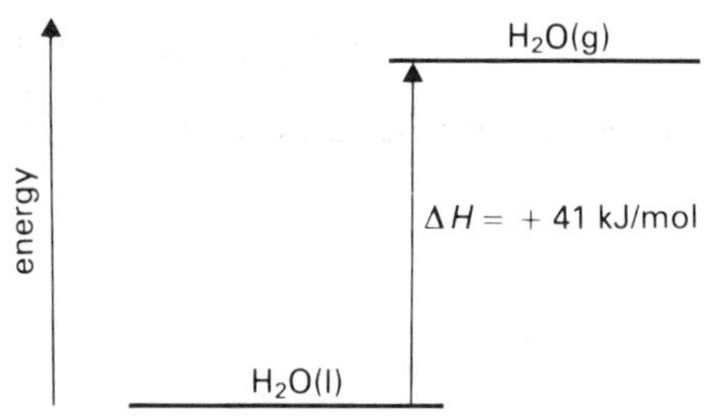

Figure 38.4

All **vaporization** and **melting** processes are endothermic. The *heat of vaporization* of water is 41 kJ/mol.

$$H_2O(l) \longrightarrow H_2O(g); \quad \Delta H = +41\,\text{kJ/mol}$$

heat energy taken in

The energy level diagram is shown in figure 38.4. 1 mol of steam at 100 °C contains 41 kJ more energy than 1 mol of water at 100 °C.

38.3 Measuring heats of reaction in solution

Heats of *combustion* reactions are measured by allowing the heat from the reaction to be absorbed by water, as shown in figure 37.8 on p. 281.

Heats of reaction *in solution* are measured by allowing the heat from the reaction to be absorbed by or taken from the solution in an insulated container (see figure 38.5). In exothermic reactions, the heat given out is absorbed by the solution. The temperature rises. The insulation reduces the amount of heat energy 'lost' to the air. In endothermic reactions, the heat is taken from the solution. The temperature falls. Insulation reduces the amount of heat energy absorbed from the air. The temperature change is used to calculate the heat of reaction.

If solutions are dilute, their specific heat capacities and densities are about the same as those of water. The density of water is about 1 g/cm^3, so the mass of a solution has about the same numerical value as its volume. The specific heat capacity is about 4.2 J/(g °C).

A useful equation is

*Heat transferred (J) = Mass of solution (g) × Temperature change (°C) × 4.2 J/(g °C)

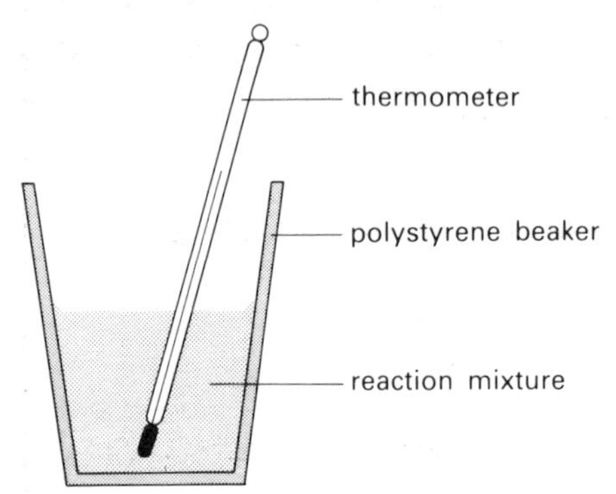

Figure 38.5 Measuring a heat of reaction in solution

Experiment 38a
The heat of a displacement reaction

copper(II) sulphate

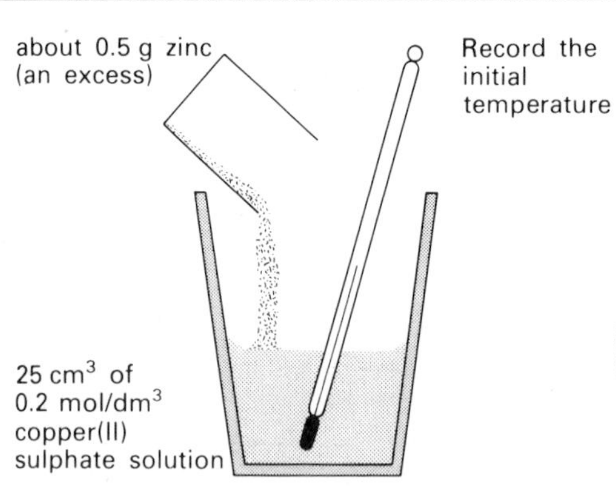

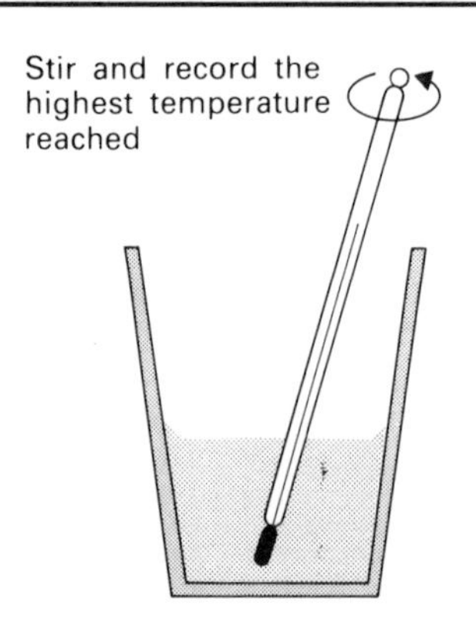

Figure 38.6

Zinc displaces copper from copper sulphate solution:

$$Zn(s) + CuSO_4(aq) \longrightarrow Cu(s) + ZnSO_4(aq)$$

or

$$Zn(s) + Cu^{2+}(aq) \longrightarrow Cu(s) + Zn^{2+}(aq)$$

The procedure for the experiment is shown in figure 38.6.

Results
Typical results are:

Initial temperature of solution = 19 °C
Final temperature of solution = 28 °C
Rise in temperature = 9 °C

Discussion

Is the reaction exothermic or endothermic?

The blue colour of copper sulphate disappears during the reaction. Why? Excess zinc is used. Why?

The energy absorbed by the solution is

$$25\,g \times 9\,°C \times 4.2\,J/(g\,°C) = 945\,J$$

The number of moles of copper ions used is

$$\frac{25}{1000}\,dm^3 \times 0.2\,mol/dm^3 = 0.005\,mol$$

Using 0.005 mol of copper ions, 945 J of heat energy is given out. Using one mole of copper ions (the amount shown in the equation), the heat of reaction is

$$\frac{945\,J}{0.005\,mol} \times 1\,mol = 189\,000\,J$$

$$= 189\,kJ$$

The reaction is exothermic, so

$$\Delta H = -189\,kJ/mol$$

Write an equation for the reaction, showing the heat of reaction. Draw an energy level diagram.

The data book value for the heat of reaction is 216 kJ/mol.

$$\Delta H = -216\,kJ/mol$$

Why is the experimental result low? How could the experiment be improved?

Figure 38.7 shows the *heats of neutralization* when different acids and alkalis react together. Despite the fact that the equations look different, the heats of neutralization are all very similar. This suggests that the same reaction is taking place in each mixture.

Reaction	**ΔH (kJ/mol)**
$HCl(aq) + NaOH(aq) \longrightarrow NaCl(aq) + H_2O(l)$	−57.9
$HCl(aq) + KOH(aq) \longrightarrow KCl(aq) + H_2O(l)$	−57.8
$HNO_3(aq) + NaOH(aq) \longrightarrow NaNO_3(aq) + H_2O(l)$	−57.6
$HNO_3(aq) + KOH(aq) \longrightarrow KNO_3(aq) + H_2O(l)$	−57.7
$HBr(aq) + NaOH(aq) \longrightarrow NaBr(aq) + H_2O(l)$	−57.6

Figure 38.7

If all acid solutions are ionized and contain hydrogen ions, $H^+(aq)$, and all alkaline solutions contain hydroxide ions, $OH^-(aq)$, the reaction taking place in each mixture is:

$$H^+(aq) + OH^-(aq) \longrightarrow H_2O(l); \qquad \Delta H \approx -57\,kJ/mol$$

The heat of neutralization is the energy change when one mole of aqueous hydrogen ions reacts with alkali to give one mole of water.

What energy change would you predict for the neutralization of sulphuric acid with sodium hydroxide?

$$H_2SO_4(aq) + 2NaOH(aq) \longrightarrow Na_2SO_4(aq) + 2H_2O(l)$$

Questions

1 Write instructions or draw a series of diagrams for an experiment to find the heat of a neutralization reaction.

2 When 4 g of ammonium nitrate dissolves in 50 cm^3 of water, the temperature falls by 6 °C.
(a) Is the reaction exothermic or endothermic?
(b) How much energy is absorbed from the water?
(c) How many moles of ammonium nitrate are used? (The mass of 1 mol of ammonium nitrate is 80 g.)
(d) What is the heat of solution per mole of ammonium nitrate?
(e) Draw an energy level diagram for the reaction

$$NH_4NO_3(s) + Water \longrightarrow NH_4NO_3(aq)$$

3 When 0.01 mol of magnesium reacts with 100 cm^3 of 1 mol/dm^3 hydrochloric acid, the temperature rises by 11 °C.
(a) Is the reaction exothermic or endothermic?
(b) How much energy is absorbed by the solution?
(c) What is the heat of reaction per mole of magnesium?
(d) Write an equation for the reaction, showing the energy change.
(e) Draw an energy level diagram.

Summary

1 Copy the following sentences, choosing the correct words from the pairs in italics.
(a) Exothermic reactions *release/absorb* energy *to/from* the surroundings. The reaction mixture becomes *hot/cold.* ΔH, the heat of reaction, is *positive/negative.*
(b) Endothermic reactions *release/absorb* energy *to/from* the surroundings. The reaction mixture becomes *hot/cold.* ΔH, the heat of reaction, is *positive/negative.*

2 Write down the two headings, 'exothermic' and 'endothermic'. Under the correct headings, list examples of reactions or processes you have met, giving equations. You should include examples of displacement, neutralization, combustion and vaporization.

39 Nuclear energy

Question

Look back to chapter 17.
(a) What are the particles in the nucleus?
(b) What mass and charge do they have?
(c) What is the **atomic number** of an atom?
(d) What is the **mass number**?
(e) How much do the electrons contribute to the mass of an atom?

39.1 Radioactivity

A chance discovery in 1896 led to the discovery of radioactivity. Some wrapped photographic plates that had been stored near a uranium salt were black when they were developed, as if they had been exposed to light. A. H. Becquerel suggested that some kind of radiation from the salt had penetrated the wrapping. More experiments showed that the air around the salt became ionized so the conductivity of the air increased. Marie and Pierre Curie called the effect *radioactivity*, and found that all uranium and thorium compounds are radioactive. Two new elements that were extracted from pitchblende – polonium and radium – proved to be even more radioactive than uranium.

These observations were important for two reasons: they played a part in the developing story of the structure of the atom and they opened up a new area of chemistry – nuclear chemistry. Nuclear chemistry has given us useful materials for industry and for the diagnosis and treatment of disease, and energy for generating electricity. It has also given us energy for destruction.

A simple picture of the structure of the atom is described in chapter 17.

Dalton's atomic theory, outlined in chapter 3, included the idea that all atoms of the same element are alike. We now know that this is not true. In 1919 Aston showed that an element can have atoms with different masses. They are called **isotopes**.

Most naturally-occurring elements have several isotopes – usually two or three, although tin, for example, has ten. Very many more isotopes can be made artificially.

Hydrogen has three isotopes, with masses 1, 2 and 3. They all contain the same number of protons and electrons, so they have the same chemical properties. The different masses are caused by different numbers of neutrons. Naturally-occurring hydrogen contains 99.9 per cent of hydrogen-1.

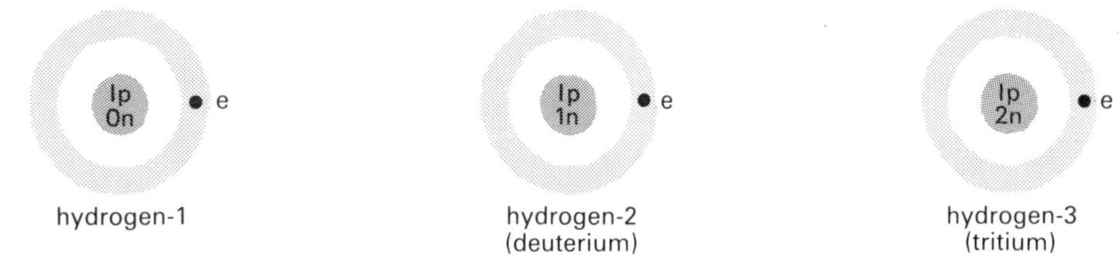

Figure 39.1 Different kinds of hydrogen atoms: protons, neutrons and electrons are shown by p, n and •

Isotopes are rather like chocolate liqueurs – they are all the same on the outside but they can have different centres. The electron structures are the same but the nuclei are different.

The composition of an isotope can be shown in a shorthand way, using the element symbol:

$$\text{mass number}-{}^{A}_{Z}\text{X}-\text{symbol of element}$$
(atomic number — Z)

So hydrogen-1 can be written as ${}^{1}_{1}H$, hydrogen-2 as ${}^{2}_{1}H$ and hydrogen-3 as ${}^{3}_{1}H$.

$^{35}_{17}Cl$, or chlorine-35 $^{37}_{17}Cl$, or chlorine-37

Figure 39.2 The two natural isotopes of chlorine

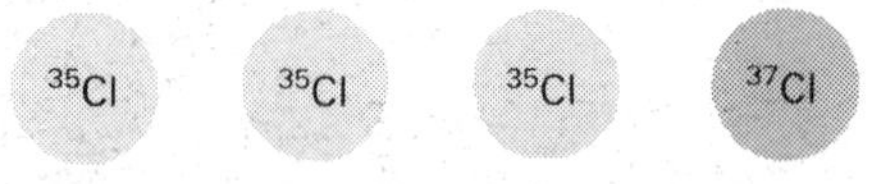

^{35}Cl ^{35}Cl ^{35}Cl ^{37}Cl

Figure 39.3 In naturally-occurring chlorine, there are three atoms of ^{35}Cl for every one of ^{37}Cl

Have you wondered why the relative atomic mass of chlorine given in table 1 on p. 336 is 35.5? Chlorine has two isotopes: chlorine-35, $^{35}_{17}Cl$, and chlorine-37, $^{37}_{17}Cl$. Naturally-occurring chlorine contains 75 per cent of chlorine-35 and 25 per cent of chlorine-37. On average, out of every four chlorine atoms, three are chlorine-35 and one is chlorine-37. The average relative atomic mass is

$$\frac{(3 \times 35) + 37}{4} \quad \text{or} \quad \frac{(75 \times 35) + (25 \times 37)}{100} = 35.5$$

Most relative atomic masses are not whole numbers because most elements have several naturally-occurring isotopes.

Questions

1 Copy and complete the table in figure 39.4.

Isotope	Number of protons	Number of neutrons	Number of electrons
$^{1}_{1}H$			
$^{4}_{2}He$			
$^{12}_{6}C$			
$^{212}_{82}Pb$			
$^{238}_{92}U$			

Figure 39.4

2 Silicon (atomic number 14) has three naturally occurring isotopes, with masses 28, 29 and 30.
(a) Write the symbols for the three isotopes.
(b) For each isotope, work out the numbers of protons and neutrons.
(c) Natural silicon contains 93% of silicon-28, 5% of silicon-29 and 2% of silicon-30. What is the approximate average atomic mass?
3 Oxygen has three isotopes: $^{16}_{8}O$, $^{17}_{8}O$ and $^{18}_{8}O$. How many different kinds of oxygen molecules, O_2, are possible?

If a nucleus is unstable it is radioactive. Unstable nuclei are particularly common in heavy elements like uranium, radium and thorium. In the atoms of these elements, the number of neutrons is much greater than the number of protons. Some light elements, including potassium, also have very small amounts of naturally-occurring unstable isotopes. Most artificial isotopes are unstable.

Radioactive isotopes eject particles from their nuclei to make them more stable. Two types of particle can be emitted – **alpha (α) particles** and **beta (β) particles**. They are often accompanied by energy in the form of gamma (γ) radiation. Each time a particle is ejected, disintegration of an atom has occurred.

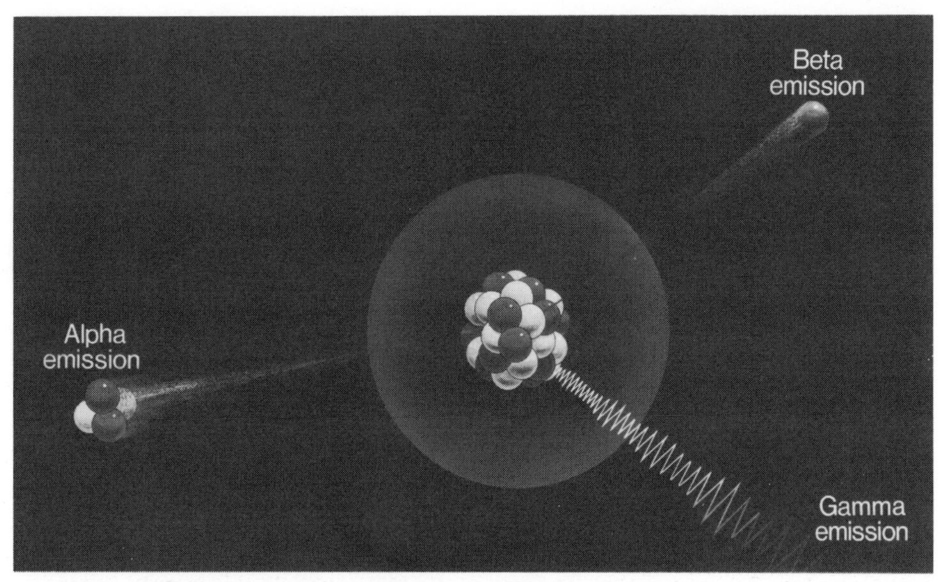

Figure 39.5 The particles that can be emitted by a radioactive atom as it decays

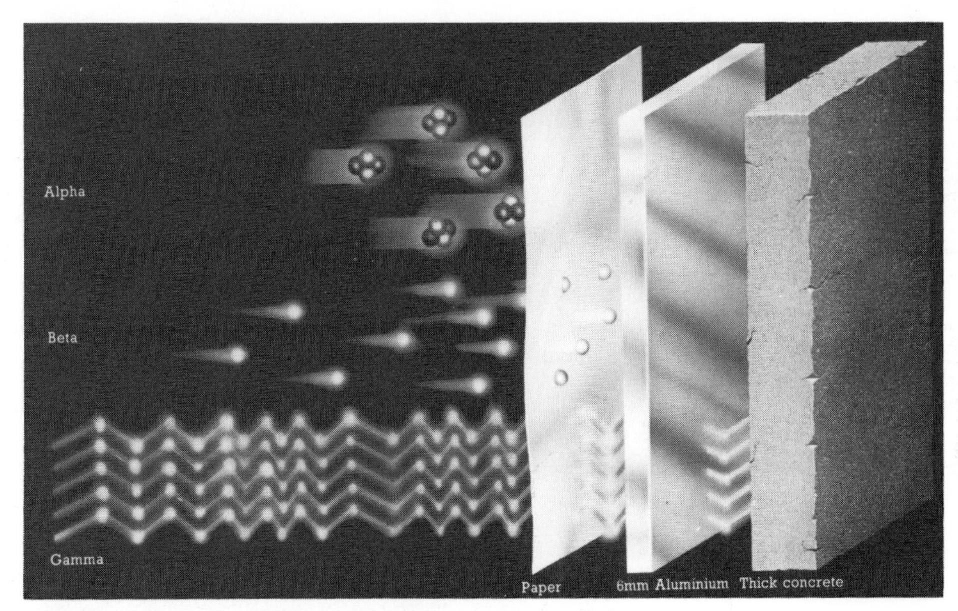

Figure 39.6 Penetrating powers of alpha, beta and gamma radiation

The penetrating powers of the different types of radiation are shown in figure 39.6. In air, alpha particles can travel only a few centimetres before they are stopped. Beta particles may travel several metres.

Unlike a 'conventional' chemical reaction, when a nuclear change takes place, the atom changes to one of a different element.

An alpha particle consists of two neutrons and two protons, tightly bound together. It is the same as a helium atom, but without the electrons, i.e. ${}^{4}_{2}He^{2+}$ When an atom ejects an alpha particle, its nucleus changes. The **mass number** decreases by 4 and the **atomic number** by 2. A new element is formed. For example

$${}^{232}_{90}Th \longrightarrow {}^{228}_{88}Ra + \alpha$$

thorium radium

A beta particle is a fast-moving electron. It is not an electron from outside the nucleus, but one formed when a neutron changes into a proton within the nucleus:

$$n \longrightarrow p^{+} + e^{-}$$

Loss of a beta particle causes no change in the mass number, but the atomic number increases by 1 because a proton has replaced a neutron. Again, a new element is formed. For example

$$^{228}_{88}\text{Ra} \longrightarrow {}^{228}_{89}\text{Ac} + \beta$$

radium actinium

As a radioactive element decays away, a new (daughter) element is formed. If the daughter product is radioactive, this also decays to give a further (granddaughter) product. Heavy elements decay through a series of steps until a stable isotope (usually of lead) is reached. Figure 39.7 shows a decay series which starts with thorium-232.

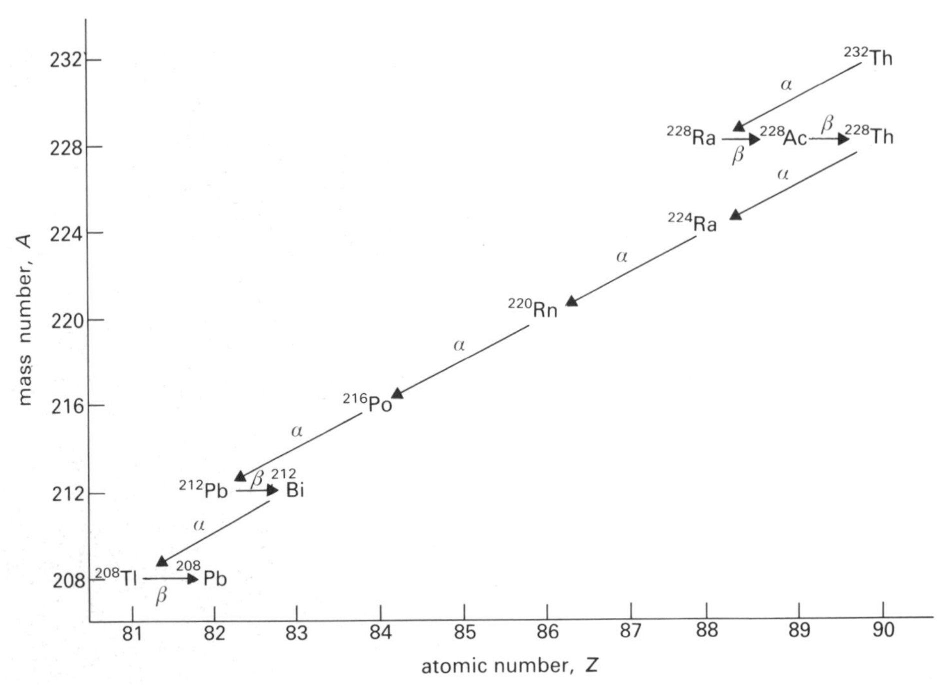

Figure 39.7 The decay series of thorium-232. Find out the names of the elements in it

Questions

1 The following isotopes decay by ejecting an alpha particle: $^{238}_{92}$U, $^{226}_{88}$Ra, $^{228}_{90}$Th, $^{218}_{84}$Po. Write equations to show the products formed. Find the name of each element.

2 The following isotopes decay by ejecting a beta particle: $^{212}_{82}$Pb, $^{225}_{88}$Ra, $^{213}_{83}$Bi, $^{234}_{91}$Pa. Write equations to show the products formed. Find the name of each element.

3 Trace the part of the decay series in figure 39.8, which starts with $^{238}_{92}$U. The first five steps are shown.

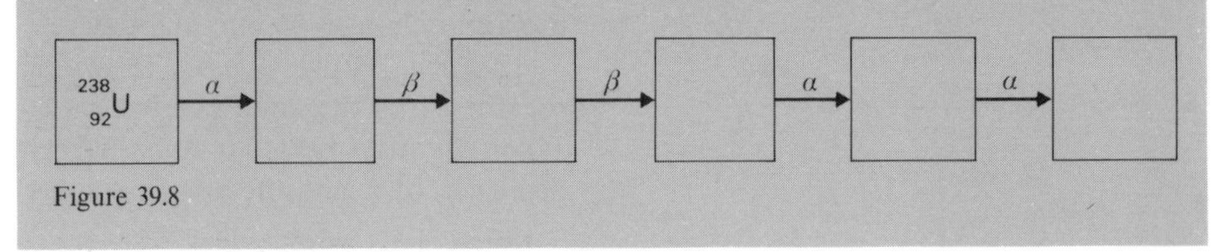

Figure 39.8

The particles emitted by radioactive isotopes are counted using a Geiger–Müller tube connected to a scaler.

In experiment 39a, the rate of decay of protactinium-234 is investigated. Protactinium-234 is a granddaughter product of uranium-238. A solution of uranyl nitrate contains this isotope along with isotopes of thorium and uranium. Use the decay series in figure 39.8 to predict what other isotopes might be present.

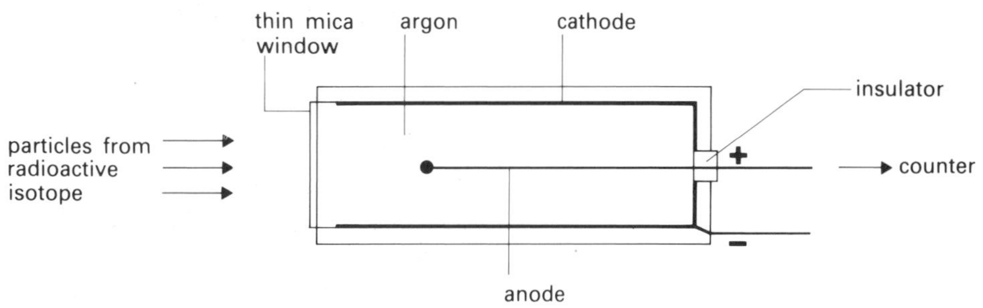

Figure 39.9 A Geiger–Müller tube: the counter detects pulses of electricity as particles ionize the argon in the tube

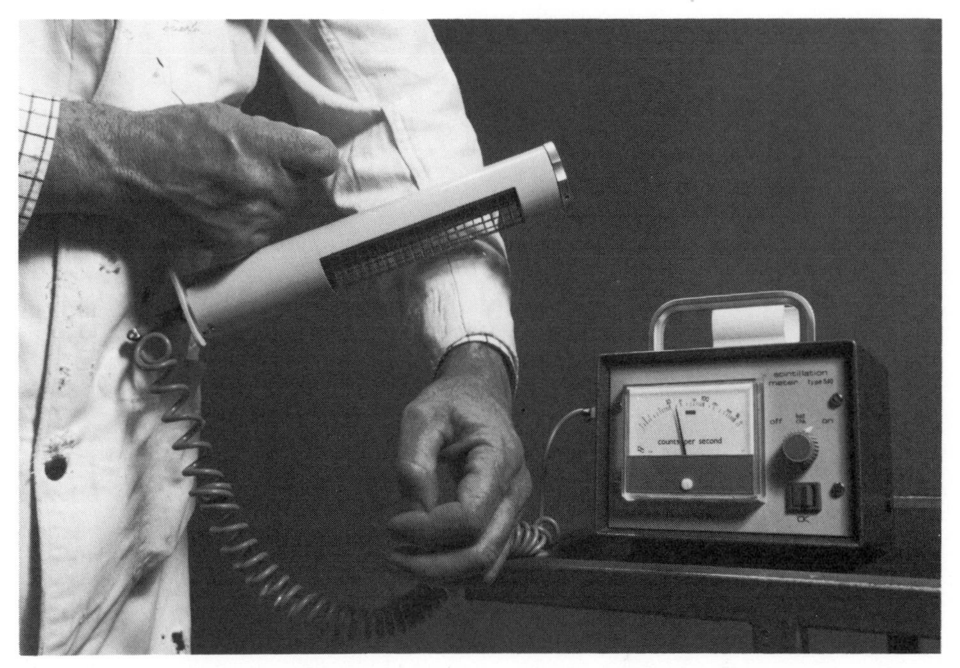

Figure 39.10 A worker using a Geiger–Müller tube to check that his clothes have not been contaminated by radioactive materials

Protactinium chloride is extracted from uranyl nitrate solution using an organic solvent.

Experiment 39a
The decay of protactinium-234

uranyl nitrate solution

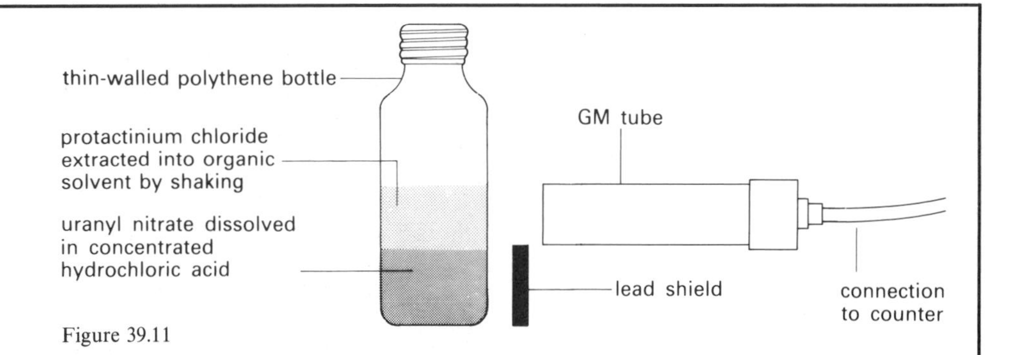

Figure 39.11

The apparatus for this experiment is shown in figure 39.11. The bottle is shaken to extract the protactinium into the organic layer. The count rate is measured at regular intervals. The background count is recorded and subtracted from the experimental readings.

Results

Figure 39.12 shows a graph of the count rate (corrected for the background radiation) against the time at the start of the 10 s count periods. Find the time

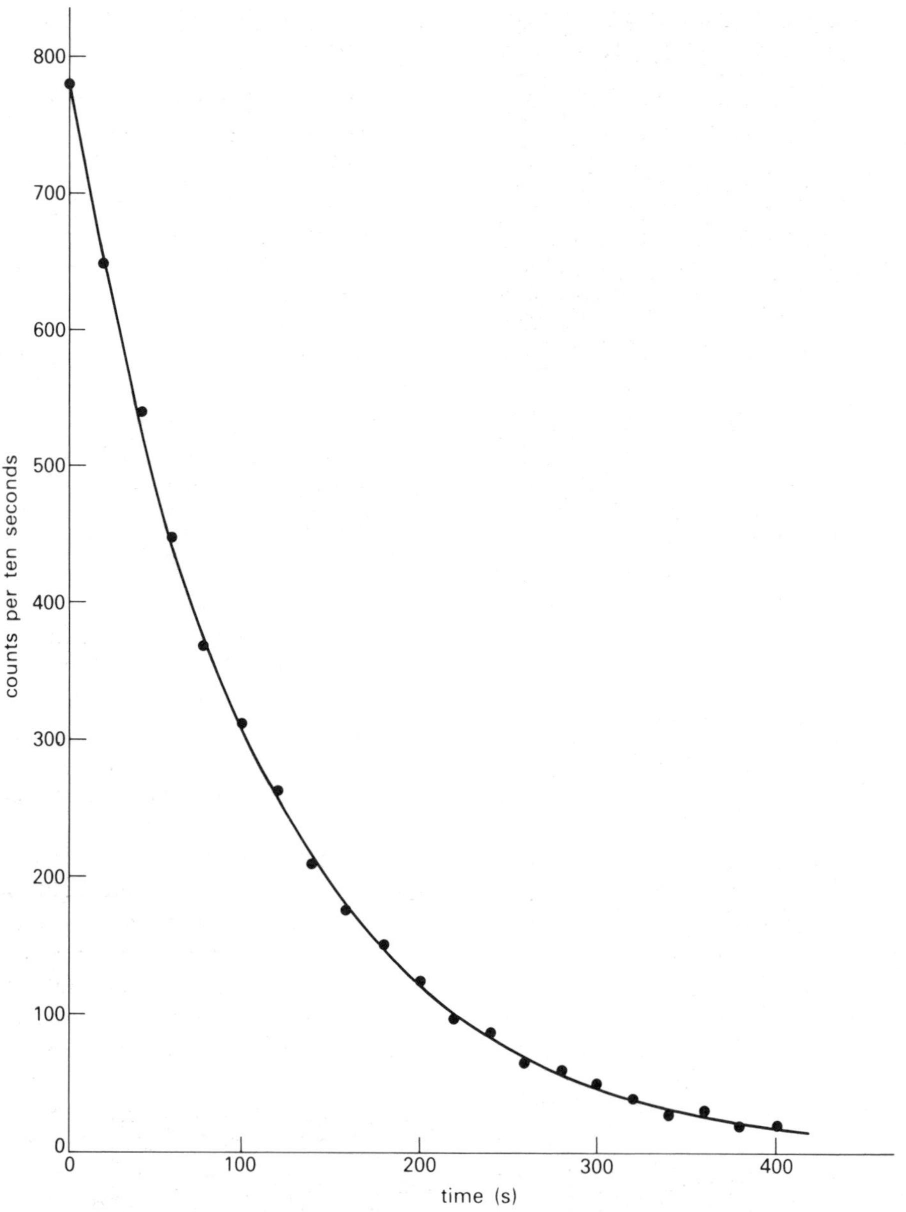

Figure 39.12

for the count rate to fall to half its starting rate. Find the time for it to fall to half again, then half again. What do you notice about the times? What is the average?

Discussion
Although all isotopes disintegrate at different rates, the pattern of decay is the same. Each isotope has a fixed *half-life*, which is the time for the count rate to fall by half. This is constant for each isotope, no matter what the initial count rate.

The half-life of an isotope is a measure of its rate of decay. It is the time for half of the atoms of the isotope to decay away. Half-lives can be as short as a fraction of a second or as long as millions of years. For a given isotope, they are constant whatever the conditions. They are not affected by changes in temperature or concentration, the presence of a catalyst or in what compound the element is included. This is an important difference between nuclear processes and ordinary chemical reactions.

Figure 39.14 The sample of wood that was used for radiocarbon dating of King Arthur's Table, Winchester. It was found to date from the thirteenth century

Questions

1 Figure 39.13 shows corrected count rates for the decay of iodine-128.
(a) Draw a graph of the results.
(b) Work out the half-life of iodine-128.

Time (minutes)	0	10	20	30	40	50	60	70	80	90
Corrected count rate (counts/second)	120	90	69	54	42	33	25	19	15	13

Figure 39.13

2 The half-life of sodium-24 is 15 hours.
(a) How much is left after 30 hours?
(b) How long will it take for the amount of sodium-24 to fall to $\frac{1}{16}$ of its original amount?

The dating of organic remains like bones and wood uses the fact that the half-life of carbon-14 is about 5700 years. While they are alive, plants and animals absorb carbon dioxide that contains a constant, but very small, amount of carbon-14. It is radioactive. After the plant or animal dies, the radioactive carbon-14 decays away and no more is absorbed to replace it. The age of a specimen is estimated by comparing the amount of carbon-14 present with the amount that would be expected if it were still absorbing carbon dioxide.

Radiation can be hazardous at high levels because it affects the growth of living cells. Figure 39.15 shows some highly active materials being handled in industry.

Despite the hazards, radiation is useful for the treatment of some diseases. Tumours can be destroyed using controlled doses of gamma radiation from cobalt-60. Surgical instruments and syringes are sterilized after they have been packed by exposing them to very high levels of gamma radiation. At these levels of radiation, all micro-organisms are killed.

Figure 39.15 Samples of plutonium being handled in glove boxes for protection against radiation

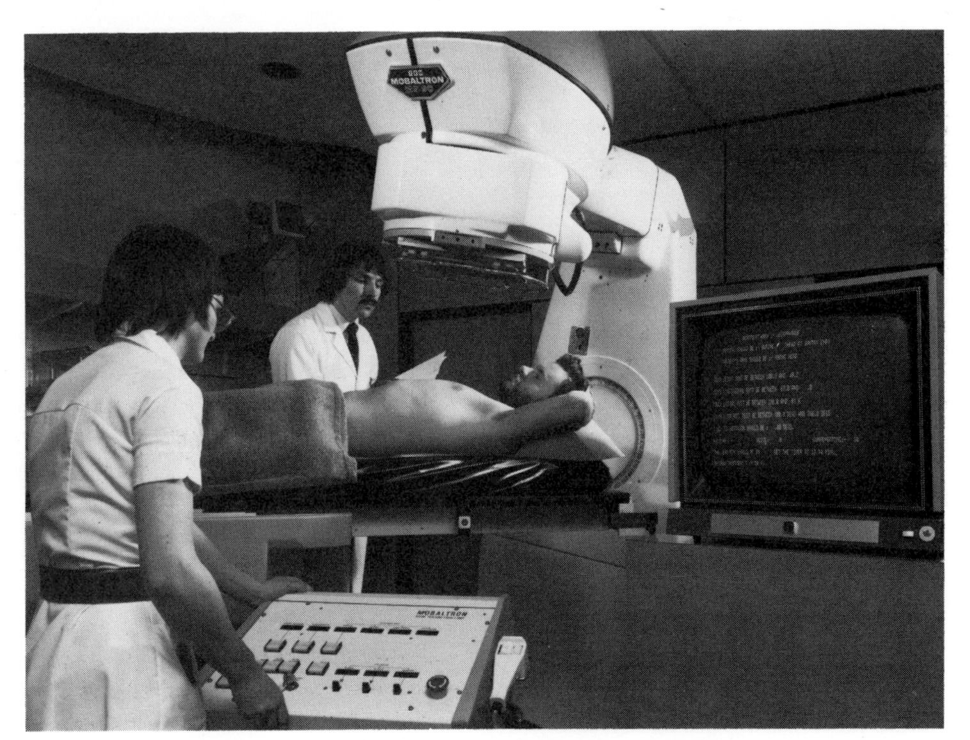

Figure 39.16 Positioning a patient for cobalt-60 radiation therapy

Radioactivity is easily detected, so radioactive atoms are useful as *tracers.* They can be followed in physical, chemical and biological processes without interfering with the processes. Chapter 27 describes how a tracer can be used to demonstrate dynamic equilibrium. Radioactive atoms are introduced with a non-active carrier of the same substance. Tracers are widely used in medical diagnosis and in industry. Two examples are shown in figure 39.17.

Figure 39.17 Radioactive tracers

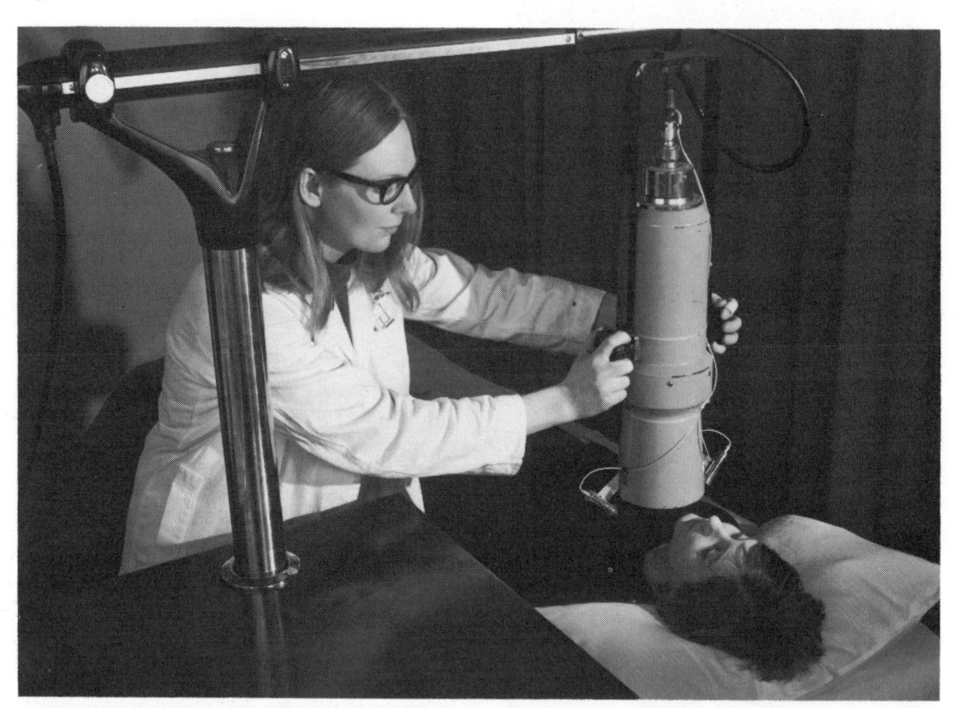

(a) Checking for abnormal activity of the thyroid gland by measuring its uptake of iodine-131

(b) Following the movement of waterborne waste in an estuary using a waterproof detector

Figure 39.18 Using a gauge that measures caesium-137 radiation to find the thickness of a pipe's walls

Whether radiation passes through something depends on the type and thickness of the material. Automatic thickness gauges and level detectors use this fact. In figure 39.18, an automatic thickness gauge is being used to check the thickness of the walls of a pipe.

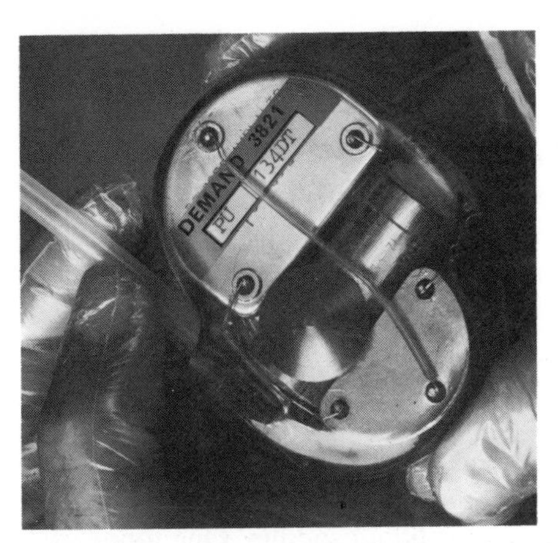

Figure 39.19 A heart pacemaker, showing the small nuclear battery, which is designed to last for 20 years

39.2 Using nuclear energy

Radioactive atoms release energy. Heat from the decay of a small amount of plutonium-238 is converted into electricity in miniature batteries in heart *pacemakers* (see figure 39.19). A heart pacemaker can be implanted in a patient's chest to keep the heart beating steadily by giving it small, regular electric shocks. Pacemakers that use nuclear batteries run for many years before they need replacing.

Vast amounts of energy are released when large atoms split. This is called *nuclear fission.* Fission of the small amount of uranium-235 in natural uranium is the source of energy in some nuclear power stations.

When a nucleus of uranium-235 absorbs a stray neutron, it becomes unstable and splits into two smaller nuclei. At the same time, it releases energy and three neutrons (see figure 39.20). These cause more uranium atoms to split and, each time, more energy and neutrons are released. A *chain reaction* occurs. In a fraction of a second a vast amount of energy is released.

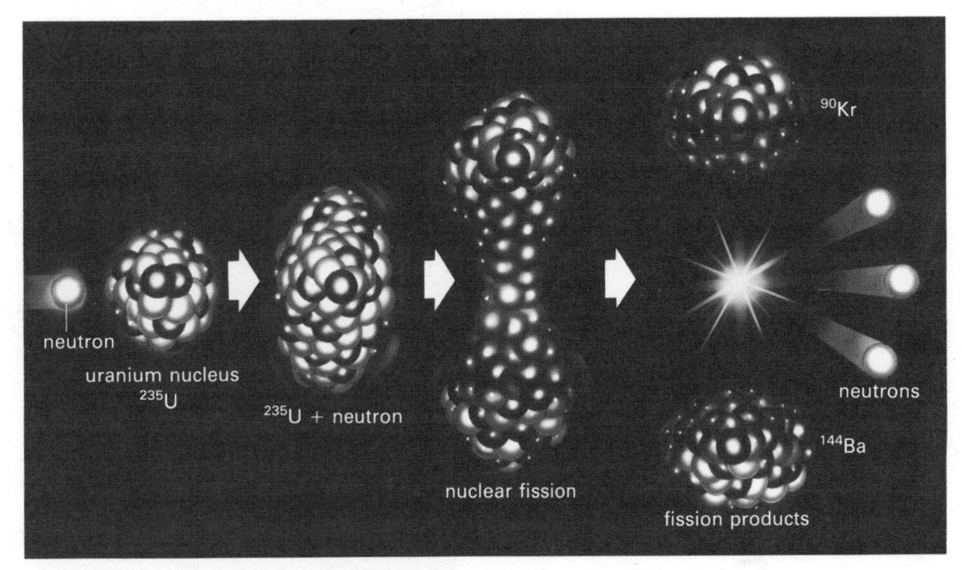

Figure 39.20 Diagram showing the fission of a nucleus of uranium-235 as it absorbs a neutron

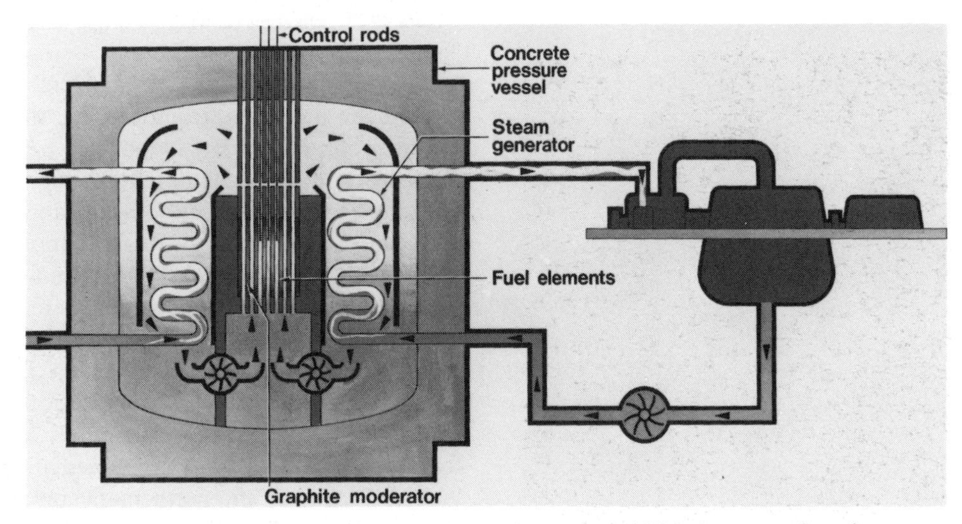

Figure 39.21 Diagram of an advanced gas-cooled nuclear reactor (AGR): the arrows show the flows of carbon dioxide and of steam

In the core of a nuclear reactor, rods containing uranium are surrounded by a material which can absorb some neutrons, slowing them down and generally keeping the reactor under control. In some reactors, graphite rods are used for this. They are raised and lowered between the fuel rods to control the heat output of the reactor. Heat from the rods is taken away with a flow of carbon dioxide gas and used to generate steam as in a normal power station. In some American-built reactors, water is used to slow down the neutrons *and* act as the coolant.

Early reactors used natural uranium containing about 1 per cent of uranium-235. Later ones, using 'enriched' uranium with up to 3 per cent of uranium-235, are more efficient and run at higher temperatures. Only a small amount of uranium is used up. Even so, one tonne of uranium gives as much energy as 20 000 tonnes of coal!

Reactors in which neutrons are slowed down by moderators are called *thermal reactors*. There are 16 thermal reactors now operating in the UK, and two more will be completed before 1990.

Another type of reactor is the *fast reactor*. In a fast reactor, the neutrons are not moderated. The fuel is plutonium and the coolant is liquid sodium. One tonne of fuel in a fast reactor can give as much energy as 1 000 000 tonnes of coal. The power station at Dounreay in the north of Scotland was the world's first fast reactor to be used for generating electricity. It has been in operation since 1975.

Plutonium which can be used in a fast reactor is produced as a by-product of thermal reactors. 99 per cent of natural uranium is uranium-238. When this isotope absorbs a neutron in a thermal reactor, it does not split like uranium-235. Instead, it ejects a beta particle and changes to plutonium-239 (see figure 39.22). Write equations to show these changes.

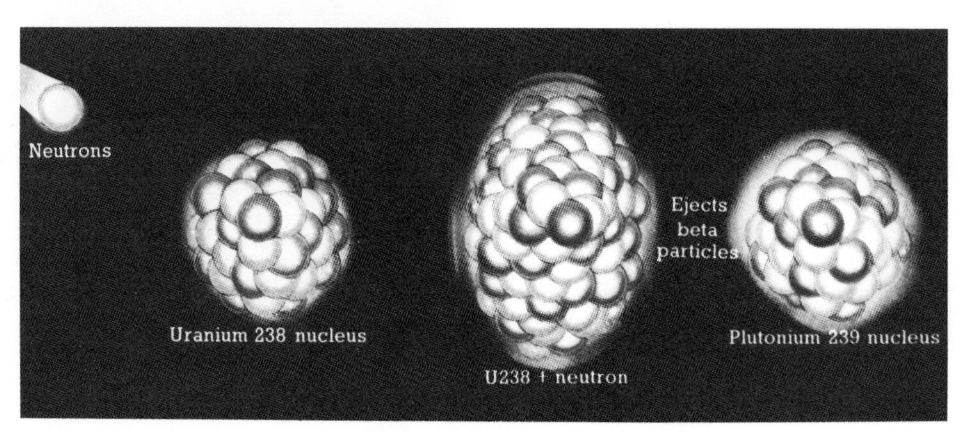

Figure 39.22 Diagram showing the formation of a nucleus of plutonium-239

Nuclear power also has its dangers. There is serious concern about each of these three main problems: reactor safety, the storage of nuclear wastes and the possible proliferation of nuclear weapons. We cannot live in a world without risks, so we have to try to understand and manage the risks we have chosen to expose ourselves to. Burning coal has its dangers, too. At the moment, the use of coal as a fuel causes many more deaths and more ill health than nuclear power does. Coal mining is dangerous and the dust and sulphur dioxide released from coal-fired power stations pollute the atmosphere and cause disease.

Fears about reactor safety were raised more dramatically in March 1979, when there was an accident in a reactor at Three Mile Island in the United States. Human error contributed to the seriousness of the accident, which illustrates the fact that the problems are not purely technical. Fortunately, the accident was controlled and there was no immediate loss of life, but the reactor was put out of action. Engineers and mathematicians have given much thought to the problems of reactor safety. They have tried to predict the main dangers and estimate the chances of an accident occurring. Finally, someone has to assess whether the benefits of cheap power outweigh the risks.

When spent nuclear fuel is unloaded from a reactor, it is highly radioactive. It is stored in tanks of water while the activity decays. After several months it is possible to reprocess the fuel rods to recover uranium and plutonium for recycling

Questions

1 Are there any ways in which nuclear fuels are like fossil fuels? What are the main differences?
2 List some of the arguments for and against increasing the number of nuclear power stations to help solve the energy problem.

as fuel. This involves the added risk of transporting highly radioactive material from the power station to the reprocessing works. Even after processing, waste remains that will be a danger to the environment for up to 25 000 years. Fortunately, the volume of the waste is relatively small – about two cubic metres per year from a large power station. The search still goes on for a permanent and safe solution to the problem of disposing of this waste. One suggestion is to convert it to a glassy form. Cylinders of solid waste could then be stored deep underground in regions which are geologically stable.

Nuclear reactors are not bombs, but they produce material that could be used to make bombs. International agencies have been set up with the task of monitoring the use of uranium and plutonium, but this is becoming increasingly difficult as the use of nuclear power becomes more widespread. One fear is that terrorists might make crude bombs with stolen plutonium and use them for blackmail.

Summary

1 Radioactivity

Copy the following paragraphs, filling in the missing words.
(a) Isotopes are atoms of the same element but with different numbers of .... Chlorine (atomic number 17) has two isotopes. One is chlorine-37; its symbol is ....

(b) Some isotopes are radioactive. Radioactive elements eject particles – alpha or beta particles – from their .... Alpha particles are the same as ... nuclei. They have a mass of ... and a charge of .... When an atom loses an alpha particle, the mass number decreases by ... and the atomic number by .... An atom of a different element is formed.

(c) Beta particles are the same as .... They are formed from ... in the nucleus. They have a mass of ... and a charge of .... When an atom loses a beta particle, the mass number does not change but the atomic number ... by .... An atom of a different element is formed.

(d) Often, alpha and beta decay is accompanied by an energetic form of radiation, called ... radiation.

(e) Radioactive decay is detected using a ... – ... tube. The half-life of an isotope is the time for the number of its atoms to fall by .... Each isotope has its own half-life, which is not affected by changes in ..., ... or ..., unlike the rates of normal chemical reactions.

2 Uses of radioactivity

Here are some uses of radioactive isotopes:

* dating vegetable and animal remains
* sterilizing surgical instruments
* treating tumours
* checking the thickness of paper during manufacture
* detecting liquid levels
* detecting abnormalities in organs of the body
* following chemical reactions
* checking root growth in plants
* following silt movement in rivers
* detecting leaks in water pipes
* pacemaker batteries.

For each use, write a sentence explaining why or how they are used for this purpose. If you can, say which isotope might be used.

3 Nuclear power

Answer the following questions about nuclear reactors.

(a) In a *thermal* reactor,

(i) what are the fuel rods made of, and what happens in them?

(ii) what may the control rods be made of, and what do they do?

(iii) what coolant can be used, and what does it do?

(iv) what could be used for control and cooling at the same time in some reactors?

(b) What is the difference between a *thermal* and a *fast* reactor?

Review questions

1 The apparatus shown below was used in an experiment to determine the molar heat of vaporization of pure trichlorotrifluoroethane, $C_2Cl_3F_3$, known commercially as 'Arklone'. Arklone is a non-toxic, non-flammable, colourless liquid.

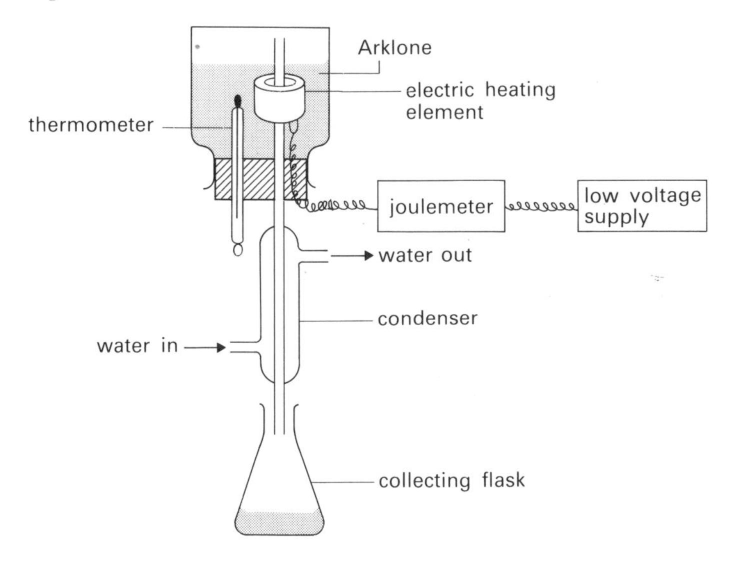

300 g of Arklone was placed in the apparatus and it was heated electrically. On reaching a temperature of 48 °C, no further rise in temperature occurred and Arklone distilled into the collecting flask. During the experiment 94 g of Arklone was collected for the use of 15 kJ of electrical energy.
(a) The relative molecular mass of Arklone is 187.5. Calculate the number of moles (g-molecules) of Arklone collected in the flask.
(b) Calculate the molar heat of vaporization, ΔH_{vap} of Arklone.
(c) If during the course of the experiment so much liquid had evaporated from the upper vessel that the thermometer bulb had become surrounded by vapour instead of liquid, would the temperature recorded by the thermometer have been lower, the same, or higher than when the bulb was in the liquid? Give a reason for your answer.
(d) What evidence is there from this experiment that Arklone is a molecular compound?
(e) Suggest a possible industrial or domestic use for Arklone. **(L)**

2 When wood chippings are heated in the absence of air the wood decomposes. The following products can be obtained from this process: carbon; pyroligneous acid and tar (which together make up the liquid product); wood gas. The table shows the substances present in pyroligneous acid.

Substances in pyroligneous acid	Boiling point (°C)
Methanol	65
Water	100
Ethanoic acid	118

(a) Draw a labelled diagram of the apparatus you would use to carry out this decomposition in the laboratory. Your apparatus should provide a means of condensing the liquid products and collecting the wood gas formed.
(b) Pyroligneous acid is a mixture of liquids. Name the process by which they could be separated.
(c) The carbon residue is graphite.
(i) Name the other allotrope of carbon.
(ii) Using your knowledge of the structure of the allotropes of carbon, explain briefly why graphite is the less dense of the two allotropes.
(d) Methanol is now made from carbon monoxide and hydrogen by the reaction

$$CO(g) + 2H_2(g) \rightleftharpoons CH_3OH(g); \qquad \Delta H = -93\,kJ/mol$$

Sketch an energy level diagram to represent the reaction shown above. **(L)**

3 This question is about the energy that can be obtained from chemical reactions. The reaction between zinc and copper(II) sulphate solution can be investigated in the apparatus shown below.

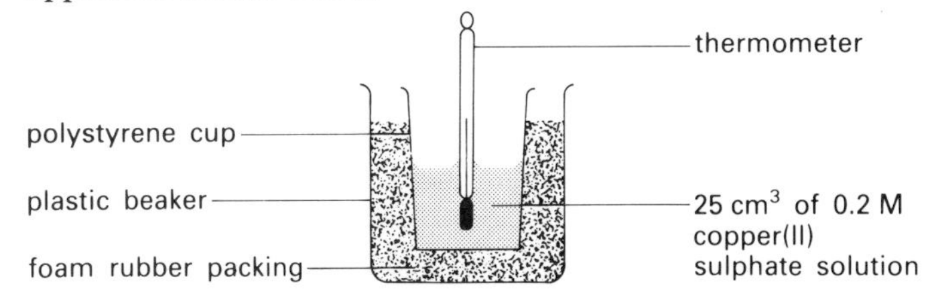

An excess of powdered zinc was added to 25 cm³ of a solution containing 0.005 mol of $Cu^{2+}(aq)$ ions (g-ions of copper). The mixture was stirred with the thermometer and the maximum temperature reached was recorded. The temperature of the mixture was found to rise by 10 °C.
(a) What would you see when stirring the mixture of copper(II) sulphate solution and zinc?

(b) Write an equation for this reaction, showing state symbols.
(c) What is the purpose of the foam rubber packing between the cup and beaker?
(d) (i) From the information given, calculate the heat change during the experiment. Show your working and state the units you use. (4.2 J (1 cal) are required to raise the temperature of 1 cm^3 of this solution by 1 °C.)
(ii) Calculate ΔH for the reaction per mole (g-ion) of $Cu^{2+}(aq)$ ions. Show your working and give the correct sign for ΔH.
(e) In the above calculation, the heat absorbed by the polystyrene cup is ignored. Explain why this can be done without making the answer too inaccurate.
(f) Describe, or draw a labelled diagram of, an electrochemical cell in which the oxidation of hydrogen or some other fuel can be used to produce electrical energy. **(L)**

4 (a) Read the following passage carefully and answer the questions below:

According to recent research, amounts of lead in the air of cities have increased dramatically. Most of the lead comes from the exhaust fumes of motor vehicles using petrol that contains lead compounds. Oil companies have for many years added these lead compounds as anti-knock agents because they are the cheapest and most practical way of increasing the octane rating of the fuel. Without them, the oil industry would have to use different and more expensive plant to improve the octane rating by adding aromatics and iso-paraffins. There is at present no cheap chemical to replace these lead compounds.

However, lead is only one of a number of pollutants put out by the automobile engine. Carbon monoxide accounts for 6% of exhaust gases, oxides of nitrogen 0.04%, sulphur dioxide 0.006%. In the UK alone, about 7 million tonnes of carbon monoxide and 250 000 tonnes of oxides of nitrogen are discharged into the atmosphere every year.

(i) Why are lead compounds added to petrol?
(ii) How could the oil industry avoid the use of lead compounds in petrol?
(iii) How would the price of petrol be affected by the removal of lead compounds from petrol?
(iv) What other pollutants are found in exhaust fumes besides lead compounds?
(v) Apart from lead, which is the most common pollutant in exhaust fumes and what proportion does it make up?
(vi) Why is lead in the atmosphere thought to be dangerous?

(b) (i) Which *one* of the following fuels produces the least smoke per gram under normal conditions of combustion: coal, natural gas, coke, wood, petrol?
(ii) Which *one* of the following produces the least amount of heat per gram of fuel: coke, wood, natural gas?
(iii) Which *one* of the following fuels is cheapest to buy in this country, per kilogram: petrol, coal, coke?
(iv) From what is smokeless fuel made?

(c) Write an essay on 'The energy problem'. In it, discuss: (i) the ability of fossil fuels to satisfy our future energy needs, (ii) how we should change our use of them to make reserves last longer and (iii) alternative sources of energy and their drawbacks. **(WMEB)**

5 The decay of a sample of a radioactive isotope was studied using a suitable Geiger–Müller tube and scaler. Readings of scaler counts against time were plotted to give the graph shown below. A count of background radiation in the laboratory was made in a separate experiment and found to be 10 counts per minute on average.

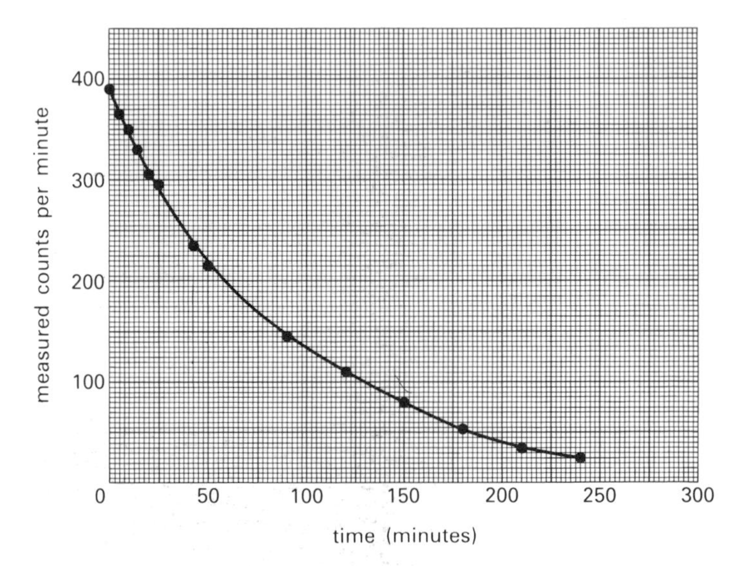

(a) At first, readings were taken every five minutes, but towards the end of the experiment they were taken every thirty minutes. Suggest a reason for this.
(b) It is usual to wear gloves when working with radioactive materials. What danger do gloves give protection against? State one danger they do not give protection against.
(c) After what time would you consider this particular sample to be radioactively harmless? Give reasons to support your answer.
(d) From the decay curve, find the value of the half-life of this isotope.
(e) What mass of a radioactive isotope having a half-life of 30 minutes would remain after 90 minutes if the initial mass of the isotope was 0.010 g?
(f) Uranium-238 decays by alpha emission to form an element X which decays by beta emission to form Y. The element Y decays by beta emission to form Z.
(i) Complete the decay series relationships by writing the mass number and atomic number of X, Y and Z.

$$^{238}_{92}U \longrightarrow X + ^{4}_{2}He$$
$$X \longrightarrow Y + ^{0}_{-1}e$$
$$Y \longrightarrow Z + ^{0}_{-1}e$$

(ii) What is the relationship between Z and $^{238}_{92}U$? **(L)**

THEME L

Chemistry and electricity

Cell city which packs the potential to produce electricity from chemical reactions

40 Electrolysis

40.1 Electrolytes

Substances which contain **ions** conduct electricity and decompose when they are liquid. They are **electrolytes**. Electrolytes can be molten **salts** or aqueous solutions of salts, or aqueous solutions of **acids** or **alkalis**.

Salts have **giant structures** of ions. When they are molten or dissolved in water, the ions are free to move and the liquid conducts electricity (see chapter 20).

Pure anhydrous acids, like dry hydrogen chloride gas, glacial ethanoic acid and solid citric acid, are molecular compounds. They cannot conduct electricity. They react with water giving ions, e.g.

$$HCl(g) + H_2O(l) \longrightarrow H_3O^+(aq) + Cl^-(aq)$$

All acids react with water to give aqueous hydrogen ions (*oxonium* ions). The solutions conduct electricity (see chapter 24).

During **electrolysis**, ions migrate towards the electrodes. The migration can be seen if the ions are coloured.

Experiment 40a
The migration of ions in copper(II) chromate(VI)

copper(II) chromate(VI) solution

The U-tube in figure 40.1 contains copper(II) chromate(VI) solution made more dense by adding urea. The clear colourless solution around the graphite electrodes is dilute hydrochloric acid.

Results
After about half an hour, the solution around the cathode (the negative electrode) becomes green–blue. The solution around the anode becomes yellow.

Discussion
Copper(II) chromate(VI) contains copper ions, Cu^{2+}, and chromate(VI) ions, CrO_4^{2-}. In solution, copper(II) ions are blue–green and migrate to the cathode. Chromate(VI) ions are yellow and migrate to the anode. Are these the directions you would expect?

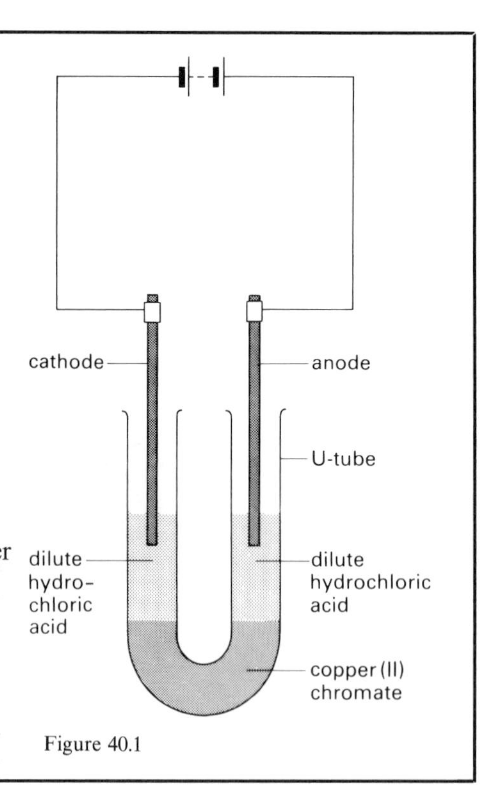

Figure 40.1

Figure 40.2 illustrates another experiment which shows the migration of ions. After a short time, a purple streak due to manganate(VII) ions spreads towards the anode.

In electrolysis, positive ions always migrate to the **cathode** and negative ions move to the **anode**.

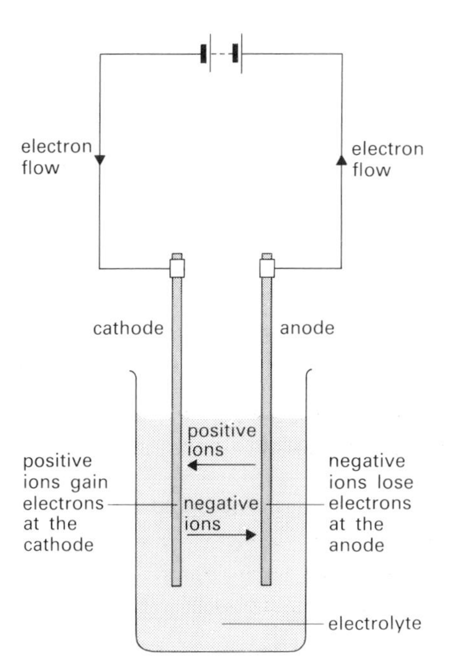

Figure 40.3 Flows of ions in an electrolysis cell

Figure 40.2 Demonstrating the migration of manganate(VII) ions

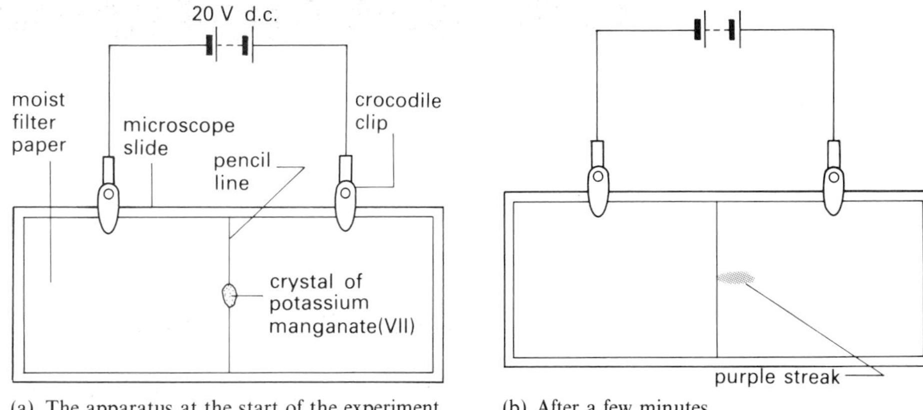

(a) The apparatus at the start of the experiment

(b) After a few minutes

Question

If the experiment in figure 40.2 were repeated using crystals of (a) copper sulphate and (b) nickel chloride, what would you expect to see? The colours of the aqueous ions are: Cu^{2+}(aq), blue; Ni^{2+}(aq), green; SO_4^{2-}(aq), colourless; Cl^-(aq), colourless.

Chemical changes take place at the electrodes. At the cathode, positive ions gain electrons and become atoms or molecules. At the anode, negative ions lose electrons and become atoms or molecules. The electrons flow into the circuit (see figure 40.3).

40.2 Electrolysis of molten salts using carbon electrodes

The electrolysis of molten lead(II) bromide and sodium chloride is described in chapters 2 and 20. The compounds are decomposed into their elements. The rules for the electrolysis of molten salts are:

At the *cathode*:

* metals are deposited.

At the *anode*:

* non-metals are released.

The electrode reactions are summarized in figure 40.4.

Electrolyte	At the cathode	At the anode
Lead(II) bromide, $PbBr_2$(l)	$Pb^{2+} + 2e^- \longrightarrow Pb$ (electrons from the cathode)	$2Br^- \longrightarrow Br_2 + 2e^-$ (electrons given up to the anode)
Sodium chloride, NaCl(l)	$Na^+ + e^- \longrightarrow Na$	$2Cl^- \longrightarrow Cl_2 + 2e^-$

Figure 40.4

Sodium, magnesium and aluminium are produced on a large scale in industry by the electrolysis of molten compounds. Two of these processes are described in chapter 13.

40.3 Electrolysis of salt solutions using carbon or platinum electrodes

Experiment 40b
Electrolysis of some aqueous salt solutions

lead(II) nitrate solution
copper(II) chloride solution

The apparatus in figure 40.5 is used. The cell and the test-tubes are filled with the electrolyte.

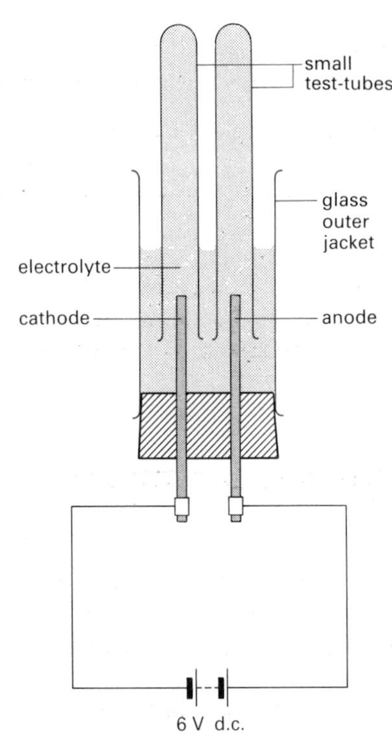

Figure 40.5

Results
Figure 40.6 shows some typical results.

Electrolyte	At the cathode	At the anode
Copper(II) chloride solution	A salmon pink solid collects around the carbon rod. It is **copper**	A yellow–green gas collects. It bleaches indicator paper. It is **chlorine**
Sodium chloride solution	A clear, colourless gas collects. The gas burns with a squeaky pop. It is **hydrogen**	**Chlorine** is produced
Potassium bromide solution	**Hydrogen** is produced	A dark red colour forms around the carbon rod. It is due to **bromine**
Lead(II) nitrate solution	A grey flaky solid collects around the carbon rod. It is **lead**	A clear, colourless gas collects. It relights a glowing splint. It is **oxygen**
Potassium sulphate solution	A clear colourless gas collects. The gas burns with a squeaky pop. It is **hydrogen**	**Oxygen** is produced
Zinc sulphate solution	**Hydrogen** is produced. Also, there is some grey solid around the rod. It is **zinc**	**Oxygen** is produced

Figure 40.6

Discussion
The pattern with solutions is not so simple as with molten salts. From some electrolytes, metal is deposited at the cathode. With others, hydrogen is released. Where does the hydrogen come from? Where are lead and copper in the activity series? Where are sodium and potassium? Is there a connection between a metal's position in the activity series and the cathode product?

At the anode, the expected non-metal is released from some electrolytes, but, with others, oxygen is released. What do chlorine and bromine have in common?

If the electrolyte is an aqueous solution of a salt, water can become involved in the electrolysis. When potassium sulphate solution is electrolysed, the products are hydrogen and oxygen. It is the water which is electrolysed, and not the salt.

The rules for the electrolysis of aqueous salts with carbon or platinum electrodes are:

At the *cathode*:

* if the metal in the salt is low in the activity series (e.g. silver, copper or lead), the metal is deposited

Questions

1 What would be the products of electrolysis of solutions of
(a) nickel(II) chloride,
(b) copper(II) nitrate,
(c) magnesium bromide and
(d) potassium iodide?
2 Why are both hydrogen and zinc formed at the cathode when zinc sulphate solution is electrolysed?

* if the metal is high in the activity series (e.g. sodium or potassium), hydrogen is evolved.

At the *anode*:

* if the non-metal ion in the salt is a halide ion (e.g. chloride, bromide or iodide), then the halogen is released
* if the non-metal is not a halogen, oxygen from the water is given off.

The electrode reactions for some electrolytes are summarized in figure 40.7.

Electrolyte	At the cathode	At the anode
Copper(II) chloride solution, $CuCl_2(aq)$	$Cu^{2+}(aq) + 2e^- \longrightarrow Cu(s)$	$2Cl^-(aq) \longrightarrow Cl_2(g) + 2e^-$
Potassium bromide solution, $KBr(aq)$	$2H^+(aq) + 2e^- \longrightarrow H_2(g)$	$2Br^-(aq) \longrightarrow Br_2(aq) + 2e^-$
Lead(II) nitrate solution $Pb(NO_3)_2(aq)$	$Pb^{2+}(aq) + 2e^- \longrightarrow Pb(s)$	$4OH^-(aq) \longrightarrow O_2(g) + 2H_2O(l) + 4e^-$

Figure 40.7

The electrolysis of aqueous sodium chloride is a very important industrial process. It leads to useful products like chlorine, hydrogen and sodium hydroxide, and, indirectly, to sodium hypochlorite, which is used for bleach. The process is described in chapter 22.

40.4 Electrolysis of salt solutions using metal (active) electrodes

When copper(II) sulphate is electrolysed using copper electrodes, fresh copper is deposited on the cathode. However, the copper anode becomes thinner and eventually disintegrates in the electrolyte. Copper atoms in the anode become copper ions in solution.

The rules for the electrolysis of aqueous salts using metal electrodes are:

At the *cathode*:

* the changes are the same as with carbon or platinum electrodes (see section 40.3).

At the *anode*:

* the metal anode dissolves as ions are formed.

Some electrode reactions are summarized in figure 40.8.

Electrolyte	Electrodes	At the cathode	At the anode
Copper(II) sulphate solution, $CuSO_4(aq)$	Copper	$Cu^{2+}(aq) + 2e^- \longrightarrow Cu(s)$	$Cu(s) \longrightarrow Cu^{2+}(aq) + 2e^-$
Silver nitrate solution, $AgNO_3(aq)$	Silver	$Ag^+(aq) + e^- \longrightarrow Ag(s)$	$Ag(s) \longrightarrow Ag^+(aq) + e^-$

Figure 40.8

The purest copper – at least 99.98 per cent pure – is made by *electrolytic refining*. Very thin sheets of pure copper are suspended in a tank containing warm

Figure 40.9 Electrolytic refining of copper: lowering a new set of thin copper cathodes into the tank

copper(II) sulphate solution and dilute sulphuric acid. The copper sheets act as cathodes. Blocks, or thicker sheets, of impure copper in between the thin sheets act as the anodes. During electrolysis, pure copper builds up on the cathodes and the anodes slowly dissolve. It takes about two weeks for the anodes to dissolve. Impurities settle as a sludge in the tank. These include silver and gold, which are recovered from the sludge.

Electroplating can be carried out in the laboratory using the simple apparatus shown in figure 40.10. An even, attractive, non-flaky coating is only obtained when the metal object is very clean and electrolysis is slow.

Many metal objects we use are electroplated to protect them from corrosion and also to make them look attractive. Metals used for plating are usually nickel, silver, chromium, tin and, sometimes, gold.

Tin plate for 'tin' cans is produced by passing steel strip continuously through a tank that contains a plating solution and anode blocks of tin. The steel sheet is in the tank for only a few seconds, but while it is there it acts as the cathode and becomes plated with tin. The coating may be about 0.0004–0.002 mm thick. What is the plating solution likely to be? Why is tin plate used for 'tin' cans? What else is tin plate used for?

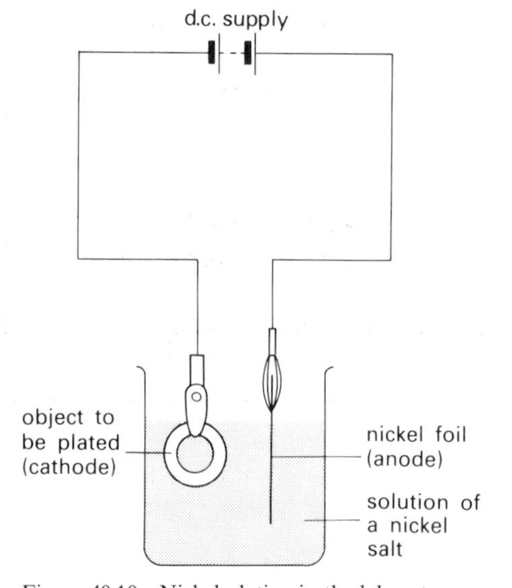

Figure 40.10 Nickel plating in the laboratory

Questions

1 Describe how you would electroplate a key with silver in the laboratory.

2 List objects you have seen or used which are plated. What metals have been used for the base and the plating in each object?

3 The letters EPNS are found on some cutlery and ornaments. What does the abbreviation stand for?

4 Before steel is plated, it is *pickled* in sulphuric acid and washed with *deionized* water. Explain the two words in italics.

Figure 40.11 Electric kettles are plated with copper and then with nickel and chromium

40.5 Electrolysis of dilute acids

Experiment 40c
Electrolysing some common acid solutions

dilute acids

The apparatus used is the same as for experiment 40b.

Results
Typical results are shown in figure 40.12.

Electrolyte	At the cathode	At the anode
Hydrochloric acid, $HCl(aq)$	A clear, colourless gas collects. It burns with a squeaky pop. It is **hydrogen**	A yellow–green gas collects. It bleaches indicator paper. It is **chlorine**
Sulphuric acid, $H_2SO_4(aq)$	**Hydrogen** is produced	A clear, colourless gas collects. It relights a glowing splint. It is **oxygen**
Nitric acid, $HNO_3(aq)$	**Hydrogen** is produced	**Oxygen** is produced
Hydrobromic acid, $HBr(aq)$	**Hydrogen** is produced	A dark red colour forms around the carbon rod. It is due to **bromine**

Figure 40.12

Discussion
Hydrogen is released at the cathode when any dilute acid is electrolysed. What ion must be present in all acid solutions? Does this support the theory about acids that is described in chapter 24?

Suggest a rule for predicting the anode product. Remember that, as in aqueous solutions of salts, water has a part in the electrolysis of acids.

The rules for the electrolysis of dilute acids are:

At the *cathode*:

* hydrogen is released.

At the *anode*:

* the changes are the same as for aqueous salt solutions with carbon electrodes.

Some electrode reactions are summarized in figure 40.13.

Electrolyte	At the cathode	At the anode
Hydrochloric acid, $HCl(aq)$	$2H^+(aq) + 2e^- \longrightarrow H_2(g)$	$2Cl^-(aq) \longrightarrow Cl_2(g) + 2e^-$
Sulphuric acid, $H_2SO_4(aq)$	$2H^+(aq) + 2e^- \longrightarrow H_2(g)$	$4OH^-(aq) \longrightarrow O_2(g) + 2H_2O(l) + 4e^-$

Figure 40.13

Question

(a) What ion is present in all alkaline solutions?
(b) What anode product would you expect from all alkalis?
(c) What products would you expect at the cathode and the anode in the electrolysis of sodium hydroxide solution?

40.6 How much reaction takes place during electrolysis?

Amounts of electricity are measured in coulombs (C). The current flowing in a circuit is the rate of flow of electricity, measured in coulombs per second. The practical unit of current is the ampere (A).

✻ $1\,A = 1\,C/s$

✻ Amount of electricity (C) = Current (A) × Time (s)

The amounts of electricity needed to deposit one mole of lead, copper and silver are found in experiments 40d and 40e.

Experiment 40d
How much electricity is needed to deposit one mole of lead?

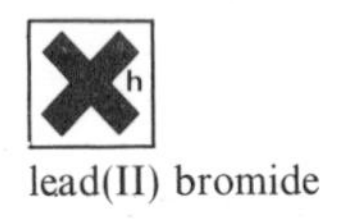

lead(II) bromide

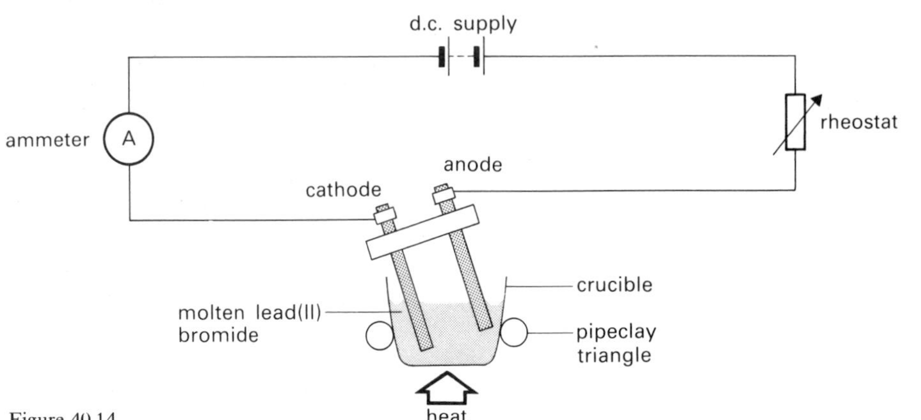

Figure 40.14

The apparatus in figure 40.14 is used. A steady current of 1 A is maintained using a rheostat (a variable resistor). After electrolysis, the molten lead(II) bromide is poured away, leaving the lead bead in the crucible. When cool, any lead(II) bromide left on the bead can be chipped away and the lead weighed.

Results
Typical results are:

Electrolysis time = 600 s
Current = 1 A (=1 C/s)
Mass of lead = 0.633 g

What would you expect to see during the experiment? Why should the experiment be carried out in a fume cupboard?

Discussion
The amount of electricity used is

$$600\,s \times 1\,C/s = 600\,C$$

So 600 C of electricity liberates 0.633 g of lead. The amount of electricity needed to liberate 1 g of lead is therefore

$$\frac{600\,C}{0.633\,g}$$

and the amount of electricity needed to liberate 1 mol (207 g) of lead is

$$\frac{600\,C}{0.633\,g} \times 207\,g/mol = 196\,200\,C/mol$$

Suggest possible sources of error in the experiment. Why are the best results obtained when the electrodes are tilted?

Experiment 40e
How much electricity is needed to deposit one mole of copper and silver?

copper(II) sulphate solution

silver nitrate solution

propanone

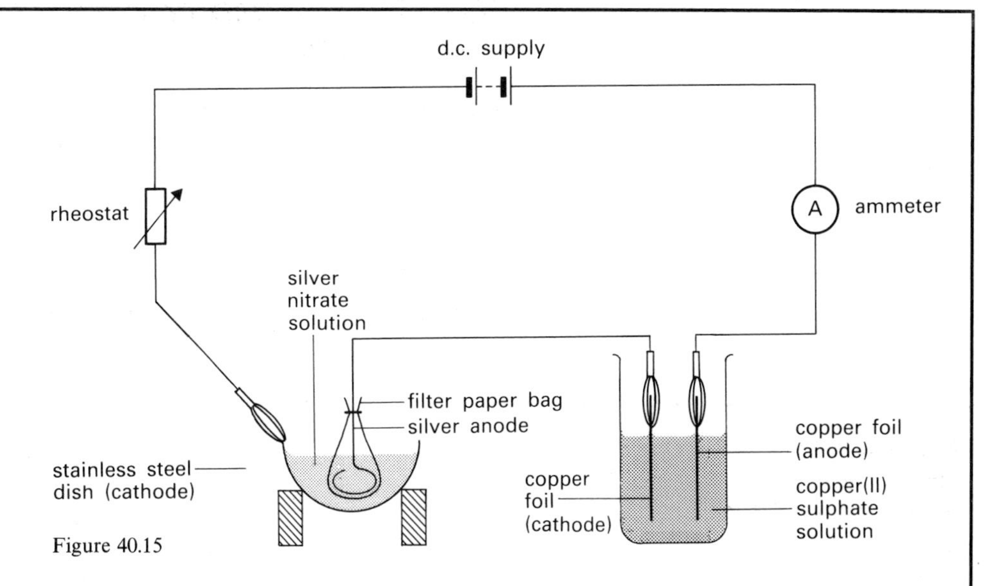

Figure 40.15

Figure 40.15 shows a copper and a silver cell in series.

In the copper cell, the copper that is liberated plates the cathode. How and why should the copper foil cathode be cleaned before use? Is it necessary to clean the anode? Before electrolysis, the cathode is weighed. After electrolysis, it is washed with water, then with propanone. It is allowed to dry in the air before the final weighing.

In the silver cell, because silver does not adhere well to silver foil, it is allowed to collect in a weighed stainless steel dish. Why is a bag tied round the anode? After electrolysis, the solution is carefully poured away and the silver washed with water, then propanone. It is dried in the air and weighed.

Results

Typical results are:

Electrolysis time	= 4500 s
Current	= 0.1 A (=0.1 C/s)
Initial mass of copper cathode	= 2.359 g
Final mass of cathode	= 2.506 g
Mass of copper deposited	= 0.147 g
Initial mass of dish	= 52.270 g
Final mass of dish	= 52.769 g
Mass of silver deposited	= 0.499 g

Discussion

The amount of electricity used is

$$4500\,\text{s} \times 0.1\,\text{C/s} = 450\,\text{C}$$

So 450 C of electricity liberates 0.147 g of copper. Therefore, the amount of electricity needed to liberate 1 mol (64 g) of copper would be

$$\frac{450\,\text{C}}{0.147\,\text{g}} \times 64\,\text{g/mol} = 195\,918\,\text{C/mol}$$

450 C of electricity also liberates 0.499 g of silver. The amount of electricity needed to liberate 1 mol (108 g) of silver would be

$$\frac{450\,\text{C}}{0.499\,\text{g}} \times 108\,\text{g/mol} = 97\,395\,\text{C/mol}$$

What are the main sources of error in this experiment? Why was the current kept so low?

Questions

1 An aluminium ion has a charge of 3+.
(a) Write an equation for the formation of aluminium at a cathode.
(b) How much electricity is needed to release 1 mol of aluminium?
2 When copper(II) sulphate solution is electrolysed using copper electrodes, the gain in mass of the cathode is the same as the loss in mass of the anode. Explain why.

Figure 40.16 shows the results of experiments 40d and 40e. The amounts of electricity needed to liberate one mole of lead and copper are about the same, but only half as much is needed to obtain one mole of silver. Why?

Element	Number of coulombs of electricity needed to liberate one mole
Lead	196 200
Copper	195 918
Silver	97 395

Figure 40.16

At the cathode, metal ions gain electrons to become atoms:

$$Pb^{2+}(aq) + 2e^- \longrightarrow Pb(s)$$

$$Cu^{2+}(aq) + 2e^- \longrightarrow Cu(s)$$

$$Ag^+(aq) + e^- \longrightarrow Ag(s)$$

From these equations, one mole of silver ions needs one mole of electrons to become one mole of silver atoms, and one mole of lead or copper ions needs two moles of electrons to become one mole of metal atoms.

Accurate experiments show that 96 500 C of electricity are needed to liberate one mole of silver. It is the amount of electricity carried by one mole of electrons. This is the **Faraday constant**, F, 96 500 C/mol.

Any ion, M^+, with a single positive charge, like Ag^+ or H^+, needs 96 500 C of electricity to deposit one mole of atoms at the cathode. Any ion, M^{2+}, with a double positive charge, like Cu^{2+} or Pb^{2+}, needs $2 \times 96\,500$ C (193 000 C) to deposit one mole of atoms.

Similarly, at the anode, one mole of any ion with a single negative charge needs the flow of 96 500 C of electricity to liberate one mole of atoms.

40.7 Calculations using the Faraday constant

In any 'electrolysis calculation', we may know

* the charge on the ions involved
* the amount of product formed
* the amount of electricity used.

If any two of these are known, the other can be calculated.

Example 1

When molten magnesium chloride is electrolysed by a current of 0.2 A flowing for 1930 s, 0.048 g of magnesium is formed. What is the charge on a magnesium ion?

The amount of electricity used is

$$1930\,\text{s} \times 0.2\,\text{C/s} = 386\,\text{C}$$

So the number of moles of electrons used is

$$\frac{386\,\text{C}}{96\,500\,\text{C/mol}} = \frac{4}{1000}\,\text{mol} = 0.004\,\text{mol}$$

The mass of 1 mol of magnesium is 24 g, so the number of moles of magnesium formed is

$$\frac{0.048\,\text{g}}{24\,\text{g/mol}} = 0.002\,\text{mol}$$

0.004 mol of electrons are needed to liberate 0.002 mol of magnesium. So 2 mol of electrons are needed to liberate 1 mol of magnesium. The charge on a magnesium ion is 2+.

Example 2
How much copper is deposited at the cathode when copper(II) sulphate is electrolysed using a current of 0.1 A for 965 s?

'Copper(II)' means that the copper ion is Cu^{2+} (see chapter 41). The amount of electricity used is

$$965\,\text{s} \times 0.1\,\text{C/s} = 96.5\,\text{C}$$

So the number of moles of electrons used is

$$\frac{96.5\,\text{C}}{96\,500\,\text{C/mol}} = 0.001\,\text{mol}$$

At the cathode:

$$Cu^{2+}(aq) + 2e^- \longrightarrow Cu(s)$$

2 mol of electrons are needed to liberate 1 mol of copper, so 0.001 mol of electrons will liberate 0.0005 mol of copper. The mass of 0.0005 mol of copper is

$$0.0005\,\text{mol} \times 64\,\text{g/mol} = 0.032\,\text{g}$$

So 0.032 g of copper will be deposited at the cathode.

Questions

Ion charges are shown in brackets.

1 How many coulombs of electricity are needed to liberate
(a) 0.1 mol of copper atoms (2+)?
(b) 0.5 mol of aluminium atoms (3+)?

2 How many moles of atoms of the following elements are released by 965 C of electricity: copper (2+), silver (1+), bromine (1−)?

3 Chromium chloride is electrolysed using chromium electrodes. A current of 0.1 A flows for 2894 s. The increase in mass of the cathode is 0.052 g.
(a) How many coulombs of electricity are used?
(b) How many moles of electrons are transferred?
(c) How many moles of chromium are liberated?
(d) How many moles of electrons are needed to liberate one mole of chromium?
(e) What is the charge on the chromium ion?

4 A current of 0.2 A is passed through dilute sulphuric acid for 9650 s.
(a) Write an equation for the formation of hydrogen atoms at the cathode.
(b) How many coulombs of electricity are used?
(c) How many moles of electrons are transferred?
(d) How many moles of hydrogen atoms are formed?
(e) How many moles of hydrogen molecules are formed?
(f) What volume of hydrogen is released at room temperature?

5 Humphry Davy discovered sodium by electrolysing molten sodium hydroxide. Use the following steps to find how long a current of 1 A must flow to give 0.23 g of sodium.
(a) How many moles of sodium are formed?
(b) How many moles of electrons are needed? (The charge on a sodium ion is 1+.)
(c) How many coulombs of electricity are needed?
(d) For what time must the current of 1 A flow to give the amount of electricity needed?

Summary

1 Copy and complete the table in figure 40.17 to summarize what happens at the electrodes during electrolysis. List any industrial applications of each type of electrolysis.

Electrolyte type	Electrodes	At the cathode	At the anode	Industrial applications
Molten salts	Carbon or platinum			
Aqueous salt solutions	Carbon or platinum			
Aqueous salt solutions	Metal (active)			
Aqueous acid solutions	Carbon or platinum			

Figure 40.17

2 Write equations for the following electrode reactions:
(a) magnesium being formed at the cathode from molten magnesium chloride,
(b) hydrogen being formed at the cathode from aqueous sodium nitrate,
(c) silver being formed at the cathode from aqueous silver nitrate,
(d) iodine being formed at the anode from aqueous potassium iodide,
(e) oxygen being formed at the anode from aqueous sodium sulphate,
(f) nickel dissolving at the anode from a nickel foil anode.
3 Write sets of rules for carrying out calculations to find (a) charges on ions, (b) amounts of product formed during electrolysis and (c) amounts of electricity needed to cause certain amounts of change. Use the examples on p. 318–9 to help you.

41 Oxidation and reduction

41.1 Redox

Burning magnesium is fun. It can also be useful because the reaction gives out much light. It used to be used in flash bulbs. When magnesium burns it combines with oxygen:

Magnesium(s) + Oxygen(g) $\longrightarrow$ Magnesium oxide(s)

In this reaction, the magnesium has gained oxygen. It has been oxidized.

Reduction is the opposite of **oxidation**. Reduction is used in industry to extract metals from their ores. In the blast furnace, iron oxide is reduced to iron by carbon monoxide:

Iron(III) oxide(s) + Carbon monoxide(g) $\longrightarrow$ Iron(s) + Carbon dioxide(g)

The iron oxide has lost oxygen. It has been reduced. The carbon monoxide has gained oxygen. It has been oxidized.

These reactions are often called **redox reactions**, because reduction and oxidation must always go together.

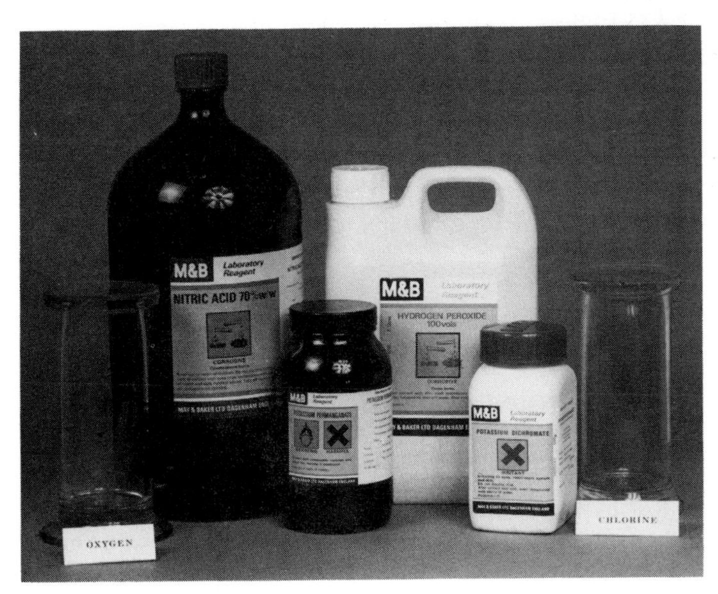

Figure 41.1 Some common oxidizing agents

Figure 41.2 Some common reducing agents

Question

Which substance is oxidized and which is reduced in the reaction of
(a) steam with hot magnesium?
(b) copper(II) oxide with hydrogen?
(c) aluminium with iron(III) oxide in the thermit process?
(d) carbon dioxide with carbon to form carbon monoxide?

Question

Write symbol equations to show the transfer of electrons in the reaction of
(a) sodium with chlorine.
(b) zinc with oxygen.
(c) aluminium with sulphur.
(d) calcium with bromine.

41.2 Electron transfer

When magnesium reacts with oxygen, the magnesium atoms become magnesium ions. Each magnesium atom gives up two electrons:

$$Mg \longrightarrow Mg^{2+} + 2e^-$$

The electrons are taken by the oxygen. Each oxygen atom turns into an oxide ion by gaining two electrons:

$$O + 2e^- \longrightarrow O^{2-}$$

Electrons have been transferred from magnesium atoms to oxygen atoms. Atoms have turned into ions.

Magnesium atoms also turn into ions when they react with other non-metals, including chlorine, bromine and sulphur.

$$Mg \longrightarrow Mg^{2+} + 2e^-$$

$$S + 2e^- \longrightarrow S^{2-}$$

In all its reactions with non-metals, magnesium loses electrons and its atoms turn into positive ions. All these reactions are examples of redox reactions. The magnesium is oxidized by loss of electrons. The non-metals are reduced by gain of electrons. The words 'oil rig' may help you remember this:

Oxidation	**R**eduction
Is	**I**s
Loss of electrons	**G**ain of electrons

This way of looking at redox reactions can also be used to explain reactions in solution. Iron metal dissolves in dilute hydrochloric acid to form iron(II) chloride:

$$Fe(s) + 2HCl(aq) \longrightarrow FeCl_2(aq) + H_2(g)$$

A simpler equation shows what has happened to the iron atoms:

$$Fe(s) \longrightarrow Fe^{2+}(aq) + 2e^-$$

The iron atoms have been *oxidized* to iron(II) ions.

If the solution of iron(II) chloride is left open to the air, it gradually turns yellow as it is oxidized further, to iron(III) chloride:

$$Fe^{2+}(aq) \longrightarrow Fe^{3+}(aq) + e^-$$

Iron(II) ions can also be oxidized to iron(III) ions by bubbling chlorine gas through the solution:

$$2Fe^{2+}(aq) \longrightarrow 2Fe^{3+}(aq) + 2e^-$$

$$Cl_2(g) + 2e^- \longrightarrow 2Cl^-(aq)$$

The number of electrons lost by the iron(II) ions must equal the number of electrons gained by the chlorine. So one chlorine molecule will oxidize two iron(II) ions.

Question

Split each of the following equations into half-equations, to show the gain and loss of electrons.

(a) $2Fe^{2+}(aq) + Br_2(aq) \longrightarrow 2Fe^{3+}(aq) + 2Br^-(aq)$

(b) $Zn(s) + Cu^{2+}(aq) \longrightarrow Zn^{2+}(aq) + Cu(s)$

(c) $2Fe^{3+}(aq) + S^{2-}(aq) \longrightarrow 2Fe^{2+}(aq) + S(s)$

When lead metal reacts with bromine, electrons are transferred from lead atoms to bromine atoms:

$$Pb \longrightarrow Pb^{2+} + 2e^-$$

$$Br_2 + 2e^- \longrightarrow 2Br^-$$

Electrolysis of molten lead(II) bromide reverses this change. The cathode turns lead ions into lead atoms by giving them back the electrons they lost:

$$Pb^{2+} + 2e^- \longrightarrow Pb$$

The anode turns bromide ions back to bromine by taking away the electrons they gained:

$$2Br^- \longrightarrow Br_2 + 2e^-$$

Lead ions are *reduced* at the cathode. The cathode is a *reducing agent.* Bromide ions are *oxidized* at the anode. The anode is an *oxidizing agent.*

41.3 Oxidation numbers

Oxidation numbers are used to show how many electrons have been gained or lost by an element. In figure 41.3, movement *up* the diagram towards more positive numbers is oxidation. Movement *down* the diagram towards more negative numbers is reduction.

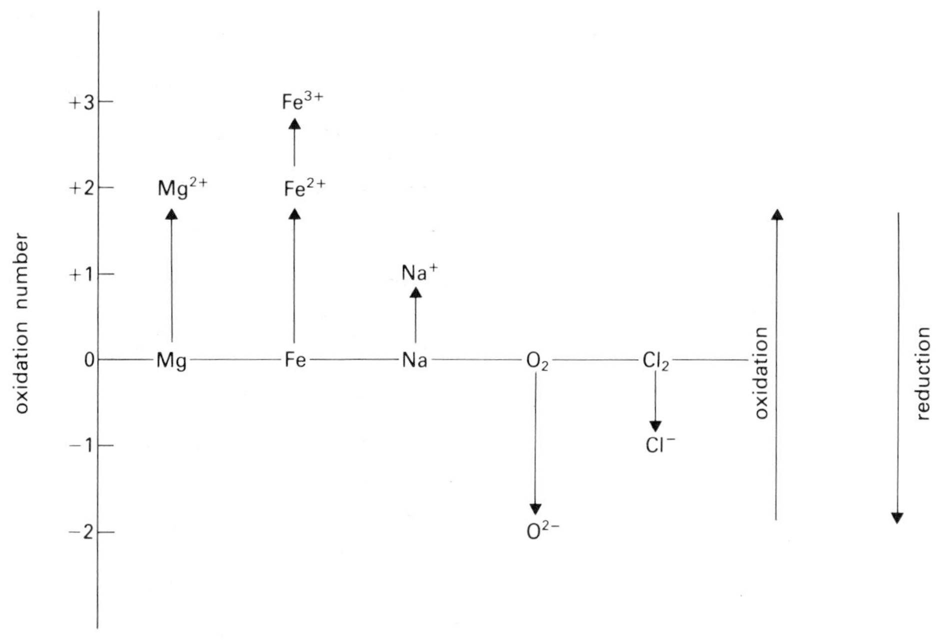

Figure 41.3 Oxidation number diagram

The oxidation numbers of the *elements* are zero. In a simple *ion*, the oxidation number of the element is equal to the charge on the ion.

Oxidation numbers are used to name chemical compounds. In iron(II) chloride, the Roman number II shows that iron is in the oxidation state +2. Iron has been oxidized by the loss of two electrons to form iron(II) chloride.

The compound with the formula $KMnO_4$ is called potassium manganate(VII). In this compound, manganese has used seven electrons to combine with oxygen in the MnO_4^- ion. Manganese has an oxidation number of +7. Potassium manganate(VII) is a powerful oxidizing agent in acid solution. When it reacts, it turns to manganese(II) ions. The change from +7 to +2 shows that the manganese has been reduced. The manganese has gained five electrons.

Questions

1 Write the symbols for the simple metal ions in (a) copper(I) oxide, (b) iron(III) nitrate, (c) cobalt(II) chloride and (d) lead(II) iodide.

2 What is the oxidation number of chromium in
(a) chromium(II) chloride?
(b) potassium dichromate(VI)?
(c) chromium(III) oxide?
(d) potassium chromate(VI)?

RULES

OXIDATION NUMBERS

1/ The oxidation numbers of uncombined elements are zero.

2/ In simple ions the oxidation number of the element is the charge on the ion.

3/ The sum of the oxidation numbers in a molecule is zero.

4/ The sum of the oxidation numbers in a complex ion equals the charge on the ion.

5/ Some elements have fixed oxidation numbers in all their compounds:

METALS		NON-METALS	
sodium	+1	hydrogen	+1
potassium	+1	halogens	−1
magnesium	+2	oxygen	−2
calcium	+2	(except in peroxides)	
aluminium	+3		

Figure 41.4 Working out oxidation numbers

There is a set of rules for working out the oxidation numbers of elements in molecules and in the more complicated ions. These are shown in figure 41.4. To use the rules, you need to know that certain elements have fixed oxidation numbers in all their compounds.

In a *molecule*, the sum of the oxidation numbers is zero. For example, in carbon dioxide each oxygen atom has oxidation number −2. The oxidation number of carbon must be +4 if the total is to be zero.

In sulphuric acid, the oxidation numbers of hydrogen and oxygen are as given in figure 41.4, so the oxidation number of sulphur must be +6.

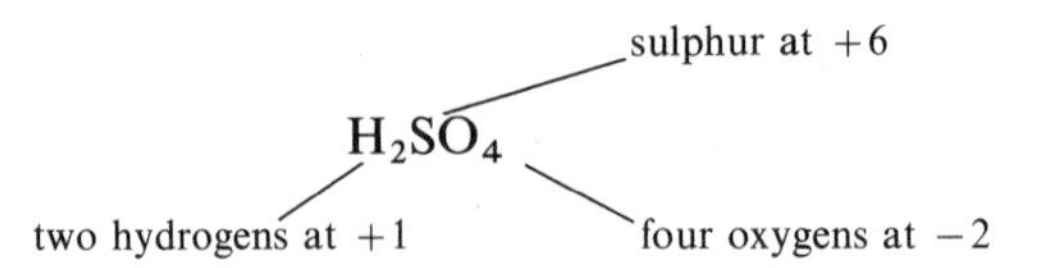

In an *ion*, the sum of the oxidation numbers is equal to the charge on the ion. In the manganate(VII) ion, MnO_4^-, there are four oxygens each with oxidation number −2. The oxidation number of manganese must therefore be +7 so that the sum is −1.

An oxidation number diagram can help to make sense of the chemistry of an element. Figure 41.5 is an oxidation number diagram for sulphur. It shows a number of sulphur compounds. Sulphur dioxide is oxidized when it is converted to sulphur trioxide or to the sulphate ion. It is reduced when it turns to sulphur. It is neither oxidized nor reduced when it reacts with water to form sulphurous acid, H_2SO_3.

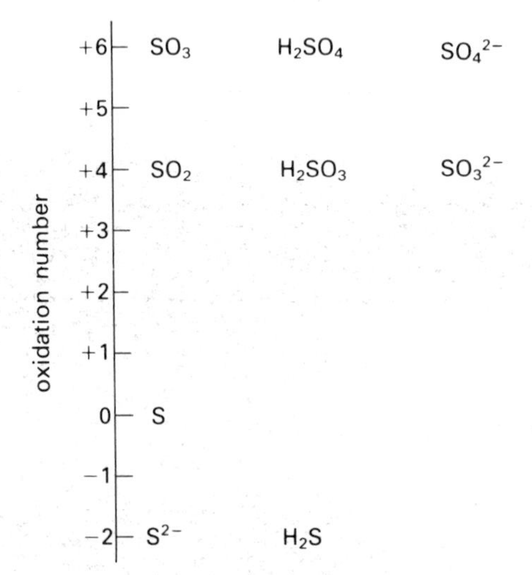

Figure 41.5 Oxidation number diagram for sulphur

Question

Work out the oxidation number of nitrogen in each of these compounds and then plot the formulae on an oxidation number diagram similar to figure 41.5: ammonia, NH_3; nitric acid, HNO_3; nitrogen dioxide, NO_2; nitrogen monoxide, NO; dinitrogen oxide, N_2O; the nitrite ion, NO_2^-; the nitrate ion, NO_3^-.

41.4 Oxidizing and reducing agents

An *agent* is someone or something which gets things done. In spy stories, the dirty work is done by secret agents. In chemistry, redox reactions involve oxidizing and reducing agents.

It is easy to get into a mental muddle when using these terms. When an *oxidizing* agent reacts, it is *reduced*. When a *reducing* agent reacts, it is *oxidized*. This is illustrated by the reaction of magnesium with chlorine, shown in figure 41.6. The magnesium is oxidized by loss of electrons. It is oxidized *by* the chlorine. So chlorine is the oxidizing agent. At the same time, the chlorine is reduced *by* the magnesium. The magnesium is the reducing agent.

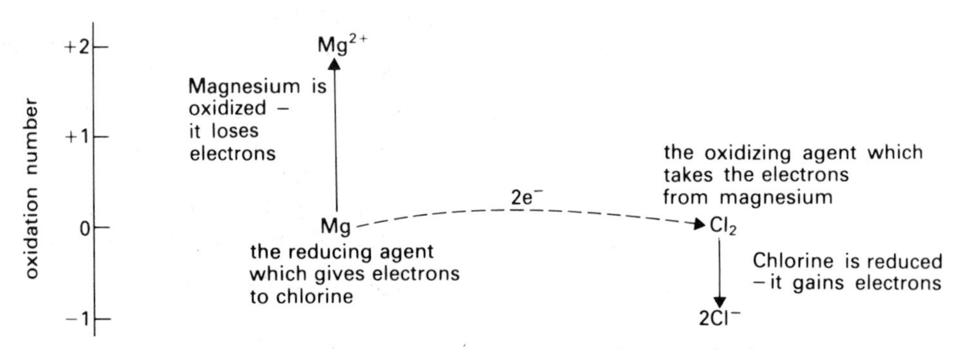

Figure 41.6 Oxidation number diagram for the reaction of magnesium with chlorine

There are three common substances which change colour when they are oxidized. They can be used to test for oxidizing agents. The details are given in figure 41.7.

Test	Results	Explanation
Add a solution of potassium iodide	Solution turns brown, and grey specks may be seen	I^- oxidized to I_2, which is sparingly soluble
or Test with starch–iodide paper	Paper turns blue–black	I^- oxidized to I_2, which then reacts with starch
Bubble $H_2S(g)$ into the solution *or* Add $H_2S(aq)$ *or* Add $Na_2S(aq)$	Yellow precipitate, which may look milky at first	S^{2-} ions oxidized to sulphur, which is insoluble
Add a fresh solution of iron(II) sulphate	Changes from very pale green to yellow–brown	Fe^{2+} oxidized to Fe^{3+} – the formation of Fe^{3+} may be confirmed by adding NaOH(aq)

Figure 41.7 Tests for oxidizing agents

Similarly, there are three common substances which change colour when they are reduced. They can be used to detect reducing agents, as shown in figure 41.8.

Test	Results	Explanation
Add a solution of potassium manganate(VII), acidified with dilute H_2SO_4	Purple solution is decolorized	Purple MnO_4^- ions are reduced to very pale pink Mn^{2+} ions
Add potassium dichromate(VI) solution, acidified with dilute H_2SO_4	Orange solution turns green	Orange $Cr_2O_7^{2-}$ ions are reduced to green Cr^{3+} ions
Add aqueous bromine	Orange–red solution is decolorized	Br_2 is reduced to colourless Br^- ions

Figure 41.8 Tests for reducing agents

41.5 Redox and the activity series

The displacement reactions of metals (see p. 84) and of the halogens (see p. 108) are all redox reactions. A more reactive metal displaces a less reactive one. A more reactive halogen displaces a less reactive one. This can be explained in terms of redox and electron transfer.

When metal atoms react, they lose electrons to form positive ions. Metals supply electrons. They are reducing agents. The higher a metal is in the activity series, the more easily it gives up electrons. The higher a metal is in the series, the greater its strength as a reducing agent.

Zinc is higher in the activity series than copper, so zinc has a greater tendency to give away electrons than copper. There is a reaction when zinc is added to a solution of copper(II) sulphate. The more reactive zinc displaces copper:

$$Zn(s) \longrightarrow Zn^{2+}(aq) + 2e^-$$

$$Cu^{2+}(aq) + 2e^- \longrightarrow Cu(s)$$

> **Question**
>
> Write half equations to show the electron transfer reactions which happen when
> (a) magnesium is added to zinc sulphate solution.
> (b) copper is added to silver nitrate solution.
> (c) bromine is added to potassium iodide solution.

When non-metal elements react, the atoms gain electrons and form negative ions. Non-metals are oxidizing agents. Of the three common halogens, chlorine is the most reactive and iodine is the least reactive. Chlorine will 'grab' electrons from bromide ions or iodide ions because of its greater reactivity as an oxidizing agent:

$$Cl_2(aq) + 2e^- \longrightarrow 2Cl^-(aq)$$

$$2Br^-(aq) \longrightarrow Br_2(aq) + 2e^-$$

Summary

The main ideas about oxidation and reduction are summarized in the following statements. Illustrate each statement with an example or further explanation.

1 Oxidation

* can be the gain of oxygen
* involves a loss of electrons
* involves an increase in oxidation numbers.

2 Reduction

* can be the loss of oxygen
* involves a gain of electrons
* involves a decrease in oxidation numbers.

3 Oxidizing agents

* accept electrons
* are reduced during the reaction
* can be detected using simple chemical tests
* can vary in their reactivity.

4 Reducing agents

* provide electrons
* are oxidized during the reaction
* can be detected using simple chemical tests
* can vary in their reactivity.

5 Redox reactions

* involve oxidation and reduction at the same time
* transfer electrons between atoms so that the total number of electrons lost is the same as the total number of electrons gained.

42 Electricity from chemical reactions

42.1 Simple cells

An electric current is a flow of electrons. Redox reactions involve electron transfer. Put like this, it seems obvious to try to use redox reactions to produce electricity.

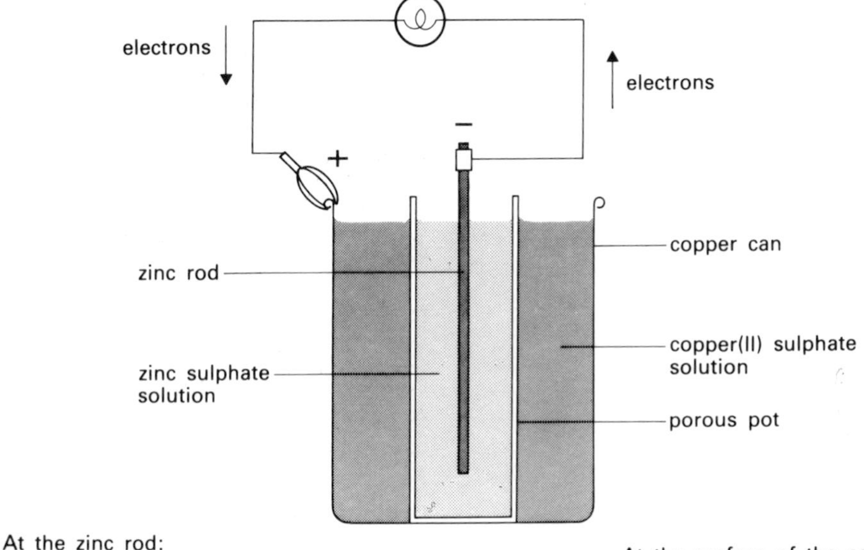

Figure 42.1 A Daniell cell

Figure 42.1 shows the chemical *cell* invented by J. F. Daniell (1790–1845). The 'porous pot' keeps the two solutions apart without breaking the circuit. The pot is soaked in a solution of ions (such as sulphuric acid) so that it is a conductor. There is no reaction until the terminals of the cell are connected by a wire. Then zinc atoms on the surface of the zinc rod turn into zinc ions by giving up electrons. The electrons flow through the wire to the copper, where they join up with copper ions and turn them into atoms. In this way, the chemical reaction produces a flow of electrons in the wire. One disadvantage of the Daniell cell is that, in time, copper ions diffuse through the porous pot into the inner container, and this wastes some of the stored energy.

The apparatus in figure 42.1 is one example of a simple chemical cell. Two or more cells joined together form a *battery*. In everyday speech, the term 'battery' is now used for both cells and batteries (see figure 42.2).

A Daniell cell cannot be stored and is easily spilled. It is not much use for powering torches and radios. The commonest cell for everyday use is the zinc–carbon ('dry') cell (which is 'dry' because the electrolyte, ammonium chloride, is made into a paste). Figure 42.3 shows a cross section of such a cell. The chemistry of these cells is more complicated than that of the Daniell cell. However, the same change happens at the negative electrode. Zinc atoms turn into zinc ions:

$$Zn(s) \longrightarrow Zn^{2+}(aq) + 2e^-$$

Questions

1 Would replacing the zinc rod and zinc sulphate solution in a Daniell cell by a magnesium rod and magnesium sulphate solution make the cell more or less powerful?
2 Why does diffusion of copper ions through the porous pot waste some of the stored chemical energy? What happens to copper ions that diffuse through the pot?
3 Why is there a limit to the amount of electricity which can be produced by a Daniell cell? What will cause the cell to run down?

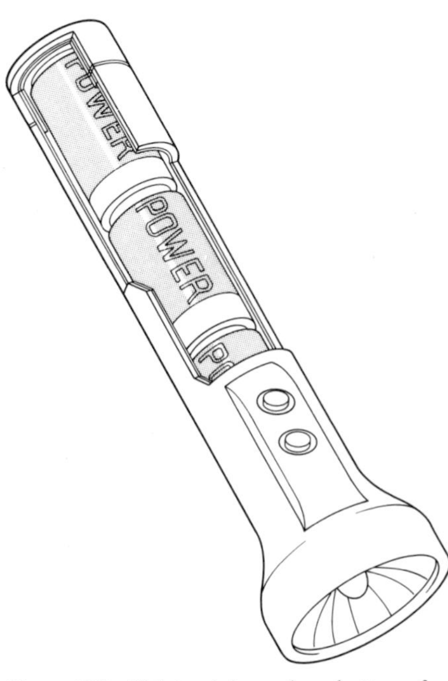

Figure 42.2 This torch is run by a battery of three cells. We usually call each cell a 'battery', which is confusing in textbooks but not in everyday life

Figure 42.4 A selection of modern batteries: the tiny battery on the right is for use in watches or hearing aids; the one next to it is for use in cameras

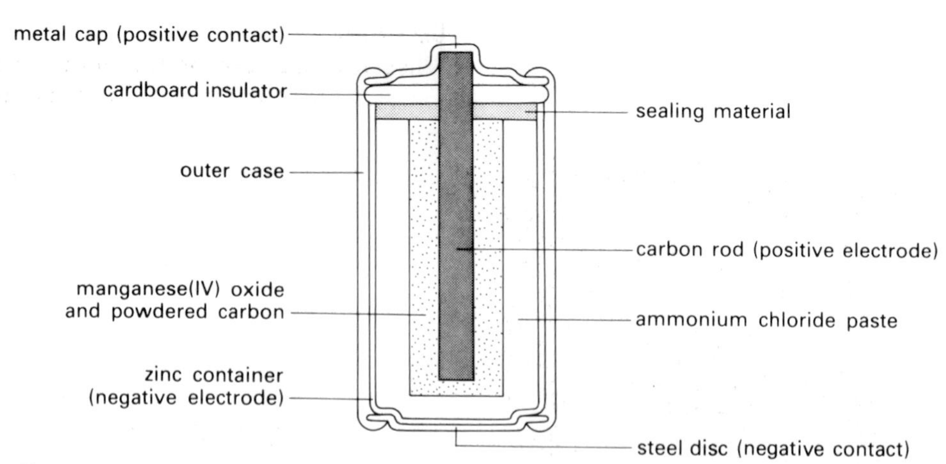

Figure 42.3 Cross section of a dry cell

In time, the zinc gets eaten away. Even with a steel outer case, the battery may then start to leak. The liquid that leaks out is very corrosive to any metals in the torch or radio. This is why old batteries should be removed and thrown away immediately after they have run down.

New types of cell have been developed for use in calculators, cameras and heart pacemakers. Some examples of these are shown in figure 42.4.

42.2 Rechargeable cells

Cells and batteries are expensive because they are made with expensive chemicals. They soon run down if they are used to power motors. One solution is to design cells which can be recharged. These cells can then be used to store electricity, collected from the mains or other source with the use of a battery charger. The commonest *rechargeable cell* is the lead–acid type, which is the basis of car batteries.

Experiment 42a
Making a lead–acid cell

dilute sulphuric acid

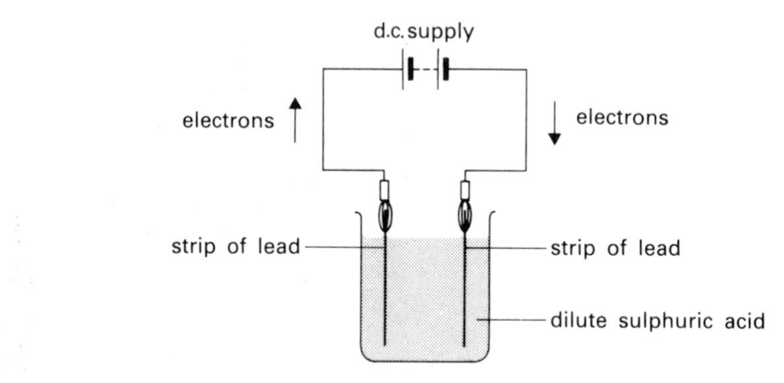

As the cell is charged:
$Pb(s) + 2H_2O(l) \rightarrow PbO_2(s) + 4H^+(aq) + 4e^-$

As the cell is charged:
$4H^+(aq) + 4e^- \rightarrow 2H_2(g)$

Figure 42.5 Charging a simple lead–acid cell

The apparatus shown in figure 42.5 will act as a simple lead–acid cell. The cell is charged by connecting it to a low voltage power supply. The diagram shows the chemical changes which take place as the cell is charged up.

Results
The simple cell used in this experiment will light a torch bulb for a few seconds. It can then be recharged again and again. It is possible to explore the connection between the length of the charging time and how long the bulb will stay alight.

Discussion
Which chemicals are formed on the electrodes as the cell charges up? What chemicals form on the plates as it runs down (see figure 42.6)?

Figure 42.7

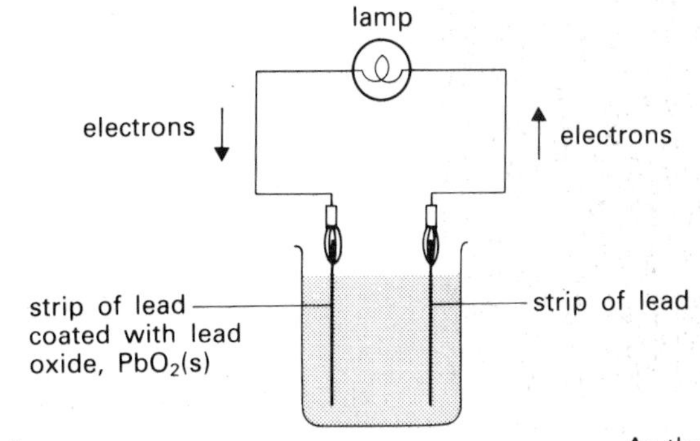

As the cell discharges:
$PbO_2(s) + 4H^+(aq) + 2e^- \rightarrow Pb^{2+}(aq) + 2H_2O(l)$

As the cell discharges:
$Pb(s) \rightarrow Pb^{2+}(aq) + 2e^-$

Figure 42.6 Discharging a simple lead–acid cell

Figure 42.7 shows the plates of a commercial lead–acid cell. Why do you think it is much more efficient than the home-made cell?

The state of charge of a lead–acid cell can be checked by using a hydrometer to measure the density of the sulphuric acid. Why does the density of the acid change as the cell is charged or discharged? Why do you think that lead–acid batteries are used to *propel* milk floats, but not motor cars?

Lead–acid cells are heavy, and each cell only produces 2 V. A car battery (see figure 42.8) usually consists of six cells together, giving 12 V. The battery is used to drive the starter motor and other electrical equipment. It is recharged by the dynamo or alternator when the engine is running.

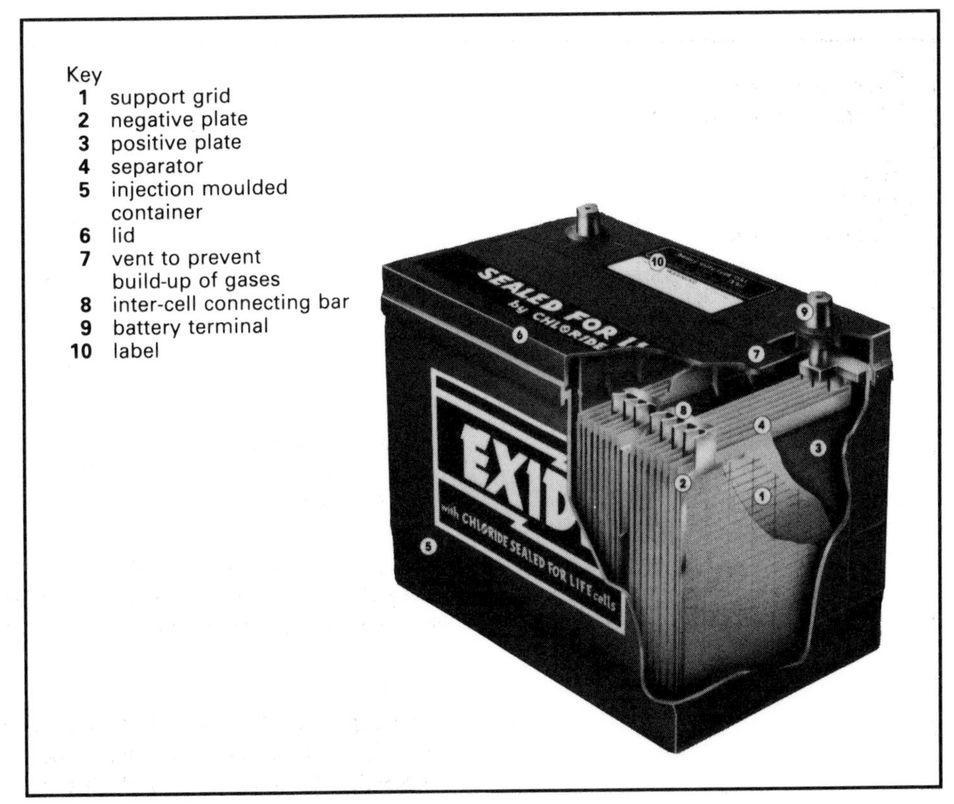

Figure 42.8 The parts of a car battery: you can see two of the six cells

Figure 42.9 Charging the battery of an experimental electrically-driven car

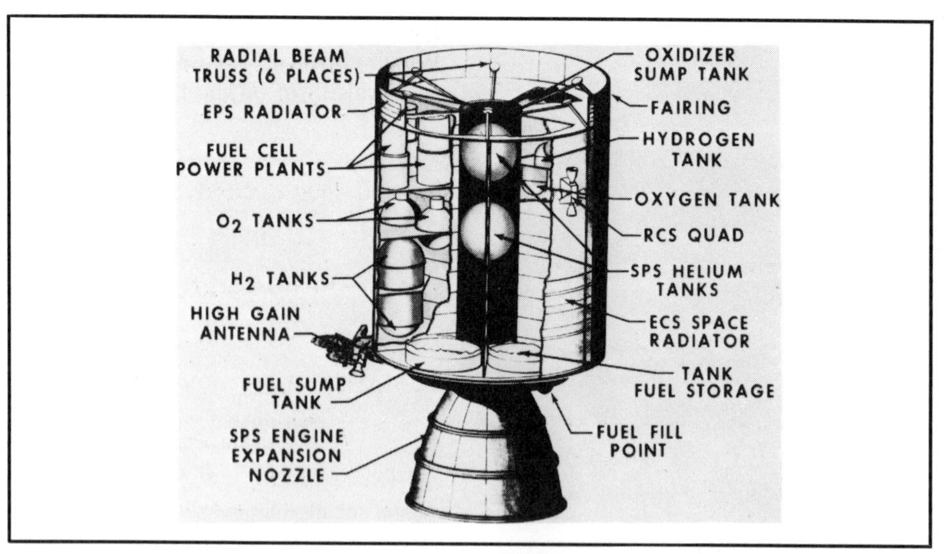

Figure 42.10 An Apollo service module, showing the fuel cells and the tanks for hydrogen and oxygen

Many new car batteries are 'sealed-for-life'. Their condition can only be checked by measuring their voltage.

Attempts are being made to develop new rechargeable cells that are lighter and produce higher voltages. It is hoped to produce cells which could be used in cars and other vehicles (see figure 42.9).

42.3 Fuel cells

Another solution to the problem of making a cell which will produce cheap electricity is the *fuel cell*. The combustion of fuels is a redox reaction. The idea is to produce electricity directly from the fuel in a cell. This is much more efficient than burning the fuel to get heat energy, which is then used to raise steam, which in turn is used to drive turbines to generate electricity (see chapter 37).

Unfortunately, no one has yet developed a reliable fuel cell which will run on the cheapest fuels. The most famous example of the use of fuel cells was the use of hydrogen–oxygen cells in the spacecraft of the Apollo moon programme (see figure 42.10).

Experiment 42b
Making a simple fuel cell

dilute sodium hydroxide

The apparatus shown in figure 42.11 can be used to demonstrate the principles of the fuel cell. Electrolysis of the dilute sodium hydroxide solution produces oxygen at the anode and hydrogen at the cathode.

Figure 42.11

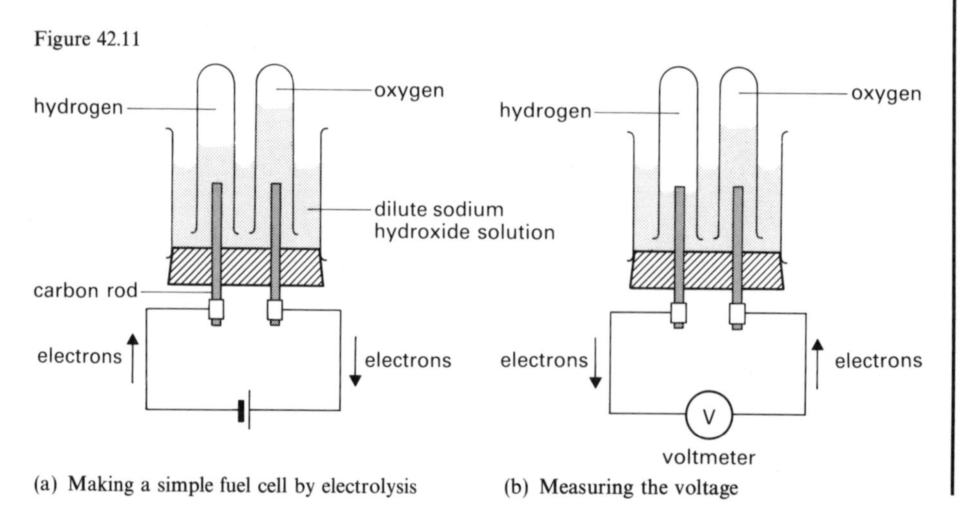

(a) Making a simple fuel cell by electrolysis (b) Measuring the voltage

Results
When electrolysis is stopped, a voltmeter is connected across the electrodes. The reading is about 1 V.

Discussion
At the oxygen electrode:

$$O_2(g) + 2H_2O(l) + 4e^- \longrightarrow 4OH^-$$

At the hydrogen electrode:

$$2H_2(g) \longrightarrow 4H^+(aq) + 4e^-$$

Add these two equations together to find the overall reaction in the cell.

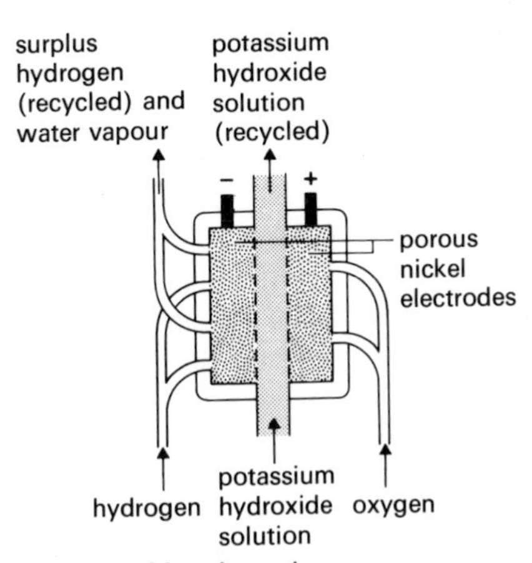

At the positive electrode:
$O_2(g) + 2H_2O(l) + 4e^- \longrightarrow 4OH^-(aq)$

At the negative electrode:
$2H_2(g) + 4OH^-(aq) \longrightarrow 4H_2O(l) + 4e^-$

Figure 42.12 The Bacon fuel cell

The simple cell in experiment 42b is not a true fuel cell because hydrogen and oxygen cannot be supplied continuously to its electrodes to maintain the source of energy. Figure 42.12 shows a diagram of one type of true fuel cell.

Experimental cells, using oxygen with fuels such as propane or methanol, have been designed, but the practical problems have yet to be overcome.

Summary

In all electric cells, redox reactions provide the flow of electricity. However, there are practical differences between different kinds of electric cells.

The table in figure 42.13 includes four types of cell that have been described in this chapter. Copy the table. For each type of cell, list its advantages, limitations and any practical uses.

Cell type	Advantages	Limitations	Practical uses
Simple wet cell, e.g. Daniell cell			
Simple dry cell, e.g. torch cell			
Rechargeable cell, e.g. lead–acid cell			
Fuel cell, e.g. hydrogen–oxygen fuel cell			

Figure 42.13

Review questions

1 (a) The apparatus below was used for the electrolysis of molten potassium iodide (K^+I^-).

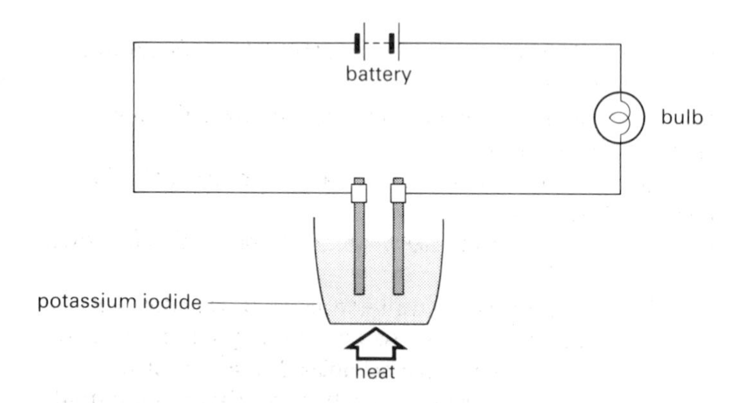

(i) What was the purpose of the bulb in the circuit?
(ii) Explain why electrolysis did not take place until the potassium iodide was molten.
(iii) The electrolysis of molten potassium iodide produced potassium at the negative electrode and iodine at the positive electrode. Explain the changes that took place at the negative electrode and the positive electrode.
(iv) State *two* ways of speeding up the electrolysis of molten potassium iodide.

(b) Electrolysis of an aqueous solution of sodium chloride was carried out with carbon electrodes. Draw a diagram of an apparatus that could be used for the electrolysis of an aqueous solution of sodium chloride and to collect the gaseous products of electrolysis.
(c) Electrolysis of an aqueous solution of sodium chloride with carbon electrodes produces a gas at both positive and negative electrodes. An aqueous solution of sodium chloride contains Na^+(aq), Cl^-(aq), H^+(aq) and OH^-(aq) ions.

(i) Which ion is present in excess in all alkaline solutions?
(ii) Which ion is discharged at the negative electrode during electrolysis of sodium chloride solution?
(iii) Which gas is produced at the negative electrode during electrolysis of sodium chloride solution?
(iv) A small amount of universal indicator was added to the sodium chloride solution before electrolysis. During the electrolysis of sodium chloride solution, the solution around the negative electrode went purple (alkaline). Why did this happen? (The addition of universal indicator does not affect the products of electrolysis.)

(EAEB)

2 Passing electricity through molten calcium bromide produces a metallic bead at the negative electrode and a red–brown gas at the positive electrode.
(a) What is the name given to the splitting up of an electrolyte such as calcium bromide by electricity?
(b) Name (i) a suitable material for the electrodes, (ii) the metal formed and (iii) the red–brown gas.
(c) Draw a labelled diagram of the apparatus set up for this experiment. (You are supplied with a battery, connecting wires, a bulb, two electrodes, a crucible containing calcium bromide, a pipeclay triangle, a tripod and a Bunsen burner.)
(d) Solid calcium bromide contains Ca^{2+} and Br^- ions.

(i) Write the chemical formula for calcium bromide.
(ii) Why does solid calcium bromide not conduct electricity?

(e) After the electricity had passed for some time the bulb started to glow brighter and the bulb continued to glow even when the apparatus had cooled to room temperature. Explain why this happened.
(f) When washing the crucible at the end of the experiment, a colourless gas was produced. The gas burned with a squeaky pop when tested with a lighted splint.

(i) Name the gas formed.
(ii) Explain how this gas was formed. **(EAEB)**

3 (a) Consider the following substances: lead(II) bromide, sugar, hydrogen chloride, copper, sodium chloride, ethanol, potassium iodide, sulphur. From these substances name those that

(i) conduct electricity in the solid state,
(ii) conduct electricity *both* in the liquid state (i.e. when molten) *and* in aqueous solution,
(iii) do not conduct electricity themselves but form conducting liquids when dissolved in water,
(iv) do not readily conduct electricity under any of these conditions.

You may list any substance more than once if appropriate.
(b) An aqueous solution of calcium hydroxide is electrolysed between carbon electrodes.

(i) What gas would you *expect* to be produced at the anode (positive electrode)?

It is observed that, during the electrolysis, the mass of the anode decreases and the solution round it becomes milky.

(ii) Suggest an explanation for these observations.

(c) A strip of moistened filter paper is laid on a microscope slide. A drop of silver nitrate solution is placed near one end of the paper and a drop of potassium iodide solution near the other end. Using inert electrodes, the apparatus is connected to a suitable d.c. supply. After some time, a pale yellow streak appears as shown below.

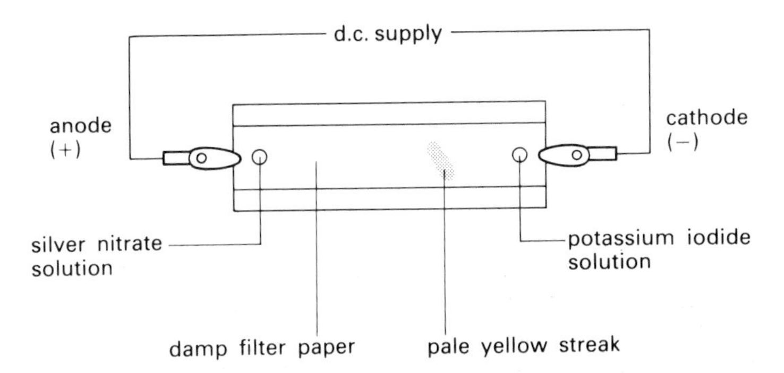

(i) Give the formulae of the ions present in the solutions of silver nitrate and potassium iodide.
(ii) Name the compound responsible for the pale yellow streak and write the ionic equation for its formation.
(iii) Explain the process leading to the formation of the streak and explain why the streak appears nearer the cathode than the anode. **(CLES)**

4

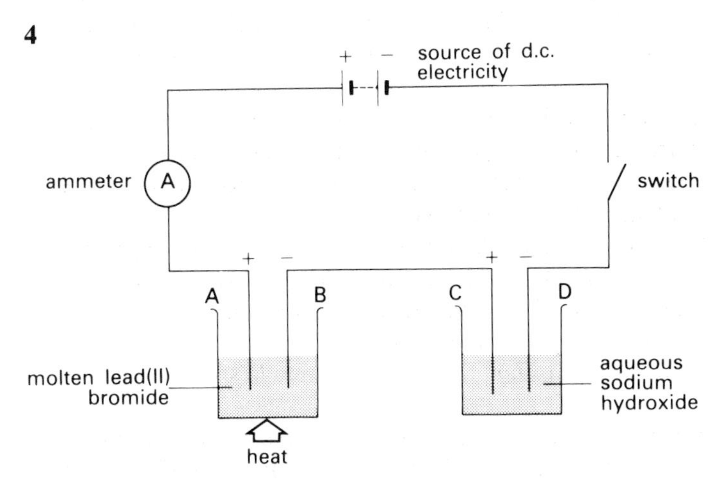

A current of 0.4 amperes is passed for 50 minutes through two cells containing respectively (i) molten lead(II) bromide and (ii) aqueous sodium hydroxide, as shown in the diagram.
(a) (i) Write ionic equations for the reactions occurring at the inert electrodes A and B.
(ii) State the products liberated at the inert electrodes C and D.
(b) (i) How many coulombs of electricity are used in the experiment?
(ii) How many moles of electrons does this represent?
(iii) How many *moles* of product are liberated at electrode B?
(iv) How many *grams* of product are liberated at electrode A? **(L)**

5 The apparatus shown in the figure below was used to investigate the electrolysis of concentrated hydrochloric acid and to determine the charge on the hydrogen ion. A current of 0.6 A (amperes) was passed for 16 min; the volume of

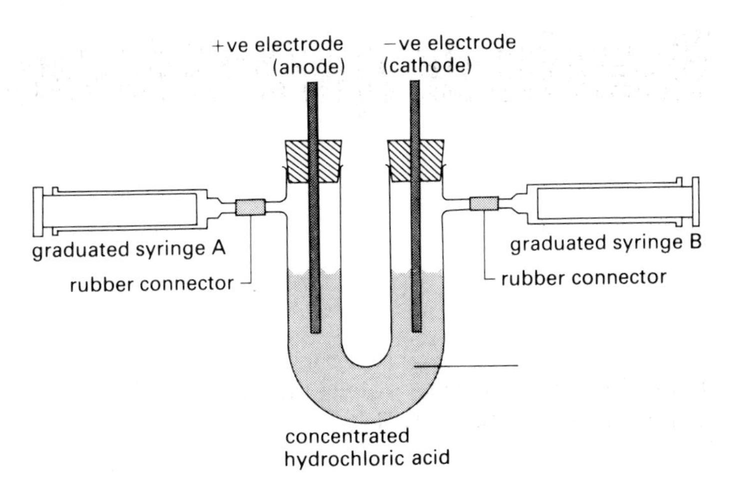

hydrogen collected was 72 cm^3 and the volume of chlorine collected was 60 cm^3.
(a) *Name* the gas collecting in syringe A. Describe *one* chemical test for this gas.
(b) Why are the electrodes made of carbon and not of a metal such as iron?
(c) Explain why the volume of chlorine collected is less than the volume of hydrogen.
(d) Use the numerical information given to calculate the number of unit charges carried by a hydrogen ion. (Assume that 1 mole of hydrogen gas occupies 24 dm^3 under the conditions of the experiment and that the Faraday constant is 96 000 coulombs per mole of electrons.)
(e) The experiment was repeated using dilute sulphuric acid and platinum electrodes, instead of concentrated hydrochloric acid and carbon electrodes. The same current was passed for the same length of time.
(i) What products are set free at the electrodes?
(ii) What volumes of gases are obtained? **(CLES)**

6 The industrial production of aqueous sodium hydroxide (NaOH) from a saturated sodium chloride solution (brine) involves two separate processes:

Stage 1: The electrolysis
Brine is electrolysed using graphite anodes placed 2 mm above a cathode consisting of a moving pool of mercury.
Chlorine is produced at the anode and sodium is formed at the cathode where it dissolves in the mercury forming a mixture known as sodium-amalgam. A current of 32 000 amperes is used.

Stage 2: The production of aqueous sodium hydroxide
The sodium-amalgam from the electrolysis is passed into water where the sodium reacts to form sodium hydroxide solution.
(a) From where might the brine for this process be obtained?
(b) Write a balanced equation including state symbols for the reaction of sodium with water in stage 2.
(c) The sodium hydroxide solution produced in the second process contains 40 g of sodium hydroxide in every 100 cm^3 of solution. What is the concentration of this solution in mol/dm^3?

(d) Calculate to the *nearest second* how long it will take to produce 1 mole of atoms of sodium metal in the electrolytic process. (The sodium ion is Na^+.)
(e) Write a balanced ionic equation (or equations) to show how chlorine molecules are produced at the anode in the first process. (Use the symbol e^- to represent an electron or mole of electrons.) **(L)**

7 (a) State the meanings of the terms *oxidation* and *reduction*.
(b) (i) When coke or charcoal is burned in an open fire, a light blue flame is often seen on the surface of the fire. Name the gas which is burning, explain how it is produced in the fire and give the equation for its combustion.
(ii) Copper was first obtained by heating copper(II) carbonate with charcoal. How is the carbonate converted into copper in this way? Give the equations.
(iii) By roasting zinc sulphide with charcoal in air, zinc can be obtained. Suggest, giving equations, the reactions which occur.
(c) From the reactions in (b) give *one* example of *oxidation* and *one* example of *reduction*.
(d) Zinc and copper form an alloy, brass, which contains one-third zinc and two-thirds copper by mass. Assuming that pure zinc sulphide was completely converted into zinc, calculate the mass of zinc sulphide required to convert 130 g copper into brass. **(AEB)**

8 (a) Ions are formed when atoms gain or lose electrons. Illustrate this statement by giving ionic equations to show the formation of the following four ions from their atoms: potassium, sulphide, oxide, calcium.
(b) Define *oxidation* and *reduction* in terms of electron transfer.
(c) For each of the four ions in (a), state whether it has been formed by oxidation or by reduction. **(O)**

9 (a) Oxidation was originally the term used to describe a chemical process which could be brought about by the action of oxygen. For example, magnesium can be converted to magnesium oxide by heating the metal in oxygen. Write an equation for this reaction and explain why it is now referred to as a *redox* reaction.
(b) For each of the following reactions write the equation, state what would be seen, and state whether the reaction is an example of a redox reaction, an acid–base reaction or a precipitation reaction: the reactions occurring when
(i) chlorine is bubbled through a solution of potassium bromide,
(ii) solutions of ammonia and sulphuric acid are mixed,
(iii) solutions of silver nitrate and hydrochloric acid are mixed
(iv) chlorine is passed over hot iron. **(JMB)**

10 Describe how you would demonstrate in the laboratory that redox reactions take place when (a) bromine water is added to a solution of Fe^{2+} ions, (b) chlorine gas is bubbled into a solution of potassium iodide, (c) zinc is added to dilute hydrochloric acid and (d) copper is added to a solution containing Ag^+ ions. Use oxidation state diagrams to explain the changes which take place in these reactions. **(L)**

11 Describe how you would bring about the changes of oxidation state shown in the diagram below.

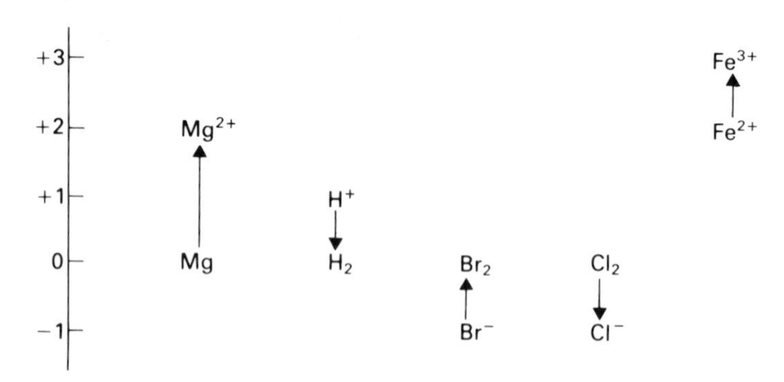

State what you would see as the changes took place and mention any chemical tests you might use to confirm that you had produced the required products. Write equations for the reactions and show that electrons are being transferred. **(L)**

12 Draw a labelled diagram of the apparatus you would use to demonstrate that it is possible to obtain electrical energy directly from a chemical reaction. (Give the equation for the reaction you choose.) Explain how you would find out experimentally the maximum amount of electrical energy available from the cell you have described? What are the advantages and disadvantages of your cell as an everyday source of electricity? **(L)**

13 Oxidation–reduction reactions occur when (a) an aqueous electrolyte is being electrolysed, and (b) when an electrical cell is supplying current. Illustrate this statement by discussing one example of each kind. **(L)**

Reference section

The standard kilogram is a Pt/Ir cylinder measuring 39 mm × 39 mm. It is kept at Sèvres, in France

Tables of data

Table 1 Physical properties of some elements

This table gives data for the elements in the shaded boxes on the periodic table on p. 97. Their *states at room temperature* are shown by 's', 'l' or 'g'. The abbreviation 'sub' indicates that the element sublimes. The densities of gases are at 25 °C.

Element	**Symbol**	**Atomic number**	**Relative atomic mass**	**State**	**Melting point (°C)**	**Boiling point (°C)**	**Density (g/cm^3)**
aluminium	Al	13	27	s	660	2350	2.70
argon	Ar	18	40	g	−189	−186	0.00166
arsenic	As	33	75	s	613 (sub)		5.78
barium	Ba	56	137	s	710	1640	3.59
beryllium	Be	4	9	s	1285	2470	1.85
boron	B	5	11	s	2030	3700	2.47
bromine	Br	35	80	l	−7	59	3.12
calcium	Ca	20	40	s	840	1490	1.53
carbon (diamond)	C	6	12	s	3550	4827	3.53
carbon (graphite)	C	6	12	s	3720 (sub)		2.25
chlorine	Cl	17	35.5	g	−101	−34	0.00299
chromium	Cr	24	52	s	1860	2600	7.19
cobalt	Co	27	59	s	1494	2900	8.80
copper	Cu	29	64	s	1084	2580	8.93
fluorine	F	9	19	g	−220	−188	0.00158
gallium	Ga	31	70	s	30	2070	5.91
germanium	Ge	32	73	s	959	2850	5.32
gold	Au	79	197	s	1064	2850	19.28
helium	He	2	4	g	−270	−269	0.00017
hydrogen	H	1	1	g	−259	−253	0.00008
iodine	I	53	127	s	114	184	4.95
iron	Fe	26	56	s	1540	2760	7.87
krypton	Kr	36	84	g	−157	−153	0.00346
lead	Pb	82	207	s	327	1760	11.34
lithium	Li	3	7	s	180	1360	0.53
magnesium	Mg	12	24	s	650	1100	1.74
manganese	Mn	25	55	s	1250	2120	7.47
mercury	Hg	80	201	l	−39	357	13.55
neon	Ne	10	20	g	−249	−246	0.00084
nickel	Ni	28	59	s	1455	2150	8.91
nitrogen	N	7	14	g	−210	−196	0.00117
oxygen	O	8	16	g	−219	−183	0.00133
phosphorus (white)	P	15	31	s	44	280	1.82
platinum	Pt	78	195	s	1772	3720	21.45
potassium	K	19	39	s	63	777	0.86
rubidium	Rb	37	85	s	39	705	1.53

Element	Symbol	Atomic number	Relative atomic mass	State	Melting point (°C)	Boiling point (°C)	Density (g/cm^3)
scandium	Sc	21	45	s	1540	2800	2.99
selenium	Se	34	79	s	220	685	4.81
silicon	Si	14	28	s	1410	2620	2.33
silver	Ag	47	108	s	962	2160	10.50
sodium	Na	11	23	s	98	900	0.97
sulphur (monoclinic)	S	16	32	s	115	445	1.96
sulphur (orthorhombic)	S	16	32	s			2.07
tin	Sn	50	119	s	232	2720	7.28
titanium	Ti	22	48	s	1670	3300	4.51
uranium	U	92	238	s	1135	4000	19.05
vanadium	V	23	51	s	1920	3400	6.09
xenon	Xe	54	131	g	−112	−108	0.0055
zinc	Zn	30	65	s	420	913	7.14

Table 2 Physical properties of some inorganic compounds

This table shows the *formulae, melting points* and *boiling points* of some inorganic compounds. Their *states at room temperature* are shown by 's', 'l' or 'g'. The *solubility* of each compound in water at room temperature is broadly classified as 'i' (insoluble), 'sps' (sparingly soluble), 's' (soluble) or 'vs' (very soluble). The abbreviation 'r' indicates that the compound reacts with water. Some compounds can have hydrated forms. This is shown by 'h'. These forms will usually dehydrate before the melting point. The abbreviation 'dec' means that the compound decomposes on heating; the abbreviation 'sub' means that it sublimes. Unless stated otherwise, all the compounds are white or colourless.

Compound	Formula	State	Melting point (°C)	Boiling point (°C)	Solubility	Notes
aluminium chloride	$AlCl_3$	s	sub		r	
aluminium oxide	Al_2O_3	s	2015	2980	i	
ammonium chloride	NH_4Cl	s	sub		s	
ammonium nitrate	NH_4NO_3	s	170	dec	vs	
ammonium sulphate	$(NH_4)_2SO_4$	s	dec		s	
barium chloride	$BaCl_2$	s	963	1560	s	h
barium oxide	BaO	s	1923	2000	r	
barium sulphate	$BaSO_4$	s	1580		i	
beryllium chloride	$BeCl_2$	s	410	492	s	
beryllium oxide	BeO	s	2540	4120	i	
boron trichloride	BCl_3	g	−107	12	r	
boron hydride	B_2H_6	g	−165	−92	r	
boron oxide	B_2O_3	s	460	1860	i	
caesium chloride	$CsCl$	s	645	1300	vs	
calcium carbonate	$CaCO_3$	s	dec		i	
calcium chloride	$CaCl_2$	s	782	2000	s	h
calcium hydroxide	$Ca(OH)_2$	s	dec		sps	
calcium nitrate	$Ca(NO_3)_2$	s	561	dec	vs	h

Compound	Formula	State	Melting point (°C)	Boiling point (°C)	Solubility	Notes
calcium oxide	CaO	s	2600	3000	r	
carbon monoxide	CO	g	−205	−191	i	
carbon dioxide	CO_2	g	sub		sps	
carbon disulphide	CS_2	l	−112	46	i	
chlorine monoxide	Cl_2O	g	−20		r	
chlorine dioxide	ClO_2	g	−60		r	
chromium(III) chloride	$CrCl_3$	s	1150	1300	i	h, green
chromium(III) oxide	Cr_2O_3	s	2435	4000	i	green
cobalt(II) chloride	$CoCl_2$	s	730	1050	s	h, red
copper(II) chloride	$CuCl_2$	s	620	dec	s	h, green
copper(II) nitrate	$Cu(NO_3)_2$	s	114	dec	vs	h, green
copper(I) oxide	Cu_2O	s	1235		i	red
copper(II) oxide	CuO	s	1326		i	black
copper(II) sulphate	$CuSO_4$	s	dec		s	h, blue
hydrogen bromide	HBr	g	−87	−67	vs	
hydrogen chloride	HCl	g	−114	−85	vs	
hydrogen fluoride	HF	g	−93	20	s	
hydrogen iodide	HI	g	−51	−35	s	
hydrogen oxide (water)	H_2O	l	0	100		
hydrogen peroxide	H_2O_2	l	0	150	vs	
hydrogen sulphide	H_2S	g	−85	−60	sps	
iron(II) chloride	$FeCl_2$	s	677	sub	s	yellow–green
iron(III) chloride	$FeCl_3$	s	307	dec	s	h, orange
iron(III) oxide	Fe_2O_3	s	1565		i	red
iron(II) sulphate	$FeSO_4$	s	dec		s	pale green
iron(II) sulphide	FeS	s	1196	dec	i	black
lead(II) bromide	$PbBr_2$	s	370	914	i	
lead(II) chloride	$PbCl_2$	s	501	950	sps	
lead(II) nitrate	$Pb(NO_3)_2$	s	dec		s	
lead(II) oxide	PbO	s	886	1472	i	yellow
lead(IV) oxide	PbO_2	s	dec		i	brown
lead(II) sulphate	$PbSO_4$	s	1170		i	
lead(II) sulphide	PbS	s	1114		i	black
lithium chloride	$LiCl$	s	614	1382	s	
lithium hydride	LiH	s	680		r	
lithium oxide	Li_2O	s	1700		r	
magnesium chloride	$MgCl_2$	s	714	1418	s	h
magnesium nitrate	$Mg(NO_3)_2$	s	89		vs	h
magnesium oxide	MgO	s	2800	3600	i	
manganese(II) chloride	$MnCl_2$	s	650	1190	s	h, pink
manganese(IV) oxide	MnO_2	s	dec		i	black
manganese(II) sulphate	$MnSO_4$	s	700	dec	s	h, pink
mercury(II) chloride	$HgCl_2$	s	276	302	sps	
mercury(II) oxide	HgO	s	dec		i	red
nickel(II) chloride	$NiCl_2$	s	1001	sub	s	yellow
nickel(II) oxide	NiO	s	1990		i	green–black
nickel(II) sulphate	$NiSO_4$	s			s	h, green

Compound	Formula	State	Melting point (°C)	Boiling point (°C)	Solubility	Notes
nitric acid	HNO_3	l	−42	83	vs	
nitrogen trichloride	NCl_3	l	−40	71	i	
nitrogen hydride (ammonia)	NH_3	g	−78	−34	vs	
nitrogen oxide	NO	g	−163	−151	sps	
nitrogen dioxide	NO_2	g	−11	21	s	brown
phosphorus trichloride	PCl_3	l	−112	76	r	
phosphorus pentachloride	PCl_5	s	dec		r	
phosphorus hydride (phosphine)	PH_3	g	−133	−90	i	
phosphorus pentoxide	P_4O_{10}	s	sub		r	
potassium bromide	KBr	s	730	1435	s	
potassium chloride	KCl	s	776	1500	s	
potassium hydroxide	KOH	s	360	1322	vs	
potassium iodide	KI	s	686	1330	vs	
potassium manganate(VII)	$KMnO_4$	s	dec		s	purple
potassium nitrate	KNO_3	s	334	dec	vs	
rubidium chloride	RbCl	s	715	1390	s	
silicon tetrachloride	$SiCl_4$	l	−70	58	r	
silicon hydride (silane)	SiH_4	g	−185	−112	i	
silicon dioxide (quartz)	SiO_2	s	1610	2230	i	
silver bromide	AgBr	s	432	dec	i	cream
silver chloride	AgCl	s	455	1550	i	
silver iodide	AgI	s	558	1506	i	yellow
silver nitrate	$AgNO_3$	s	212	dec	vs	
sodium bromide	NaBr	s	755	1390	s	
sodium carbonate	Na_2CO_3	s	851	dec	s	h
sodium chloride	NaCl	s	808	1465	s	
sodium hydroxide	NaOH	s	318	1390	s	
sodium nitrate	$NaNO_3$	s	307	dec	s	
sodium oxide	Na_2O	s	sub		r	
sodium sulphate	Na_2SO_4	s	890		s	h
sodium thiosulphate	$Na_2S_2O_3$	s	dec		vs	h
sulphur monochloride	S_2Cl_2	l	−80	136	r	yellow
sulphur dichloride	SCl_2	l	−78	dec	r	red
sulphur dioxide	SO_2	g	−75	−10	vs	
sulphur trioxide	SO_3	l	−17	43	r	
sulphuric acid	H_2SO_4	l	10	330	vs	
zinc chloride	$ZnCl_2$	s	283	732	vs	
zinc oxide	ZnO	s	1975		i	
zinc sulphate	$ZnSO_4$	s	740	dec	vs	h

Table 3 Physical properties of some organic compounds

This table shows the *formulae*, *melting points* and *boiling points* of some organic compounds, which are grouped together as far as possible according to their type. Their *states at room temperature* are shown by 's', 'l' or 'g'. The *heat of combustion* is the energy released when one mole of the compound is completely burned.

Compound	Formula	State	Melting point (°C)	Boiling point (°C)	Heat of combustion (kJ/mol)
Alkanes					
methane	CH_4	g	−182	−161	−890
ethane	C_2H_6	g	−183	−88	−1560
propane	C_3H_8	g	−188	−42	−2220
butane	C_4H_{10}	g	−138	−0.5	−2877
pentane	C_5H_{12}	l	−130	36	−3509
hexane	C_6H_{14}	l	−95	69	−4195
decane	$C_{10}H_{22}$	l	−30	174	−6778
hexadecane	$C_{16}H_{34}$	l	18	287	
eicosane	$C_{20}H_{42}$	s	37	344	
Alkenes					
ethene	C_2H_4	g	−169	−104	−1411
propene	C_3H_6	g	−185	−48	−2058
Alcohols					
methanol	CH_3OH	l	−98	65	−726
ethanol	C_2H_5OH	l	−114	78	−1366
propan-1-ol	C_3H_7OH	l	−126	97	−2017
butan-1-ol	C_4H_9OH	l	−89	118	−2675
Carboxylic acids					
methanoic (formic) acid	HCO_2H	l	9	101	−270
ethanoic (acetic) acid	CH_3CO_2H	l	17	118	−873
propanoic acid	$C_2H_5CO_2H$	l	−21	141	−1574
octadecanoic (stearic) acid	$C_{17}H_{35}CO_2H$	s	71	375	
Esters					
ethyl ethanoate (acetate)	$CH_3CO_2C_2H_5$	l	−84	77	−2238
ethyl propanoate	$C_2H_5CO_2C_2H_5$	l	−74	99	−2890
methyl ethanoate	$CH_3CO_2CH_3$	l	−98	57	−1593
methyl propanoate	$C_2H_5CO_2CH_3$	l	−87	80	−2246
Halogen-containing compounds					
chloromethane	CH_3Cl	g	−98	−24	
trichloromethane (chloroform)	$CHCl_3$	l	−63	61	
tetrachloromethane	CCl_4	l	−23	77	
1,1,1-trichloroethane	CH_3CCl_3	l	−30	74	
Miscellaneous					
glucose	$C_6H_{12}O_6$	s	146	dec	−964
naphthalene	$C_{10}H_8$	s	80	218	−5149
propanone (acetone)	CH_3COCH_3	l	−95	56	−1821
sucrose	$C_{12}H_{22}O_{11}$	s	186	dec	−5645

Table 4 Heats of fusion and vaporization of some elements and compounds

In this table, substances are grouped together according to their type. The *heat of fusion* is the energy needed to change one mole of a substance from a solid into liquid at its melting point. The *heat of vaporization* is the energy needed to change one mole of a substance from a liquid into a gas at its boiling point.

Substance	Symbol or formula	Melting point (°C)	Boiling point (°C)	Heat of fusion (kJ/mol)	Heat of vaporization (kJ/mol)
Metal elements					
calcium	Ca	840	1490	8.7	149.8
iron	Fe	1540	2760	15.4	351.0
lead	Pb	327	1760	4.8	179.5
magnesium	Mg	650	1100	8.9	128.7
potassium	K	63	777	2.3	77.5
silver	Ag	962	2160	11.3	255.1
sodium	Na	98	900	2.6	89.0
zinc	Zn	420	913	7.4	115.3
Non-metal elements					
bromine	Br_2	−7	59	5.3	15.0
chlorine	Cl_2	−101	−34	3.2	10.2
hydrogen	H_2	−159	−253	0.06	0.5
iodine	I_2	114	184	7.9	20.9
neon	Ne	−249	−246	0.33	1.8
oxygen	O_2	−219	−183	0.22	3.4
phosphorus	P_4	44	280	0.63	12.4
sulphur	S_8	115	445	1.41	9.6
Compounds containing metal and non-metal elements					
lithium chloride	LiCl	614	1382	13.4	150.6
potassium bromide	KBr	730	1435	29.3	155.6
potassium chloride	KCl	776	1500	25.5	163.1
silver chloride	AgCl	455	1550	13	183
sodium chloride	NaCl	808	1465	28.9	170.2
sodium iodide	NaI	651	1304	22.2	159.4
zinc chloride	$ZnCl_2$	283	732	23	129
Compounds containing non-metal elements only					
ammonia	NH_3	−78	−34	5.6	23
ethanol	C_2H_5OH	−114	78	5.0	39
hydrogen chloride	HCl	−114	−85	2.0	16
methane	CH_4	−182	−161	0.9	8.2
propanone (acetone)	CH_3COCH_3	−95	56	5.7	30
sulphur dioxide	SO_2	−75	−10	7.4	25
tetrachloromethane	CCl_4	−23	77	2.5	30
water	H_2O	0	100	6.0	41

Table 5 Densities of some common materials

Material	Density (g/cm^3)
balsa wood	0.2
brass	8.4
brick	1.5–1.8
cedar wood	0.55
concrete	2.2–2.4
Duralumin	2.8
glass (Pyrex)	2.23
mahogany	0.8
marble	2.7
mild steel	7.9
nylon	1.12–1.17
pine	0.5
polystyrene (not expanded)	1.04–1.09
polythene	0.91–0.96
stainless steel	7.8

Table 6 Solubilities of some salts in water

In the table, the solubilities are given as the mass (in grams) of the salt which can be dissolved in 100 g of water to give a saturated solution at the temperature shown.

		Temperature (°C)						
Salt	**Formula**	**0**	**10**	**20**	**40**	**60**	**80**	**100**
ammonium chloride	NH_4Cl	29.4	33.3	37.2	45.8	55.2	65.6	77.3
copper(II) sulphate	$CuSO_4.5H_2O$	14.3	17.4	20.7	28.5	40.0	55.0	75.4
potassium chloride	KCl	28.1	31.2	34.2	40.0	45.8	51.3	56.3
potassium bromide	KBr	53.5	59.5	65.2	75.5	85.5	95.0	104
potassium nitrate	KNO_3	13.3	20.9	31.6	63.9	110	169	246
sodium chloride	$NaCl$	35.7	35.8	36.0	36.6	37.3	38.4	39.8

Table 7 Solubilities of some gases in water

The solubility of a gas is shown as the volume in cm^3 which will dissolve in 1 cm^3 of water. The gas volumes are those that the gas would have at 0 °C. The solubilities are given at three temperatures. (Room temperature is about 20 °C.)

Gas	Formula	Temperature (°C) 10	20	30
ammonia	NH_3	870	680	530
argon	Ar	0.041	0.032	0.028
carbon dioxide	CO_2	1.16	0.848	0.652
chlorine	Cl_2	3.09	2.26	1.77
helium	He	0.0091	0.0086	0.0084
hydrogen	H_2	0.0195	0.0182	0.0170
hydrogen chloride	HCl	475	442	412
hydrogen sulphide	H_2S	3.28	2.51	1.97
nitrogen	N_2	0.0183	0.0152	0.0133
oxygen	O_2	0.037	0.030	0.026
sulphur dioxide	SO_2	56.6	39.4	27.2

Table 8 Charges on some common ions

Positive ions (cations)

Charge	Cation	Symbol
1+	copper(I)	Cu^+
	hydrogen	H^+
	lithium	Li^+
	potassium	K^+
	silver	Ag^+
	sodium	Na^+
2+	calcium	Ca^{2+}
	copper(II)	Cu^{2+}
	iron(II)	Fe^{2+}
	lead(II)	Pb^{2+}
	magnesium	Mg^{2+}
	nickel(II)	Ni^{2+}
	zinc	Zn^{2+}
3+	aluminium	Al^{3+}
	iron(III)	Fe^{3+}

Negative ions (anions)

Charge	Anion	Symbol
1−	bromide	Br^-
	chloride	Cl^-
	hydroxide	OH^-
	iodide	I^-
	nitrate	NO_3^-
2−	carbonate	CO_3^{2-}
	oxide	O^{2-}
	sulphate	SO_4^{2-}
	sulphide	S^{2-}
	sulphite	SO_3^{2-}
3−	nitride	N^{3-}
	phosphate	PO_4^{3-}

Units

A physical quantity is expressed as a number with a unit. Below are some notes about the units used in this book. In general, for every quantity there is a *base unit*, which may have a *prefix*.

Prefixes commonly used are:

* kilo (k), which means $\times 1000$
* deci (d), which means $\times 0.1$ or $\times \frac{1}{10}$
* centi (c), which means $\times 0.01$ or $\times \frac{1}{100}$
* milli (m), which means $\times 0.001$ or $\times \frac{1}{1000}$

For example,

1 km = 1000 m
1 dm = 0.1 m or $\frac{1}{10}$ m

Physical quantity	Unit	Notes
length	m (metre) cm (centimetre) mm (millimetre) nm (nanometre)	The nanometre is used for expressing the size of atomic particles. 1 m = 1 000 000 000 nm 1 m = 100 cm 1 cm = 10 mm 1 mm = 1 000 000 nm
volume	dm^3 cm^3 l (litre)	The unit of volume in everyday use is the litre (l). 1 l = 1 dm^3 On some glassware, volume is shown in millilitres (ml). 1 ml = 1 cm^3 1 dm^3 = 1000 cm^3
mass	g (gram) kg (kilogram)	1 kg = 1000 g
time	s (second)	Although the second is the basic unit of time, it is sometimes more convenient to measure times of reactions in minutes.
temperature	°C (degree Celsius)	The Celsius scale is the one most commonly used in the laboratory. The internationally recommended unit is the kelvin (K). On the kelvin scale 0 °C = 273 K 100 °C = 373 K
amount of substance	mol (mole)	The definition of the mole is included in the glossary.
energy	J (joule) kJ (kilojoule)	1 kJ = 1000 J
power	W (watt)	Power is the rate of transfer of energy. 1 W = 1 J/s
amount of electricity (electric charge)	C (coulomb)	The rate of flow of electric charge is the current. 1 A = 1 C/s
current	A (ampere)	

Names of chemicals

The chemical names used throughout this book are the ones which are likely to be most familiar. Changes in names are taking place to comply with international recommendations on the naming of substances. Some equivalent names are shown in the following list.

Traditional name	**Newer, recommended name**
acetic acid	ethanoic acid
acetone	propanone
acetylene	ethyne
carbon tetrachloride	tetrachloromethane
chloroform	trichloromethane
ethyl acetate	ethyl ethanoate
ethylene	ethene
formic acid	methanoic acid
hypochlorite ion	chlorate(I) ion
nitrate ion	nitrate(V) ion
nitric acid	nitric(V) acid
nitrite ion	nitrate(III) ion
nitrous acid	nitric(III) acid
permanganate ion	manganate(VII) ion
stearic acid	octadecanoic acid
sulphate ion	sulphate(VI) ion
sulphite ion	sulphate(IV) ion
sulphuric acid	sulphuric(VI) acid
sulphurous acid	sulphuric(IV) acid
styrene	phenylethene
toluene	methylbenzene
xylene	dimethylbenzene

Glossary

acid An acid dissolves in water to give a solution with pH below 7. Acid solutions turn litmus red, form salts when neutralized by a base, react with carbonates to form carbon dioxide and give off hydrogen when they react with the more reactive metals. An acid is a substance which contains hydrogen in its formula and reacts with water to form aqueous hydrogen ions. In this, and other reactions, an acid acts as a proton donor.

addition reaction A reaction in which molecules add together to form a single product. *Addition polymerization* is used to make polymers from compounds with double bonds.

alcohol An organic compound containing the reactive group —O—H. Ethanol is an alcohol. It has the formula C_2H_5OH.

alkali An alkali is a base which is soluble in water. A solution of an alkali has a pH above 7 and turns litmus blue. An alkali will neutralize an acid to form a salt. A solution is alkaline if it contains hydroxide ions.

alkali metal A metal in Group I of the periodic table – one of lithium, sodium, potassium, rubidium and caesium.

alkaline earth metal A metal in Group II of the periodic table – one of beryllium, magnesium, calcium, strontium and barium.

alkane Alkanes are hydrocarbons found in crude oil. They have the general formula C_nH_{2n+2}. They are saturated compounds.

alkene The alkenes are hydrocarbons with the general formula C_nH_{2n}. The best known example is ethene, C_2H_4. The alkenes have double bonds in their formulae and are unsaturated.

allotropy The existence of more than one form of an element in the same physical state. Carbon, sulphur and phosphorus are common elements which have *allotropes*.

alpha particle A particle which can be emitted from the nucleus of a radioactive atom. It is a helium nucleus and consists of two protons and two neutrons.

amino acid Amino acids are the monomers from which proteins are made. Each amino acid molecule has an amine group, $—NH_2$, and an acid group, $—CO_2H$.

amorphous Describes a form of a substance which has no regular crystalline structure or shape. Plastic sulphur is an amorphous form of sulphur.

amphoteric The property of an oxide or hydroxide to react as a base or an acid. Amphoteric oxides react with acids to form salts and with bases to form salts, too. Zinc oxide is amphoteric.

anhydrous Means 'without water'. Crystalline substances which contain no water of crystallization are said to be anhydrous. Anhydrous copper(II) sulphate is white.

anion A negatively charged ion which migrates to the anode during electrolysis.

anode The positive electrode in electrolysis. Electrons flow out from the anode to the power pack or battery.

atom The smallest particle of an element. All the atoms of the same element have the same atomic number, which means that they all have the same number of protons in the nucleus.

atomic number The atomic number of an element is the number of protons in the nucleus of its atoms. In the periodic table, the elements are arranged in order of atomic number.

Avogadro constant The number of particles in a mole. The value of the constant is approximately 6×10^{23} per mole.

base The chemical opposite of an acid. A base will neutralize an acid to form a salt. Common bases include the oxides and hydroxides of metals. Ammonia is also a base. A base is a proton acceptor.

beta particle A particle which can be emitted from the nucleus of a radioactive atom. It is an electron formed in the nucleus.

boiling A liquid boils when bubbles of vapour start to form within the liquid on heating. The temperature at which a liquid boils is called the *boiling point*. The boiling point varies with pressure. Raising the pressure raises the boiling point. Boiling points are usually measured at atmospheric pressure.

carbohydrates A family of energy foods including starch and sugars. These compounds consist of carbon, hydrogen and oxygen. An example is glucose, $C_6H_{12}O_6$. In the formulae of carbohydrates, the hydrogen and oxygen atoms are present in the same ratio as in water.

carboxylic acid An organic acid with the reactive group

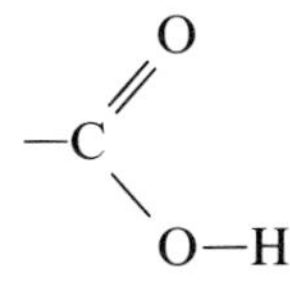

It is the hydrogen atom bonded to oxygen which is acidic.

catalyst A substance which speeds up a chemical reaction but is not used up in the reaction.

cathode The negative electrode in electrolysis. Electrons flow into the cathode from the power pack or battery.

cation A positively charged ion which migrates to the cathode during electrolysis.

chromatography A method for separating and analysing mixtures. The substances are separated as they are carried over a stationary phase (paper, or a powder) by a moving liquid or gas. The result is a *chromatogram.*

combustion Burning (usually in air). The combustion of fuels is used to produce heat and light.

compound A pure substance consisting of two, or more, elements which are chemically joined.

condensation The change of state from vapour, or gas, to liquid.

condensation reaction A reaction in which molecules are joined together by splitting off a small molecule such as water. Nylon and polyesters are polymers made by *condensation polymerization.*

cracking A process used in oil refining to split large hydrocarbon molecules into smaller ones. The reaction takes place in the gas phase and involves heat and catalysts.

crystallization The process of forming crystals. This often refers to the formation of crystals by the cooling or evaporation of a solution.

dehydration The term can have two meanings. It can mean the removal of water contained in a substance. The water of crystallization in hydrated copper(II) sulphate can be removed by heating or using concentrated sulphuric acid. Concentrated sulphuric acid is a *dehydrating agent.* Dehydration can also mean the removal of the elements hydrogen and oxygen from a substance, forming water. Ethanol can be dehydrated by passing the vapour over aluminium oxide. Ethene and water are formed.

detergent A substance which helps to get things clean. There are two main types: soaps, and synthetic detergents made from oil.

diffusion This is a spreading out and mixing process, seen mainly in gases and liquids. The particles of one substance mingle with, and move through, the particles of another. Diffusion goes on until the mixture is uniform.

distillation A method of purifying liquids. The process involves evaporation of the liquid followed by condensation of the vapour. The condensed liquid is called the *distillate.*

electrolysis This happens when a compound is split up by an electric current. The substance split up is called the *electrolyte.* Electrolysis will only happen when the electrolyte is molten or dissolved in water.

electrolyte A substance which, when liquid or in solution, conducts electricity and is decomposed by the current.

electron One of the types of particle in the atom. Electrons are located outside the nucleus. They have negligible mass and a single negative electric charge. The symbol e^- is used to represent an electron or one mole of electrons.

element Substances which cannot be decomposed into simpler substances are called elements. All the atoms of an element have the same atomic number.

endothermic This describes a process which takes in heat energy.

enzyme A protein which acts as a catalyst. Enzymes are specific in their action and are most effective at a particular temperature and pH.

equilibrium The state of balance in a reversible reaction when neither the forward nor the backward reaction is complete. The reaction appears to have stopped. At equilibrium in most reactions, the forward and backward reactions are still occurring but at equal rates. This is *dynamic equilibrium.* At equilibrium reactants and products are all present and their concentrations are constant. The position of equilibrium can be changed by changing the reaction conditions. The sign '$\rightleftharpoons$' in an equation shows that a reaction is in equilibrium.

ester A compound formed by a condensation reaction between an acid and an alcohol. The process is called *esterification.*

evaporation The change of state from liquid to vapour, e.g. the change of water into steam. Evaporation can take place below the boiling point.

exothermic This describes a process which gives out heat.

Faraday constant The electric charge on one mole of electrons. The approximate value of the constant is 96 500 C/mol.

fermentation The conversion of sugars to ethanol by enzymes present in yeast or bacteria. The reaction is most efficient at about 40 °C. The equation for the fermentation of glucose is:

$$C_6H_{12}O_6 \longrightarrow 2C_2H_5OH + CO_2$$

filtration The process used to separate insoluble solids from liquids. In the laboratory, this is done by using porous paper in a funnel. The liquid which passes through the paper is called the *filtrate.*

fixation Nitrogen in the air is unreactive and free. Fixation is used to describe processes, such as the Haber process, which combine nitrogen with other elements. The compounds so formed are used as fertilizers.

formula A shorthand way of describing a chemical substance using symbols. The *empirical formula* is the simplest formula. It shows the ratio between the numbers of each type of atom in a molecule or giant structure. The empirical formula of ethanoic acid is CH_2O. The *molecular formula* shows how many of each type of atom there are in the molecule. For ethanoic acid, this is $C_2H_4O_2$. The *structural formula* shows how the atoms are arranged. The structural formula for ethanoic acid is:

```
      H     O
      |    //
   H—C—C
      |    \
      H     O—H
```

fractional distillation The use of distillation to separate a mixture of liquids with different boiling points. Common applications are the separation of the following mixtures: liquid air, ethanol and water, and crude oil.

freezing The change of state from liquid to solid. A pure substance freezes at a definite temperature, called its *freezing point* (which is the same temperature as its melting point).

giant structure A crystal structure in which all the particles are strongly linked together by a network of bonds extending throughout the crystal.

group A vertical column of elements in the periodic table.

halogen An element in group VII of the periodic table – one of fluorine, chlorine, bromine, iodine and astatine.

heat of reaction The energy change which accompanies a reaction when the amounts of substances shown in the equation are used. The energy changes associated with some specific processes are given special names, e.g. the *heats of combustion, neutralization* and *vaporization.* These are defined in chapters 36, 37 and 38.

homologous series A series of organic compounds of the same type with the same general formula. There is a constant structure difference between one member and the next. The alkanes methane (CH_4), ethane (C_2H_6) and propane (C_3H_8) are part of a homologous series. The constant structure difference is $—CH_2—$.

hydrated Means 'with water'. Substances containing water of crystallization are hydrated. Hydrated copper(II) sulphate, $CuSO_4.5H_2O$, is blue.

hydrocarbon A compound consisting of carbon and hydrogen only.

hydrolysis A reaction in which the molecules of a compound are split into smaller molecules by reaction with water. Heat and catalysts are often used to speed up the reaction. Examples are the breakdown of fats to soap, and of starch to glucose.

indicator A substance which changes colour in different conditions. Indicators are commonly used to detect whether a solution is acidic or alkaline and to show up the 'end point' in titrations. Full range (universal) indicator is a mixture of indicators which has different colours in solutions of different pH.

ion An atom, or group of atoms, with an electric charge. Metal ions are positively charged. Non-metal ions are negatively charged.

isomers Compounds with the same molecular formula but different structures.

isotopes Atoms of the same element which have different mass numbers because they have different numbers of neutrons in their nuclei.

mass number The total number of protons and neutrons in the nucleus of an atom.

melting The change of state from solid to liquid. A pure substance melts at a constant temperature called its *melting point* (which is the same as its freezing point).

molar solution A molar solution contains one mole of the dissolved substance in one litre ($1\,dm^3$) of the solution. Molar solutions are labelled 1.0 M or $1.0\,mol/dm^3$.

mole One mole of any substance contains the same number of particles as one mole of any other substance. (It must be stated whether the particles are atoms, molecules, or ions.) To be precise, the mole is the amount of substance which contains the same number of particles as there are atoms in 12.000 g of carbon-12.

molecule A group of atoms joined together. Most non-metals are molecular (e.g. Cl_2, S_8, O_2 and H_2). Most compounds of non-metals with other non-metals are also molecular (e.g. H_2O, CO_2, CH_4 and NH_3).

monomer A small molecule which can be polymerized to make a polymer.

neutron One of the types of particle in the nucleus of an atom. It has a relative mass of 1 and no electric charge. The symbol for the neutron is n. The hydrogen atom has no neutrons.

nucleus The tiny central core of an atom which contains the protons and neutrons. The mass of an atom is concentrated in the nucleus.

oxidation At its simplest, oxidation involves combination with oxygen. During oxidation, an atom, molecule or ion loses electrons. An element is oxidized when its oxidation number becomes more positive, or less negative. Oxidation is brought about by *oxidizing agents.*

oxidation number A number to show the extent to which an element has been oxidized or reduced. Oxidation numbers are worked out according to the rules given on p. 324.

period A horizontal row of elements in the periodic table. The first three are short periods, the remainder are long periods.

physical properties Examples of physical properties are colour, melting point, density and electrical conductivity. These properties are to do with changes which do not involve one substance changing into another.

polymer A long-chain molecule made by polymerization.

polymerization A reaction by which a very large number of small molecules is joined together to make a large molecule. Polymerization may occur by addition or by condensation.

precipitate An insoluble solid which separates out from a solution during a reaction. Precipitation reactions are used to make insoluble salts.

protein Proteins are essential nutrients. They are compounds of carbon, hydrogen, oxygen, nitrogen and other elements. The molecules consist of long chains of amino acids.

proton One of the types of particle in the nucleus of an atom. It has a relative mass of 1 and a single positive electric charge. The symbol is p. A hydrogen ion is simply a proton.

pure Something is chemically pure if it is a single substance not mixed with anything else.

redox reaction A reaction involving reduction and oxidation.

reduction This is the opposite of oxidation and at its simplest it involves the removal of oxygen. During reduction, an atom, molecule, or ion gains electrons. An element is reduced when its oxidation number becomes more negative or less positive. Reduction is brought about by *reducing agents.*

relative atomic mass The relative mass of an atom on a scale on which an atom of carbon-12 is 12.000.

relative molecular mass The relative mass of a molecule on the same scale as relative atomic mass.

reversible reaction A reaction which can go forwards or backwards depending on the conditions. Many reversible reactions can reach a state of *equilibrium.*

salt An ionic compound formed when an acid is neutralized by a base. In the formula of a salt, the hydrogen of the parent acid is replaced by metal ions. For example, magnesium sulphate, $MgSO_4$, is a salt of sulphuric acid, H_2SO_4.

saturated compound In the molecules of a saturated compound, all of the bonds are single bonds, so there is no spare bonding. Alkanes are saturated hydrocarbons.

saturated solution A solution is saturated when it contains as much of the dissolved substance as possible at a particular temperature.

solute A solute is a substance which dissolves in a *solvent* to make a *solution.* Brine is a solution of salt (the solute) in water (the *solvent*).

solvent A liquid which can be used to dissolve things. Water is the commonest solvent. White spirit and ethanol are also often used as solvents.

state symbols These symbols are included in equations to show the states of the reactants and products: (s) – solid, (l) – liquid, (g) – gas and (aq) – aqueous (meaning 'dissolved in water').

sublimation This is the change of state from solid to vapour and back from vapour to solid without passing through the liquid state. The process can be used to purify substances such as iodine. Solid carbon dioxide is called 'dry ice' because it sublimes and does not melt to a liquid.

substitution reaction A reaction in which an atom, or group of atoms, in a molecule is replaced by a different atom, or group of atoms.

surface tension A measure of the attraction between molecules in the surface of a liquid. The surface tension of water is high compared with other molecular liquids.

synthesis Joining things together to make something more complicated. Compounds are synthesized from elements, e.g. ammonia is synthesized from nitrogen and hydrogen. Manufactured substances such as plastics are often described as *synthetic.*

thermoplastic A plastic which softens on heating but hardens again on cooling. Examples are polythene, polystyrene and pvc.

thermoset A plastic resin which sets hard when heated. Once formed in a mould, a thermosetting plastic cannot be remelted. One example is Bakelite.

titration A technique for investigating the volumes of solutions which react together. One solution, in a burette, is usually run into a fixed volume of another solution until the 'end point' is reached, as shown by an indicator.

transition metal A metal in the middle block of elements in the long periods of the periodic table. Examples are chromium, iron, nickel and copper.

unsaturated compound There are double or triple bonds in the molecules of unsaturated compounds. This means that there is spare bonding and the molecules can undergo addition reactions. Ethene, the simplest alkene, is an unsaturated compound.

valency The valency of an element is a number which shows its ability to combine with other elements. In molecules, the valency gives the number of covalent bonds which the atoms of the element can form. In ionic compounds, the valency gives the charge on the ions of the element.

vaporization The change of state from liquid to vapour. The term is usually used to describe the change at the boiling point of the liquid.

viscosity The viscosity of a liquid shows how easily it flows. Runny liquids have low viscosities. Liquids which are thick and sticky like treacle and tar are highly viscous.

volatile Liquids and solids which easily turn to vapour are volatile. Petrol is a volatile liquid. Substances which are hard to vaporize, such as iron or salt, are involatile.

water of crystallization The water molecules trapped in a crystalline substance. Hydrated copper(II) sulphate has five moles of water of crystallization per mole of copper sulphate. Its formula is $CuSO_4.5H_2O$.

Index

The page numbers in **bold** type refer to entries in the glossary.

acetic acid, *see* ethanoic acid
acetylene, *see* ethyne
acid–base theory, 170
acids
  as catalysts, 229, 234, 246–8
  definitions of, **346**
  effect on indicators, 148–50
  electrolysis of, 310, 315
  mineral, 151
  reactions with bases, 149–50, 156–7
  reactions with carbonates, 149–50
  reactions with metals, 82, 100–1, 149–50
  strength of, 149, 171, 233
  theories of, 170–2, 293
activation energy, 200–1
activity series for metals, 82–6, 87, 312–13
addition reactions, 227, 237, **346**
air
  composition of, 34–7
  in breathing, 44–6
  in burning, 42–4, 222
  in rusting, 46–7
  liquefaction of, 50–1
  pollution of, 48–9, 225, 280–1, 304
  reactions with metals, 80
  solubility of, 64–5
  uses of, 89–92, 213, 257
alcohol, *see* ethanol
alcohols, 229–31, 283, 340, **346**
alkalis, 100
  action on ammonium salts and proteins, 252–4
  definitions of, 156, **346**
  manufacture of, 157–60
  theory of, 172, 293
  uses of, 67, 156, 158, 160
alkali metals, *see* Group I
alkaline earth metals, *see* Group II
alkanes, **346**
  comparison with alkenes, 225–6
  heats of combustion of, 283
  names and formulae of, 219, 340
  physical properties of, 340
  reactions of, 220
alkenes, 225–8, 340, **346**
allotropy, 132–7, **346**
alloys, 76, 124–7
alpha particles, 296–8, **346**
aluminium
  alloys of, 124
  extraction of, 87–9
  physical properties of, 77
  reactions of, 83–4
  uses of, 6, 77
aluminium oxide, 88–9, 230
aluminium sulphate, 26
amino acids, 254–5, **346**
ammonia
  diffusion of, 20
  manufacture of, 256–8
  physical properties of, 259
  reactions of, 108–9, 172, 259–61, 263
  reactions producing, 252–4, 259
  tests for, 155, 253
  uses of, 156, 159, 258, 262
ammonium chloride, 20, 108, 184, 202–3, 253
ammonium nitrate, 262–4
ammonium salts, 167
ammonium sulphate, 262–4
amphoteric hydroxides, 168, **346**
amphoteric oxides, **346**
anions, 343, **346**
anode, 15, 311–13, 315, **346**
argon
  discovery of, 36–7
  separation from air, 51
  uses of, 54, 299
asbestos, 138
atomic mass, *see* relative atomic mass
atomic number, 97, 112–13, 295, 297, 336–7, **346**
atomic theory, 23–4, 96, 110–13
atoms, 23–4, 27, 96, **346**
  in crystals, 118–23, 128–30, 132, 134, 136–9
  moles of, 176–8
  structure of, 110–13, 295–6
  symbols for, 24, 97, 295–6
Avogadro constant, 177, **346**

barium, 100
barium chloride, 100
barium hydroxide, 100
barium sulphate, 100–1, 118, 164
bases
  definitions of, 150, 156–7, **346**
  reactions of, 150, 157, 171–2
  solubility of, 162
  uses of, 156
batteries, 327–30
Benedict's solution, 101, 245, 247
beta particles, 296–8, **346**
biuret test, 251–2
blast furnace, 89–91, 125
blood, 43, 45, 199, 251
boiling, 7–8, 271, **346**
boiling point, 4, 7–8, 271, **346**
  and structure, 122, 128, 130–1, 140–1, 275
  tables of values, 336–41
bonding
  covalent, 128–31, 136–9, 142–3
  hydrogen, 143
  ionic, 140–3
  metallic, 122–3
Bragg, W. H. and W. L., 119
brass, 124
breathing, 44–6, 52
bromine
  bonding, 25
  diffusion of, 20
  molecules, 24, 106
  reactions of, 107–9, 209, 220, 227, 323, 325
  sources of, 7, 104
  structure of, 106, 128
  uses of, 105
bronze, 124
Brownian motion, 22
brown ring test, 169
burette, 186, 188
burning, 17, 42–9, 220, 222, 277, 280–3, **347**
butane, 130, 219, 281–2

calcium, 40, 80–1, 83, 87, 100, 202
calcium carbonate
  decomposition of, 27–8, 85–6, 209–10
  natural forms of, 2, 70, 100, 118
  reaction with acids, 70, 149–50, 166
calcium chloride, 26, 88, 100, 154, 159–60
calcium fluoride, 88, 118, 140–1
calcium hydrogencarbonate, 70
calcium hydroxide, 81, 100, 156, 253
calcium oxide, 25, 27–8, 40, 90, 100, 253
calcium sulphate, 25, 69, 71, 100–1, 162
carbohydrates, 46, 244–50, **346**
carbonates, 100
  decomposition of, 85–6
  reactions with acids, 149–50, 171
  solubility of, 162
  tests for, 167, 169
carbon
  allotropy of, 134–7
  bonding, 25
  cycle, 245–6
  organic chemistry of, 218–40
  reactions of, 17, 39, 103
  structure of, 103, 134–7
  uses of, 88–91, 134–7, 301, 327–8
carbon dioxide
  as a product of burning, 17, 39, 42–3, 45–6
  in the air, 7, 36–7, 45
  physical properties of, 43

preparation of, 154
reactions of, 70, 103, 166
structure of, 137
test for, 18, 43, 155
carbonic acid, 166
carbon monoxide, 43, 89, 91, 251
carboxylic acids, 231–4, 340, **347**
cast iron, 79
catalase, 199
catalysts
effect on reaction rates, 196–8, **347**
enzymes as, 199–200
theory of, 200–1
transition elements as, 101, 198
use in industry, 198–9, 213, 223–5, 236–7, 256–7, 260
use in the laboratory, 38, 230
cathode, 15, 311–13, 315, **347**
cations, 343, **347**
cells, 327–31
cellulose, 244, 246, 249
centrifuge, 8, 165
chalk, 2, 69–71
changes of state, 271
charcoal, 17, 242–3
chlorides, 169
as salts, 150, 161
formulae of, 98
of metals, 100–1, 108–9
of non-metals, 103
solubility of, 162
test for, 109, 169
chlorine
bonding, 25
isotopes of, 296
manufacture of, 88, 157–8
molecules, 24, 106–7
physical properties of, 105
preparation of, 104
reactions of, 62, 100–1, 107–9, 140, 207–8, 220, 225, 322, 324
structure of, 106
test for, 155
uses of, 67, 105
chlorophyll, 10, 245
chromatography, 10, 247–8, **347**
chromium, 77–8, 83, 125, 314
citric acid, 148, 150, 170, 233
coal
as a fuel, 48, 277–80, 285, 304
chemicals from, 10–11, 279–80
gas from, 221, 279–80
resources, 285
cobalt chloride, 14, 253
coke, 11, 42–3, 87, 89, 91, 159, 279–80
combining power, 25–6, 129
combustion, **347**, *see also* burning *and* heat
compounds, 17–19, 24–7, 60–2, 96, **347**
ionic, 27
of metals with non-metals, 26–7, 140–2
of non-metals, 25–6, 128–31, 137–9, 143
physical properties of, 337–43
concentration
effect on equilibria, 208–9
effect on reaction rates, 193–5, 201
units of, 185–6
condensation, 7, 11, **347**
condensation reactions, 234, 238–9, **347**
contact process, 210–14
copper
extraction of, 87, 91–2
physical properties of, 77
purification of, 313–14
reactions of, 34, 39, 42, 80, 82–4, 101, 149
structure of, 122
uses of, 6, 77, 124, 262, 314, 327
copper(II) carbonate, 85–6
copper chlorides, 101, 312–13
copper(II) chromate(VI), 310
copper(I) oxide, 101, 245
copper(II) oxide, 34, 39, 101, 149, 179–80, 218, 253, 260
copper(II) sulphate
anhydrous, 14, **346**
decomposition of, 14
electrolysis of, 313–14, 317–19
formula of, 26
physical properties of, 25
uses of, 82, 314, 327
covalent bonding, *see* bonding
cracking, 220, 223–6, 230, **347**
crude oil
chemicals from, 10–11, 13, 223–6
fuels from, 48, 221, 223–5, 277, 279
refining, 11, 223–5
resources, 285
crystallization, **347**
crystals
cleavage of, 141–2
from melts, 120–1, 132–3
from solutions, 4, 8, 63–4, 84, 132–3, 161, 163
in rocks, 2, 118–19
structure of, 118–43
cyclohexane, 131

Dalton, J., 23–4, 96, 111
Daniell cell, 327
Davy, H., 16, 96–7
decay series, 298
dehydration, 231, **347**
density, tables of, 336–7, 342
detergents, 67–9, **347**
diamond, 134–7
diatomic molecules, 24
1,2-dibromoethane, 227
diffusion, 20, **347**
dislocations, 123
displacement series
for halogens, 108, 325–6
for metals, 84, 325–6
distillation, 6–7, **347**
Down's cell, 87–8
dry cell, 327–8
dynamic equilibrium, 205–7, **347**

electricity
from cells, 327–31
generation of, 78, 278, 285–8
units of, 316–19, 344
electrolysis, 14, 96, **347**
calculations, 316–19
electroplating, 314
extraction of metals by, 87–9
of molten salts, 15–16, 142, 311
of sodium chloride, 87–8, 142, 157
of solutions, 14, 157, 170, 312–13, 315
theory of, 88–9, 142, 310–13, 315, 323
to purify copper, 313–14
electrolyte, 26, 310–11, **347**
electron microscope, 23
electrons
in atoms, 111–13, 128–9, 131, 295–7, **347**
in electrolysis, 311, 313, 315, 318–19
in metals, 122–3
transfer of, 140, 322–3, 324–6, 327, 331
electroplating, 314
elements
abundance of, 16
atoms of, 110–13
definition of, **347**
discovery of, 96
in the periodic table, 97–109
physical properties of, 336–7
symbols for, 16, 24, 97, 336–7
endothermic processes, 291–2, **347**
end-point, 187
energy
alternative sources of (geothermal, hydroelectric, solar, tidal, wind), 285–9
forms of, 270
from burning fuels, 277–85
nuclear, *see* nuclear power
units of, 272–3, 344
energy changes
and bonding, 284–5
and chemical reactions, 290–3
during changes of state, 270–5
energy level diagrams, 284, 290–2
enzymes, 199–200, 246–8, 251, **347**
equations
calculations from, 181, 183–5, 188
determination by experiment, 180–8
rules for writing, 28
symbol, 27–8
word, 14
equilibrium, 204–10, 213, **347**
esters, 233–4, 239, 340, **347**
ethane, 130, 227
ethanoic acid
formation of, 232
occurrence and uses of, 148
reactions of, 149, 233–4
strength of, 149, 233
structure of, 150, 232
ethanol
as a fuel, 281–3
from fermentation, 8–9
manufacture of, 225
physical properties of, 25
reactions of, 230–2, 234
structure of, 130, 274
uses of, 229–30
ethene, 225–7, 229, 235–7
ethyl ethanoate, 233–4
ethyne, 52
evaporation, 3–4, 8, 66, **347**
exothermic processes, 17, 260–1, 277, 284, 290–2, **347**

Faraday constant, 318, **347**
fats, 46, 67, 227, 234
Fehling's solution, 101, 245
fermentation, 8, 52, 199, 250, **347**

fertilizers, 256, 259, 262–5, 280
filtration, 4, 163, 165, **347**
fire extinguishers, 44
fire triangle, 44
flame tests, 100, 168
fluorine, 104, 128–9
food, 244, 264–5, *see also* carbohydrates, fats *and* protein
formic acid, *see* methanoic acid
formulae, **348**
  determination by experiment, 178–80
  empirical, *see* simplest
  general, 219, 226
  in the periodic table, 98
  molecular, 24–5, 129–30, 150, 219, **348**
  of ionic compounds, 26, 337–41
  of non-metal compounds, 25, 337–41
  simplest, 178, 180, **348**
  structural, 129–30, 219, 226, **348**
fossil fuels, 277, 285
fountain experiment, 65
fractional distillation, 8–9, 11, 50–1, 223–4, **348**
Frasch process, 4–5
freezing, 6, 271–2, 275, **348**
fuel cells, 330–1
fuels, 42–3, 45–6, 48, 52, 277–83, 285–7, 304
full range indicator, 148–9
fusion, *see* heat

galvanizing, 79
gamma radiation, 296–7, 301
gases
  moles of, 183–5
  particles in, 21
  preparation of, 153–4
  solubility of, 64–5, 343
  tests for, 155
Geiger–Müller tube, 206, 298–9
germanium, 16
giant structures, 122, 128, 136–40, 275, **348**
glucose, 46, 244–50
glycerol, 68
glycine, 254
gold, 77, 82, 87, 99, 314
grain, 120–5
granite, 2
graphite, 125, 134–7, 303–4
groups, 98, 99–109, 112, **348**
Group I, 99–100, 112, **346**
Group II, 99, 100–1, **346**
Group IV, 99, 103, 137
Group VII, 103–6, 132, 220, 313, 325–6, **348**
Group VIII, *see* noble gases
gypsum, 71

Haber process, 256–8
half-life, 300–1
halogens, *see* Group VII
hardness of water, *see* water
heat
  of combustion, 282–3, 291, 340
  of fusion, 341
  of neutralization, 293
  of reaction, 290–3, **348**
  of vaporization, 271, 274–5, 292, 341
helium, 16, 37, 51, 54–5, 111, 288, 297

homologous series, 219, **348**
hydrated compounds, **348**
hydrides, 103, 106, 108
hydrobromic acid, 315
hydrocarbons, 219, 223–7, 235, 237, **348**
hydrochloric acid, 101, 104
  as a catalyst, 246–8
  electrolysis of, 315
  preparation of, 152
  reactions of, 149–50, 169, 193
  salts of, 150, 161
  uses of, 151
hydrogen, 16
  atomic structure of, 111–13, 171, 295
  bonding in, 25, 284
  isotopes of, 288, 295
  manufacture of, 157, 257
  molecules, 24, 129–30
  preparation of, 81–2, 154
  reactions of, 17, 27, 107, 196, 227, 256
  reactions producing, 100–1, 150, 257
  test for, 155
  uses of, 54, 158, 281, 329–30
hydrogen bonding, *see* bonding
hydrogen bromide, 106–8
hydrogen chloride
  acidity in different solvents, 170–1
  bonding in, 129–30
  diffusion of, 20
  physical properties of, 106
  preparation of, 151–2
  reactions of, 101, 106–8
  solubility of, 65
  structure of, 130–1, 150
  test for, 155
hydrogen iodide, 106–8
hydrogen peroxide, 9, 38, 196–7, 199
hydrogen sulphide, 155, 212, 325
hydrolysis, 234, 246–8, 255, 348
hydroxides
  as bases, 156–7, 172
  decomposition of, 86, 167
  formulae of, 100
  precipitation of, 167–8
  reactions producing, 81, 100
  reactions with acids, 150
  solubility of, 162

ice, structure of, 61, 143
indicators, 39, 148–9, 163, 187–8, **348**
industry
  economics and efficiency, 3–4, 212, 214, 221, 225
  location of, 158, 160, 214, 221
  use of water by, 60, 66, 278, 304
inert gases, *see* noble gases
iodine
  equilibria involving, 204–5, 207–8
  extraction of, 8–9, 104
  molecules, 24, 106
  physical properties of, 131
  reactions of, 107–9, 207, 245
  structure of, 106, 132
  sublimation of, 10
  uses of, 106, 302
iodine monochloride, 207–8
ion exchange, 71
ionic bonding, *see* bonding
ionic equations, 140–1, 164

ions, 26–7, 70, **348**
  and electrolysis, 310–11, 313, 315, 318–19
  in crystals, 140–3, 161
  in solution, 157–8, 164–5, 171–2, 293
  of metals and non-metals, 27, 98, 100–1, 106, 113, 140, 161
  table of charges, 343
  tests for, 167–9
iron
  alloys of, 77
  conversion to steel, 52, 90–1
  extraction of, 87, 89–91
  physical properties of, 77
  reactions of, 39, 80, 82–3, 101, 107, 182, 322
  rusting of, 46–7
  uses of, 125–7, 257
iron chlorides, 101, 107, 322
iron(III) oxide, 39, 83, 101
iron(II) sulphate, 101, 151, 325
isomers, 219, **348**
isotopes, 295–6, 298, 300, 304, **348**

joule, 272, 344

kilojoule, 272–3, 344
krypton, 37, 51, 54

Lavoisier, A. L., 35, 96, 150
law of conservation of mass, 19
law of constant composition, 18
lead
  alloys of, 78
  extraction of, 87, 91
  physical properties of, 77
  reactions of, 82–4, 323
  structure of, 120–1
  uses of, 77–8, 328–9
lead–acid cell, 328–9
lead(II) bromide, 15, 26, 311, 316, 323
lead(II) carbonate, 85–6
lead(II) nitrate, 84, 85–6, 312–13
lead(II) oxide, 85, 167
lead(II) sulphide, 118
Le Chatelier's principle, 208–10
limestone
  in formation of hard water, 69–71
  quarry, 2
  thermal decomposition of, 15–16, 27–8
  uses of, 89–91, 159
limewater, 17, 100, 155
liquids, 21, 128, 271–5
lithium, 80–1, 99–100, 111
lithium chloride, 100
lithium hydroxide, 100
litmus, 148–9, 156–7, 203

magnesium
  extraction of, 87
  reactions of, 18, 39–40, 42, 62, 80, 82–4, 100, 149, 321–2, 324
  source of, 7
  uses of, 47
magnesium chloride, 100
magnesium hydroxide, 62, 100, 156
magnesium oxide, 18, 39, 62, 80, 100, 161, 178–9
magnesium sulphate, 84, 100, 164

maltose, 245, 248
manganese, 83, 262
manganese(IV) oxide, 38, 104, 197–8
marble, 2, 70
margarine, 198, 227, 234
mass number, 295, 297, **348**
melting, 5, 292, **348**
melting point, 4–5, 270–1, **348**
  and purity, 5
  and structure, 122, 128, 130–1, 136–7, 140–1
  of metals, 78, 122
  table of values, 336–41
Mendeléev, D. I., 95, 96
mercury, 35, 38, 78, 84
mercury(II) chloride, 84
mercury oxide, 35, 38
metals
  activity series of, 82–6
  bonding in, 122–3
  extraction of, 87–92
  in the periodic table, 98, 99–102
  ions of, 113, 140, 311–13, 322–3, 325–6
  oxides of, 40–1, 80, 82, 85–6, 100–1, 149–50, 157, 162, 197
  physical properties of, 16, 76–9, 122–7
  reactions of, 39–40, 80–6, 149–50, 171
  resources, 6
  structure of, 120–7
  uses of, 75–9
methanoic acid, 232
methane
  bonding in, 129–30
  burning of, 28, 277
  from sewage, 286
  hydrogen from, 257
  in natural gas, 221–2
  molecules of, 25, 129–30
  reaction with chlorine, 220
methanol, 230
methyl orange, 187–8
2-methylpropane, 130, 219
mica, 139
minerals, 100, 117–19, 138–9, 161–2
mixtures, 17
molar solutions, 185, 348
molecular mass, *see* relative molecular mass
molecules, 24, 27, 128–34, 137, 142–3, 150, 219, 284, **348**
moles, 176–89, 344, **348**
  in solution, 185–9
  of electrons, 318–19
  of gases, 183–5
  of particles, 176–8
molybdenum, 125
monomers, 237, 249, **348**

names of compounds
  inorganic, 18, 323, 345
  organic, 226, 230, 232, 345
nanometre, 22, 344
naphthalene, 131
natural gas, 212, 221–2, 277, 279, 285
neon, 37, 51, 54, 128
neutralization, 150, 156–7, 163, 172, 293
neutrons, 111–13, 295–8, 303–4, **348**
nickel, 77–8, 125, 198, 227, 257, 314
nitrates
  as fertilizers, 262–4
  decomposition of, 38, 85–6
  in sewage, 72
  of alkali metals, 100
  solubility of, 162
  tests for, 167, 169
nitric acid, 148, 151, 315
  manufacture of, 260–1
  uses of, 152, 263
nitrogen
  bonding, 25
  cycle, 255
  fixation, 255, 257–8, **348**
  in ammonia and proteins, 253
  in fertilizers, 262–4
  in the air, 34–7, 51, 65
  molecules, 24
  oxides of, 255, 260–1
  physical properties of, 39
  problem, 256
  reactions of, 40, 256
  uses of, 53
nitrogen dioxide, 85–6, 155, 261
noble gases, 37, 54–5, 103, 112–13
non-metals
  in the periodic table, 98, 103–6
  ions of, 113, 140, 311–13, 322–3, 325–6
  oxides of, 40–1, 103, 152
  physical properties of, 16
  reactions of, 39–40
  structure of, 128–37
nuclear power, 78, 278–9, 285
  from fission, 303–5
  from fusion, 287–8
nucleus, 111, **348**
nylon, 217, 238–9

oil, *see* crude oil, olive oil *and* vegetable oils
olive oil, 23
ores of metals, 87–91, 100–1
organic acids, 231–4, 340, **347**
organic chemistry, 218–40
oxidation, **348**
  in terms of electrons, 322–6
  in terms of oxidation numbers, 323–4
  in terms of oxygen, 40, 42–3, 47, 80, 92, 232, 260–1, 321
oxidation numbers, 323–4, **349**
oxides
  acid–base properties of, 40, 103, 149–50, 152, 157
  as catalysts, 197
  extraction of metals from, 87–91
  formation of, 39–40, 80, 82, 85–6
  formulae of, 100, 103
  solubility of, 162
oxidizing agents, 321, 323, 324–6, **348**
oxonium ion, 171
oxygen
  bonding in, 25, 129, 284
  in breathing, 45–6
  in burning, 17, 27, 39–40, 42–4, 321–2
  in rusting, 47
  in solution, 64–5
  in the air, 34–7, 51, 246
  molecules, 24, 129
  physical properties of, 39
  preparation of, 38
  reactions with metals, 80, 84, 100–1
  reactions with non-metals, 103, 213
  test for, 38, 155
  uses of, 52, 72, 90–1, 329–30

paper making, 249
paraffin, 11, 42
particles
  collisions between, 200–1
  diffusion of, 20
  in compounds, 24–5
  in elements, 23–4
  in solids, liquids and gases, 21–3, 270–1, 275
  size of, 23
percentage composition, 264
periods, 97, 112, **349**
periodic table
  and electron configuration, 112–13
  discovery of, 95–8
  metals in, 99–102, 113
  non-metals in, 99, 113
petrol, 11, 42–3, 223–4
petroleum, *see* crude oil
phenolphthalein, 187
phosphates, 72
phosphoric acid, 229, 231
phosphorus, 24, 39, 262
phosphorus pentoxide, 39
photosynthesis, 7, 246, 277
pH scale, 148–9
physical properties, 98, 336–43, **349**
pipette, 186, 188
plants, 7–10, 246, 255, 262
plastics, *see also* polymers
  disposal of, 48–9
  from natural polymers, 249
  shaping of, 239–40
  structure of, 137, 236–9
  uses of, 235–6, 239
platinum, 77, 196, 260–1, 312
plutonium, 301, 303–5
pollution (causes and prevention)
  of air, 48–9, 89, 91, 213–14, 225, 280–1, 304
  of water, 60, 65, 72, 160, 302
polyester, 239
polymers, 137, 237–40, 249, 254, **349**
polymerization, 223, 225, 236–9, **349**
polythene, 137, 235–7, 239
polystyrene, 225, 237, 239
polyvinyl chloride, *see* pvc
potassium
  reactions of, 80–3, 99–100
  structure of, 122
  uses of, 262
potassium bromide, 106, 108–9, 312–13
potassium carbonate, 85–6
potassium chlorate, 63–4, 196
potassium chloride, 100, 106, 108–9
potassium dichromate, 232, 325
potassium hydroxide, 100, 156–7
potassium iodide, 106, 108–9, 325
potassium manganate(VII), 104, 311, 323, 325
potassium nitrate, 38, 85–6, 104
potassium nitrite, 38, 85
potassium sulphate, 312
precipitates, 109, 164–9, 186–7, **349**